Algorithms and Computation in Mathematics • Volume 14

Editors

Manuel Bronstein   Arjeh M. Cohen
Henri Cohen   David Eisenbud
Bernd Sturmfels

Alicia Dickenstein
Ioannis Z. Emiris (Editors)

Solving Polynomial Equations

Foundations, Algorithms, and Applications

With 46 Figures

 Springer

Editors

Alicia Dickenstein

Departamento de Matemática
Facultad de Ciencias Exactas y Naturales
Universidad de Buenos Aires
Ciudad Universitaria – Pab. I
C1428EGA Buenos Aires, Argentina
e-mail: alidick@dm.uba.ar

Ioannis Z. Emiris

Department of Informatics and Telecommunications
National Kapodistrian University of Athens
Panepistimiopolis GR-15784, Greece
e-mail: emiris@di.uoa.gr

Mathematics Subject Classification (2000): 14QXX, 68W30, 65H10, 13PXX, 11C08, 12F10, 12Y05, 13P05, 13P10, 32A27, 52A39, 52B11, 52B20, 68Q25, 68U07

ISSN 1431-1550

ISBN 978-3-642-06361-9   e-ISBN 978-3-540-27357-8

Springer is a part of Springer Science+Business Media

springeronline.com

© Springer-Verlag Berlin Heidelberg 2005
Softcover reprint of the hardcover 1st edition 2005

Cover design: *design & production* GmbH, Heidelberg

Printed on acid-free paper   46/sz - 5 4 3 2 1 0

To my family.
A.D.

To my parents.
I.Z.E.

Preface

The subject of this book is the solution of polynomial equations, that is, systems of (generally) non-linear algebraic equations. This study is at the heart of several areas of mathematics and its applications. It has provided the motivation for advances in different branches of mathematics such as algebra, geometry, topology, and numerical analysis. In recent years, an explosive development of algorithms and software has made it possible to solve many problems which had been intractable up to then and greatly expanded the areas of applications to include robotics, machine vision, signal processing, structural molecular biology, computer-aided design and geometric modelling, as well as certain areas of statistics, optimization and game theory, and biological networks. At the same time, symbolic computation has proved to be an invaluable tool for experimentation and conjecture in pure mathematics. As a consequence, the interest in effective algebraic geometry and computer algebra has extended well beyond its original constituency of pure and applied mathematicians and computer scientists, to encompass many other scientists and engineers. While the core of the subject remains algebraic geometry, it also calls upon many other aspects of mathematics and theoretical computer science, ranging from numerical methods, differential equations and number theory to discrete geometry, combinatorics and complexity theory.

The goal of this book is to provide a general introduction to modern mathematical aspects in computing with multivariate polynomials and in solving algebraic systems. It is aimed to upper-level undergraduate and graduate students, and researchers in pure and applied mathematics and engineering, interested in computational algebra and in the connections between computer algebra and numerical mathematics. Most chapters assume a solid grounding in linear algebra while for several of them a basic knowledge of Gröbner bases, at the level of [CLO97] is expected. Gröbner bases have become a basic standard tool in computer algebra and the reader may consult any other textbook such as [AL94, BW93, CLO98, GP02], or the introductory chapter in [CCS99]. Below we discuss briefly the content of each chapter and some of their prerequisites.

The book describes foundations, recent developments and applications of Gröbner and border bases, residues, multivariate resultants, including toric elimination theory, primary decomposition of ideals, multivariate polynomial factorization, as well as homotopy continuation methods. While some of the chapters are introductory in nature, others present the state-of-the-art in symbolic techniques in polynomial system solving, including effective and algorithmic methods in algebraic geometry and computational algebra, complexity issues, and applications. We also discuss several numeric and symbolic-numeric methods. This is not a standard textbook in that each chapter is independent and, largely, self-contained. However, there are strong links between the different chapters as evidenced by the many cross-references. While the reader gains the advantage of being able to access the book at many different places and of seeing the interplay of different views of the same concepts, we should note that, because of the different needs and traditions, some notations inevitably vary between different chapters. We have tried to note this in the text whenever it occurs. The single bibliography and index underline the unity of the subject.

The first chapter gives an introduction to the notions of residues and resultants, and the interplay between them, starting with the univariate case and synthesizing different approaches. The sections on univariate residues and resultants could be used in an undergraduate course on complex analysis, abstract algebra, or computational algebra as an introduction to more advanced topics and to illustrate the interdependence of different areas of mathematics. The multivariate sections, on the other hand, directed to graduate students and researchers, are intended as an introduction to concepts which are widely used in current research and applications.

The second chapter puts the accent on linear algebra methods to deal with polynomial systems: the multiplication maps in the quotient algebra by a polynomial ideal are linear and allow for the use of eigenvalues and eigenvectors, duality, etc. Applications to Galois theory, factoring, and primary decomposition are offered. The first sections require, besides standard linear algebra, some background on computational algebraic geometry (for instance, the first five chapters of [CLO97]). Some acquaintance with local rings (as in Chapter 4 of [CLO98]) would also be helpful. Known basic facts about field extensions and Galois theory are assumed in the last part.

The third chapter also elaborates on the concepts in the first two chapters, and combines them with numerical methods for polynomial system solving and several applications. The tools and methods developed are used to solve problems arising in implicitization of rational surfaces, determination of the position of a camera or a parallel robot, molecular conformations, and blind identification in signal processing. The required background is very similar to that needed for the first sections of Chapter 2.

Chapter 4 is devoted to laying the algebraic foundations for border bases of ideals, an extension of the theory of Gröbner bases, yielding more flexible bases of the quotient algebras. Border bases yield a connection between computer algebra and numerical mathematics. An application to design of experiments in statistics is included.

The fifth chapter concentrates on various techniques for computing primary decomposition of ideals. This machinery is applied to study an interesting class of ideals coming from Bayesian networks, establishing an important link between algebraic geometry and the emerging field of algebraic statistics. Besides Gröbner bases, the readers are expected to have a casual understanding of the algebra-geometry dictionary between ideals in polynomial rings and their zero set. Many propositions that can be found in the literature are stated without proof, but the chapter contains several accessible exercises dealing with the structure and decomposition of polynomial ideals.

Chapter 6 studies the inherent complexity of polynomial system solving when working with the dense encoding of the input polynomials and under the model of straight-line programs, i.e., when polynomials are not given by their monomials but by evaluation programs. Being a brief survey of algebraic complexity applied to computational algebraic geometry, there is not much background required, though knowledge of basic notions of algebraic geometry and commutative algebra would be helpful. The chapter is mostly self-contained; when necessary, basic bibliography supplements are indicated.

Chapter 7 is devoted to the study of sparse systems of polynomial equations, i.e., algebraic equations with a specific monomial structure, presenting a comprehensive state-of-the-art introduction to the field. Combinatorial and discrete geometry, together with matrices of special structure, are ingredients of the presentation of toric (or sparse) elimination theory. The chapter focuses on applications to geometric modelling and computer-aided design. It also provides the tools for exploiting the structure of algebraic systems which may arise in different applications. Some basic knowledge of discrete geometry for polyhedral objects in arbitrary dimension is assumed. This chapter will be of particular interest to graduate students and researchers in theoretical computer science or applied mathematics wishing to combine discrete and algebraic geometry.

Chapter 8 deals with numerical algebraic geometry, a term coined some years ago to describe a new field, which bears the same relation to algebraic geometry as numerical linear algebra does to linear algebra. Modern homotopy methods to describe solution components of polynomial systems are presented. The prerequisites include a basic course in numerical analysis, in particular Newton's method for nonlinear systems. Because of the numerical flavor of the proposed methods, this chapter is expected to be particularly appealing to engineers.

Lastly, Chapter 9 gives a complete overview of old and recent methods for the important problem of approximate factorization of a multivariate polynomial, in other words, the complex irreducible decomposition of a hypersurface. The main techniques rely on approximate numerical computation but the results are exact and certified. It is addressed to students and researchers with some basic knowledge of commutative algebra, algebraic numbers and holomorphic functions of several variables.

This book grew out of Course Notes prepared for the CIMPA Graduate School on Systems of Polynomial Equation that we organized in Buenos Aires, in July 2003. We take this opportunity to thank CIMPA for the funding and the academic support to carry out this activity. We are also grateful for the support from the following institutions: International Centre for Theoretical Physics (ICTP, Italy), Consejo Nacional de Investigaciones Científicas y Técnicas (CONICET, Argentina), Institut National de Recherche en Informatique et en Automatique (INRIA, France), PROSUL Programme from CNPq (Brazil), Délégation régionale de coopération Française au Chili, and Universidad de Buenos Aires (Argentina). We also thank ECOS-Sud, whose project A00E02 between INRIA and Universidad de Buenos Aires provided the initial framework for our collaboration. Special thanks go to Gregorio Malajovich and Alvaro Rittatore, who co-organized with us the I Latin American Workshop on Polynomial Systems which followed the School. Finally, we would like to thank deeply all the speakers and all the participants.

December 2004   *Alicia Dickenstein and Ioannis Z. Emiris*

Contents

1

Introduction to residues and resultants

Eduardo Cattani [*1] and Alicia Dickenstein [**2]

[1] Department of Mathematics and Statistics - University of Massachusetts,
Amherst, MA 01003, USA, cattani@math.umass.edu
[2] Departamento de Matemática - FCEyN - Universidad de Buenos Aires, Ciudad
Universitaria - Pab. I - (1428) Buenos Aires, Argentina, alidick@dm.uba.ar

Summary. This chapter is an expanded version of the lecture notes prepared by the second-named author for her introductory course at the CIMPA Graduate School on Systems of Polynomial Equations held in Buenos Aires, Argentina, in July 2003. We present an elementary introduction to residues and resultants and outline some of their multivariate generalizations. Throughout we emphasize the application of these ideas to polynomial system solving.

1.0 Introduction

This chapter is an introduction to the theory of residues and of resultants. These are very classical topics with a long and distinguished history. It is not our goal to present a full historical account of their development but rather to introduce the basic notions in the one-dimensional case, to discuss some of their applications -in particular, those related to polynomial system solving- and present their multivariate generalizations. We emphasize in particular the applications of residues to duality theory and the explicit computation of resultants which, in turn, results in the explicit elimination of variables.

Most readers are probably familiar with the classical theory of local residues which was introduced by Augustin-Louis Cauchy in 1825 as a powerful tool for the computation of integrals and for the summation of infinite series. Perhaps less familiar is the fact that given a meromorphic form $(H(z)/P(z))dz$ on the complex plane, its global residue, i.e. the sum of local residues at the zeros of P, defines an easily computable linear functional on the quotient algebra $\mathcal{A} := \mathbb{C}[z]/\langle P(z)\rangle$ whose properties encode many important features of this algebra. As in Chapters 2 and 3, it is through the study of this algebra, and its multivariate generalization, that we make the connection with the roots of the associated polynomial system.

[*] Partially supported by NSF Grant DMS-0099707.
[**] Partially supported by Action A00E02 of the ECOS-SeTCIP French-Argentina bilateral collaboration, UBACYT X052 and ANPCYT 03-6568, Argentina.

The basic definitions and properties of the univariate residue are reviewed in Section 1.1 and we discuss some nice applications in Section 1.2. Although there are many different possible definitions of the residue, we have chosen to follow the classical integral approach for the definition of the local residue. Alternatively, one could define the global residue by its algebraic properties and use ring localization to define the local residue. We indicate how this is done in a particular case.

In Section 1.5 we study multidimensional residues. Although, as coefficients of certain Laurent expansions, they are already present in the work of Jacobi [Jac30], the first systematic treatment of bivariate residue integrals is the 1887 memoir of Poincaré [Poi87], more than 60 years after the introduction of univariate residues. He makes the very interesting observation that geometers were long stopped from extending the one-dimensional theory because of the lack of geometric intuition in 4 dimensions (referring to $\mathbb{C}^2$). The modern theory of residues and the duality in algebraic geometry is due to Leray and Grothendieck. There have been many developments since the late 70's: in the algebro-geometric side with the work of Grothendieck (cf. [Har66]); in analytic geometry where we may mention the books by Griffiths and Harris [GH78] and Arnold, Varchenko and Guseĭn -Zadé [AGZV85]; in commutative algebra with the work of Scheja and Storch [SS75, SS79], Kunz [Kun86], and Lipman [Lip87]; and in the analytic side with the residual currents approach pioneered by Coleff and Herrera [CH78]. In the 90's the possibility of implementing symbolic computations brought about another important expansion in the theory and computation of multidimensional residues and its applications to elimination theory as pioneered by the Krasnoyarsk school [AY83, BKL98, Tsi92]. It would, of course, be impossible to fully present all these approaches to the theory of residues or to give a complete account of all of its applications. Indeed, even a rigorous definition of multivariate residues would take us very far afield. Instead we will attempt to give an intuitive idea of this notion, explain some of its consequences, and describe a few of its applications. In analogy with the one-variable case we will begin with an "integral" definition of local residue from which we will define the total residue as a sum of local ones. The reader who is not comfortable with integration of differential forms should not despair since, as in the univariate case, we soon show how one can give a purely algebraic definition of global, and then local, residues using Bezoutians. We also touch upon the geometric definition of Arnold, Varchenko and Guseĭn-Zadé.

In Sections 1.3 and 1.4 we discuss the definition and application of the univariate resultant. This is, again, a very classical concept which goes back to the work of Euler, Bézout, Sylvester and Cayley. It was directly motivated by the problem of elimination of variables in systems of polynomial equations. While the idea behind the notion of the resultant is very simple, its computation leads to very interesting problems such as the search for determinantal formulas. We recall the classical Sylvester and Bezoutian matrices in Section 1.4.

The rebirth of the classical theory of elimination in the last decade owes much to the work of Jouanolou [Jou79, Jou91, Jou97] and of Gelfand, Kapranov and Zelevinsky [GKZ94], as well as to the possibility of using resultants not only as a computational tool to solve polynomial systems but also to study their complexity aspects. In particular, homogeneous and multi-homogeneous resultants are essential tools in the implicitization of surfaces. We discuss the basic constructions and properties in Section 1.6. We refer to [Stu93, Stu98], [Stu02, Ch. 4] and to Chapters 2, 3, and 7 in this book for further background and applications. A new theoretical tool in elimination theory yet to be fully explored is the use of exterior algebra methods in commutative algebra (starting with Eisenbud and Schreyer [ESW03] and Khetan [Khe03, Khe]).

In the last section of this chapter we recall how the resultant appears naturally as the denominator of the residue and apply this to obtain a normal form algorithm for the computation of resultants which, as far as we know, has not been noted before.

Although many of the results in this chapter, including those in the last section, are valid in much greater generality, we have chosen to restrict most of the exposition to the affine and projective cases. We have tried to direct the reader to the appropriate references.

For further reading we refer to a number of excellent books on the topics treated here: [AY83, AGZV85, CLO98, GKZ94, GH78, EM, Tsi92].

1.1 Residues in one variable

1.1.1 Local analytic residue

We recall that, given a holomorphic function $h(z)$ with an isolated singularity at a point ξ in $\mathbb{C}$, we may consider its Laurent expansion

$$h(z) = \sum_{n=1}^{\infty} \frac{b_n}{(z - \xi)^n} + \overline{h}(z),$$

where $\overline{h}$ is holomorphic in a neighborhood of ξ, and define the *residue* of h at ξ as

$$\operatorname{res}_\xi(h) = b_1. \tag{1.1}$$

The classical Residue Theorem tells us that the residue is "what remains after integrating" the differential form $(1/2\pi i)\, h(z)\, dz$ on a small circle around ξ. Precisely:

$$\operatorname{res}_\xi(h) = \frac{1}{2\pi i} \int_{|z-\xi|=\delta} h(z)dz,$$

for any sufficiently small positive δ and where the circle $\{|z - \xi| = \delta\}$ is oriented counter-clockwise.

Remark 1.1.1. As defined in (1.1), the residue depends on the choice of local coordinate z. Associating the residue to the meromorphic 1-form $h(z)\,dz$ makes it invariant under local change of coordinates. We will, however, maintain the classical notation, $\mathrm{res}_\xi(h)$ rather than write $\mathrm{res}_\xi(h(z)dz)$.

We can also think of the residue of a holomorphic function h at ξ as a linear operator $\mathrm{res}_\xi[h] : \mathcal{O}_\xi \to \mathbb{C}$, which assigns to any holomorphic function f defined near ξ the complex number

$$\mathrm{res}_\xi[h](f) := \mathrm{res}_\xi(f \cdot h).$$

Suppose h has a pole at ξ of order m, Then, the action of $\mathrm{res}_\xi[h]$ maps

$$1 \longmapsto b_1$$
$$z - \xi \longmapsto b_2$$
$$\vdots \quad \vdots \quad \vdots$$
$$(z - \xi)^{m-1} \longmapsto b_m$$

and for any $k \geq m$, $(z - \xi)^k \mapsto 0$ since $(z - \xi)^k \cdot h$ is holomorphic at ξ. These values suffice to characterize the residue map $\mathrm{res}_\xi[h]$ in this case: indeed, given f holomorphic near ξ, we write

$$f(z) = \sum_{j=0}^{m-1} \frac{f^{(j)}(\xi)}{j!}(z - \xi)^j + (z - \xi)^m g(z),$$

with g holomorphic in a neighborhood of ξ. Therefore

$$\mathrm{res}_\xi[h](f) = \sum_{j=0}^{m-1} \frac{f^{(j)}(\xi)}{j!}\mathrm{res}_\xi[h]((z - \xi)^j) = \sum_{j=0}^{m-1} \frac{b_{j+1}}{j!}f^{(j)}(\xi). \qquad (1.2)$$

Note, in particular, that the residue map $\mathrm{res}_\xi[h]$ is then the evaluation at ξ of a constant coefficient differential operator and that it carries the information of the principal part of h at ξ.

1.1.2 Residues associated to polynomials

In this notes we will be interested in the algebraic and computational aspects of residues and therefore we shall restrict ourselves to the case when $h(z)$ is a rational function $h(z) = H(z)/P(z)$, $H, P \in \mathbb{C}[z]$. Clearly, $\mathrm{res}_\xi(h) = 0$ unless $P(\xi) = 0$. It is straightforward to check the following basic properties of residues:

- If ξ is a simple zero of P, then

$$\mathrm{res}_\xi\left(\frac{H(z)}{P(z)}\right) = \frac{H(\xi)}{P'(\xi)}. \qquad (1.3)$$

- If ξ is a root of P of multiplicity m, then

$$\text{res}_\xi \left(\frac{H(z)P'(z)}{P(z)} \right) = m \cdot H(\xi). \tag{1.4}$$

Since $(P'(z)/P(z))dz = d(\ln P(z))$ wherever a logarithm $\ln P$ of P is defined, the expression above is often called the (local) logarithmic residue.

Given a polynomial $P \in \mathbb{C}[z]$, its polar set $Z_P := \{\xi \in \mathbb{C} : P(\xi) = 0\}$ is finite and we can consider the total sum of local residues

$$\text{res} \left(\frac{H}{P} \right) = \sum_{\xi \in Z_P} \text{res}_\xi (H/P),$$

where $H \in \mathbb{C}[z]$. We will be particularly interested in the global residue operator.

Definition 1.1.2. *The global residue* $\text{res}_P : \mathbb{C}[z] \to \mathbb{C}$ *is the sum of local residues:*

$$\text{res}_P(H) = \sum_{\xi \in Z_P} \text{res}_\xi (H/P)$$

Remark 1.1.3. We may define the sum of local residues over the zero set of P for any rational function h which is regular on Z_P. Moreover, if we write $h = H/Q$, with $Z_P \cap Z_Q = \emptyset$, then by the Nullstellensatz, there exist polynomials R, S such that $1 = RP + SQ$. It follows that the total sum of local residues

$$\sum_{\xi \in Z_P} \text{res}_\xi (h/P) = \text{res}_P(HS),$$

coincides with the global residue of the polynomial HS.

Let $R > 0$ be large enough so that Z_P be contained in the open disk $\{|z| < R\}$. Then, for any polynomial H the rational function $h = H/P$ is holomorphic for $|z| > R$ and has a Laurent expansion $\sum_{n \in \mathbb{Z}} e_n z^n$ valid for $|z| > R$. The residue of h at infinity is defined as

$$\text{res}_\infty(h) := -e_{-1}. \tag{1.5}$$

Note that integrating term by term the Laurent expansion, we get

$$\text{res}_\infty(h) = -\frac{1}{2\pi i} \int_{|z|=R} h(z)dz.$$

Since by the Residue Theorem,

$$\text{res}_P(H) = \frac{1}{2\pi i} \int_{|z|=R} \frac{H(z)}{P(z)} dz,$$

we easily deduce

Proposition 1.1.4. *Let* $P, H \in \mathbb{C}[z]$. *Then* $\mathrm{res}_P(H) = -\mathrm{res}_\infty(H/P)$.

Remark 1.1.5. We note that the choice of sign in (1.5) is consistent with Remark 1.1.1: If $h = H/P$ is holomorphic for $|z| > R$, then we may regard h as being holomorphic in a punctured neighborhood of the point at infinity in the Riemann sphere $S^2 = \mathbb{C} \cup \{\infty\}$. Taking $w = 1/z$ as local coordinate at infinity we have: $h(z)dz = -(h(1/w)/w^2)dw$ and

$$\mathrm{res}_0(-(h(1/w)/w^2)) = -e_{-1}. \tag{1.6}$$

Note also that Proposition 1.1.4 means that the sum of the local residues of the extension of the meromorphic form $(H(z)/P(z))\,dz$ to the Riemann sphere is zero.

Proposition 1.1.6. *Given* $P, H \in \mathbb{C}[z]$, $\mathrm{res}_P(H)$ *is linear in* H *and is a rational function of the coefficients of* P *with coefficients in* $\mathbb{Q}$.

Proof. The first statement follows from the definition of $\mathrm{res}_P(H)$ and the linearity of the local residue. Thus, in order to prove the second statement it suffices to consider $\mathrm{res}_P(z^k)$, $k \in \mathbb{N}$. Let $d = \deg P$, $P(z) = \sum_{j=0}^d a_j z^j$, $a_d \neq 0$. Then, if follows from Proposition 1.1.4 and (1.6) that

$$\mathrm{res}_P(z^k) = \mathrm{res}_0\left(\frac{(1/w)^k}{w^2\,P(1/w)}\right) = \mathrm{res}_0\left(\frac{1}{w^{k+2-d}\,P_1(w)}\right),$$

where $P_1(w) = \sum_{j=0}^d a_j w^{d-j}$. Note that $P_1(0) = a_d \neq 0$ and therefore $1/P_1(w)$ is holomorphic near 0. Hence

$$\mathrm{res}_P(z^k) = \begin{cases} 0 & \text{if } k+2-d \leq 0 \\ \frac{1}{\ell!}\frac{d^\ell}{dw^\ell}\left(\frac{1}{P_1}\right)(0) & \text{if } \ell := k+1-d \geq 0 \end{cases} \tag{1.7}$$

Now, writing $P_1 = a_d(1 + \sum_{j=0}^{d-1}\frac{a_j}{a_d}w^{d-j})$, the expression $\frac{1}{\ell!}\frac{d^\ell}{dw^\ell}\left(\frac{1}{P_1}\right)(0)$ may be computed as the w^ℓ coefficient of the geometric series

$$\frac{1}{a_d}\sum_{r=0}^{\infty}\left(-\sum_{j=0}^{d-1}\frac{a_j}{a_d}w^{d-j}\right)^r \tag{1.8}$$

and the result follows.

In fact, we can extract from (1.7) and (1.8) the following more precise dependence of the global residue on the coefficients of P.

Corollary 1.1.7. *Given a polynomial* $P = \sum_{j=0}^d a_j z^j \in \mathbb{C}[z]$ *of degree* d *and* $k \geq d-1$, *there exists a polynomial with integer coefficients* C_k *such that*

$$\mathrm{res}_P(z^k) = \frac{C_k(a_0, \ldots, a_d)}{a_d^{k-d+2}}.$$

In particular, when P, H *have coefficients in a subfield* $\mathbf{k}$, *it holds that* $\mathrm{res}_P(H) \in \mathbf{k}$.

We also deduce from (1.7) a very important vanishing result:

Theorem 1.1.8. (Euler-Jacobi vanishing conditions) *Given polynomials $P, H \in \mathbb{C}[z]$ satisfying $\deg(H) \leq deg(P) - 2$, the global residue*

$$\operatorname{res}_P(H) = 0.$$

We note that, in view of (1.3), when all the roots of P are simple, Theorem 1.1.8 reduces to the following algebraic statement: For every polynomial $H \in \mathbb{C}[z]$, with $\deg H < \deg P - 1$,

$$\sum_{\xi \in Z_P} \frac{H(\xi)}{P'(\xi)} = 0. \qquad (1.9)$$

The following direct proof of this statement was suggested to us by Askold Khovanskii. Let $d = \deg(P)$, $Z_P = \{\xi_1, \ldots, \xi_d\}$, and $P(z) = a_d \prod_{i=1}^{d}(z - \xi_i)$. Let L_i be the Lagrange interpolating polynomial

$$L_i(z) = \frac{\prod_{j \neq i}(z - \xi_j)}{\prod_{j \neq i}(\xi_i - \xi_j)}.$$

For any polynomial H with $\deg(H) \leq d - 1$,

$$H(z) = \sum_{i=1}^{d} H(\xi_i)\, L_i(z).$$

So, if $\deg(H) < d - 1$, the coefficient of z^{d-1} in this sum should be 0. But this coefficient is precisely

$$\sum_{i=1}^{d} H(\xi_i) \frac{1}{\prod_{j \neq i}(\xi_i - \xi_j)} = a_d \sum_{i=1}^{d} \frac{H(\xi_i)}{P'(\xi_i)}.$$

Since $a_d \neq 0$, statement (1.9) follows.

Since, clearly, $\operatorname{res}_P(G.P) = 0$, for all $G \in \mathbb{C}[z]$, the global residue map res_P descends to $\mathcal{A} := \mathbb{C}[z]/\langle P \rangle$, the quotient algebra by the ideal generated by P. On the other hand, if $\deg P = d$, then $\mathcal{A}$ is a finite dimensional $\mathbb{C}$-vector space of dimension $\deg(P)$, and a basis is given by the classes of $1, z, \ldots, z^{d-1}$. As in 2 we will denote by $[H]$ the class of H in the quotient $\mathcal{A}$. It follows from (1.7) and (1.8) that, as a linear map,

$$\operatorname{res}_P : \mathcal{A} \to \mathbb{C}$$

is particularly simple:

$$\operatorname{res}_P([z^k]) = \begin{cases} 0 & \text{if } 0 \leq k \leq d - 2, \\ \frac{1}{a_d} & \text{if } k = d - 1. \end{cases} \qquad (1.10)$$

The above observations suggest the following "normal form algorithm" for the computation of the global residue $\operatorname{res}_P(H)$ for any $H \in \mathbb{C}[z]$:

1) Compute the remainder $r(z) = r_{d-1}z^{d-1} + \cdots + r_1 z + r_0$ in the Euclidean division of H by $P = a_d z^d + \cdots + a_0$.
2) Then, $\mathrm{res}_P(H) = \frac{r_{d-1}}{a_d}$.

We may also use (1.10) to reverse the local-global direction in the definition of the residue obtaining, in the process, an algebraic definition which extends to polynomials with coefficients in an arbitrary algebraically-closed field $\mathbb{K}$ of characteristic zero. We illustrate this construction in the case of a polynomial $P(z) = \sum_{j=0}^{d} a_j z^j \in \mathbb{K}[z]$ with simple zeros. Define a linear map $L: \mathbb{K}[z]/\langle P \rangle \to \mathbb{K}$ as in (1.10). Let $Z_P = \{\xi_1, \ldots, \xi_d\} \subset \mathbb{K}$ be the zeros of P and $L_1, \ldots, L_d$ be the interpolating polynomials. For any $H \in \mathbb{K}[z]$ we set:

$$\mathrm{res}_{\xi_i}(H/P) := L([H.L_i]).$$

One can then check that the defining property (1.3) is satisfied. We will discuss another algebraic definition of the univariate residue in Section 1.2.1 and we will discuss the general passage from the global to the local residue in Section 1.5.3. We conclude this section by remarking on another consequence of Theorem 1.1.8. Suppose $P_1, P_2 \in \mathbb{C}[z]$ are such that their set of zeros Z_1, Z_2 are disjoint. Then, for any $H \in \mathbb{C}[z]$ such that

$$\deg H \leq \deg P_1 + \deg P_2 - 2$$

we have that

$$\sum_{\xi \in Z_1 \cup Z_2} \mathrm{res}_\xi \left(\frac{H}{P_1 P_2} \right) = 0$$

and, therefore

$$\mathrm{res}_{P_1}(H/P_2) = \sum_{\xi \in Z_1} \mathrm{res}_\xi \left(\frac{H}{P_1 P_2} \right) = -\sum_{\xi \in Z_2} \mathrm{res}_\xi \left(\frac{H}{P_1 P_2} \right) = -\mathrm{res}_{P_2}(H/P_1).$$

$$(1.11)$$

We denote the common value by $\mathrm{res}_{\{P_1, P_2\}}(H)$. Note that it is skew-symmetric on P_1, P_2. This is the simplest manifestation of a *toric* residue ([Cox96, CCD97]). We will discuss a multivariate generalization in Section 1.5.6.

1.2 Some applications of residues

1.2.1 Duality and Bezoutian

The global residue may be used to define a dualizing form in the algebra $\mathcal{A}$. We give, first of all, a proof of this result based on the local properties of the residue and, after defining the notion of the Bezoutian, we will give an algebraic construction of the dual basis.

Theorem 1.2.1. *For $P \in \mathbb{C}[z]$, let $\mathcal{A} = \mathbb{C}[z]/\langle P \rangle$. The pairing $\mathcal{A} \times \mathcal{A} \to \mathbb{C}$*

$$([H_1], [H_2]) \mapsto \operatorname{res}_P(H_1 \cdot H_2)$$

is non degenerate, i.e.

$$\operatorname{res}_P(H_1 \cdot H_2) = 0 \quad \text{for all} \quad H_2 \quad \text{if and only if} \quad H_1 \in \langle P \rangle.$$

Proof. Let $d = \deg P$ and denote by $\xi_1, \ldots, \xi_r$ the roots of P, with respective multiplicities $m_1, \ldots, m_r$. Assume, for simplicity, that P is monic. Suppose $\operatorname{res}_P(H_1 \cdot H_2) = 0$ for all H_2. Given $i = 1 \ldots, r$, let $G_i = \prod_{j \neq i}(z - \xi_j)^{m_j}$. Then, for any $\ell \leq m_i$,

$$0 = \operatorname{res}_P(H_1 \cdot (z - \xi_i)^\ell G_i) = \operatorname{res}_{\xi_i}(H_1/(z - \xi_i)^{m_i - \ell})$$

which, in view of (1.1.1), implies that $(z - \xi_i)^{m_i}$ divides H_1. Since these factors of P are pairwise coprime, it follows that $H_1 \in \langle P \rangle$, as desired.

As before, we denote by $\mathbb{K}$ an algebraically-closed field of characteristic zero.

Definition 1.2.2. *Let $P \in \mathbb{K}[z]$ be a polynomial of degree d. The Bezoutian associated to P is the bivariate polynomial*

$$\Delta_P(z, w) := \frac{P(z) - P(w)}{z - w} = \sum_{i=0}^{d-1} \Delta_i(z) w^i \in \mathbb{K}[z, w].$$

Proposition 1.2.3. *The classes $[\Delta_0(z)], \ldots, [\Delta_{d-1}(z)] \in \mathcal{A} = \mathbb{K}[z]/\langle P \rangle$ give the dual basis of the standard basis $[1], [z], \ldots, [z^{d-1}]$, relative to the non-degenerate pairing defined by the global residue.*

Proof. We note, first of all, that

$$P(z) - P(w) = \left(\sum_{i=0}^{d-1} \Delta_i(z) w^i \right) (z - w) = \sum_{i=0}^{d} (z \Delta_i(z) - \Delta_{i-1}(z)) w^i,$$

where it is understood that $\Delta_{-1}(z) = \Delta_d(z) = 0$. Writing $P(w) = \sum_{i=0}^{d} a_i w^i$ and comparing coefficients we get the following recursive definition of $\Delta_i(z)$:

$$z \Delta_i(z) = \Delta_{i-1}(z) - a_i, \tag{1.12}$$

with initial step: $z \Delta_0(z) = P(z) - a_0$. We now compute $\operatorname{res}_P([z^j] \cdot [\Delta_i(z)])$. Since $\deg \Delta_i = d - 1 - i$, $\deg(z^j \Delta_i(z)) = d - 1 - i + j$. Hence, if $i > j$, $\deg(z^j \Delta_i(z)) \leq d - 2$ and, by Theorem 1.1.8,

$$\operatorname{res}_P([z^j] \cdot [\Delta_i(z)]) = 0 \text{ for } i > j.$$

If $i = j$, then $\deg(z^j \Delta_j) = d - 1$ and it is easy to check from (1.12) that its leading coefficient is a_d, the leading coefficient of P. Hence

$$\mathrm{res}_P([z^j] \cdot [\Delta_j(z)]) \;=\; \mathrm{res}_P(a_d z^{d-1}) = 1.$$

Finally, we consider the case $i < j$. The relations (1.12) give:

$$z^j \Delta_i(z) \;=\; z^{j-1} z \Delta_i(z) \;=\; z^{j-1}(\Delta_{i-1}(z) - a_i)$$

and, therefore

$$\mathrm{res}_P(z^j \Delta_i(z)) = \mathrm{res}_P(z^{j-1} \Delta_{i-1}(z))$$

given that $\mathrm{res}_P(a_i z^{j-1}) = 0$ since $j - 1 \le d - 2$. Continuing in this manner we obtain

$$\mathrm{res}_P(z^j \Delta_i(z)) = \cdots = \mathrm{res}_P(z^{j-i} \Delta_0(z)) = \mathrm{res}_P(z^{j-i-1} P(z)) = 0.$$

Remark 1.2.4. Note that Proposition 1.2.3 provides an algebraic proof of Theorem 1.2.1. Indeed, we have shown that Theorem 1.2.1 only depends on the conditions (1.10) that we used in the algebraic characterization of the global residue. We may also use Proposition 1.2.3 to give an alternative algebraic definition of the global residue. Let $\Phi \colon \mathcal{A} \times \mathcal{A} \to \mathcal{A}$ denote the bilinear symmetric form defined by the requirement that $\Phi([z^i], [\Delta^j]) = \delta_{ij}$. Then, the global residue map $\mathrm{res} \colon \mathcal{A} \to \mathbb{K}$ is defined as the unique linear map such that $\Phi(\alpha, \beta) = \mathrm{res}(\alpha \cdot \beta)$, for $\alpha, \beta \in \mathcal{A}$.

Remark 1.2.5. The recursive relations (1.12) are exactly those defining the classical *Horner polynomials* $H_{d-i}(z) = a_d z^{i-1} + a_{d-1} z^{i-2} + \cdots + a_{d-i+1}$, associated to the polynomial $P(z) = \sum_{j=0}^{d} a_j z^j$.

1.2.2 Interpolation

Definition 1.2.6. *Let* $Z := \{\xi_1, \ldots, \xi_r\} \subset \mathbb{K}$ *be a finite set of points together with multiplicities* $m_1, \ldots, m_r \in \mathbb{N}$. *Let* $d = m_1 + \cdots + m_r$ *and* $h \in \mathbb{K}[z]$. *A polynomial* $H \in \mathbb{K}[z]$ *is said to interpolate* h *over* Z *if* $\deg H \le d - 1$ *and* $H^{(j)}(\xi_i) = h^{(j)}(\xi_i)$ *for all* $j = 1, \ldots, m_i - 1$.

Proposition 1.2.7. *Let* $Z \subset \mathbb{K}$ *and* $h \in \mathbb{K}[z]$ *be as above. Let* $P(z) := \prod_{i=1}^{r}(z - \xi_i)^{m_i}$. *Then* H *interpolates* h *over* Z *if and only if* $[H] = [h]$ *in* $\mathcal{A} = \mathbb{K}[z]/\langle P \rangle$, *i.e. if* H *is the remainder of dividing* h *by* P.

Proof. If we write $h = Q \cdot P + H$, with $\deg H < d$, then

$$h^{(j)}(\xi_i) \;=\; \sum_{k=0}^{j} c_k Q^{(k)}(\xi_i) P^{(k-j)}(\xi_i) + H^{(j)}(\xi_i),$$

for suitable coefficients $c_k \in \mathbb{K}$. Since $P^{(\ell)}(\xi_i) = 0$ for $\ell = 0, \ldots, m_i - 1$, it follows that H interpolates h. On the other hand, it is easy to check that the interpolating polynomial is unique and the result follows.

Lemma 1.2.8. *With notation as above, given $h \in \mathbb{K}[z]$, the interpolating polynomial H of h over Z equals*

$$H(w) = \sum_{i=1}^{d-1} c_i(h)w^i \quad \text{where } c_i(h) = \operatorname{res}_P(h \cdot \Delta_i).$$

Proof. This is a straightforward consequence of the fact that $\operatorname{res}_p(z^j \cdot \Delta_i(z)) = \delta_{ij}$. For the sake of completeness, we sketch a proof for the complex case using the integral representation of the residue.

For any $\epsilon > 0$ and any w with $|P(w)| < \epsilon$, we have by the Cauchy integral formula

$$h(w) = \frac{1}{2\pi i} \int_{|P(z)|=\epsilon} \frac{h(z)}{z-w}\, dz = \frac{1}{2\pi i} \int_{|P(z)|=\epsilon} \frac{h(z)}{P(z)-P(w)} \Delta_P(z,w)dz.$$

Denote $\Gamma := \{|P(z)| = \epsilon\}$; for any $z \in \Gamma$ we have the expansion

$$\frac{1}{P(z)-P(w)} = \frac{1}{P(z)} \frac{1}{1 - \frac{P(w)}{P(z)}} = \sum_{n\geq 0} \frac{P(w)^n}{P(z)^{n+1}},$$

which is uniformly convergent over Γ. Then,

$$h(w) = \sum_{n\geq 0} \left(\frac{1}{2\pi i} \int_\Gamma \frac{h(z)\,\Delta_P(z,w)}{P(z)^{n+1}} \right) P(w)^n, \tag{1.13}$$

and so, isolating the first summand we get

$$h(w) = \operatorname{res}_P(h(z)\,\Delta_P(z,w)) + Q(w)\,P(w). \tag{1.14}$$

Finally, call $H(w) := \operatorname{res}_P(h(z)\,\Delta_P(z,w))$. It is easy to check that $H = 0$ or $\deg(H) \leq d-1$, and by linearity of the residue operator, $H(w) = \sum_{i=1}^{d-1} c_i(h)\, w^i$, as desired.

1.2.3 Ideal membership

Let again $P(z) = \sum_{i=0}^d a_i z^i \in \mathbb{C}[z]$. While in the univariate case is trivial, it is useful to observe that Theorem 1.2.1 allows us to derive a *residual* system of d linear equations in the coefficients of all polynomials $H(z) = \sum_{j=1}^m h_j z^j$ of degree less than or equal to m, whose vanishing is equivalent to the condition that $H \in \langle P \rangle$.

Such a system can be deduced from any basis $B = \{\beta_0, \ldots, \beta_{d-1}\}$ of $\mathcal{A} = \mathbb{C}[z]/\langle P \rangle$. We can choose for instance the canonical basis of monomials $\{[z^j], j = 0, \ldots, d-1\}$, or the dual basis $\{[\Delta_k(z)], k = 0, \ldots, d-1\}$. Theorem 1.2.1 means that $H \in \langle P \rangle$, i.e. $[H] = 0$ if and only if

$$\mathrm{res}_P([H] \cdot \beta_i) \;=\; \sum_{j=0}^{m} h_j \mathrm{res}_P([z^j]\beta_i) \;=\; 0 \quad \forall\, i = 0,\dots,d-1.$$

Suppose $m \geq d$, when B is the monomial basis, the first $d \times d$ minor of the $d \times m$ matrix of the system is triangular, while if B is the dual basis given by the Bezoutian, this minor is simply the identity.

If $H \in \langle P \rangle$, we can obtain the quotient $Q(z) = H(z)/P(Z) \in \mathbb{C}[z]$ from equations (1.13), (1.14). Indeed, we have:

$$Q(w) = \sum_{n \geq 1} \mathrm{res}[P^{n+1}](H(z)\, \Delta_P(z,w)) P(w)^{n-1}.$$

By Theorem 1.1.8, the terms in this sum vanish when $n \geq \dfrac{\deg(H) + 1}{d}$.

1.2.4 Partial fraction decomposition

We recall the *partial fraction decomposition* of univariate rational functions. This is a very important classical result because of its usefulness in the computation of integrals of rational functions.

Let $P, H \in \mathbb{K}[z]$ with $\deg(H) + 1 \leq \deg(P) = d$. Let $\{\xi_1,\dots,\xi_r\}$ be the zeros of P and let $m_1,\dots,m_r$ denote their multiplicities. Then the rational function $H(z)/P(z)$ may be written as:

$$\frac{H(z)}{P(z)} \;=\; \sum_{i=1}^{r} \left(\frac{A_{i1}}{(z - \xi_i)} + \cdots + \frac{A_{im_i}}{(z - \xi_i)^{m_i}} \right) \tag{1.15}$$

for appropriate constants $A_{ij} \in \mathbb{K}$.

There are, of course, many elementary proofs of this result. Here we would like to show how it follows from the Euler-Jacobi vanishing Theorem 1.1.8. The argument below also gives a simple formula for the coefficients in (1.15) when P has only simple zeros.

For any $z \notin \{\xi_1,\dots,\xi_r\}$ we consider the auxiliary polynomial $P_1(w) = (z - w)P(w) \in \mathbb{K}[w]$. Its zeros are ξ_i, with multiplicity m_i, $i = 1,\dots,r$, and z with multiplicity one. On the other hand, $\deg H \leq \deg P_1 - 2$, and therefore Theorem 1.1.8 gives:

$$0 \;=\; \mathrm{res}_{P_1}(H) \;=\; \mathrm{res}_z(H/P_1) + \sum_{i=1}^{r} \mathrm{res}_{\xi_i}(H/P_1).$$

Since P_1 has a simple zero at z, we have $\mathrm{res}_z(H/P_1) = H(z)/P_1'(z) = -H(z)/P(z)$ and, therefore

$$\frac{H(z)}{P(z)} \;=\; \sum_{i=1}^{r} \mathrm{res}_{\xi_i}\left(\frac{H(w)}{(z-w)P(w)} \right).$$

In case P has simple zeros we have $\mathrm{res}_{\xi_i}(H/P_1) = H(\xi_i)/P_1'(\xi_i)$ which gives:

$$\frac{H(z)}{P(z)} = \sum_{i=1}^{r} \frac{(H(\xi_i)/P'(\xi_i))}{(z - \xi_i)}.$$

In the general case, it follows from (1.2) that

$$\mathrm{res}_{\xi_i}(H/P_1) = \sum_{j=0}^{m_i-1} k_j \frac{d^j(H(w)/(z-w))}{dw^j}(\xi_i) = \sum_{j=0}^{m_i-1} \frac{a_j}{(z-\xi_i)^{j+1}}$$

for suitable constants k_j and a_j.

We leave it as an exercise for the reader to compute explicit formulas for the coefficients A_{ij} in (1.15).

1.2.5 Computation of traces and Newton sums

Let $P(z) = \sum_{i=0}^{d} a_i z^i \in \mathbb{C}[z]$ be a polynomial of degree d, $\{\xi_1, \ldots, \xi_r\}$ the set of zeros of P, and $m_1, \ldots, m_r$ their multiplicities. As always, we denote by $\mathcal{A}$ the $\mathbb{C}$-algebra $\mathcal{A} = \mathbb{C}[z]/\langle P \rangle$. We recall (cf. Theorem 2.1.4 in Chapter 2) that for any polynomial $Q \in \mathbb{C}[z]$, the eigenvalues of the multiplication map

$$M_Q : \mathcal{A} \to \mathcal{A} ; \quad [H] \mapsto [Q \cdot H]$$

are the values $Q(\xi_i)$. In particular, using (1.4), the trace of M_Q may be expressed in terms of global residues:

$$\mathrm{tr}(M_Q) = \sum_i m_i Q(\xi_i) = \mathrm{Res}_P(Q \cdot P').$$

Theorem 1.2.9. *The pairing* $\mathcal{A} \times \mathcal{A} \to \mathbb{C}$

$$([g_1], [g_2]) \mapsto \mathrm{tr}(M_{g_1 g_2}) = \mathrm{Res}_P(g_1 \cdot g_2 \cdot P')$$

is non degenerate only when all zeros of P are simple. More generally, the trace $\mathrm{tr}(M_{g_1 g_2}) = 0$ *for all g_2 if and only if $g_1(\xi_i) = 0$, for all $i = 1, \ldots, r$ or, equivalently, if and only if $g_1 \in \sqrt{\langle P \rangle}$.*

Proof. Fix $g_1 \in \mathbb{C}[z]$. As $\mathrm{tr}(M_{g_1 g_2}) = \mathrm{res}_P(g_1 \cdot P' \cdot g_2)$, it follows from Theorem 1.2.1 that the trace of $g_1 \cdot g_2$ vanishes for all g_2 if and only if $g_1 P' \in \langle P \rangle$. But this happens if and only if g_1 vanishes over Z_P, since the multiplicity of P' at any zero p of P is one less than the multiplicity of P at p.

As $\mathrm{tr}_P(Q)$ is linear in Q, all traces can be computed from those corresponding to the monomials z^k; i.e. the *power sums* of the roots:

$$S_k := \sum_{i=1}^{r} m_i \xi_i^k = \mathrm{res}_P(z^k \cdot P'(z)).$$

It is well known that the S_k's are rational functions of the elementary symmetric functions on the zeros of P, i.e. the coefficients of P, and conversely (up to the choice of a_d). Indeed, the classical *Newton identities* give recursive relations to obtain one from the other. It is interesting to remark that not only the power sums S_k can be expressed in terms of residues, but that we can also use residues to obtain the Newton identities. The proof below is an adaptation to the one-variable case of the approach followed by Aïzenberg and Kytmanov [AK81] to study the multivariate analogues.

Lemma 1.2.10. (Newton identities) *For all* $\ell = 0, \ldots, d-1$,

$$(d - \ell)a_\ell = - \sum_{j>\ell}^{d} a_j S_{j-\ell} \tag{1.16}$$

Proof. The formula (1.16) follows from computing:

$$\mathrm{res}\left(\frac{P'(z)}{z^\ell P(z)} P(z) \right) \quad ; \quad \ell \in \mathbb{N}$$

in two different ways:

i) As $\mathrm{res}\left(\dfrac{P'(z)}{z^\ell} \right) = \mathrm{res}_0\!\left(\dfrac{P'(z)}{z^\ell} \right) = \ell a_\ell.$

ii) Expanding it as a sum:

$$\sum_{j=0}^{d} a_j \, \mathrm{res}\left(\frac{P'(z)\, z^j}{z^\ell P(z)} \right) = \sum_{j<\ell} a_j \mathrm{res}\left(\frac{P'(z)}{z^{\ell-j} P(z)} \right) + \sum_{j\geq\ell} a_j \mathrm{res}\left(\frac{P'(z) z^{j-\ell}}{P(z)} \right)$$

The terms in the first sum vanish by Theorem 1.1.8 since $\deg(z^{\ell-j}P(z)) \geq \deg(P'(z)) + 2$, while the second sum may be expressed as $\sum_{j\geq\ell} a_j S_{j-\ell}$. Since $S_0 = d$, the identity (1.16) follows.

1.2.6 Counting integer points in lattice tetrahedra

Let $\mathcal{P} \subset \mathbb{R}^n$ be a polytope with integral vertices and let $\mathcal{P}^\circ$ denote its interior. For any $t \in \mathbb{N}$, call

$$L(\mathcal{P}, t) := \#(t \cdot \mathcal{P}) \cap \mathbb{Z}^n \quad ; \quad L(\mathcal{P}^\circ, t) := \#(t \cdot \mathcal{P}^\circ) \cap \mathbb{Z}^n,$$

the number of the lattice points in the dilated polyhedron $t \cdot \mathcal{P}$ and in its dilated interior. Ehrhart [Ehr67] proved that these are polynomial functions of degree n. They are known as the Ehrhart polynomials associated to $\mathcal{P}$ and $\mathcal{P}^\circ$. Moreover, he determined the two leading coefficients and the constant term in terms of the volume of the polytope, the normalized volume of its boundary and its Euler characteristic. The other coefficients are not as easily accessible, and a method of computing these coefficients was unknown until quite recently (cf. [Bar94, KK93, Pom93]). There is a remarkable relation between these two polynomials, the Ehrhart-Macdonald reciprocity law:

$$L(\mathcal{P}^\circ, t) = (-1)^n L(\mathcal{P}, t).$$

In [Bec00], Matthias Beck shows how to express these polynomials in terms of (multidimensional) residues. In the particular case when $\mathcal{P}$ is a tetrahedron, this is just a rational one-dimensional residue. We illustrate Beck's approach by sketching a proof of Ehrhart-Macdonald reciprocity in the case of a tetrahedron.

Fix $\alpha_1, \ldots, \alpha_n \in \mathbb{N}$ and consider the tetrahedron with vertices at the origin and at the points $(0, \ldots, \alpha_i, \ldots, 0)$:

$$\Sigma = \{(x_1, \ldots, x_n) \in \mathbb{R}^n_{\geq 0} : \sum_{k=1}^n \frac{x_k}{\alpha_k} \leq 1\}.$$

Clearly,

$$\Sigma^\circ = \{(x_1, \ldots, x_n) \in \mathbb{R}^n_{>0} : \sum_{k=1}^n \frac{x_k}{\alpha_k} < 1\}.$$

Let $A := \prod_{i=1}^n \alpha_i$, $A_k := \prod_{i \neq k} \alpha_i$, $k = 1, \ldots, n$. Then,

$$L(\Sigma, t) = \#\{m \in \mathbb{Z}^n_{\geq 0} : \sum_{k=1}^n \frac{m_k}{\alpha_k} \leq t\}$$

$$= \#\{m \in \mathbb{Z}^n_{\geq 0} : \sum_{k=1}^n m_k A_k \leq t\, A\}$$

$$= \#\{m \in \mathbb{Z}^{n+1}_{\geq 0} : \sum_{k=1}^n m_k A_k + m_{n+1} = t\, A\}.$$

So, we can interpret $L(\Sigma, t)$ as the coefficient of z^{tA} in the series product:

$$(1 + z^{A_1} + z^{2A_1} + \ldots) \ldots (1 + z^{A_n} + z^{2A_n} + \ldots)(1 + z + z^2 + \ldots),$$

i.e. as the coefficient of z^{tA} in the Taylor expansion at the origin of

$$\frac{1}{(1 - z^{A_1}) \ldots (1 - z^{A_n})(1 - z)}.$$

Thus,

$$L(\Sigma, t) = \mathrm{res}_0 \left(\frac{z^{-tA-1}}{(1 - z) \cdot \prod_{i=1}^n (1 - z^{A_i})} \right)$$

$$= 1 + \mathrm{res}_0 \left(\frac{z^{-tA} - 1}{z \cdot (1 - z) \cdot \prod_{i=1}^n (1 - z^{A_i})} \right).$$

For $t \in \mathbb{Z}$, let us denote by $f_t(z)$ the rational function

$$f_t(z) := \frac{z^{tA} - 1}{z \cdot (1 - z) \cdot \prod_{i=1}^{n}(1 - z^{A_i})}.$$

Note that for $t > 0$, $\mathrm{res}_0(f_t) = -1$, while for $t < 0$, $\mathrm{res}_\infty(f_t) = 0$. In particular, denoting by Z the set of non-zero, finite poles of f_t, we have for $t > 0$:

$$L(\Sigma, t) = 1 + \mathrm{res}_0(f_{-t}(z)) = 1 - \sum_{\xi \in Z} \mathrm{res}_\xi(f_{-t}(z)). \qquad (1.17)$$

Since $L(\Sigma, t)$ is a polynomial, this identity now holds for every t.

Similarly, we compute that

$$L(\Sigma^\circ, t) = \#\{m \in \mathbb{Z}_{>0}^{n+1} : \sum_{k=1}^{n} m_k A_k + m_{n+1} = t\,A\}.$$

That means that $L(\Sigma^\circ, t)$ is the coefficient of w^{tA} in the series product:

$$(w^{A_1} + w^{2A_1} + \dots) \dots (w^{A_n} + w^{2A_n} + \dots)(w + w^2 + \dots)$$

or, in terms of residues:

$$L(\Sigma^\circ, t) = \mathrm{res}_0 \left(\frac{w^{A_1} \dots w^{A_n} (w^{-tA} - 1)}{(1 - w^{A_1}) \dots (1 - w^{A_n})(1 - w)} \right).$$

The change of variables $z = 1/w$ now yields

$$\begin{aligned}
L(\Sigma^\circ, t) &= (-1)^n \, \mathrm{res}_\infty \left(\frac{z^{tA} - 1}{z\,(1 - z^{A_1}) \dots (1 - z^{A_n})\,(1 - z)} \right) \\
&= (-1)^n \mathrm{res}_\infty(f_t(z)). \qquad (1.18)
\end{aligned}$$

The Ehrhart-Macdonald reciprocity law now follows from comparing (1.17) and (1.18), and using the fact that for $t > 0$, $\mathrm{res}_0(f_t) = -1$.

1.3 Resultants in one variable

1.3.1 Definition

Fix two natural numbers d_1, d_2 and consider generic univariate polynomials of these degrees and coefficients in a field $\mathbf{k}$:

$$P(z) = \sum_{i=0}^{d_1} a_i z^i, \quad Q(z) = \sum_{i=0}^{d_2} b_i z^i. \qquad (1.19)$$

The system $P(z) = Q(z) = 0$ is, in general, overdetermined and has no solutions. The following result is classical:

Theorem 1.3.1. *There exists a unique (up to sign) irreducible polynomial*

$$\text{Res}_{d_1,d_2}(P,Q) = \text{Res}_{d_1,d_2}(a_0,\ldots,a_{d_1},b_0,\ldots,b_{d_2}) \in \mathbb{Z}[a_0,\ldots,a_{d_1},b_0,\ldots,b_{d_2}],$$

called the resultant of P and Q, which verifies that for any specialization of the coefficients a_i, b_i in $\mathbf{k}$ with $a_{d_1} \neq 0, b_{d_2} \neq 0$, the resultant vanishes if and only if the polynomials P and Q have a common root in any algebraically closed field $\mathbb{K}$ containing $\mathbf{k}$.

Geometrically, the hypersurface $\{(a,b) \in \mathbb{K}^{d_1+d_2+2} : \text{Res}_{d_1,d_2}(a,b) = 0\}$ is the projection of the incidence variety $\{(a,b,z) \in \mathbb{K}^{d_1+d_2+3} : \sum_{i=0}^{d_1} a_i z^i = \sum_{i=0}^{d_2} b_i z^i = 0\}$; that is to say, the variable z is eliminated. Here, and in what follows, $\mathbb{K}$ denotes an algebraically closed field.

A well known theorem of Sylvester allows us to compute the resultant as the determinant of a matrix of size $d_1 + d_2$, whose entries are 0 or a coefficient of either P or Q. For instance, when $d_1 = d_2 = 2$, the resultant is the following polynomial in 6 variables $(a_0, a_1, a_2, b_0, b_1, b_2)$:

$$b_2^2 a_0^2 - 2b_2 a_0 a_2 b_0 + a_2^2 b_0^2 - b_1 b_2 a_1 a_0 - b_1 a_1 a_2 b_0 + a_2 b_1^2 a_0 + b_0 b_2 a_1^2$$

and can be computed as the determinant of the 4×4 matrix:

$$M_{2,2} := \begin{pmatrix} a_0 & 0 & b_0 & 0 \\ a_1 & a_0 & b_1 & b_0 \\ a_2 & a_1 & b_2 & b_1 \\ 0 & a_2 & 0 & b_2 \end{pmatrix}. \tag{1.20}$$

Let us explain how one gets this result. The basic idea is to linearize the problem in order to use the eliminant polynomial par excellence: the determinant. Note that the determinant of a square homogeneous linear system $A \cdot x = 0$ allows to eliminate x: the existence of a non trivial solution $x \neq 0$ of the system, is equivalent to the fact that the determinant of A (a polynomial in the entries of A) vanishes.

Assume $\deg(P) = d_1, \deg(Q) = d_2$. A first observation is that P and Q have a common root if and only if they have a common factor of positive degree (since $P(z_0) = 0$ if and only if $z - z_0$ divides P). Moreover, the existence of such a common factor is equivalent to the existence of polynomials g_1, g_2 with $\deg(g_1) \leq d_2 - 1, \deg(g_2) \leq d_1 - 1$, such that $g_1 P + g_2 Q = 0$. Denote by S_ℓ the space of polynomials of degree ℓ and consider the map

$$\begin{aligned} S_{d_2-1} \times S_{d_1-1} &\longrightarrow S_{d_1+d_2-1} \\ (g_1, g_2) &\longmapsto g_1 P + g_2 Q \end{aligned} \tag{1.21}$$

This defines a $\mathbb{K}$-linear map between two finite dimensional $\mathbb{K}$-vector spaces of the same dimension $d_1 + d_2$, which is surjective (and therefore an isomorphism) if and only if P and Q do not have any common root in $\mathbb{K}$. Denote by M_{d_1,d_2}

the matrix of this linear map in the monomial bases. It is called the *Sylvester matrix* associated to P and Q. Then

$$\text{Res}_{d_1,d_2}(P,Q) = \pm \det(M_{d_1,d_2}). \tag{1.22}$$

The sign in this last equality cannot be determined, but the positive sign is taken by convention.

Note that for $d_1 = d_2 = 2$ we obtain the matrix $M_{2,2}$ in (1.20). The general shape of the Sylvester matrix is:

$$\begin{pmatrix}
a_0 & & & & b_0 & & & \\
a_1 & a_0 & & & b_1 & b_0 & & \\
& a_1 & \ddots & & & b_1 & \ddots & \\
\vdots & & \ddots & a_0 & \vdots & & \ddots & b_0 \\
& \vdots & & a_1 & & \vdots & & b_1 \\
a_{d_1} & & & & b_{d_2} & & & \\
& a_{d_1} & & \vdots & & b_{d_2} & & \vdots \\
& & \ddots & & & & \ddots & \\
& & & a_{d_1} & & & & b_{d_2}
\end{pmatrix}$$

where the blank spaces are filled with zeros.

Note that setting $a_{d_1} = 0$ but $b_{d_2} \neq 0$, the determinant of the Sylvester matrix equals b_{d_2} times the determinant of the Sylvester matrix M_{d_1-1,d_2} (in $a_0, \ldots, a_{d_1-1}, b_0, \ldots, b_{d_2}$). We deduce that when $\deg(P) = d_1' < d_1$ and $\deg(Q) = d_2$, the restriction of the (d_1, d_2) resultant polynomial to the closed set $(a_{d_1} = \cdots = a_{d_1'+1} = 0)$ of polynomials of degrees d_1', d_2 factorizes as $\text{Res}_{d_1,d_2}(P,Q) = b_{d_2}^{d_1-d_1'} \text{Res}_{d_1',d_2}(P,Q)$.

What happens if we specialize both P and Q to polynomials of respective degrees smaller than d_1 and d_2? Then, the last row of the Sylvester matrix is zero and so the resultant vanishes, but in principle P and Q do not need to have a common root in K. One way to recover the equivalence between the vanishing of the resultant and the existence of a common root is the following.

Given P, Q as in (1.19), consider the homogenizations P^h, Q^h defined by

$$P^h(z,w) = \sum_{i=0}^{d_1} a_i z^i w^{d_1-i}, \quad Q^h = \sum_{i=0}^{d_2} b_i z^i w^{d_2-i}.$$

Then, P, Q can be recovered by evaluating at $w = 1$ and $(z_0, 1)$ is a common root of P^h, Q^h if and only if $P(z_0) = Q(z_0) = 0$. But also, on one hand $P^h(0,0) = Q^h(0,0) = 0$ for any choice of coefficients, and on the other P^h, Q^h have the common root $(1,0)$ when $a_{d_1} = b_{d_2} = 0$. The space obtained as the classes of pairs $(z,w) \neq (0,0)$ after identification of (z,w) with $(\lambda z, \lambda w)$ for any $\lambda \in \mathbb{K} - \{0\}$, denoted $\mathbb{P}^1(\mathbb{K})$, is called the projective line over $\mathbb{K}$. Since for homogeneous polynomials as P^h it holds that $P^h(\lambda z, \lambda w) = \lambda^{d_1} P^h(z, w)$ (and similarly for Q^h), it makes sense to speak of their zeros in $\mathbb{P}^1(\mathbb{K})$. So, we could restate Theorem 1.3.1 saying that for any specialization of the coefficients of P and Q, the resultant vanishes if and only if their homogenizations have a common root in $\mathbb{P}^1(\mathbb{K})$. As we have already remarked, when $\mathbb{K} = \mathbb{C}$, the projective space $\mathbb{P}^1(\mathbb{C})$ can be identified with the Riemann sphere, a compactification of the complex plane, where the class of the point $(1, 0)$ is identified with the point at infinity.

1.3.2 Main properties

It is interesting to realize that many properties of the resultant can be derived from its expression (1.22) as the determinant of the Sylvester matrix:

i)   The resultant Res_{d_1, d_2} is homogeneous in the coefficients of P and Q separately, with respective degrees d_2, d_1. So, the degree of the resultant in the coefficients of P is the number of roots of Q, and vice-versa.

ii)   The resultants Res_{d_1, d_2} and Res_{d_2, d_1} coincide up to sign.

iii)   There exist polynomials $A_1, A_2 \in \mathbb{Z}[a_0, \ldots, b_{d_2}][z]$ with $\deg(A_1) = d_2 - 1, \deg(A_2) = d_1 - 1$ such that

$$\mathrm{Res}_{d_1, d_2}(P, Q) = A_1 P + A_2 Q. \tag{1.23}$$

Let us sketch the proof of property iii). If we add to the first row in the Sylvester matrix z times the second row, plus z^2 times the third row, and so on, the first row becomes

$$P(z) \quad zP(z) \quad \ldots \quad z^{d_2-1}P(z) \quad Q(z) \quad zQ(z) \quad z^{d_1-1}Q(z)$$

but the determinant is unchanged. Expanding along this modified first row, we obtain the desired result.

Another important classical property of the resultant $\mathrm{Res}_{d_1, d_2}(P, Q)$ is that it can be written as a product over the zeros of P or Q:

Proposition 1.3.2. (Poisson formula) *Let P, Q polynomials with respective degrees d_1, d_2 and write $P(z) = a_{d_1} \prod_{i=1}^r (z - p_i)^{m_i}$, $Q(z) = b_{d_2} \prod_{j=1}^s (z - q_j)^{n_j}$. Then*

$$\mathrm{Res}_{d_1, d_2}(P, Q) = a_{d_1}^{d_2} \prod_{i=1}^r Q(p_i)^{m_i} = (-1)^{d_1 d_2} b_{d_2}^{d_1} \prod_{j=1}^s P(q_i)^{n_i}$$

Proof. Again, one possible way of proving the Poisson formula is by showing that

$$\mathrm{Res}_{d_1, d_2}((z - p)P_1, Q) = Q(p)\mathrm{Res}_{d_1-1, d_2}(P_1, Q),$$

using the expression of the resultant as the determinant of the Sylvester matrix, and standard properties of determinants. The proof would be completed by induction, and the homogeneity of the resultant.

Alternatively, one could observe that $R'(a, b) := a_{d_1}^{d_2} \prod_{i=1}^{r} Q(p_i)^{m_i}$ depends polynomially on the coefficients of Q and, given the equalities

$$R'(a, b) = a_{d_1}^{d_2} b_{d_2}^{d_1} \prod_{i,j} (p_i - q_j)^{m_i n_j} = (-1)^{d_1 d_2} b_{d_2}^{d_1} \prod_{j=1}^{s} P(q_i)^{n_i},$$

on the coefficients of P as well. Since the roots are unchanged by dilation of the coefficients, we see that, as Res_{d_1, d_2}, the polynomial R' has degree $d_1 + d_2$ in the coefficients $(a, b) = (a_0, \dots, b_{d_2})$. Moreover, $R'(a, b) = 0$ if and only if there exists a common root, i.e. if and only if $\mathrm{Res}_{d_1, d_2}(a, b) = 0$. This holds in principle over the open set $(a_{d_1} \neq 0, b_{d_2} \neq 0)$ but this implies that the loci $\{R' = 0\}$ and $\{\mathrm{Res}_{d_1, d_2} = 0\}$ in $\mathbb{K}^{d_1 + d_2 + 2}$ agree. Then, the irreducibility of Res_{d_1, d_2} implies the existence of a constant $c \in \mathbb{K}$ such that $\mathrm{Res}_{d_1, d_2} = c \cdot R'$. Evaluating at $P(z) = 1, Q(z) = z^{d_2}$, the Sylvester matrix M_{d_1, d_2} reduces to the identity $I_{d_1 + d_2}$ and we get $c = 1$.

We immediately deduce

Corollary 1.3.3. *Assume $P = P_1 \cdot P_2$ with $\deg(P_1) = d_1'$, $\deg(P_2) = d_1''$ and $\deg(Q) = d_2$. Then,*

$$\mathrm{Res}_{d_1' + d_1'', d_2}(P, Q) = \mathrm{Res}_{d_1', d_2}(P_1, Q) \, \mathrm{Res}_{d_1'', d_2}(P_2, Q).$$

There are other determinantal formulas to compute the resultant, coming from suitable generalizations of the map (1.21), which are for instance described in [DD01]. In case $d_1 = d_2 = 3$, the Sylvester matrix $M_{3,3}$ is 6×6. Denote $[ij] := a_i b_j - a_j b_i$, for all $i, j = 0, \dots, 3$. The resultant $\mathrm{Res}_{3,3}$ can also be computed as the determinant of the following 3×3 matrix:

$$B_{3,3} := \begin{pmatrix} [03] & [02] & [01] \\ [13] & [03] + [12] & [02] \\ [23] & [13] & [03] \end{pmatrix}, \tag{1.24}$$

or as minus the determinant of the 5×5 matrix

$$\begin{pmatrix} a_0 & 0 & b_0 & 0 & [01] \\ a_1 & a_0 & b_1 & b_0 & [02] \\ a_2 & a_1 & b_2 & b_1 & [03] \\ a_3 & a_2 & b_3 & b_3 & 0 \\ 0 & a_3 & 0 & b_3 & 0 \end{pmatrix}.$$

Let us explain how the matrix $B_{3,3}$ was constructed and why $\mathrm{Res}_{3,3} = \det(B_{3,3})$. We assume, more generally, that $d_1 = d_2 = d$.

Definition 1.3.4. *Let P, Q polynomials of degree d as in (1.19). The Bezoutian polynomial associated to P and Q is the bivariate polynomial*

$$\Delta_{P,Q}(z,y) = \frac{P(z)Q(y) - P(y)Q(z)}{z - y} = \sum_{i,j=0}^{d-1} c_{ij} z^i y^j.$$

The $d \times d$ matrix $B_{P,Q} = (c_{ij})$ is called the Bezoutian matrix associated to P and Q.

Note that $\Delta_{P,1} = \Delta_P$ defined in (1.2.2) and that each coefficient c_{ij} is a linear combination with integer coefficients of the brackets $[k, \ell] = a_k b_\ell - a_\ell b_k$.

Proposition 1.3.5. *With the above notations,*

$$\text{Res}_{d,d}(a, b) = \det(B_{P,Q}). \tag{1.25}$$

Proof. The argument is very similar to the one presented in the proof of Poisson's formula. Call $R' := \det(B_{P,Q})$. This is a homogeneous polynomial in the coefficients (a, b) of the same degree $2d = d + d$ as the resultant. Moreover, if $\text{Res}_{d,d}(a, b) = 0$, there exists $z_0 \in \mathbb{K}$ such that $P(z_0) = Q(z_0) = 0$, and so, $\Delta_{P,Q}(y, z_0) = \sum_{i=0}^{d-1} \left(\sum_{j=0}^{d-1} c_{ij} z_0^j \right) y^i$ is the zero polynomial. This shows that $R'(a, b) = 0$ since the non trivial vector $(1, z_0, \ldots, z_0^{d-1})$ lies in the kernel of the Bezoutian matrix $B_{P,Q}$. By Hilbert's Nullstellensatz, the resultant divides a power of R'. Using the irreducibility of $\text{Res}_{d,d}$ plus a particular specialization to adjust the constant, we get the desired result.

The Bezoutian matrices are more compact and practical experience seems to indicate that these matrices are numerically more stable than the Sylvester matrices.

1.4 Some applications of resultants

1.4.1 Systems of equations in two variables

Suppose that we want to solve a polynomial system in two variables $f(z, y) = g(z, y) = 0$ with $f, g \in \mathbb{K}[z, y]$. We can "hide the variable y in the coefficients" and think of $f, g \in \mathbb{K}[y][z]$. Denote by d_1, d_2 the respective degrees in the variable z. Then, the resultant $\text{Res}_{d_1, d_2}(f, g)$ with respect to the variable z will give us back a polynomial (with integer coefficients) in the coefficients, i.e. we will have a polynomial in y, which vanishes on every y_0 for which there exists z_0 with $f(z_0, y_0) = g(z_0, y_0) = 0$. So, we can eliminate the variable z from the system, detect the second coordinates y_0 of the solutions, and then try to recover the full solutions (z_0, y_0).

Assume for instance that $f(z, y) = z^2 + y^2 - 10$, $g(z, y) = z^2 + 2y^2 + zy - 16$. We write

$$f(z, y) = z^2 + 0z + (y^2 - 10), \quad g(z, y) = z^2 + yz + (2y^2 - 16).$$

Then, $\mathrm{Res}_{2,2}(f,g)$ equals

$$\mathrm{Res}_{2,2}((1,0,y^2-10),(1,y,2y^2-16)) = -22y^2+2y^4+36 = 2(y+3)(y-3)(y^2-2).$$

For each of the four roots $y_0 = -3,3,\sqrt{2},-\sqrt{2}$, we replace $g(z,y_0) = 0$ and we need to solve $z = \frac{y_0^2-6}{y_0}$. Note that $f(z,y_0) = 0$ will also be satisfied due to the vanishing of the resultant. So, there is precisely one solution z_0 for each y_0. The system has $4 = 2 \times 2$ real solutions.

It is easy to deduce from the results and observations made in Section 1.3 the following extension theorem.

Theorem 1.4.1. *Write $f(z,y) = \sum_{i=1}^{d_1} f_i(y)z^i$, $g(z,y) = \sum_{i=1}^{d_2} g_i(y)z^i$, with $f_i, g_i \in \mathbb{K}[y]$, and f_{d_1}, g_{d_2} non zero. Let y_0 be a root of the resultant with respect to z, $\mathrm{Res}_{d_1,d_2}(f,g) \in \mathbb{K}[y]$. If either $f_{d_1}(y_0) \neq 0$ or $g_{d_2}(y_0) \neq 0$, there exists $z_0 \in \mathbb{K}$ such that $f(z_0,y_0) = g(z_0,y_0) = 0$.*

Assume now that $f(z,y) = yz-1$, $g(z,y) = y^3-y$. It is immediate to check that they have two common roots, namely $\{f = g = 0\} = \{(1,1),(-1,-1)\}$. Replace g by the polynomial $\tilde{g} := g + f$. Then, $\{f = \tilde{g} = 0\} = \{f = g = 0\}$ but now both $f, \tilde{g}$ have positive degree 1 with respect to the variable z. The resultant with respect to z equals

$$\mathrm{Res}_{1,1}(f,\tilde{g}) = \det \begin{pmatrix} y & -1 \\ y & y^3 - y - 1 \end{pmatrix} = y^2(y^2 - 1).$$

Since both leading coefficients with respect to z are equal to the polynomial y, Theorem 1.4.1 asserts that the two roots $y_0 = \pm 1$ can be extended. On the contrary, the root $y_0 = 0$ cannot be extended.

Consider now $f(z,y) = yz^2 + z - 1$, $g(z,y) = y^3 - y$ and let us again consider f and $\tilde{g} := g + f$, which have positive degree 2 with respect to z. In this case, $y_0 = 0$ is a root of $\mathrm{Res}_{2,2}(f,\tilde{g}) = y^4(y^2 - 1)^2$. Again, $y_0 = 0$ annihilates both leading coefficients with respect to z. But nevertheless it can be extended to the solution $(0,1)$.

So, two comments should be made. The first one is that finding roots of univariate polynomials is in general not an algorithmic task! One can try to detect the rational solutions or to approximate the roots numerically if working with polynomials with complex coefficients. The second one is that even if we can obtain the second coordinates explicitly, we have in general a sufficient but not necessary condition to ensure that a given partial solution y_0 can be extended to a solution (z_0, y_0) of the system, and an ad hoc study may be needed.

1.4.2 Implicit equations of curves

Consider a parametric plane curve $\mathcal{C}$ given by $z = f(t), y = g(t)$, where $f, g \in \mathbb{K}[t]$, or more precisely,

$$\mathcal{C} = \{(z,y) \in \mathbb{K}^2 : z = f(t), \, y = g(t) \text{ for some } t \in \mathbb{K}\}.$$

Having this parametric expression allows one to "follow" or "travel along" the curve, but it is hard to detect if a given point in the plane is in $\mathcal{C}$. One can instead find an implicit equation $f \in \mathbb{K}[z,y]$, i.e. a bivariate polynomial f such that $\mathcal{C} = \{f = 0\}$. This amounts to eliminating t from the equations $z - f(t) = y - g(t) = 0$ and can thus be done by computing the resultant with respect to t of these polynomials.

This task could also be solved by a Gröbner basis computation. But we propose the reader to try in any computer algebra system the following example suggested to us by Ralf Fröberg. Consider the curve $\mathcal{C}$ defined by $z = t^{32}$, $y = t^{48} - t^{56} - t^{60} - t^{62} - t^{63}$. Then the resultant $\mathrm{Res}_{32,63}(t^{32} - z, t^{48} - t^{56} - t^{60} - t^{62} - t^{63} - y)$ with respect to t can be computed in a few seconds, giving the answer $f(z,y)$ we are looking for. It is a polynomial of degree 63 in z and degree 32 in y with 257 terms. On the other side, a Gröbner basis computation seems to be infeasible.

For a plane curve $\mathcal{C}$ with a rational parametrization; i.e.

$$\mathcal{C} = \{(p_1(t)/q_1(t), p_2(t)/q_2(t)) : q_1(t) \neq 0, q_2(t) \neq 0\},$$

where $p_i, q_i \in \mathbb{K}[t]$, the elimination ideal

$$I_1 := \langle q_1(t)z - p_1(t), q_2(t)y - p_2(t)\rangle \cap \mathbb{K}[z,y]$$

defines the Zariski closure of $\mathcal{C}$ in $\mathbb{K}^2$. We can obtain a generator of I_1 with a resultant computation that eliminates t. For example, let

$$\mathcal{C} = \left\{ \left(\frac{t^2 - 1}{(1 + 2t)^2}, \frac{t + 1}{(1 + 2t)(1 - t)} \right), t \neq 1, -1/2 \right\}.$$

Then $\overline{\mathcal{C}} = \mathcal{V}(I_1)$ is the zero locus of

$$f(z,y) = \mathrm{Res}_{2,2}((1 + 2t)^2 z - (t^2 - 1), (1 + 2t)(1 - t))y - (t + 1))$$

which equals

$$27y^2 z - 18yz + 4y + 4z^2 - z.$$

We leave it to the reader to verify that $\mathcal{C}$ is not Zariski closed.

One could also try to implicitize non planar curves. We show a general classical trick in the case of the space curve $\mathcal{C}$ with parametrization $x = t^2, y = t^3, z = t^5$. We have 3 polynomials $x - t^2, y - t^3, z - t^5$ from which we want to eliminate t. Add two new indeterminates u, v and compute the resultant

$$\mathrm{Res}_{2,5}(x - t^2, u(y - t^3) + v(z - t^5)) = (-y^2 + x^3)u^2 + (2x^4 - 2yz)uv + (-z^2 + x^5)v^2.$$

Then, since the resultant must vanish for all specializations of u and v, we deduce that

$$\mathcal{C} = \{-y^2 + x^3 = 2x^4 - 2yz = -z^2 + x^5 = 0\}.$$

1.4.3 Bézout's theorem in two variables

Similarly to the construction of $\mathbb{P}^1(\mathbb{K})$, one can define the projective plane $\mathbb{P}^2(\mathbb{K})$ (and in general projective n-space) as the complete variety whose points are identified with lines through the origin in $\mathbb{K}^3$. We may embed $\mathbb{K}^2$ in $\mathbb{P}^2(\mathbb{K})$ as the set of lines through the points $(x, y, 1)$. Again, it makes sense to speak of the zero set in $\mathbb{P}^2(\mathbb{K})$ of homogeneous polynomials (i.e. polynomials $f(x, y, z)$ such that $f(\lambda x, \lambda y, \lambda z) = \lambda^d f(x, y, z)$, for $d = \deg(f)$).

Given two homogeneous polynomials $f, g \in \mathbb{K}[x, y, z]$ without common factors, with $\deg(f) = d_1$, $\deg(g) = d_2$, a classical theorem of Bézout asserts that they have $d_1 \cdot d_2$ common points of intersection in $\mathbb{P}^2(\mathbb{K})$, counted with appropriate intersection multiplicities. A proof of this theorem using resultants is given for instance in [CLO97]. The following weaker version suffices to obtain such nice consequences as Pascal's Mystic Hexagon theorem [CLO97, Sect. 8.7] (see Corollary 1.5.15 for a proof using multivariable residues).

Theorem 1.4.2. *Let $f, g \in \mathbb{K}[x, y, z]$ be homogeneous polynomials, without common factors, and of respective degrees d_1, d_2. Then $(f = 0) \cap (g = 0)$ is finite and has at most $d_1 \cdot d_2$ points.*

Proof. Assume $(f = 0) \cap (g = 0)$ have more than $d_1 \cdot d_2$ points, which we label $p_0, \ldots, p_{d_1 d_2}$. Let L_{ij} be the line through p_i and p_j for $i, j = 0, \ldots, d_1 d_2$. Making a linear change of coordinates, we can assume that $(0, 0, 1) \notin (f = 0) \cup (g = 0) \cup (\cup_{ij} L_{ij})$. Write $f = \sum_{i=0}^{d_1} a_i z^i$, $g = \sum_{j=0}^{d_2} b_j z^j$, as polynomials in z with coefficients $a_i, b_j \in \mathbb{K}[x, y]$. Since $f(0, 0, 1) \neq 0$, $g(0, 0, 1) \neq 0$ and f and g do not have any common factor, it is straightforward to verify from the expression of the resultant as the determinant of the Sylvester matrix, that the resultant $\mathrm{Res}_{d_1, d_2}(f, g)$ with respect to z is a non zero homogeneous polynomial in x, y of total degree $d_1 \cdot d_2$. Write $p_i = (x_i, y_i, z_i)$. Then, $\mathrm{Res}_{d_1, d_2}(f, g)(x_i, y_i) = 0$ for all $i = 0, \ldots, d_1 \cdot d_2$. The fact that $(0, 0, 1)$ does not lie in any of the lines L_{ij} implies that the $(d_1 d_2 + 1)$ points (x_i, y_i) are distinct, and we get a contradiction.

1.4.4 GCD computations and Bézout identities

Let P, Q be two univariate polynomials with coefficients in a field $\mathbf{k}$. Assume they are coprime, i.e. that their greatest common divisor $\mathrm{GCD}(P, Q) = 1$. We can then find polynomials $h_1, h_2 \in \mathbf{k}[z]$ such that the *Bézout identity* $1 = h_1 P + h_2 Q$ is satisfied, by means of the Euclidean algorithm to compute $\mathrm{GCD}(P, Q)$. A we have already remarked, $\mathrm{GCD}(P, Q) = 1$ if and only if P and Q do not have any common root in any algebraically field $\mathbb{K}$ containing $\mathbf{k}$. If d_1, d_2 denote the respective degrees, this happens precisely when $\mathrm{Res}_{d_1, d_2}(P, Q) \neq 0$. Note that since the resultant is an integer polynomial in the coefficients, $\mathrm{Res}_{d_1, d_2}(P, Q)$ also lies in $\mathbf{k}$. Moreover, by property iii) in Section 1.3.2, one deduces that

$$1 = \frac{A_1}{\mathrm{Res}_{d_1,d_2}(P,Q)}P + \frac{A_2}{\mathrm{Res}_{d_1,d_2}(P,Q)}Q. \tag{1.26}$$

So, it is possible to find h_1, h_2 whose coefficients are rational functions with integer coefficients evaluated in the coefficients of the input polynomials P, Q, and denominators equal to the resultant. Moreover, these polynomials can be explicitly obtained from the proof of (1.23). In particular, the coefficients of A_1, A_2 are particular minors of the Sylvester matrix M_{d_1,d_2}.

This has also been extended to compute $\mathrm{GCD}(P,Q)$ even when P and Q are not coprime (and the resultant vanishes), based on the so called *subresultants*, which are again obtained from particular minors of M_{d_1,d_2}. Note that $\mathrm{GCD}(P,Q)$ is the (monic polynomial) of least degree in the ideal generated by P and Q (i.e. among the polynomial linear combinations $h_1 P + h_2 Q$). So one is led to study non surjective specializations of the linear map (1.21). In fact, the dimension of its kernel equals the degree of $\mathrm{GCD}(P,Q)$, i.e. the number of common roots of P and Q, counted with multiplicity.

Note that if $1 \le d_2 \le d_1$ and $C = \sum_{i=0}^{d_1-d_2} c_i z^i$ is the quotient of P in the Euclidean division by Q, the remainder equals

$$R = P - \sum_{i=0}^{d_1-d_2} c_i(z^i Q).$$

Thus, subtracting from the first column of M_{d_1,d_2} the linear combination of the columns corresponding to $z^i Q, i = 0, \ldots, d_1 - d_2$, with respective coefficients c_i, we do not change the determinant but we get the coefficients of R in the first column. In fact, it holds that

$$R_{d_1,d_2}(P,Q) = a_{d_1}^{d_2-\deg(R)} R_{\deg(R),d_2}(R,Q).$$

So, one could describe an algorithm for computing resultants similar to the Euclidean algorithm. However, the Euclidean remainder sequence to compute greatest common divisors has a relatively bad numerical behavior. Moreover, it has bad specialization properties when the coefficients depend on parameters. Collins [Col67] studied the connections between subresultants and Euclidean remainders, and he proved in particular that the polynomials in the two sequences are pairwise proportional. But the subresultant sequence has a good behavior under specializations and well controlled growth of the size of the coefficients. Several efficient algorithms have been developed to compute subresultants [LRD00].

1.4.5 Algebraic numbers

A complex number α is said to be algebraic if there exists a polynomial $P \in \mathbb{Q}[z]$ such that $P(\alpha) = 0$. The algebraic numbers form a subfield of $\mathbb{C}$. This can be easily proved using resultants.

Lemma 1.4.3. *Let* $P, Q \in \mathbb{Q}[z]$ *with degrees* d_1, d_2 *and let* $\alpha, \beta \in \mathbb{C}$ *such that* $P(\alpha) = Q(\beta) = 0$. *Then,*

i) $\alpha + \beta$ *is a root of the polynomial* $u_+(z) = \mathrm{Res}_{d_1, d_2}(P(z - y), Q(y)) = 0$,
ii) $\alpha \cdot \beta$ *is a root of the polynomial* $u_\times(z) = \mathrm{Res}_{d_1, d_2}(y^{d_1} P(z/y), Q(y))$,
iii) for $\alpha \neq 0$, α^{-1} *is a root of the polynomial* $u_{-1}(z) = \mathrm{Res}_{d_1, d_2}(zy - 1, P(y))$,

where the resultants are taken with respect to y.

The proof of Lemma 1.4.3 is immediate. Note that even if P (resp. Q) is the minimal polynomial annihilating α (resp. β), i.e. the monic polynomial with minimal degree having α (resp. β) as a root, the roots of the polynomial $u_\times$ are all the products $\alpha_i \cdot \beta_j$ where α_i (resp. β_j) is any root of P (resp. Q), which need not be all different, and so $u_\times$ need not be the minimal polynomial annihilating $\alpha \cdot \beta$. This happens for instance in case $\alpha = \sqrt{2}, P(z) = z^2 - 2, \beta = \sqrt{3}, Q(z) = z^2 - 3$, where $u_\times(z) = (z^2 - 6)^2$.

1.4.6 Discriminants

Given a generic univariate polynomial of degree d, $P(z) = a_0 + a_1 z + \cdots + a_d z^d$, $a_d \neq 0$, it is also classical the existence of an irreducible polynomial $D_d(P) = D_d(a_0, \ldots, a_d) \in \mathbb{Z}[a_0, \ldots, a_d]$, called the *discriminant* (or d-discriminant) whose value at a particular set of coefficients (with $a_d \neq 0$) is non-zero if and only if the corresponding polynomial of degree d has only simple roots. Equivalently, $D_d(a_0, \ldots, a_n) = 0$ if and only if there exists $z \in \mathbb{C}$ with $P(z) = P'(z) = 0$.

Geometrically, the discriminantal hypersurface

$$\{a = (a_0, \ldots, a_d) \in \mathbb{C}^{d+1} : D_d(a) = 0\}$$

is the projection over the first $(d + 1)$ coordinates of the intersection of the hypersurfaces $\{(a, z) \in \mathbb{C}^{d+2} : a_0 + a_1 z + \cdots + a_d z^d = 0\}$ and $\{(a, z) \in \mathbb{C}^{d+2} : a_1 + 2a_2 z + \cdots + d a_d z^{d-1} = 0\}$, i.e. the variable z is eliminated.

The first guess would be that $D_d(P)$ equals the resultant $\mathrm{Res}_{d,d-1}(P, P')$, but it is easy to see that in fact $\mathrm{Res}_{d,d-1}(P, P') = (-1)^{d(d-1)/2} a_d D_d(P)$. In case $d = 2$, $P(z) = az^2 + bz + c$, $D_2(a, b, c)$ is the well known discriminant $b^2 - 4ac$. When $d = 6$ for instance, D_6 is an irreducible polynomial of degree 10 in the coefficients $(a_0, \ldots, a_6)$ with 246 terms.

The extremal monomials and coefficients of the discriminant have very interesting combinatorial descriptions. This notion has important applications in singularity theory and number theory. The distance of the coefficients of a given polynomial to the discriminantal hypersurface is also related to the numerical stability of the computation of its roots. For instance, consider the Wilkinson polynomial $P(z) = (z + 1)(z + 2) \ldots (z + 19)(z + 20)$, which clearly has 20 real roots at distance at least 1 from the others, and is known to be numerically unstable. The coefficients of P are very close to the coefficients of a polynomial with a multiple root. The polynomial $Q(z) = P(z) + 10^{-9} z^{19}$, obtained by a "small perturbation" of one of the coefficients of P, has only 12 real roots and 4 pairs of imaginary roots, one of which has imaginary part close to $\pm 0.88i$. Consider then the parametric family of polynomials $P_\lambda(z) = P(z) + \lambda z^{19}$ and note that $P(z) = P_0$ and $Q(z) = P_{10^{-9}}$. Thus, for some intermediate value of λ, two complex roots merge to give a double real root and therefore that value of the parameter is a zero of the discriminant $D(\lambda) = D_{20}(P_\lambda)$.

1.5 Multidimensional residues

In this section we will extend the theory of residues to the several variables case. As in the one-dimensional case we will begin with an "integral" definition of local residue from which we will define the total residue as a sum of local ones. We will also indicate how one can give a purely algebraic definition of global, and then local, residues using Bezoutians. We shall also touch upon the geometric definition of Arnold, Varchenko and Guseĭn-Zadé [AGZV85].

Let $\mathbb{K}$ be an algebraically closed field of characteristic zero and let $I \subset \mathbb{K}[x_1, \ldots, x_n]$ be a zero-dimensional ideal. We denote by $Z(I) = \{\xi_1, \ldots, \xi_s\} \subset \mathbb{K}^n$ the variety of zeros of I. We will assume, moreover, that I is a *complete intersection* ideal, i.e. that it has a presentation of the form $I = \langle P_1, \ldots, P_n \rangle$, $P_i \in \mathbb{K}[x_1, \ldots, x_n]$. For simplicity, we will denote by $\langle P \rangle$ the ordered n-tuple $\{P_1, \ldots, P_n\}$. As before, let $\mathcal{A}$ be the finite dimensional commutative algebra $\mathcal{A} = \mathbb{K}[x_1, \ldots, x_n]/I$. Our goal is to define a linear map

$$\mathrm{res}_{\langle P \rangle} : \mathcal{A} \to \mathbb{K}$$

whose properties are similar to the univariate residue map. In particular, we would like it to be dualizing in the sense of Theorem 1.2.1 and to be compatible with local maps $\mathrm{res}_{\langle P \rangle, \xi} : \mathcal{A}_\xi \to \mathbb{K}$, $\xi \in Z(I)$.

1.5.1 Integral definition

In case $\mathbb{K} = \mathbb{C}$, given $\xi \in Z(I)$, let $\mathcal{U} \subset \mathbb{C}^n$ be an open neighborhood of ξ containing no other points of $Z(I)$, and let $H \in \mathbb{C}[x_1, \ldots, x_n]$. We define the local *Grothendieck* residue

$$\mathrm{res}_{\langle P \rangle, \xi}(H) = \frac{1}{(2\pi i)^n} \int_{\Gamma_\xi(\epsilon)} \frac{H(x)}{P_1(x) \cdots P_n(x)} \, dx_1 \wedge \cdots \wedge dx_n, \qquad (1.27)$$

where $\Gamma_\xi(\epsilon)$ is the real n-cycle $\Gamma_\xi(\epsilon) = \{x \in \mathcal{U} : |P_i(x)| = \epsilon_i\}$ oriented by the n-form $d(\arg(P_1)) \wedge \cdots \wedge d(\arg(P_n))$. For almost every $\epsilon = (\epsilon_1, \ldots, \epsilon_n)$ in a neighborhood of the origin, $\Gamma_\xi(\epsilon)$ is smooth and by Stokes' Theorem the integral (1.27) is independent of ϵ. The choice of the orientation form implies that $\mathrm{res}_{\langle P \rangle, \xi}(H)$ is skew-symmetric on $P_1, \ldots, P_n$. We note that this definition makes sense as long as H is holomorphic in a neighboorhood of ξ. If $\xi \in Z(I)$ is a point of multiplicity one then the Jacobian

$$J_{\langle P \rangle}(\xi) := \det\left(\frac{\partial P_i}{\partial x_j}(\xi)\right)$$

is non-zero, and

$$\operatorname{res}_{\langle P \rangle, \xi}(H) = \frac{H(\xi)}{J_{\langle P \rangle}(\xi)}. \tag{1.28}$$

This identity follows from making a change of coordinates $y_i = P_i(x)$ and iterated integration.

It follows from Stokes's theorem that if $H \in I_\xi$, the ideal defined by I in the local ring defined by ξ (cf. Section 2.1.3 in Chapter 2), then $\operatorname{res}_{\langle P \rangle, \xi}(H) = 0$ and therefore the local residue defines a map $\operatorname{res}_{\langle P \rangle, \xi} \colon \mathcal{A}_\xi \to \mathbb{C}$. We then define the global residue map as the sum of local residues

$$\operatorname{res}_{\langle P \rangle}(H) := \sum_{\xi \in Z(I)} \operatorname{res}_{\langle P \rangle, \xi}(H)$$

which we may view as a map $\operatorname{res}_{\langle P \rangle} \colon \mathcal{A} \to \mathbb{C}$. We may also define the global residue $\operatorname{res}_{\langle P \rangle}(H_1/H_2)$ of a rational function regular on $Z(I)$, i.e. such that H_2 does not vanish on $Z(I)$. At this point one may be tempted to replace the local cycles $\Gamma_\xi(\epsilon)$ by a global cycle

$$\Gamma(\epsilon) := \{x \in \mathbb{C}^n : |P_i(x)| = \epsilon_i\}$$

but $\Gamma(\epsilon)$ need not be compact and integration might not converge. However, if the map

$$(P_1, \ldots, P_n) \colon \mathbb{C}^n \to \mathbb{C}^n$$

is proper, then $\Gamma(\epsilon)$ is compact and we can write

$$\operatorname{res}_{\langle P \rangle}(H) := \frac{1}{(2\pi i)^n} \int_{\Gamma(\epsilon)} \frac{H(x)}{P_1(x) \cdots P_n(x)} \, dx_1 \wedge \cdots \wedge dx_n.$$

The following two theorems summarize basic properties of the local and global residue map.

Theorem 1.5.1 (Local and Global Duality). *Let* $I = \langle P_1, \ldots, P_n \rangle \subset \mathbb{C}[x_1, \ldots, x_n]$ *be a complete intersection ideal and* $\mathcal{A} = \mathbb{C}[x_1, \ldots, x_n]/I$. *Let* $\mathcal{A}_\xi$ *be the localization at* $\xi \in Z(I)$. *The pairings*

$$\mathcal{A}_\xi \times \mathcal{A}_\xi \to \mathbb{C} \quad ; \quad ([H_1], [H_2]) \mapsto \operatorname{res}_{\langle P \rangle, \xi}(H_1 \cdot H_2)$$

and

$$\mathcal{A} \times \mathcal{A} \to \mathbb{C} \quad ; \quad ([H_1], [H_2]) \mapsto \operatorname{res}_{\langle P \rangle}(H_1 \cdot H_2)$$

are non-degenerate.

Theorem 1.5.2 (Local and Global Transformation Laws). *Let $I = \langle P_1, \ldots, P_n \rangle$ and $J = \langle Q_1, \ldots, Q_n \rangle$ be zero-dimensional ideals such that $J \subset I$. Let*

$$Q_j(x) = \sum_{i=1}^{n} a_{ij}(x) P_i(x).$$

Denote by $A(x)$ the $n \times n$-matrix $(a_{ij}(x))$, then for any $\xi \in Z(I)$,

$$\operatorname{res}_{\langle P \rangle, \xi}(H) = \operatorname{res}_{\langle Q \rangle, \xi}(H \cdot \det(A)). \tag{1.29}$$

Moreover, a similar formula holds for global residues

$$\operatorname{res}_{\langle P \rangle}(H) = \operatorname{res}_{\langle Q \rangle}(H \cdot \det(A)).$$

Remark 1.5.3. We refer the reader to [Tsi92, Sect. 5.6 and 8.4] for a proof of the duality theorems and to [Tsi92, Sect. 5.5 and 8.3] for full proofs of the transformation laws. The local theorems are proved in [GH78, Sect. 5.1] and extended to the global case in [TY84]; a General Global Duality Law is discussed in [GH78, Sect. 5.4] Here we will just make a few remarks about Theorem 1.5.2.

Suppose that $\xi \in Z(I)$ is a simple zero and that $\det(A(\xi)) \neq 0$. Then, since

$$J_{\langle Q \rangle}(\xi) := J_{\langle P \rangle}(\xi) \cdot \det(A(\xi))$$

we have

$$\operatorname{res}_{\langle P \rangle, \xi}(H) = \frac{H(\xi)}{J_{\langle P \rangle}(\xi)} = \frac{H(\xi) \cdot \det(A(\xi))}{J_{\langle Q \rangle}(\xi)} = \operatorname{res}_{\langle Q \rangle, \xi}(H \cdot \det(A)),$$

as asserted by (1.29). The case of non-simple zeros which are common to both I and J is dealt-with using a perturbation technique after showing that when the input $\{P_1, \ldots, P_n\}$ depends smoothly on a parameter so does the residue. Finally, one shows that if $\xi \in Z(J) \backslash Z(I)$, then $\det(A) \in J_\xi$ and the local residue $\operatorname{res}_{\langle Q \rangle, \xi}(H \cdot \det(A))$ vanishes.

1.5.2 Geometric definition

For the sake of completeness, we include a few comments about the geometric definition of the residue of Arnold, Varchenko and Guseĭn-Zadé [AGZV85]. Here, the starting point is the definition of the residue at a simple zero $\xi \in Z(I)$ as in (1.28). Suppose now that $\xi \in Z(I)$ has multiplicity μ. In a sufficiently small neighborhood $\mathcal{U}$ of ξ in $\mathbb{C}^n$ we can consider the map

$$P = (P_1, \ldots, P_n) : \mathcal{U} \to \mathbb{C}^n.$$

By Sard's theorem, almost all values $y \in P(\mathcal{U})$ are regular and at such points the equation $P(x) - y = 0$ has exactly μ simple roots $\eta_1(y), \ldots, \eta_\mu(y)$. Consider the map

$$\phi(y) := \sum_{i=1}^{\mu} \frac{H(\eta_i(y))}{J_{\langle P \rangle}(\eta_i(y))} \,.$$

It is shown in [AGZV85, Sect. 5.18] that $\phi(y)$ extends holomorphically to $0 \in \mathbb{C}^n$. We can then define the local residue $\mathrm{res}_{\langle P \rangle, \xi}(H)$ as the value $\phi(0)$. A continuity argument shows that both definitions agree.

1.5.3 Residue from Bezoutian

In this section we generalize to the multivariable case the univariate approach discussed in Section 1.2.1. This topic is also discussed in Section 3.3 of Chapter 3. We will follow the presentation of [BCRS96] and [RS98] to which we refer the reader for details and proofs. We note that other purely algebraic definitions of the residue may also be found in [KK87, Kun86, SS75, SS79].

Let $\mathbb{K}$ be an algebraically closed field $\mathbb{K}$ of characteristic zero and let $\mathcal{A}$ be a finite-dimensional commutative $\mathbb{K}$ algebra. Recall that $\mathcal{A}$ is said to be a *Gorenstein* algebra if there exists a linear form $\ell \in \widehat{\mathcal{A}} := \mathrm{Hom}_{\mathbb{K}}(\mathcal{A}, \mathbb{K})$ such that the bilinear form

$$\phi_\ell : \mathcal{A} \times \mathcal{A} \to \mathbb{K} \quad ; \quad \phi_\ell(a, b) := \ell(a \cdot b)$$

is non-degenerate. Given such a dualizing linear form ℓ, let $\{a_1, \ldots, a_r\}$ and $\{b_1, \ldots, b_r\}$ be ϕ_ℓ-dual bases of $\mathcal{A}$, and set

$$B_\ell := \sum_{i=1}^{r} a_i \otimes b_i \in \mathcal{A} \otimes \mathcal{A}.$$

B_ℓ is independent of the choice of dual bases and is called a *generalized Bezoutian*. It is characterized by the following two properties:

- $(a \otimes 1)B_\ell = (1 \otimes a)B_\ell$, for all $a \in \mathcal{A}$, and
- If $\{a_1, \ldots, a_r\}$ is a basis of $\mathcal{A}$ and $B_\ell = \sum_i a_i \otimes b_i$, then $\{b_1, \ldots, b_r\}$ is a basis of $\mathcal{A}$ as well.

It is shown in [BCRS96, Th. 2.10] that the correspondence $\ell \mapsto B_\ell$ is an equivalence between dualizing linear forms on $\mathcal{A}$ and generalized Bezoutians in $\mathcal{A} \otimes \mathcal{A}$.

As in Section 1.2.5 we can relate the dualizing form, the Bezoutian and the computation of traces. The dual $\widehat{\mathcal{A}}$ may be viewed as a module over $\mathcal{A}$ by $a \cdot \lambda(b) := \lambda(a \cdot b)$, $a, b \in \mathcal{A}, \lambda \in \widehat{\mathcal{A}}$. A dualizing form $\ell \in \widehat{\mathcal{A}}$ generates $\widehat{\mathcal{A}}$ as an $\mathcal{A}$-module. Moreover, it defines an isomorphism $\mathcal{A} \to \widehat{\mathcal{A}}$, $a \mapsto \ell(a\bullet)$. In particular there exists a unique element $J_\ell \in \mathcal{A}$ such that $\mathrm{tr}(M_q) = \ell(J_\ell \cdot q)$, where $M_q : \mathcal{A} \to \mathcal{A}$ denotes multiplication by $q \in \mathcal{A}$. On the other hand, if $\{a_1, \ldots, a_r\}$ and $\{b_1, \ldots, b_r\}$ are ϕ_ℓ-dual bases of $\mathcal{A}$, then

$$M_q(a_j) = q \cdot a_j = \sum_{i=1}^{r} \phi_\ell(q \cdot a_j, b_i) a_i$$

and therefore

$$\mathrm{tr}(M_q) = \sum_{i=1}^{r} \phi_\ell(q \cdot a_i, b_i) = \sum_{i=1}^{r} \ell(q \cdot a_i \cdot b_i) = \ell\Big(q \cdot \sum_{i=1}^{r} a_i b_i\Big)$$

from which it follows that

$$J_\ell = \sum_{i=1}^{r} a_i \cdot b_i. \tag{1.30}$$

Note that, in particular,

$$\ell(J_\ell) = \sum_{i=1}^{r} \ell(a_i \cdot b_i) = r = \dim(\mathcal{A}). \tag{1.31}$$

Suppose now that $I \subset \mathbb{K}[x_1, \ldots, x_n]$ is a zero-dimensional complete intersection ideal. We may assume without loss of generality that I is generated by a regular sequence $\{P_1, \ldots, P_n\}$. The quotient algebra $\mathcal{A} = \mathbb{K}[x_1, \ldots, x_n]/I$ is a Gorenstein algebra. This can be done by defining directly a dualizing linear form (global residue or Kronecker symbol) or by defining an explicit Bezoutian as in [BCRS96, Sect. 3]:

Let

$$\partial_j P_i := \frac{P_i(y_1, \ldots, y_{j-1}, x_j, \ldots, x_n) - P_i(y_1, \ldots, y_j, x_{j+1}, \ldots, x_n)}{x_j - y_j} \tag{1.32}$$

and set

$$\Delta_{\langle P \rangle}(x, y) = \det(\partial_j P_i) \in \mathbb{K}[x, y]. \tag{1.33}$$

We shall also denote by $\Delta_{\langle P \rangle}(x, y)$ its image in the tensor algebra

$$\mathcal{A} \otimes \mathcal{A} \cong \mathbb{K}[x, y]/\langle P_1(x), \ldots, P_n(x), P_1(y), \ldots, P_n(y) \rangle. \tag{1.34}$$

Remark 1.5.4. In the analytic context, the polynomials $\partial_j P_i$ are the coefficients of the so called Hefer expansion of P_i. We refer to [TY84] for the relationship between Hefer expansions and residues.

Theorem 1.5.5. *The element $\Delta_{\langle P \rangle}(x, y) \in \mathcal{A} \otimes \mathcal{A}$ is a generalized Bezoutian.*

This is Theorem 3.2 in [BCRS96]. It is easy to check that $\Delta_{\langle P \rangle}$ satisfies the first condition characterizing generalized Bezoutians. Indeed, given the identification (1.34), it suffices to show that $[f(x)] \cdot \Delta_{\langle P \rangle}(x, y) = [f(y)] \cdot \Delta_{\langle P \rangle}(x, y)$ for all $[f] \in \mathcal{A}$. This follows directly from the definition of $\Delta_{\langle P \rangle}$. The proof of the second property is much harder. Becker et al. show it by reduction to the local case where it is obtained through a deformation technique somewhat similar to that used in the geometric case in [AGZV85].

We denote by τ the Kronecker symbol; that is, the dualizing linear form associated with the Bezoutian $\Delta_{\langle P \rangle}$. As we shall see below, for $\mathbb{K} = \mathbb{C}$, the Kronecker symbol agrees with the global residue. In order to keep the context clear, we will continue to use the expression Kronecker symbol throughout this section.

If H_1/H_2 is a rational function such that H_2 does not vanish on $Z(I)$, then $[H_2]$ has an inverse $[G_2]$ in $\mathcal{A}$ and we define $\tau(H_1/H_2) := \tau([H_1] \cdot [G_2])$.

If $\{[x^\alpha]\}$ is a monomial basis of $\mathcal{A}$ and we write

$$\Delta_{\langle P \rangle}(x,y) = \sum x^\alpha \Delta_\alpha(y)$$

then $\{[x^\alpha]\}$ and $\{[\Delta_\alpha(x)]\}$ are dual basis and it follows from (1.30) and (1.34) that

$$J_{\langle P \rangle}(x) := J_\tau(x) = \sum_\alpha x^\alpha \Delta_\alpha(x) = \Delta_{\langle P \rangle}(x,x).$$

Since $\lim_{y \to x} \partial_j P_i(x,y) = \dfrac{\partial P_i}{\partial x_j}$ it follows that $J_{\langle P \rangle}(x)$ agrees with the standard Jacobian of the polynomials $P_1, \ldots, P_n$. As we did in Section 1.1.2 for univariate residues, we can go from the global Kronecker symbol to local operators. Let $Z(I) = \{\xi_1, \ldots, \xi_s\}$ and let

$$I = \cap_{\xi \in Z(I)} I_\xi$$

be the primary decomposition of I as in Section 2.1.3 of Chapter 2. Let $\mathcal{A}_\xi = \mathbb{K}[x_1, \ldots, x_n]/I_\xi$, we have an isomorphism:

$$\mathcal{A} \cong \prod_{\xi \in Z(I)} \mathcal{A}_\xi.$$

We recall (cf. [CLO98, Sect. 4.2]) that there exist idempotents $e_\xi \in \mathbb{K}[x_1, \ldots, x_n]$ such that, in $\mathcal{A}$, $\sum_{\xi \in Z(I)} e_\xi = 1$, $e_{\xi_i} e_{\xi_j} = 0$ if $i \neq j$, and $e_\xi^2 = 1$. These generalize the interpolating polynomials we discussed in Section 1.1.2. We can now define

$$\tau_\xi([H]) := \tau(e_\xi \cdot [H])$$

and it follows easily that the global Kronecker symbol is the sum of the local ones. In analogy with the global case, we may define the local Kronecker symbol $\tau_\xi([H_1/H_2])$ of a rational function H_1/H_2, regular at ξ as $\tau_\xi([H_1] \cdot [G_2])$, where $[G_2]$ is the inverse of $[H_2]$ in the algebra $\mathcal{A}_\xi$. The following proposition shows that in the case of simple zeros and $\mathbb{K} = \mathbb{C}$, the Kronecker symbol agrees with the global residue defined in Section 1.5.1.

Proposition 1.5.6. *Suppose that $J_{\langle P \rangle}(\xi) \neq 0$ for all $\xi \in Z(I)$. Then*

$$\tau(H) = \sum_{\xi \in Z(I)} \frac{H(\xi)}{J_{\langle P \rangle}(\xi)} \tag{1.35}$$

for all $H \in \mathbb{K}[x_1, \ldots, x_n]$.

Proof. Recall that the assumption that $J_{\langle P \rangle}(\xi) \neq 0$ for all $\xi \in Z(I)$ implies that $[J_{\langle P \rangle}]$ is invertible in $\mathcal{A}$. Indeed, since $J_{\langle P \rangle}, P_1, \ldots, P_n$ have no common zeros in $\mathbb{K}^n$, the Nullstellensatz implies that there exists $G \in \mathbb{K}[x_1, \ldots, x_n]$ such that

$$G.J_{\langle P \rangle} = 1 \mod I.$$

Given $H \in \mathbb{K}[x_1, \ldots, x_n]$, consider the trace of the multiplication map $M_{H \cdot G} \colon \mathcal{A} \to \mathcal{A}$. On the one hand, we have from Theorem 2.1.4 in Chapter 2 that

$$\mathrm{tr}(M_{H \cdot G}) = \sum_{\xi \in Z(I)} H(\xi)G(\xi) = \sum_{\xi \in Z(I)} \frac{H(\xi)}{J_{\langle P \rangle}(\xi)}.$$

But, recalling the definition of the Jacobian we also have

$$\mathrm{tr}(M_{H \cdot G}) = \tau(J_{\langle P \rangle} \cdot G \cdot H) = \tau(H)$$

and (1.35) follows.

Remark 1.5.7. As in the geometric case discussed in Section 1.5.2 one can use continuity arguments to show that the identification between the Kronecker symbol and the global residue extends to the general case. We refer the reader to [RS98] for a proof of this fact as well as for a proof of the Transformation Laws in this context. In particular, Theorem 1.5.2 holds over any algebraically closed field of characteristic zero.

1.5.4 Computation of residues

In this section we would like to discuss briefly some methods for the computation of global residues; a further method is discussed in Section 3.3.1 in Chapter 3. Of course, if the zero-dimensional ideal $I = \langle P_1, \ldots, P_n \rangle$ is radical and we can compute the zeros $Z(I)$, then we can use (1.28) to compute the local and global residue. We also point out that the transformation law gives a general, though not very efficient, algorithm to compute local and global residues. Indeed, since I is a zero dimensional ideal there exist univariate polynomials $f_1(x_1), f_2(x_2), \ldots, f_n(x_n)$ in the ideal I. In particular we can write

$$f_j(x_j) = \sum_{i=1}^{n} a_{ij}(x)P_i(x)$$

and for any $H \in \mathbb{K}[x_1, \ldots, x_n]$,

$$\mathrm{res}_{\langle P \rangle}(H) = \mathrm{res}_{\langle f \rangle}(H \cdot \det(a_{ij})). \tag{1.36}$$

Moreover, the right hand side of the above equation may be computed as an iterated sequence of univariate residues. What makes this a less than desirable computational method is that even if the polynomials $P_1, \ldots, P_n$ and

$f_1, \ldots, f_n$ are very simple, the coefficients $a_{ij}(x)$ need not be so. The following example illustrates this.

Consider the polynomials

$$P_1 = x_1^2 - x_3$$
$$P_2 = x_2 - x_1 x_3^2 \qquad (1.37)$$
$$P_3 = x_3^2 - x_1^3$$

The ideal $I = \langle P_1, P_2, P_3 \rangle$ is a zero-dimensional ideal; the algebra $\mathcal{A}$ has dimension four, and the zero-locus $Z(I)$ consists of two points, the origin, which has multiplicity three, and the point $(1, 1, 1)$. Gröbner basis computations with respect to lexicographic orders give the following univariate polynomials in the ideal I:

$$f_1 = x_1^4 - x_1^3$$
$$f_2 = x_2^2 - x_2 \qquad (1.38)$$
$$f_3 = x_3^3 - x_3^2.$$

We observe that we could also have used iterated resultants to find univariate polynomials in I. However, this will generally yield higher degree polynomials. For instance, for our example (1.37) a Singular [GPS01] computation gives:

```
>resultant(resultant(P_1,P_2,x_3),resultant(P_2,P_3,x_3),x_2);
x_1^10-2*x_1^9+x_1^8
```

Returning to the polynomials (1.38), we can obtain, using the Singular command "division", a coefficient matrix $A = (a_{ij}(x))$:

$$\begin{pmatrix} x_1^2 + x_3 \; x_3^3 + (x_1^2 + x_1 + 1)x_3^2 + (x_1^2 + x_1 + x_2)x_3 + x_1^2 x_2 & (x_1 + 1)x_3 + x_1^2 \\ 0 & x_1^3 + x_2 - 1 & 0 \\ 1 & (x_1 + 1)(x_2 + x_3) + x_3^2 & x_1 + x_3 \end{pmatrix}$$

So that

$$\begin{aligned} \det(A) &= (x_2 + x_1^3 - 1)x_3^2 + (x_1^2 x_2 + x_1^5 - x_1^3 - x_1^2 - x_2 + 1)x_3 + \\ &= x_1^3 x_2 + x_1^6 - x_1^5 - x_1^3 + x_1^2 - x_1^2 x_2. \end{aligned}$$

Rather than continuing with the computation of a global residue $\operatorname{res}_{\langle P \rangle}(H)$ using (1.36) and iterated univariate residues or Bezoutians, we will refer the reader to Chapter 3 where improved versions are presented and discuss instead how we can use the multivariate Bezoutian in computations. The Bezoutian matrix $(\partial_j P_i)$ is given by

$$\begin{pmatrix} x_1 + y_1 & -x_3^2 & -(x_1^2 + x_1 y_1 + y_1^2) \\ 0 & 1 & 0 \\ -1 & -y_1(x_3 + y_3) & x_3 + y_3 \end{pmatrix}$$

and therefore

$$\Delta_{\langle P\rangle}(x,y) \;=\; x_1x_3 + x_1y_3 + x_3y_1 + y_1y_3 - x_1^2 - x_1y_1 - y_1^2 \,.$$

Computing a Gröbner basis relative to grevlex gives a monomial basis of $\mathcal{A}$ of the form $\{1, x_1, x_2, x_3\}$. Reducing $\Delta_{\langle P\rangle}(x,y)$ relative to the corresponding basis of $\mathcal{A} \otimes \mathcal{A}$ we obtain:

$$\Delta_{\langle P\rangle}(x,y) \;=\; (y_2 - y_3) + (y_3 - y_1)x_1 + x_2 + (y_1 - 1)x_3 \,.$$

Hence the dual basis of $\{1, x_1, x_2, x_3\}$ is the basis $\{x_2 - x_3, x_3 - x_1, 1, x_1 - 1\}$.

We now claim that given $H \in \mathbb{K}[x_1, \ldots, x_n]$, if we compute the grevlex normal form:

$$N(H) \;=\; \lambda_0 + \lambda_1[x_1] + \lambda_2[x_2] + \lambda_3[x_3]$$

then, $\mathrm{res}_{\langle P\rangle}(H) = \lambda_2$. More generally, suppose that $\{[x^\alpha]\}$ is a monomial basis of $\mathcal{A}$ and that $\{[\Delta_\alpha(x)]\}$ is the dual basis given by the Bezoutian, then if $[H] = \sum_\alpha \lambda_\alpha[x^\alpha]$ and $1 = \sum_\alpha \mu_\alpha[\Delta_\alpha]$,

$$\mathrm{res}_{\langle P\rangle}(H) \;=\; \sum_\alpha \lambda_\alpha \mu_\alpha \,. \tag{1.39}$$

Indeed, we have

$$\mathrm{res}_{\langle P\rangle}(H) \;=\; \mathrm{res}_{\langle P\rangle}(H \cdot 1) \;=\; \mathrm{res}_{\langle P\rangle}\Big(\sum_\alpha \lambda_\alpha x^\alpha \cdot \sum_\beta \mu_\beta \Delta_\beta\Big)$$

$$=\; \sum_{\alpha,\beta} \lambda_\alpha \mu_\beta \mathrm{res}_{\langle P\rangle}(x^\alpha \cdot \Delta_\beta) \;=\; \sum_\alpha \lambda_\alpha \mu_\alpha \,.$$

Although the computational method based on the Bezoutian allows us to compute $\mathrm{res}_{\langle P\rangle}(H)$ as a linear combination of normal form coefficients of H, it would be nice to have a method that computes the global residue as a single normal form coefficient, generalizing the univariate algorithm based on the identities (1.10). This can be done if we make some further assumptions on the generators of the ideal I. We will discuss here one such case which has been extensively studied both analytically and algebraically, following the treatment in [CDS96]. A more general algorithm will be presented in Section 1.5.6. Assume the generators $P_1, \ldots, P_n$ satisfy:

Assumption: $P_1, \ldots, P_n$ are a Gröbner basis for some term order $\prec$.

Since we can always find a weight $w \in \mathbb{N}^n$ such that $\mathrm{in}_w(P_i) = \mathrm{in}_\prec(P_i)$, $i = 1, \ldots, n$, and given that I is a zero dimensional ideal, it follows that, up to reordering the generators, our assumption is equivalent to the existence of a weight w such that:

$$\mathrm{in}_w(P_i) = c_i\, x_i^{r_i+1} \tag{1.40}$$

It is clear that in this case $\dim_\mathbb{K}(\mathcal{A}) = r_1 \cdots r_n$, and a monomial basis of $\mathcal{A}$ is given by $\{[x^\alpha] : 0 \le \alpha_i \le r_i\}$.

We point out that, for appropriately chosen term orders, our assumption leads to interesting examples.

- Suppose $\prec$ is lexicographic order with $x_n \prec \cdots \prec x_1$. In this case

$$P_i = c_i x_i^{r_i+1} + P_i'(x_i, \ldots, x_n)$$

and $\deg_{x_i}(P_i') \leq r$. This case was considered in [DS91].
- Let $\prec$ be degree lexicographic order with $x_1 \prec \cdots \prec x_n$. Then

$$P_i(x) = c_i x_i^{r_i+1} + \sum_{j=1}^{i-1} z_j \phi_{ij}(x) + \psi_i(x),$$

where $\deg(\phi_{ij}) = r_i$ and $\deg(\psi_i(x)) \leq r_i$. This case has been extensively studied by the Krasnoyarsk School (see, for example, [AY83, Ch. 21] and [Tsi92, II.8.2]) using integral methods. Some of their results have been transcribed to the algebraic setting in [BGV02] under the name of Pham systems of type II.

Note also that the polynomials in (1.37) satisfy these conditions. Indeed, for $w = (3, 14, 5)$ we have:

$$\text{in}_w(P_1) = x_1^2 , \quad \text{in}_w(P_2) = x_2 , \quad \text{in}_w(P_3) = x_3^2 \qquad (1.41)$$

The following theorem, which may be viewed as a generalization of the basic univariate definition (1.1), is due to Aïzenberg and Tsikh. Its proof may be found in [AY83, Ch. 21] and [CDS96, Th. 2.3].

Theorem 1.5.8. *Let* $P_1, \ldots, P_n \in \mathbb{C}[x_1, \ldots, x_n]$ *satisfy (1.40). Then for any* $H \in \mathbb{C}[x_1, \ldots, x_n]$ $\text{res}_{\langle P \rangle}(H)$ *is equal to the* $\dfrac{1}{x_1 \cdots x_n}$*-coefficient of the Laurent series expansion of:*

$$\frac{H(x)}{\prod_i c_i x_i^{r_i+1}} \prod_i \left(\frac{1}{1 + P_i'(x)/(c_i x_i^{r_i+1})} \right) , \qquad (1.42)$$

obtained through geometric expansions.

The following simple consequence of Theorem 1.5.8 generalizes (1.10) and is the basis for its algorithmic applications.

Corollary 1.5.9. *Let* $P_1, \ldots, P_n \in \mathbb{C}[x_1, \ldots, x_n]$ *satisfy (1.40) and let* $\{[x^\alpha] : 0 \leq \alpha_i \leq r_i\}$ *be the corresponding monomial basis of* $\mathcal{A}$. *Let* $\mu = (r_1, \ldots, r_n)$, *then*

$$\text{res}_{\langle P \rangle}([x^\alpha]) = \begin{cases} 0 & \text{if } \alpha \neq \mu \\ \frac{1}{c_1 \cdots c_n} & \text{if } \alpha = \mu \end{cases} \qquad (1.43)$$

Remark 1.5.10. A proof of (1.43) using the Bezoutian approach may be found in [BCRS96]. Hence, Corollary 1.5.9 may be used in the algebraic setting as well.

As in the univariate case, we are led to the following algorithm for computing residues when $P_1, \ldots, P_n$ satisfy (1.40).

Algorithm 1: Compute the normal form $N(H)$ of $H \in \mathbb{K}[x_1, \ldots, x_n]$ relative to any term order which refines w-degree. Then,

$$\mathrm{res}_{\langle P \rangle}(H) = \frac{a_\mu}{c_1 \cdots c_n}, \qquad (1.44)$$

where a_μ is the coefficient of x^μ in $N(H)$.

Remark 1.5.11. Given a weight w for which (1.40) holds it is easy to carry the computations in the above algorithm using the weighted orders wp (weighted grevlex) and Wp (weighted deglex) in Singular [GPS01]. For example, for the polynomials in (1.37), the Jacobian $J_{\langle P \rangle}(x) = 4x_1 x_3 - 3x_1^2$ and we get:

```
> ring R = 0, (x1, x2, x3), wp(3,14,5);
> ideal I = x1^2-x3, x2-x1*x3^2, x3^2 - x1^3;
> reduce(4*x1*x3 - 3*x1^2,std(I));
            4*x1*x3-3*x3
```

Thus, the $x_1 x_3$ coefficient of the normal form of $J_{\langle P \rangle}(x)$ is 4, i.e. $\dim_{\mathbb{K}}(\mathcal{A})$ as asserted by (1.30).

1.5.5 The Euler-Jacobi vanishing theorem

We will now discuss the multivariate extension of Theorem 1.1.8. The basic geometric assumption that we need to make is that if we embed $\mathbb{C}^n$ in a suitable compactification then the ideal we are considering has all its zeros in $\mathbb{C}^n$. Here we will restrict ourselves to the case when the chosen compactification is weighted projective space. The more general vanishing theorems are stated in terms of global residues in the torus and toric compactifications as in [Kho78a].

Let $w \in \mathbb{N}^n$ and denote by $\deg_w$ the weighted degree defined by w. We set $|w| = \sum_i w_i$. Let $I = \langle P_1, \ldots, P_n \rangle$ be a zero-dimensional complete intersection ideal and write

$$P_i(x) = Q_i(x) + P_i'(x),$$

where $Q_i(x)$ is weighted homogeneous of w-degree d_i and $\deg_w(P_i') < d_i$. We call Q_i the *leading form* of P_i. We say that I has no zeros at infinity in weighted projective space if and only if

$$Q_1(x) = \cdots = Q_n(x) = 0 \quad \text{if and only if} \quad x = 0. \qquad (1.45)$$

In the algebraic context an ideal which has a presentation by generators satisfying (1.45) is called a *strict complete intersection* [KK87].

Theorem 1.5.12 (Euler-Jacobi vanishing). *Let* $I = \langle P_1, \ldots, P_n \rangle$ *be a zero-dimensional complete intersection ideal with no zeros at infinity in weighted projective space. Then,*

$$\mathrm{res}_{\langle P \rangle}(H) = 0 \quad \text{if} \quad \deg_w(H) < \sum_{i=1}^{n} \deg_w(P_i) - |w|.$$

Proof. We begin by proving the assertion in the particular case when $Q_i(x) = x_i^{N+1}$. By linearity it suffices to prove that if x^α is a monomial with $\langle w, \alpha \rangle < N|w|$, then $\mathrm{res}_{\langle P \rangle}(x^\alpha) = 0$. We prove this by induction on $\delta = \langle w, \alpha \rangle$. If $\delta = 0$ then $x^\alpha = 1$ and the result follows from Corollary 1.5.9. Suppose then that the result holds for any monomial of degree less than $\delta = \langle w, \alpha \rangle$, if every $\alpha_i \le N$ then the result follows, again, from Corollary 1.5.9. If, on the other hand, some $\alpha_i > N$ then we can write

$$ x^\alpha = x^\beta \cdot P_i - x^\beta \cdot P_i' \,, $$

where $\beta = \alpha - (N+1)e_i$. It then follows that $\mathrm{res}_{\langle P \rangle}(x^\alpha) = -\mathrm{res}_{\langle P \rangle}(x^\beta \cdot P_i')$, but all the monomials appearing in the right-hand side have weighted degree less than δ and therefore the residue vanishes.

Consider now the general case. In view of (1.45) and the Nullstellensatz there exists N sufficiently large such that

$$ x_i^{N+1} \in \langle Q_1(x), \ldots, Q_n(x) \rangle \,. $$

In particular, we can write

$$ x_j^{N+1} = \sum_{i=1}^{n} a_{ij}(x) Q_i(x) \,, $$

where $a_{ij}(x)$ is w-homogeneous of degree $(N+1)w_j - d_i$. Let now

$$ F_j(x) = \sum_{i=1}^{n} a_{ij}(x) P_i(x) = x_j^{N+1} + F_j'(x) \,, $$

and $\deg_w(F_j') < (N+1)w_j$. Given now $H \in \mathbb{K}[x_1, \ldots, x_n]$ with $\deg_w(H) < \sum_i d_i - |w|$, we have by the Global Transformation Law:

$$ \mathrm{res}_{\langle P \rangle}(H) = \mathrm{res}_{\langle F \rangle}(\det(a_{ij}) \cdot H) \,. $$

But, $\deg_w(\det(a_{ij})) \le (N+1)|w| - \sum_i d_i$ and therefore

$$ \deg_w(det(a_{ij}) \cdot H) \le \deg_w(\det(a_{ij})) + \deg_w(H) < N|w| \,, $$

and the result follows from the previous case.

Remark 1.5.13. The Euler-Jacobi vanishing theorem is intimately connected to the continuity of the residue. The following argument from [AGZV85, Ch. 1, Sect. 5] makes the link evident. Suppose $P_1, \ldots, P_n$ have only simple zeros and satisfy (1.45). For simplicity we take $w = (1, \ldots, 1)$, the general case is completely analogous. Consider the family of polynomials

$$ \tilde{P}_i(x; t) := t^{d_i} P_i(t^{-1}x_1, \ldots, t^{-1}x_n) \,. \tag{1.46} $$

Note that $\tilde{P}_i(t \cdot x, t) = t^{d_i} P_i(x)$. In particular if $P_i(\xi) = 0$, $\tilde{P}_i(t\xi; t) = 0$ as well. Suppose now that $\deg(H) < \sum_i d_i - n$ and let $\tilde{H}(x; t)$ be defined as in (1.46). Then

$$\mathrm{res}_{\langle \tilde{P} \rangle}(\tilde{H}) = \sum_{\xi \in Z(I)} \frac{\tilde{H}(t\xi; t)}{\mathrm{Jac}_{\langle \tilde{P} \rangle}(t\xi)} = t^a \sum_{\xi \in Z(I)} \frac{H(\xi)}{\mathrm{Jac}_{\langle P \rangle}(\xi)} = t^a \mathrm{res}_{\langle P \rangle}(H),$$

where $a = \deg(H) - \deg(\mathrm{Jac}_{\langle P \rangle}(x)) = \deg(H) - (\sum_i d_i - n)$. Hence, if $a < 0$, the limit

$$\lim_{t \to 0} \mathrm{res}_{\langle \tilde{P} \rangle}(\tilde{H})$$

may exist only if $\mathrm{res}_{\langle P \rangle}(H) = 0$ as asserted by the Euler-Jacobi theorem.

We conclude this subsection with some applications of Theorem 1.5.12 to plane projective geometry (cf. [GH78, 5.2]). The following theorem is usually referred to as the Cayley-Bacharach Theorem though, as Eisenbud, Green, and Harris point out in [EGH96], it should be attributed to Chasles.

Theorem 1.5.14 (Chasles). *Let C_1 and C_2 be curves in $\mathbb{P}^2$, of respective degrees d_1 and d_2, intersecting in $d_1 d_2$ distinct points. Then, any curve of degree $d = d_1 + d_2 - 3$ that passes through all but one of the points in $C_1 \cap C_2$ must pass through the remaining point as well.*

Proof. After a linear change of coordinates, if necessary, we may assume that no point in $C_1 \cap C_2$ lies in the line $x_3 = 0$. Let $C_i = \{\tilde{P}_i(x_1, x_2, x_3) = 0\}$, $\deg P_i = d_i$. Set $P_i(x_1, x_2) = \tilde{P}_i(x_1, x_2, 1)$. Given $\tilde{H} \in \mathbb{K}[x_1, x_2, x_3]$, homogeneous of degree d, let $H \in \mathbb{K}[x_1, x_2]$ be similarly defined. We can naturally identify the points in $C_1 \cap C_2$ with the set of common zeros

$$Z = \{\xi \in \mathbb{K}^2 : P_1(\xi) = P_2(\xi) = 0\}.$$

Since $\deg H < \deg P_1 + \deg P_2 - 2$, Theorem 1.5.12 implies that $\mathrm{res}_{\langle P \rangle}(H) = 0$, but then

$$0 = \mathrm{res}_{\langle P \rangle}(H) = \sum_{\xi \in Z} \frac{H(\xi)}{\mathrm{Jac}_{\langle P \rangle}(\xi)}$$

which implies that if H vanishes at all but one of the points in Z it must vanish on the remaining one as well.

Corollary 1.5.15 (Pascal's Mystic Hexagon). *Consider a hexagon inscribed in a conic curve of $\mathbb{P}^2$. Then, the pairs of opposite sides meet in collinear points.*

Proof. Let $L_1 \ldots L_6$ denote the hexagon inscribed in the conic $Q \subset \mathbb{P}^2$, where L_i is a line in $\mathbb{P}^2$. Let ξ_{ij} denote the intersection point $L_i \cap L_j$. Consider the cubic curves $C_1 = L_1 + L_3 + L_5$ and $C_2 = L_2 + L_4 + L_6$. The intersection $C_1 \cap C_2$ consists of the nine points ξ_{ij} with i odd and j even. The cubic $Q + L(\xi_{14}\xi_{36})$, where $L(\xi_{14}\xi_{36})$ denotes the line joining the two points, passes through eight of the points in $C_1 \cap C_2$ hence must pass through the ninth point ξ_{52}. For degree reasons this is only possible if $\xi_{52} \in L(\xi_{14}\xi_{36})$ and therefore the three points are collinear.

1.5.6 Homogeneous (projective) residues

In this section we would like to indicate how the notion or residue may be extended to meromorphic forms in projective space. This is a special instance of a much more general theory of residues in toric varieties. A full discussion of this topic is beyond the scope of these notes so we will restrict ourselves to a presentation of the basic ideas, in the case $\mathbb{K} = \mathbb{C}$, and refer the reader to [GH78, TY84, PS83, Cox96, CCD97] for details and proofs.

Suppose $F_0, \ldots, F_n \in \mathbb{C}[x_0, \ldots, x_n]$ are homogeneous polynomials of degrees $d_0, \ldots, d_n$, respectively. Let $V_i = \{x \in \mathbb{P}^n : F_i(x) = 0\}$ and assume that

$$V_0 \cap V_1 \cdots \cap V_n = \emptyset. \tag{1.47}$$

This means that the zero locus of the ideal $I = \langle F_0, \ldots, F_n \rangle$ is the origin $0 \in \mathbb{C}^{n+1}$. Given any homogeneous polynomial $H \in \mathbb{C}[x_0, \ldots, x_n]$ we can define the *projective residue* of H relative to the $n+1$-tuple $\langle F \rangle = \{F_0, \ldots, F_n\}$ as:

$$\mathrm{res}^{\mathbb{P}^n}_{\langle F \rangle}(H) := \mathrm{res}_{\langle F \rangle}(H) = \mathrm{res}_{\langle F \rangle, 0}(H).$$

It is clear from the integral definition of the Grothendieck residue, that the local residue at 0 is invariant under the change of coordinates $x_i \mapsto \lambda x_i$, $\lambda \in \mathbb{C}^*$. On the other hand, if $\deg(H) = d$ we see that, for

$$\rho := \sum_{i=0}^{n} (d_i - 1),$$

$$\frac{H(\lambda \cdot x)}{F_0(\lambda \cdot x) \cdots F_n(\lambda \cdot x)} d(\lambda x_0) \wedge \cdots \wedge d(\lambda x_n) = \frac{\lambda^{d-\rho} H(x)}{F_0(x) \cdots F_n(x)} dx_0 \wedge \cdots \wedge dx_n.$$

Hence,

$$\mathrm{res}^{\mathbb{P}^n}_{\langle F \rangle}(H) = 0 \quad \text{if} \quad \deg(H) \neq \rho.$$

Being a global (and local) residue, the projective residue is a dualizing form in the algebra $\mathcal{A} = \mathbb{C}[x_0, \ldots, x_n]/I$. Moreover, since I is a homogeneous ideal, $\mathcal{A}$ is a graded algebra and the projective residue is compatible with the grading. These dualities properties are summarized in the following theorem.

Theorem 1.5.16. *The graded algebra $\mathcal{A} = \oplus \mathcal{A}_d$ satisfies:*

a) $\mathcal{A}_d = 0$ for $d > \rho := d_0 + \cdots + d_n - (n+1)$.
b) $\mathcal{A}_\rho \cong \mathbb{C}$.

c) For $0 \leq d \leq \rho$, the bilinear pairing

$$\mathcal{A}_d \times \mathcal{A}_{\rho-d} \to \mathbb{C} \; ; \quad ([H_1], [H_2]) \mapsto \operatorname{res}_{\langle F \rangle}^{\mathbb{P}^n}(H_1 \cdot H_2)$$

is non-degenerate.

Proof. The assumption (1.47) implies that $F_0, \ldots, F_n$ are a regular sequence in the ring $\mathbb{C}[x_0, \ldots, x_n]$. Computing the Poincaré series for $\mathcal{A}$ using the exactness of the Koszul sequence yields the first two assertions. See [PS83, Sect. 12] for details. A proof using residues may be found in [Tsi92, Sect. 20]. The last assertion follows from Theorem 1.5.1.

An important application of Theorem 1.5.16 arises in the study of smooth hypersurfaces $X_F = \{x \in \mathbb{P}^n : F(x) = 0\}$, of degree d, in projective space [CG80]. In this case we take $F_i = \partial F / \partial x_i$, the smoothness condition means that $\{F_0, \ldots, F_n\}$ satisfy (1.47), and the Hodge structure of X may be described in terms of the *Jacobian ideal* generated by $\{\partial F / \partial x_i\}$. Indeed, $\rho = (n+1)(d-2)$ and setting, for $0 \leq p \leq n-1$, $\delta(p) := d(p+1) - (n+1)$, we have $\delta(p) + \delta(n-1-p) = \rho$, and

$$H^{p, n-1-p}(X) \cong \mathcal{A}_{\delta(p)} \, .$$

Moreover, the pairing

$$\operatorname{res}_{\langle F \rangle}^{\mathbb{P}^n} : \mathcal{A}_{\delta(p)} \times \mathcal{A}_{\delta(n-1-p)} \to \mathbb{C}$$

corresponds to the intersection pairing

$$H^{p, n-1-p}(X) \times H^{n-1-p, p}(X) \to \mathbb{C} \, .$$

The projective residue may be related to affine residues in a different way. If we identify $\mathbb{C}^n \cong \{x \in \mathbb{P}^n : x_0 \neq 0\}$, then after a linear change of coordinates, if necessary, we may assume that for every $i = 0, \ldots, n$,

$$Z_i := V_0 \cap \cdots \cap \widehat{V}_i \cap \cdots \cap V_n \subset \mathbb{C}^n \, . \tag{1.48}$$

Let $P_i \in \mathbb{C}[x_1, \ldots, x_n]$ be the polynomial $P_i(x_1, \ldots, x_n) = F_i(1, x_1, \ldots, x_n)$ and let us denote by $\langle P^i \rangle$ the n-tuple of polynomials $P_0, \ldots, P_{i-1}, P_{i+1}, \ldots, P_n$.

Theorem 1.5.17. *For any homogeneous polynomial $H \in \mathbb{C}[x_0, \ldots, x_n]$ with* $\deg(H) \leq \rho$,

$$\operatorname{res}_{\langle F \rangle}^{\mathbb{P}^n}(H) := (-1)^i \operatorname{res}_{\langle P^i \rangle}(h/P_i) \, , \tag{1.49}$$

where $h(x_1, \ldots, x_n) = H(1, x_1, \ldots, x_n)$.

Proof. We will only prove the second, implicit, assertion that the right-hand side of (1.49) is independent of i. This statement, which generalizes the identity (1.11), is essentially Theorem 5 in [TY84]. For the main assertion we refer to [CCD97, Sect. 4], where it is proved in the more general setting of simplicial toric varieties.

Note that the assumption (1.47) implies that the rational function h/P_i is regular on Z_i and hence it makes sense to compute $\mathrm{res}_{\langle P^i \rangle}(h/P_i)$. For each $i = 0, \ldots, n$, consider the n-tuple of polynomials in $\mathbb{K}[x_1, \ldots, x_n]$: $\langle Q_i \rangle = \{P_0, \ldots, (P_i \cdot P_{i+1}), \ldots, P_n\}$, if $i < n$ and $\langle Q_n \rangle = \{P_1, \ldots, P_{n-1}, (P_n \cdot P_0)\}$. The set of common zeros of the polynomials in Q_i is $Z(Q_i) = Z_i \cup Z_{i+1}$. Hence, it follows from (1.48) that the ideal generated by the n-tuple Q_i is zero-dimensional and has no zeros at infinity. Hence, given that $\deg(H) \le \rho$, the Euler-Jacobi vanishing theorem implies that

$$
\begin{aligned}
0 &= \mathrm{res}_{\langle Q_i \rangle}(h) = \sum_{\xi \in Z_i} \mathrm{res}_{\langle Q_i \rangle, \xi}(h) \;+\; \sum_{\xi \in Z_{i+1}} \mathrm{res}_{\langle Q_i \rangle, \xi}(h) \\
&= \sum_{\xi \in Z_i} \mathrm{res}_{\langle P^i \rangle, \xi}(h/P_i) \;+\; \sum_{\xi \in Z_{i+1}} \mathrm{res}_{\langle \widehat{P^{i+1}} \rangle, \xi}(h/P_{i+1}) \\
&= \mathrm{res}_{\langle P^i \rangle}(h/P_i) \;+\; \mathrm{res}_{\langle \widehat{P^{i+1}} \rangle}(h/P_{i+1})
\end{aligned}
$$

and, consequently, the theorem follows. We should point out that the equality $\mathrm{res}_{\langle Q_i \rangle, \xi}(h) = \mathrm{res}_{\langle P^i \rangle, \xi}(h/P_i)$, which is clear from the integral definition of the local residue, may be obtained in the general case from the Local Transformation Law and the fact that $\mathrm{res}_{\langle P^i \rangle, \xi}(h/P_i)$ was defined as $\mathrm{res}_{\langle P^i \rangle, \xi}(h \cdot Q_i)$, where Q_i inverts P_i in the local algebra A_ξ^i and, consequently, the statement holds over any algebraically closed field of characteristic zero.

We can use the transformation law to exhibit a polynomial $\Delta(x)$ of degree ρ with non-zero residue. Write

$$
F_j = \sum_{i=0}^{n} a_{ij}(x)\, x_i \; ; \quad j = 0, \ldots, n\,,
$$

and set $\Delta(x) = \det(a_{ij}(x))$. Then, $\deg(\Delta) = \rho$, and

$$
\mathrm{res}_{\langle F \rangle}^{\mathbb{P}^n}(\Delta) \;=\; 1 \tag{1.50}
$$

Indeed, let $\langle G \rangle$ denote the $n + 1$-tuple $G = \{x_0, \ldots, x_n\}$. Then by the transformation law

$$
\mathrm{res}_{\langle G \rangle}^{\mathbb{P}^n}(1) \;=\; \mathrm{res}_{\langle F \rangle}^{\mathbb{P}^n}(\Delta)
$$

and a direct computation shows that the left-hand side of the above identity is equal to 1.

Putting together part b) of Theorem 1.5.16 with (1.50) we obtain the following normal form algorithm for computing the projective residue $\mathrm{res}_{\langle F \rangle}^{\mathbb{P}^n}(H)$:

Algorithm 2:   1. Compute a Gröbner basis of the ideal $\langle F_0, \ldots, F_n \rangle$.

2. Compute the normal form $N(H)$ of H and the normal form $N(\Delta)$ of Δ, with respect to the Gröbner basis.

3. The projective residue $\text{res}^{\mathbb{P}^n}_{\langle F \rangle}(H) = \dfrac{N(H)}{N(\Delta)}$.

Remark 1.5.18. There is a straightforward variant of this algorithm valid for weighted homogeneous polynomials. This more general algorithm has been used by Batyrev and Materov [BM02], to compute the Yukawa 3-point function of the generic hypersurface in weighted projective $\mathbb{P}^4_w$, $w = (1,1,2,2,2)$. This function, originally computed in [CdlOF+94] has a series expansion whose coefficients have enumerative meaning. We refer to [BM02, 10.3] and [CK99, 5.6.2.1] for more details.

We can combine Theorem 1.5.17 and Algorithm 2 to compute the global (affine) residue with respect to a zero-dimensional complete intersection ideal with no zeros at infinity in projective space. The construction below is a special case of a much more general algorithm described in [CD97] and it applies, in particular, to the weighted case as well. It also holds over any algebraically closed field $\mathbb{K}$ of characteristic zero.

Let $I = \{P_1, \ldots, P_n\} \in \mathbb{K}[x_1, \ldots, x_n]$ be polynomials satisfying (1.45). Let $d_i = \deg(P_i)$ and denote by

$$F_i(x_0, x_1, \ldots, x_n) := x_0^{d_i} P(\frac{x_1}{x_0}, \ldots, \frac{x_n}{x_0})$$

the homogenization of P_i. Let $h(x_1, \ldots, x_n) \in \mathbb{K}[x_1, \ldots, x_n]$. If $d = \deg(h) < \sum_i (d_i - 1)$, then $\text{res}_{\langle P \rangle}(h) = 0$ by the Euler-Jacobi theorem. Suppose, then that $d \geq \sum_i (d_i - 1)$, let $H \in \mathbb{K}[x_0, \ldots, x_n]$ be its homogenization, and let

$$F_0 = x_0^{d_0} ; \quad d_0 = d - \sum_{i=1}^{n}(d_i - 1) + 1 .$$

Then, $d = \sum_{i=0}^{n}(\deg(F_i) - 1)$ and it follows from Theorem 1.5.17 that

$$\text{res}^{\mathbb{P}^n}_{\langle F \rangle}(H) = \text{res}_{\langle P^0 \rangle}(h/P_0) = \text{res}_{\langle P \rangle}(h) .$$

1.5.7 Residues and elimination

One of the basic applications of residues is to elimination theory. The key idea is very simple (see also Section 3.3.1 in Chapter 3). Let $I = \langle P_1, \ldots, P_n \rangle \subset \mathbb{K}[x_1, \ldots, x_n]$ be a zero-dimensional, complete intersection ideal. Let $\xi_i = (\xi_{i1}, \ldots, \xi_{in}) \in \mathbb{K}^n$, $i = 1, \ldots, r$, be the zeros of I. Let $\mu_1, \ldots, \mu_r$ denote their respective multiplicities. Then the power sum

$$S_j^{(k)} := \sum_{i=1}^{r} \xi_{ij}^k$$

is the trace of the multiplication map $M_{x_j^k} : \mathcal{A} \to \mathcal{A}$ and, therefore, it may be expressed as a global residue:

$$S_j^{(k)} = \operatorname{tr}(M_{x_j^k}) = \operatorname{res}_{\langle P \rangle}(x_j^k \cdot J_{\langle P \rangle}(x)).$$

The univariate Newton identities of Section 1.2.5 now allow us to compute inductively the coefficients of a polynomial in the variable x_j with roots at $\xi_{1j}, \ldots, \xi_{rj} \in \mathbb{K}$ and respective multiplicities $\mu_1, \ldots, \mu_r$.

We illustrate the method with the following example. Let

$$I = \langle x_1^3 + x_1^2 - x_2, x_1^3 - x_2^2 + x_1 x_2 \rangle.$$

It is easy to check that the given polynomials are a Gröbner basis for any term order that refines the weight order defined by $w = (5, 9)$. The leading terms are $x_1^3, -x_2^2$. A normal form computation following Algorithm 1 in Section 1.5.4 yields:

$$S_1^{(1)} = -2 \; ; \; S_1^{(2)} = 4 \; ; \; S_1^{(3)} = -2 \; ; \; S_1^{(4)} = 0 \; ; \; S_1^{(5)} = 8 \; ; \; S_1^{(6)} = -20.$$

For example, the following Singular [GPS01] computation shows how the values $S_1^{(3)}$ and $S_1^{(4)}$ were obtained:

```
> ring R = 0, (x1,x2), wp(5,9);
> ideal I = x1^3 + x1^2 - x2, x1^3 - x2^2 + x1*x2;
> poly J = -6*x1^2*x2+3*x1^3-4*x1*x2+5*x1^2+x2;
> reduce(x1^3*J,std(I));
2*x1^2*x2+2*x1*x2+10*x1^2-10*x2
> reduce(x1^4*J,std(I));
-8*x1*x2-12*x1^2+12*x2
```

Now, using the Newton identities (1.16) we may compute the coefficients of a monic polynomial of degree 6 on the variable x_1 lying on the ideal:

$$a_5 = 2 \; ; \; a_4 = 0 \; ; \; a_3 = -2 \; ; \; a_2 = 0 \; ; \; a_1 = 0 \; ; \; a_0 = 0.$$

Hence, $f_1(x_1) = x_1^6 + 2x_1^5 - 2x_1^3 \in I$.

We refer the reader to [AY83, BKL98] for a fuller account of this elimination procedure. Note also that in Section 3.6 of Chapter3 there is an application of residues to the implicitization problem.

1.6 Multivariate resultants

In this section we will extend the notion of the resultant to multivariate systems. We will begin by defining the resultant of $n+1$ homogeneous polynomials in $n + 1$ variables and discussing some formulas to compute it. We will also discuss some special examples of the so-called sparse or toric resultant.

1.6.1 Homogeneous resultants

When trying to generalize resultants associated to polynomials in any number of variables, the first problem one faces is which families of polynomials one is going to study, i.e. which will be the variables of the resultant. For example, in the univariate case, fixing the degrees d_1, d_2 amounts to setting $(a_0, \ldots, a_{d_1}, b_0, \ldots, b_{d_2})$ as the input variables for the resultant Res_{d_1, d_2}. One obvious, and classical choice, in the multivariable case is again, to fix the degrees $d_0, \ldots, d_n$ of $n+1$ polynomials in n variables, which will generally define an overdetermined system. If one wants the vanishing of the resultant $\text{Res}_{d_0, \ldots, d_n}$ to be equivalent to the existence of a common root, one realizes that a compactification of affine space naturally comes into the picture, in this case projective n-space.

Consider, for instance, a bivariate linear system

$$\begin{cases} f_0(x,y) = a_{00}x + a_{01}y + a_{02} \\ f_1(x,y) = a_{10}x + a_{11}y + a_{12} \\ f_2(x,y) = a_{20}x + a_{21}y + a_{22} \end{cases} \tag{1.51}$$

We fix the three degrees equal to 1, i.e. we have nine variables a_{ij} ($i, j = 0, 1, 2$), and we look for an irreducible polynomial $\text{Res}_{1,1,1} \in \mathbb{Z}[a_{ij}, i, j = 0, 1, 2]$ which vanishes if and only the system has a solution (x, y). If such a solution (x, y) exists, then $(x, y, 1)$ would be a non-trivial solution of the augmented 3×3-linear system and consequently the determinant of the matrix (a_{ij}) must vanish. However, as the following example easily shows, the vanishing of the determinant does not imply that (1.51) has a solution. Let

$$\begin{cases} f_0(x,y) = x + 2y + 1 \\ f_1(x,y) = x + 2y + 2 \\ f_2(x,y) = x + 2y + 3 \end{cases}$$

The determinant vanishes but the system is incompatible in $\mathbb{C}^2$. On the other hand, the lines defined by $f_i(x, y) = 0$ are parallel and therefore we may view them as having a common point at infinity in projective space. We can make this precise by passing to the homogenized system

$$\begin{cases} F_0(x,y,z) = x + 2y + z \\ F_1(x,y,z) = x + 2y + 2z \\ F_2(x,y,z) = x + 2y + 3z, \end{cases}$$

which has non zero solutions of the form $(-2y, y, 0)$, i.e. the homogenized system has a solution in the projective plane $\mathbb{P}^2(\mathbb{C})$, a compactification of the affine plane $\mathbb{C}^2$.

We denote $x = (x_0, \ldots, x_n)$ and for any $\alpha = (\alpha_0, \ldots, \alpha_n) \in \mathbb{N}^{n+1}$, $|\alpha| = \alpha_0 + \cdots + \alpha_n$, $x^\alpha = x_0^{\alpha_0} \ldots x_n^{\alpha_n}$. Recall that $f = \sum_\alpha a_\alpha x^\alpha \in \mathbf{k}[x_0, \ldots, x_n]$ is called homogeneous (of degree $\deg(f) = d$) if $|\alpha| = d$ for all $|\alpha|$ with $a_\alpha \neq 0$, or equivalently, if for all $\lambda \in \mathbf{k}$, it holds that $f(\lambda\, x) = \lambda^d\, f(x)$, for all $x \in \mathbf{k}^{n+1}$.

As we already remarked in Section 1.3.1, the variety of zeros of a homogeneous polynomial is well defined over $\mathbb{P}^n(\mathbf{k}) = (\mathbf{k}^{n+1}\backslash\{0\})\,/\sim$, where we identify $x \sim \lambda x$, for all $\lambda \in \mathbf{k}\backslash\{0\}$. As before, $\mathbb{K}$ denotes the algebraic closure of $\mathbf{k}$.

The following result is classical.

Theorem 1.6.1. *Fix $d_0, \ldots, d_n \in \mathbb{N}$ and write $F_i = \sum_{|\alpha|=d_i} a_{i\alpha} x^\alpha$, $i = 1, \ldots, n$. There exists a unique irreducible polynomial*

$$\mathrm{Res}_{d_0,\ldots,d_n}(F_0, \ldots, F_n) \in \mathbb{Z}[a_{i\alpha}\,;\, i = 0, \ldots, n, |\alpha| = d_i]$$

which verifies:

(i) $\mathrm{Res}_{d_0,\ldots,d_n}(F_0, \ldots, F_n) = 0$ *for a given specialization of the coefficients in* $\mathbf{k}$ *if and only if there exists $x \in \mathbb{P}^n(\mathbb{K})$ such that $F_0(x) = \cdots = F_n(x) = 0$.*
(ii) $\mathrm{Res}_{d_0,\ldots,d_n}(x_0^{d_0}, \ldots, x_n^{d_n}) = 1$.

The resultant $\mathrm{Res}_{d_0,\ldots,d_n}$ depends on N variables, where $N = \sum_{i=0}^{n}\binom{n+d_i}{d_i}$. A geometric proof of this theorem, which is widely generalizable, can be found for instance in [Stu98]. It is based on the consideration of the incidence variety

$$\mathcal{Z} = \{((a_{i\alpha}), x) \in \mathbb{K}^N \times \mathbb{P}^n(\mathbb{K}) \,:\, \sum_{|\alpha|=d_i} a_{i\alpha} x^\alpha,\, i = 1, \ldots, n\},$$

and its two projections to $\mathbb{K}^N$ and $\mathbb{P}^n(\mathbb{K})$. In fact, $\mathcal{Z}$ is an irreducible variety of dimension $N-1$ and the fibers of the first projection is generically $1-1$ onto its image.

As we noticed above, in the linear case $d_0 = \cdots = d_n = 1$, the resultant is the determinant of the linear system. We now state the main properties of multivariate homogeneous resultants, which generalize the properties of determinants and of the univariate resultant (or bivariate homogeneous resultant) in Section 1.3.2. The proofs require more background, and we will omit them.

Main properties

i)   The resultant $\mathrm{Res}_{d_0,\ldots,d_n}$ is homogeneous in the coefficients of F_i of degree $d_0 \ldots d_{i-1} d_{i+1} \ldots d_n$, i.e. by Bézout's theorem, the number of generic common roots of $F_0 = \cdots = F_{i-1} = F_{i+1} = \cdots = F_n = 0$.
ii) The resultants $\mathrm{Res}_{d_0,\ldots,d_i\ldots,d_j,\ldots,d_n}$ and $\mathrm{Res}_{d_0,\ldots,d_j\ldots,d_i,\ldots,d_n}$ coincide up to sign.
iii) For any monomial x^γ of degree $|\gamma|$ greater than the *critical degree* $\rho := \sum_{i=0}^{n}(d_i-1)$, there exist homogeneous polynomials $A_0, \ldots, A_n$ in the variables $x_0, \ldots, x_n$ with coefficients in $\mathbb{Z}[(a_{i\alpha})]$ and $\deg(A_i) = |\gamma| - d_i$, such that

$$\mathrm{Res}_{d_0,\ldots,d_n}(F_0, \ldots, F_n) \cdot x^\gamma = A_0 F_0 + \cdots + A_n F_n. \tag{1.52}$$

Call $f_i(x_1, \ldots, x_n) = F_i(1, x_1, \ldots, x_n) \in \mathbf{k}[x_1, \ldots x_n]$ the dehomogenizations of $F_0, \ldots, F_n$. One can define the resultant

$$\mathrm{Res}_{d_0, \ldots, d_n}(f_0, \ldots, f_n) := \mathrm{Res}_{d_0, \ldots, d_n}(F_0, \ldots, F_n)$$

and try to translate to the affine setting these properties of the homogeneous resultant. We point out the following direct consequence of (1.52). Taking $\gamma = (\rho + 1, 0, \ldots, 0)$ and then specializing $x_0 = 1$, we deduce that there exist polynomials $A_0, \ldots, A_n \in \mathbb{Z}[(a_{i\alpha})][x_1, \ldots, x_n]$, with $\deg(A_i)$ bounded by $\rho + 1 - d_i = \sum_{j \neq i} d_i - n$, and such that

$$\mathrm{Res}_{d_0, \ldots, d_n}(f_0, \ldots, f_n) = A_0 f_0 + \cdots + A_n f_n. \tag{1.53}$$

As we remarked in the linear case, the resultant $\mathrm{Res}_{d_0, \ldots, d_n}(f_0, \ldots, f_n)$ can vanish even if $f_0, \ldots, f_n$ do not have any common root in $\mathbb{K}^n$ if their homogenizations $F_0, \ldots, F_n$ have a nonzero common root with $x_0 = 0$. Denote by $f_{i, d_i} = F_i(0, x_1, \ldots, x_n)$ the homogeneous component of top degree of each f_i. The corresponding version of Proposition 1.3.2 is as follows.

Proposition 1.6.2. (Homogeneous Poisson formula) *Let $F_0, \ldots, F_n$ be homogeneous polynomials with degrees $d_0, \ldots, d_n$ and let $f_i(x_1, \ldots, x_n)$ and $f_{i, d_i}(x_1, \ldots, x_n)$ as above. Then*

$$\mathrm{Res}_{d_0, \ldots, d_n}(F_0, \ldots, F_n)) = \mathrm{Res}_{d_1, \ldots, d_n}(f_{1, d_1}, \ldots, f_{n, d_n})^{d_0} \prod_{\xi \in V} f_0(\xi)^{m_\xi},$$

where V is the common zero set of $f_1, \ldots, f_n$, and m_ξ denotes the multiplicity of $\xi \in V$.

This factorization holds in the field of rational functions over the coefficients $(a_{i\alpha})$. Stated differently, the product $\prod_{\xi \in V} f_0(\xi)^{m_\xi}$ is a rational function of the coefficients, whose numerator is the irreducible polynomial $\mathrm{Res}_{d_0, \ldots, d_n}(F_0, \ldots, F_n)$ and whose denominator is the d_0 power of the irreducible polynomial $\mathrm{Res}_{d_1, \ldots, d_n}(f_{1, d_1}, \ldots, f_{n, d_n})$, which only depends on the coefficients of the monomials of highest degree $d_1, \ldots, d_n$ of $f_1, \ldots, f_n$. Note that taking $F_0 = x_0$ we get, in particular, the expected formula

$$\mathrm{Res}_{1, d_1, \ldots, d_n}(x_0, F_1, \ldots, F_n) = \mathrm{Res}_{d_1, \ldots, d_n}(f_{1, d_1}, \ldots, f_{n, d_n}). \tag{1.54}$$

Another direct consequence of Proposition 1.6.2 is the multiplicative property:

$$\mathrm{Res}_{d_0' \cdot d_0'', d_1, \ldots, d_n}(F_0' \cdot F_0'', F_1, \ldots, F_n) = \tag{1.55}$$

$$\mathrm{Res}_{d_0', d_1, \ldots, d_n}(F_0', F_1, \ldots, F_n) \cdot \mathrm{Res}_{d_0'', d_1, \ldots, d_n}(F_0'', F_1, \ldots, F_n),$$

where F_0', F_0'' are homogeneous polynomials of respective degrees d_0', d_0''. More details and applications of the homogeneous resultant to study V and the quotient ring by the ideal $\langle f_1, \ldots, f_n \rangle$ can be found in 2, Section 2.3.2.

Some words on the computation of homogeneous resultants

When trying to find explicit formulas for multivariate resultants like the Sylvester or Bézout formulas (1.22) (1.25), one starts searching for maps as (1.21) which are an isomorphism if and only if the resultant does not vanish. But this is possible only in very special cases or low dimensions, and higher linear algebra techniques are needed, in particular the notion of the determinant of a complex [GKZ94]. Given $d_0, \ldots, d_n$, the first idea to find a linear map whose determinant equals the resultant $\mathrm{Res}_{d_0,\ldots,d_n}(F_0, \ldots, F_n)$, is to consider the application

$$
\begin{aligned}
S_{\rho+1-d_0} \times \cdots \times S_{\rho+1-d_n} &\longrightarrow S_{\rho+1} \\
(G_0, \ldots, G_n) &\longmapsto G_0 F_0 + \cdots + G_n F_n,
\end{aligned}
\tag{1.56}
$$

where we denote by S_ℓ the space of homogeneous polynomials of degree ℓ and we recall that $\rho + 1 = d_0 + \cdots + d_n - n$.

For any specialization in $\mathbb{K}$ of the coefficients of $F_0, \ldots, F_n$ (with respective degrees $d_0, \ldots, d_n$), we get a $\mathbb{K}$-linear map between finite dimensional $\mathbb{K}$-vector spaces which is surjective if and only if $F_0, \ldots, F_n$ do not have any common root in $\mathbb{K}^{n+1} \setminus \{0\}$. But it is easy to realize that the dimensions are not equal, except if $n = 1$ or $d_0 = \cdots = d_n = 1$. Macaulay [Mac02, Mac94] then proposed a choice of a generically non zero maximal minor of the corresponding rectangular matrix in the standard bases of monomials, which exhibits the multivariate resultant not as a determinant but as a quotient of two determinants. More details on this can be found in Chapters 2 and 3; see also [CLO98].

We now recall the multivariate Bezoutian defined in Section 1.5 (cf. also Chapter 3).

Let $F_0, \ldots, F_n$ polynomials of degrees $d_0, \ldots, d_n$. Write $x = (x_0, \ldots, x_n)$, $y = (y_0, \ldots, y_n)$ and let $F_i(x) - F_i(y) = \sum_{j=0}^n F_{ij}(x, y)(x_j - y_j)$, where F_{ij} are homogeneous polynomials in $2(n+1)$ variables of degree $d_i - 1$. The Bezoutian polynomial $\Delta_{(F)}$ is defined as the determinant

$$
\Delta_{(F)}(x, y) = \det((F_{ij}(x, y))) = \sum_{|\alpha| \le \rho} \Delta_\alpha(x) y^\alpha.
$$

For instance, we can take as in (1.32)

$$
F_{ij}(x, y) = (F_i(y_0, \ldots, y_{j-1}, x_j, \ldots, x_n) - F_i(y_0, \ldots, y_j, x_{j+1}, \ldots, x_n)) / (x_j - y_j).
$$

This polynomial is well defined modulo $\langle F_0(x) - F_0(y), \ldots, F_n(x) - F_n(y) \rangle$. Note that the sum of the degrees $\deg(\Delta_\alpha) + |\alpha|$ equals the critical degree $\rho = \sum_i (d_i - i)$. In fact, for any specialization of the coefficients in $\mathbf{k}$ such that $R_{d_0,\ldots,d_n}(F_0, \ldots, F_n)$ is non zero, the specialized polynomials $\{\Delta_\alpha, |\alpha| = m\}$ give a system of generators (over $\mathbf{k}$) of the classes of homogeneous polynomials of degree m in the quotient $\mathbf{k}[x_0, \ldots, x_n]/\langle F_0(x), \ldots, F_n(x) \rangle$, for any $m \le \rho$.

In particular, according to Theorem 1.5.16, the graded piece of degree ρ of the quotient has dimension one and a basis is given by the coefficient

$$\Delta_0(x) = \Delta_{\langle F\rangle}(x,0).\tag{1.57}$$

On the other side, by (1.52), any homogeneous polynomial of degree at least $\rho + 1$ lies in the ideal $\langle F_0(x),\ldots,F_n(x)\rangle$.

There is a determinantal formula for the resultant $\mathrm{Res}_{d_0,\ldots,d_n}$ (as the determinant of a matrix involving coefficients of the given polynomials and coefficients of their Bezoutian $\Delta_{\langle F\rangle}$) only when $d_2 + \cdots + d_n < d_0 + d_1 + n$. In general, it is possible to find smaller Macaulay formulas than those arising from (1.56), as the quotient of the determinants of two such explicit matrices (c.f. [Jou97], [WZ94], [DD01]).

Assume for example that $n = 2$, $(d_0, d_1, d_2) = (1, 1, 2)$, and let

$$
\begin{aligned}
F_0 &= a_0 x_0 + a_1 x_1 + a_2 x_2\\
F_1 &= b_0 x_0 + b_1 x_1 + b_2 x_2\\
F_2 &= c_1 x_0^2 + c_2 x_1^2 + c_3 x_2^2 + c_4 x_0 x_1 + c_5 x_0 x_2 + c_6 x_1 x_2
\end{aligned}
$$

be generic polynomials of respective degrees $1, 1, 2$. Macaulay's classical matrix looks as follows:

$$
\begin{pmatrix}
a_0 & 0 & 0 & 0 & 0 & c_1\\
0 & a_1 & 0 & b_1 & 0 & c_2\\
0 & 0 & a_2 & 0 & b_2 & c_3\\
a_1 & a_0 & 0 & b_0 & 0 & c_4\\
a_2 & 0 & a_0 & 0 & b_0 & c_5\\
0 & a_2 & a_1 & b_2 & b_1 & c_6
\end{pmatrix}
$$

and its determinant equals $-a_0 \mathrm{Res}_{1,1,2}$. In this case, the extraneous factor a_0 is the 1×1 minor formed by the element in the fourth row, second column. On the other hand, we can exhibit a determinantal formula for $\pm\mathrm{Res}_{1,1,2}$, given by the determinant of

$$
\begin{pmatrix}
\Delta_{(1,0,0)} & a_0 & b_0\\
\Delta_{(0,1,0)} & a_1 & b_1\\
\Delta_{(0,0,1)} & a_2 & b_2
\end{pmatrix},
$$

where the coefficients Δ_γ of the Bezoutian $\Delta_{\langle F\rangle}$ are given by

$$\Delta_{(1,0,0)} = c_1(a_1 b_2 - a_2 b_1) - c_4(a_0 b_2 - a_2 b_0) + c_5(a_0 b_1 - a_1 b_0),$$

$$\Delta_{(0,1,0)} = c_6(a_0 b_1 - a_1 b_0) - c_2(a_0 b_2 - b_0 a_2)$$

and

$$\Delta_{(0,0,1)} = c_3(a_0 b_1 - b_0 a_1).$$

In fact, in this case the resultant can be also computed as follows. The generic space of solutions of the linear system $f_0 = f_1 = 0$ is generated by the vector of minors $(a_1 b_2 - a_2 b_1, -(a_0 b_2 - a_2 b_0), a_1 b_2 - a_2 b_1)$. Then

$$\mathrm{Res}_{1,1,2}(F_0, F_1, F_2) = F_2(a_1 b_2 - a_2 b_1, -(a_0 b_2 - a_2 b_0), a_1 b_2 - a_2 b_1).$$

Suppose now that $F_0 = \sum_{i=0}^n a_i x_i$ is a linear form. As in expression (1.54) one gets, using the homogeneity of the resultant, that

$$\mathrm{Res}_{1,d_1,\ldots,d_n}(F_0, F_1, \ldots, F_n) = a_0^{d_1 \cdots d_n} \mathrm{Res}_{1,d_1,\ldots,d_n}\left(x_0 + \sum_{i=1}^n \frac{a_i}{a_0} x_i, F_1, \ldots, F_n\right)$$

$$= a_0^{d_1 \cdots d_n} \mathrm{Res}_{d_1,\ldots,d_n}\left(F_1\left(-\sum_{i=1}^n \frac{a_i}{a_0} x_i, x_1, \ldots, x_n\right), \ldots, F_n\left(-\sum_{i=1}^n \frac{a_i}{a_0} x_i, x_1, \ldots, x_n\right)\right).$$

More generally, let $\ell_0, \ldots, \ell_{r-1}$ be generic linear forms and $F_r, \ldots, F_n$ be homogeneous polynomials of degree $d_r, \ldots, d_n$ on the variables $x_0, \ldots, x_n$. Write $\ell_i = \sum_{j=0}^n a_j^i x_j$ and for any subset J of $\{0, \ldots, n\}$, $|J| = r$, denote by δ_J the determinant of the square submatrix $A_J := (a_j^i), j \in J$. Obviously, $\delta_J \in \mathbb{Z}[a_j^i, j \in J]$ vanishes if and only if $\ell_0 = \cdots = \ell_{r-1} = 0$ cannot be parametrized by the variables $(x_j)_{j \notin J}$.

Assume for simplicity that $J = \{0, \ldots, r-1\}$ and let $\delta_J \neq 0$. Left multiplying by the inverse matrix of A_J, the equality $A.x^t = 0$ is equivalent to $x_k = $ k-th coordinate of $-(A_J)^{-1}.(a_j^i)_{j \notin J}(x_r, \ldots, x_n)^t$, for all $k \in J$. Call $F_j^J(x_r, \ldots, x_n), j = r, \ldots, n$, the homogeneous polynomials of degrees $d_r, \ldots, d_n$ respectively gotten from $F_j, j = r, \ldots, n$ after this substitution. Using standard properties of Chow forms (defined below), we then have

Proposition 1.6.3. *Up to sign,*

$$\mathrm{Res}_{1,\ldots,1,d_r,\ldots,d_n}(\ell_0, \ldots, \ell_{r-1}, F_r, \ldots, F_n) = \delta_J^{d_r \cdots d_n} \mathrm{Res}_{d_r,\ldots,d_n}(F_r^J, \ldots, F_n^J).$$

In case $r = n$ we moreover have

$$\mathrm{Res}_{1,\ldots,1,d_n}(\ell_0, \ldots, \ell_{n-1}, F_n) = F_n(\delta_{\{1,\ldots,n\}}, -\delta_{\{0,2,\ldots,n\}}, \ldots, (-1)^n \delta_{\{0,\ldots,n-1\}}).$$

As we have already remarked in the univariate case, resultants can, in principle, be obtained by a Gröbner basis computation using an elimination order. However, this is often not feasible in practice, while using geometric information contained in the system of equations to build the resultant matrices may make it possible to obtain the result. These matrices may easily become huge (c.f. [DD01] for instance), but they are structured. For some recent implementations of resultant computations in Macaulay2 and Maple, together with examples and applications, we also refer to [Bus03].

The unmixed case

Assume we have an unmixed system, i.e. all degrees are equal. Call $d_0 = \cdots = d_n = d$ and write $F_i(x) = \sum_{|\gamma|=d} a_{i\gamma} x^\gamma$. Then, the coefficients of each Δ_α are linear combinations with integer coefficients of the brackets $[\gamma_0, \ldots, \gamma_n] := \det(a_{i\gamma_j}, i, j = 0, \ldots, n)$, for any subset $\{\gamma_0, \ldots, \gamma_n\}$ of multi-indices of degree

d. In fact, in this equal-degree case, if $F_0, \ldots, F_n$ and $G_0, \ldots, G_n$ are homogeneous polynomials of degree d, and $G_i = \sum_{j=0}^{n} m_{ij} F_j$, $i = 0, \ldots, n$, where $M = (m_{ij}) \in \mathbf{k}^{(n+1) \times (n+1)}$, then,

$$\text{Res}_{d,\ldots,d}(G_0, \ldots, G_n) = \det(M)^{d^n} \text{Res}_{d,\ldots,d}(F_0, \ldots, F_n).$$

In particular, the resultant $\text{Res}_{d,\ldots,d}$ is invariant under the action of the group $SL(n, \mathbf{k})$ of matrices with determinant 1, and by the Fundamental Theorem of Invariant Theory, there exists a (non unique) polynomial P in the brackets such that $\text{Res}_{d,\ldots,d}(F_0, \ldots, F_n) = P([\gamma_0, \ldots, \gamma_n], |\gamma_i| = d)$. There exists a determinantal formula in terms of the coefficients of the Bezoutian as in (1.24) only if $n = 1$ or $d = 1$. In the "simple" case $n = 2$, $d = 2$, $\text{Res}_{2,2,2}$ is a degree 12 polynomial with more than 20,000 terms in the 18 coefficients of F_0, F_1, F_2, while it has degree 4 in the 20 brackets with considerably fewer terms.

Given a projective variety $X \in \mathbb{P}^N(\mathbb{K})$, of dimension n, and n generic linear forms $\ell_1, \ldots, \ell_n$, the intersection $X \cap (\ell_1 = 0) \cap \cdots \cap (\ell_n = 0)$ is finite of cardinal equal to the degree of the variety $\deg(X)$. If we take instead $(n + 1)$ generic linear forms $\ell_0, \ldots, \ell_n$ in $\mathbb{P}^N(\mathbb{K})$, the intersection $X_\ell := X \cap (\ell_0 = 0) \cap \cdots \cap (\ell_n = 0)$ is empty. The Chow form $\mathcal{C}_X$ of X is an irreducible polynomial in the coefficients of $\ell_0, \ldots, \ell_n$ verifying

$$\mathcal{C}_X(\ell_0, \ldots, \ell_n) = 0 \iff X_\ell \neq \emptyset.$$

Consider for example the twisted cubic, i.e the curve V defined as the closure in $\mathbb{P}^3(\mathbb{K})$ of the points parametrized by $(1 : t : t^2 : t^3)$, $t \in \mathbb{K}$. It can also be presented as

$$V = \{(\xi_0 : \xi_1 : \xi_2 : \xi_3) \in \mathbb{P}^3(\mathbb{K}) : \; : \; \xi_1^2 - \xi_0 \xi_2 = \xi_2^2 - \xi_1 \xi_3 = \xi_0 \xi_3 - \xi_1 \xi_2 = 0\}.$$

Given a linear form $\ell_0 = a_0 \xi_0 + a_1 \xi_1 + a_2 \xi_2 + a_3 \xi_3$ (resp. $\ell_1 = b_0 \xi_0 + b_1 \xi_1 + b_2 \xi_2 + b_3 \xi_3$), a point in V of the form $(1 : t : t^2 : t^3)$ is annihilated by ℓ_0 (resp. ℓ_1) if and only if t is a root of the cubic polynomial $f_0 = a_0 + a_1 t + a_2 t^2 + a_3 t^3$ (resp. $f_1 = b_0 + b_1 t + b_2 t^2 + b_3 t^3$). It follows that

$$\mathcal{C}_V(\ell_0, \ell_1) = \text{Res}_{3,3}(f_0, f_1).$$

In general, given n and d, denote $N = \binom{n+d}{d}$ and consider the Veronese variety $V_{n,d}$ in $\mathbb{P}^{N-1}(\mathbb{K})$ defined as the image of the Veronese map

$$\mathbb{P}^n(\mathbb{K}) \longrightarrow \mathbb{P}^{N-1}(\mathbb{K})$$
$$(t_0 : \cdots : t_n) \longmapsto (t^\alpha)_{|\alpha|=d}.$$

Given coefficients $(a_{i\alpha}, i = 0, \ldots, n, |\alpha| = d)$, denote by $\ell_i = \sum_{|\alpha|=d} a_{i\alpha} \xi_\alpha$ and $f_i = \sum_{|\alpha|=d} a_{i\alpha} t^\alpha$, $i = 0, \ldots, n$, the corresponding linear forms in the N variables ξ_α and degree d polynomials in the n variables t_i. Then,

$$\mathcal{C}_{V_{n,d}}(\ell_0, \ldots, \ell_n) = \text{Res}_{d,\ldots,d}(f_0, \ldots, f_n).$$

For the use of exterior algebra methods to compute Chow forms, and a fortiori unmixed resultants, we refer to [ESW03].

1.6.2 A glimpse of other multivariate resultants

Resultants behave quite badly with respect to specializations or give no information, and so different notions of resultants tailored for special families of polynomials are needed, together with appropriate different algebraic compactifications.

Suppose we want to define a resultant which describes the existence of a common root of three degree 2 polynomials of the form

$$f_i(x_1, x_2) = a_i x_1 x_2 + b_i x_1 + c_i x_2 + d_i \,;\; a_i, b_i, c_i, d_i \in \mathbb{K},\; i = 0, 1, 2, \quad (1.58)$$

i.e. ranging in the subvariety of the degree 2 polynomials with zero coefficients in the monomials x_1^2, x_2^2. Note that the homogenized polynomials

$$F_i(x_0, x_1, x_2) = = a_i x_1 x_2 + b_i x_0 x_1 + c_i x_0 x_2 + d_i x_0^2, \quad i = 0, 1, 2,$$

vanish at $(0, 1, 0)$ and $(0, 0, 1)$ for any choice of coefficients a_i, b_i, c_i, d_i. Therefore the homogeneous resultant $\mathrm{Res}_{2,2,2}(f_0, f_1, f_2)$ is meaningless because it is identically zero. Nevertheless, the closure in the 12 dimensional parameter space $\mathbb{K}^{12}$ with coordinates $(a_0, \ldots, d_2)$ of the vectors of coefficients for which f_0, f_1, f_2 have a common root in $\mathbb{K}^2$, is an irreducible hypersurface, whose equation is the following polynomial with 66 terms:

$$\mathrm{Res}_{(1,1),(1,1),(1,1)}(f_0, f_1, f_2) = -c_2 a_0 d_1^2 a_2 b_0 - a_1 c_2^2 b_0^2 d_1 - a_1 c_0^2 b_2^2 d_1 + a_2^2 c_1 d_0^2 b_1$$

$$+2a_0 c_1 b_2 c_2 b_1 d_0 - a_1 c_2 b_0 c_0 b_1 d_2 - a_0 c_1^2 b_2^2 d_0 + c_2 a_0^2 d_1^2 b_2 - c_2^2 a_0 b_1^2 d_0 + a_1 c_2 d_0 a_0 b_1 d_2$$

$$+c_0 a_2^2 d_1^2 b_0 + 2c_0 a_2 b_1 c_1 b_0 d_2 - 2c_2 a_0 d_1 b_2 a_1 d_0 + a_2 c_1^2 b_0 b_2 d_0 + a_1 c_2 d_0 a_2 b_0 d_1 + a_1^2 c_2 d_0^2 b_2$$

$$+a_2 c_1 d_0 a_0 b_2 d_1 - a_2^2 c_1 d_0 b_0 d_1 + a_2 c_1 d_0 a_1 b_0 d_2 - a_2 c_1 d_0^2 b_2 a_1 + c_0 a_2 d_1 b_2 a_1 d_0 - a_1 c_2 d_0^2 b_1 a_2$$

$$+ c_2 a_0 d_1 b_1 a_2 d_0 + c_2 a_0 d_1 a_1 b_0 d_2 - a_1 c_0 d_2^2 a_0 b_1 - c_0 a_2 b_1 b_0 c_2 d_1 - a_2 c_1 b_0 b_2 c_0 d_1 - c_0^2 a_2 b_1^2 d_2$$

$$-a_1 c_2 b_0 c_1 b_2 d_0 + c_2^2 a_0 b_1 b_0 d_1 + a_1 c_2 b_0^2 c_1 d_2 - a_0 c_1 b_2 c_0 b_1 d_2 + a_0 c_1 b_2^2 c_0 d_1 - 2a_1 c_0 d_2 a_2 b_0 d_1$$

$$+a_1 c_0 d_2 a_0 b_2 d_1 - c_0 a_2 d_1^2 a_0 b_2 - a_0^2 c_1 d_2 b_2 d_1 - a_1^2 c_2 d_0 b_0 d_2 - 2a_0 c_1 d_2 b_1 a_2 d_0 + c_0 a_2 d_1 a_0 b_1 d_2$$

$$-c_0 a_2^2 d_1 b_1 d_0 + c_0^2 a_2 b_1 b_2 d_1 + a_1 c_0^2 b_2 b_1 d_2 + a_0 c_1 d_2 a_2 b_0 d_1 - a_0 c_1 d_2^2 a_1 b_0 + a_2 c_1 b_0^2 c_2 d_1$$

$$+c_0 a_2 b_1^2 c_2 d_0 + a_1 c_0 d_2 b_1 a_2 d_0 + a_0 c_1 d_2 b_2 a_1 d_0 + c_2 a_0 b_1^2 c_0 d_2 - c_2 a_0 b_1 b_2 c_0 d_1 - c_0 a_2 b_1 c_1 b_2 d_0$$

$$-a_1 c_0 b_2 c_1 b_0 d_2 + 2a_1 c_0 b_2 b_0 c_2 d_1 - a_2 c_1 b_0 c_2 b_1 d_0 - a_1 c_0 b_2 c_2 b_1 d_0 + a_1 c_0 b_2^2 c_1 d_0 + a_0 c_1^2 b_2 b_0 d_2$$

$$-a_0 c_1 b_2 b_0 c_2 d_1 - c_2 a_0 b_1 c_1 b_0 d_2 - c_2 a_0^2 d_1 b_1 d_2 - a_1^2 c_0 d_2 b_2 d_0 + a_1^2 c_0 d_2^2 b_0 + a_1 c_2^2 b_0 b_1 d_0$$

$$-a_2 c_1^2 b_0^2 d_2 + a_0^2 c_1 d_2^2 b_1. \quad (1.59)$$

This polynomial is called the multihomogeneous resultant (associated to bidegrees $(1, 1)$). In Section 1.7 we will describe a method to compute it.

There are also determinantal formulas to compute this resultant, i.e. formulas that present $\mathrm{Res}_{(1,1),(1,1),(1,1)}(f_0, f_1, f_2)$ as the determinant of a matrix whose entries are coefficients of the given polynomials or of an adequate version of their Bezoutian. The smallest such formula gives the resultant as the determinant of a 2×2 matrix, as follows. Given f_0, f_1, f_2, as in (1.58) introduce two new variables y_1, y_2 and let B be the matrix:

$$B = \begin{pmatrix} f_0(x_1, x_2) & f_1(x_1, x_2) & f_2(x_1, x_2) \\ f_0(y_1, x_2) & f_1(y_1, x_2) & f_2(y_1, x_2) \\ f_0(y_1, y_2) & f_1(y_1, x_2) & f_2(y_1, x_2) \end{pmatrix}$$

Compute the Bezoutian polynomial

$$\frac{1}{(x_1 - y_1)(x_2 - y_2)} \det(B) = B_{11} + B_{12}x_2 + B_{21}y_1 + B_{22}x_2y_1,$$

where the coefficients B_{ij} are homogeneous polynomials of degree 3 in the coefficients $(a_0, \ldots, b_2)$ with tridegree $(1, 1, 1)$ with respect to the coefficients of f_0, f_1 and f_2. Moreover, they are brackets in the coefficient vectors; for instance, $B_{11} = c_1 b_0 d_2 - b_0 c_2 d_1 - c_0 b_1 d_2 + c_2 b_1 d_0 + b_2 c_0 d_1 - c_1 b_2 d_0$ is the determinant of the matrix with rows $(b_0, c_0, d_0), (b_1, c_1, d_1), (b_2, c_2, d_2)$. Finally,

$$\mathrm{Res}_{(1,1),(1,1),(1,1)}(f_0, f_1, f_2) = \det(B_{ij}).$$

These formulas go back to the pioneering work of Dixon [Dix08]. For a modern account of determinantal formulas for multihomogeneous resultants see [DE03].

Multihomogeneous resultants are special instances of *sparse (or toric) resultants*. We refer to 7 for the computation and applications of sparse resultants. The setting is as follows (cf. [GKZ94, Stu93]). We fix $n + 1$ finite subsets $A_0, \ldots, A_n$ of $\mathbb{Z}^n$. To each $\alpha \in \mathbb{Z}^n$ we associate the Laurent monomial $x_1^{\alpha_1} \ldots x_n^{\alpha_n}$ and consider consider

$$f_i = \sum_{\alpha \in A_i} a_{i\alpha} x^{\alpha}, \quad i = 0, \ldots, n.$$

For instance, one could fix lattice polytopes $P_0, \ldots, P_n$ and take $A_i = P_i \cap \mathbb{Z}^n$. In general A_i is a subset of the lattice points in its convex hull P_i. For generic choices of the coefficients $a_{i\alpha}$, the polynomials $f_0, \ldots, f_n$ have no common root. We consider then, the closure H_A of the set of coefficients for which $f_0, \ldots, f_n$ have a common root in the torus $(\mathbb{K} \setminus \{0\})^n$. If H_A is a hypersurface, it is irreducible, and its defining equation, which has integer coefficients (defined up to sign by the requirement that its content be 1), is called the sparse resultant $\mathrm{Res}_{A_0, \ldots, A_n}$. The hypersurface condition is fulfilled if the family of polytopes $P_0, \ldots, P_n$ is *essential*, i.e. if for any proper subset I of $\{0, \ldots, n\}$, the dimension of the Minkowski sum $\sum_{i \in I} P_i$ is at least $|I|$. In this case, the sparse resultant depends on the coefficients of all the polytopes; this is the case of the homogeneous resultant. When the codimension of H_A is greater than 1, the sparse resultant is defined to be the constant 1. For example, set $n = 4$ and consider polynomials of the form

$$\begin{cases} f_0 = a_1 x_1 + a_2 x_2 + a_3 x_3 + a_4 x_4 + a_5 \\ f_1 = b_1 x_1 + b_2 x_2 \\ f_2 = c_1 x_1 + c_2 x_2 \\ f_3 = b_3 x_3 + b_4 x_4 \\ f_4 = c_3 x_3 + c_4 x_4. \end{cases}$$

Then, the existence of a common root in the torus implies the vanishing of both determinants $b_1c_2 - b_2c_1$ and $b_3c_4 - b_4c_3$, i.e. the variety H_A has codimension two. In this case, the sparse resultant is defined to be 1 and it does not vanish for those vectors of coefficients for which there is a common root. Another unexpected example is the following, which corresponds to a non essential family. Set $n = 2$ and let

$$\begin{cases} f_0 = a_1x_1 + a_2x_2 + a_3 \\ f_1 = b_1x_1 + b_2x_2 \\ f_2 = c_1x_1 + c_2x_2. \end{cases}$$

In this case, the sparse resultant equals the determinant $b_1c_2 - b_2c_1$ which does not depend on the coefficients of f_0.

There are also arithmetic issues that come into the picture, as in the following simple example. Set $n = 1$ and consider two univariate polynomials of degree 2 of the form $f_0 = a_0 + b_0x^2$, $f_1 = a_1 + b_1x^2$. In this case, the sparse resultant equals the determinant $D := a_0b_1 - b_0a_1$. But if we think of f_0, f_1 as being degree 2 polynomials with vanishing x-coefficient, and we compute its univariate resultant $\text{Res}_{2,2}(f_0, f_1)$, the answer is D^2. The exponent 2 is precisely the rank of the quotient of the lattice $\mathbb{Z}$ by the lattice $2\mathbb{Z}$ generated by the exponents in f_0, f_1. As in the case of the projective resultant, there is an associated algebraic compactification $X_{A_n,...,A_n}$ of the n-torus, called the toric variety associated to the family of supports, which contains $(\mathbb{K} \setminus \{0\})^n$ as a dense open set. For essential families, the sparse resultant vanishes at a vector of coefficients if and only if the closures of the hypersurfaces $(f_i = 0), i = 0, \ldots, n$, have a common point of intersection in $X_{A_n,...,A_n}$. In the bihomogeneous example (1.58) that we considered, $A_i = \{(0,0), (1,0), (0,1), (1,1)\}$ are the vertices of the unit square in the plane for $i = 0, 1, 2$, and the corresponding toric variety is the product variety $\mathbb{P}^1(\mathbb{K}) \times \mathbb{P}^1(\mathbb{K})$.

Sparse resultants are in turn a special case of *residual resultants*. Roughly speaking, we have families of polynomials which generically have some fixed common points of intersection, and we want to find the condition under which these are the only common roots. Look for instance at the homogeneous case: for any choice of positive degrees $d_0, \ldots, d_n$, generic polynomials $F_0, \ldots, F_n$ with these degrees will all vanish at the origin $0 \in \mathbb{K}^{n+1}$, and the homogeneous resultant $\text{Res}_{d_0,...,d_n}(F_0, \ldots, F_n)$ is non zero if and only if the origin is the only common solution. This problem arises naturally when trying to find implicit equations for families of parametric surfaces with base points of codimension greater than 1. We refer to Chapter 3 and to [Bus03, BEM03] for more background and applications.

1.7 Residues and resultants

In this section we would like to discuss some of the connections between residues and resultants. We will also sketch a method, based on residues, to compute multidimensional resultants which, as far as we know, has not been made explicit before.

Suppose $P(z), Q(z)$ are univariate polynomials of respective degrees d_1, d_2 as in (1.19) and let $Z_P = \{\xi_1, \ldots, \xi_r\}$ be the zero locus of P. If Q is regular on Z_P, equivalently $\mathrm{Res}_{d_1, d_2}(P, Q) \neq 0$, then the global residue $\mathrm{res}_P(1/Q)$ is defined and the result will be a rational function on the coefficients (a, b) of P and Q. Thus, it is reasonable to expect that the denominator of this rational function (in a minimal expression) will be the resultant. This is the content of the following proposition:

Proposition 1.7.1. *For any* $k = 0, \ldots, d_1 + d_2 - 2$, *the residue* $\mathrm{res}_P(z^k/Q)$ *is a rational function of the coefficients* (a, b) *of* P, Q, *and there exists a polynomial* $C_k \in \mathbb{Z}[a, b]$ *such that*

$$\mathrm{res}_P\left(z^k/Q\right) = \frac{C_k(a, b)}{\mathrm{Res}_{d_1, d_2}(P, Q)}.$$

Proof. We have from (1.26) that

$$1 = \frac{A_1}{\mathrm{Res}_{d_1, d_2}(P, Q)} P + \frac{A_2}{\mathrm{Res}_{d_1, d_2}(P, Q)} Q,$$

with $A_1, A_2 \in \mathbb{Z}[a, b][z]$, $\deg(A_1) = d_2 - 1$, and $\deg(A_2) = d_1 - 1$. Then,

$$\mathrm{res}_P\left(z^k/Q\right) = \mathrm{res}_P\left(z^k \frac{A_2}{\mathrm{Res}_{d_1, d_2}(P, Q)}\right),$$

and we deduce from Corollary 1.1.7 that there exists a polynomial $C_k'(a, b) \in \mathbb{Z}[a, b][z]$ such that

$$\mathrm{res}_P\left(z^k/Q\right) = \frac{C_k'(a, b)}{\mathrm{Res}_{d_1, d_2}(P, Q)\, a_{d_1}^{k+1}}.$$

Thus, it suffices to show that $a_{d_1}^{k+1}$ divides $C_k'(a, b)$. But, since $k \leq d_1 + d_2 - 2$ we know from (1.11) that

$$\mathrm{res}_P\left(z^k/Q\right) = -\mathrm{res}_Q\left(z^k/P\right) = \frac{C_k''(a, b)}{\mathrm{Res}_{d_1, d_2}(P, Q)\, b_{d_2}^{k+1}},$$

for a suitable polynomial $C_k'' \in \mathbb{Z}[a, b][z]$. Since $\mathrm{Res}_{d_1, d_2}(P, Q)$ is irreducible, the result follows.

Note that according to Theorem 1.5.17, we have

$$\mathrm{res}^{\mathbb{P}^1}_{\tilde{P},\tilde{Q}}(z^k) \; = \; \mathrm{res}_P\left(z^k/Q\right) \; = \; -\mathrm{res}_Q\left(z^k/P\right),$$

where $\tilde{P}, \tilde{Q}$ denote the homogenization of P and Q, respectively. This is the basis for the generalization of Proposition 1.7.1 to the multidimensional case. The following is a special case of [CDS98, Th. 1.4].

Theorem 1.7.2. *Let* $F_i(x) = \sum_{|\alpha|=d_i} a_{i\alpha} x^\alpha \in \mathbb{C}[x_0,\ldots,x_n]$, $i = 0,\ldots,n$, *be homogeneous polynomials of degrees* $d_0,\ldots,d_n$. *Then, for any monomial* x^β *with* $|\beta| = \rho = \sum_i (d_i - 1)$, *the homogeneous residue* $\mathrm{res}^{\mathbb{P}^n}_{\langle F \rangle}(x^\beta)$ *is a rational function on the coefficients* $\{a_{i\alpha}\}$ *which can be written as*

$$\mathrm{res}^{\mathbb{P}^n}_{\langle F \rangle}(x^\beta) \; = \; \frac{C_\beta(a_{i\alpha})}{\mathrm{Res}_{d_0,\ldots,d_n}(F_0,\ldots,F_n)}$$

for a suitable polynomial $C_\beta \in \mathbb{Z}[a_{i\alpha}]$.

We sketch a proof of this result, based on [Jou97, CDS98] and the notion of the determinant of a complex [GKZ94].

Proof. We retrieve the notations in (1.56), but we consider now the application "at level ρ"

$$\begin{array}{rcl} S_{\rho-d_0} \times \cdots \times S_{\rho-d_n} \times S_0 & \longrightarrow & S_\rho \\ (G_0,\ldots,G_n,\lambda) & \longmapsto & G_0 F_0 + \cdots + G_n F_n + \lambda \Delta_0, \end{array} \qquad (1.60)$$

where Δ_0 is defined in (1.57). For any specialization in $\mathbb{K}$ of the coefficients of $F_0,\ldots,F_n$ (with respective degrees $d_0,\ldots,d_n$), we get a $\mathbb{K}$-linear map between finite dimensional $\mathbb{K}$-vector spaces which is surjective if and only if $F_0,\ldots,F_n$ do not have a common root in $\mathbb{K}^{n+1} \setminus \{0\}$, or equivalently, if and only if the resultant $\mathrm{Res}_{d_0,\ldots,d_n}(F_0,\ldots,F_n)$ is non zero. Moreover, it holds that the resultant equals the greatest common divisor of all maximal minors of the above map. Let $\mathcal{U}$ be the intersection of Zariski open set in the space of coefficients $a = (a_{i\alpha})$ of the given polynomials where all (non identically zero) maximal minors do not vanish. For $a \in \mathcal{U}$, the specialized $\mathbb{K}$-linear map is surjective and for any monomial x^β of degree ρ we can write

$$x^\beta \; = \; \sum_{i=0}^{n} A_i(a;x)\, F_i(a;x) + \lambda(a)\, \Delta_0(a;x),$$

where λ depends rationally on a. Since the residue vanishes on the first sum and takes the value 1 on Δ_0, we have that

$$\mathrm{res}^{\mathbb{P}^n}_{\langle F \rangle}(x^\beta) \; = \; \lambda(a),$$

This implies that every maximal minor which is not identically zero must involve the last column and that $\lambda(a)$ is unique. Thus, it follows from Cramer's rule that $\mathrm{res}^{\mathbb{P}^n}_{\langle F \rangle}(x^\beta)$ may be written as a rational function with denominator M for all non-identically zero maximal minors M. Consequently it may also be written as a rational function with denominator $\mathrm{Res}_{d_0,\ldots,d_n}(F_0,\ldots,F_n)$.

In fact, (1.60) can be extended to a generically exact complex

$$0 \to S_{d_0-(n+1)} \times \cdots \times S_{d_n-(n+1)} \to \cdots \to S_{\rho-d_0} \times \cdots \times S_{\rho-d_n} \times S_0 \to S_\rho \to 0,$$

which is a graded piece of the Koszul complex associated to $F_0, \ldots, F_n$, which is exact if and only if $\operatorname{Res}_{d_0,\ldots,d_n}(F_0, \ldots, F_n) \neq 0$. Moreover, the resultant equals (once we index appropriately the terms and choose monomial bases for them) the determinant of the complex. This concept goes back to Cayley [Cay48] and generalizes the determinant of a linear map between two vector spaces of the same dimension with chosen bases. For short exact sequences of finitely dimensional vector spaces V_{-1}, V_0, V_1 with respective chosen bases, the determinant of the based complex is defined as follows [GKZ94, Appendix A]. Call d_{-1} and d_0 the linear maps

$$0 \longrightarrow V_{-1} \xrightarrow{d_{-1}} V_0 \xrightarrow{d_0} V_1 \longrightarrow 0,$$

and let $\ell_i = \dim V_i$, $i = -1, 0, 1$. Thus, $\ell_0 = \ell_{-1} + \ell_1$. Denote by M_{-1} and M_0 the respective matrices of d_{-1} and d_0 in the chosen bases. Choose any subset I of $\{0, \ldots, \ell_0\}$ with $|I| = \ell_{-1}$ and let M_{-1}^I be the submatrix of M_{-1} given by all the ℓ_{-1} rows and the ℓ_{-1} columns corresponding to the index set I. Similarly, denote by M_0^I the submatrix of M_0 given by the ℓ_1 rows indexed by the complement of I and all the ℓ_1 columns. Then, it can be easily checked that $\det(M_{-1}^I) \neq 0 \iff \det(M_0^I) \neq 0$. Moreover, up to (an explicit) sign, it holds that whenever they are non zero, the quotient of determinants

$$\frac{\det(M_{-1}^I)}{\det(M_0^I)}$$

is independent of the choice of I. The determinant of the based complex is then defined to be this common value. In the case of the complex given by a graded piece of the Koszul complex we are considering, the hypotheses of [GKZ94, Appendix A, Th. 34] are fulfilled, and its determinant equals the greatest common divisor of the rightmost map (1.60) we considered in the proof of Theorem 1.7.2.

We recall that, by b) in Theorem 1.5.16, the graded piece of degree ρ in the graded algebra $\mathcal{A} = \mathbb{C}[x_0, \ldots, x_n]/\langle F_0, \ldots, F_n \rangle$, is one-dimensional. We can exploit this fact together with the relation between residues and resultants to propose a new algorithm for the computation of resultants. Given a term order $\prec$, there will be a unique standard monomial of degree ρ, the smallest monomial x^{β_0}, relative to $\prec$, not in the ideal $\langle F_0, \ldots, F_n \rangle$. Consequently, for any $H \in \mathbb{C}[x_0, \ldots, x_n]_\rho$, its normal form $N(H)$ relative to the reduced Gröbner basis for $\prec$, will be a multiple of x^{β_0}.

In particular, let $\Delta \in \mathbb{C}[x_0, \ldots, x_n]$ be the element of degree ρ and homogeneous residue 1 constructed in Section 1.5.6. We can write

$$N(\Delta) = \frac{P(a_{i\alpha})}{Q(a_{i\alpha})} \cdot x^{\beta_0}.$$

Theorem 1.7.3. *With notation as above, if $P(a_{i\alpha})$, and $Q(a_{i\alpha})$ are relatively prime*

$$\mathrm{Res}_{d_0,\ldots,d_n}(F_0,\ldots,F_n) \; = \; P(a_{i\alpha}).$$

Proof. We have:

$$1 \; = \; \mathrm{res}_{\langle F \rangle}^{\mathbb{P}^n}(\Delta) \; = \; \mathrm{res}_{\langle F \rangle}^{\mathbb{P}^n}\left(\frac{P(a_{i\alpha})}{Q(a_{i\alpha})} \cdot x^{\beta_0}\right) \; = \; \frac{P(a_{i\alpha})}{Q(a_{i\alpha})} \frac{C_{\beta_0}(a_{i\alpha})}{\mathrm{Res}_{d_0,\ldots,d_n}(F_0,\ldots,F_n)}.$$

Therefore

$$\mathrm{Res}_{d_0,\ldots,d_n}(F_0,\ldots,F_n)Q(a_{i\alpha}) \; = \; P(a_{i\alpha})C_{\beta_0}(a_{i\alpha}),$$

but since $\mathrm{Res}_{d_0,\ldots,d_n}(F_0,\ldots,F_n)$ is irreducible and coprime with $C_{\beta_0}(a_{i\alpha})$ this implies the assertion.

Remark 1.7.4. Note that Theorem 1.7.3 holds even if the polynomials F_i are not densely supported as long as the resultant $\mathrm{Res}_{d_0,\ldots,d_n}(F_0,\ldots,F_n)$ is not identically zero.

Consider the example from Section 1.6.1:

$$\begin{aligned}
F_0 &= a_0 x_0 + a_1 x_1 + a_2 x_2 \\
F_1 &= b_0 x_0 + b_1 x_1 + b_2 x_2 \\
F_2 &= c_1 x_0^2 + c_2 x_1^2 + c_3 x_2^2 + c_4 x_0 x_1 + c_5 x_0 x_2 + c_6 x_1 x_2
\end{aligned}$$

Then $\rho = 1$ and

$$\Delta \; = \; \det \begin{pmatrix} a_0 & a_1 & a_2 \\ b_0 & b_1 & b_2 \\ c_1 x_0 + c_4 x_1 + c_5 x_2 & c_2 x_1 + c_6 x_2 & c_3 x_2 \end{pmatrix}.$$

We can now read off the resultant $\mathrm{Res}_{1,1,2}(F_0, F_1, F_2)$ from the normal form of Δ relative to any Gröbner basis of $I = \langle F_0, F_1, F_2 \rangle$. For example computing relative to grevlex with $x_0 > x_1 > x_2$, we have:

$$N(\Delta) = ((a_0^2 b_1^2 c_3 - a_0^2 b_1 b_2 c_6 + a_0^2 b_2^2 c_2 + a_0 a_1 b_0 b_2 c_6 - a_0 a_2 b_1^2 c_5 +$$

$$a_0 a_1 b_1 b_2 c_5 - a_0 a_1 b_2^2 c_4 + a_0 a_2 b_0 b_1 c_6 - a_0 a_2 b_1 b_2 c_4 - 2a_0 a_1 b_0 b_1 c_3 + a_1^2 b_0^2 c_3 -$$

$$a_1^2 b_0 b_2 c_5 + a_1^2 b_2^2 c_1 - a_1 a_2 b_0^2 c_6 + a_1 a_2 b_0 b_1 c_5 + a_1 a_2 b_0 b_2 c_4 + 2a_0 a_2 b_0 b_2 c_2 -$$

$$2a_1 a_2 b_1 b_2 c_1 + a_2^2 b_0^2 c_2 - a_2^2 b_0 b_1 c_4 + a_2^2 b_1^2 c_1)/(a_0 b_1 - a_1 b_0))x_2$$

and the numerator of the coefficient of x_2 in this expression is the resultant. Its denominator is the subresultant polynomial in the sense of [Cha95], whose vanishing is equivalent to the condition $x_2 \in I$.

Theorem 1.7.3 is a special case of a more general result which holds in the context of toric varieties [CD]. We will not delve into this general setup here but will conclude this section by illustrating this computational method in the case of the sparse polynomials described in (1.58). As noted in Section 1.6.2, the homogeneous resultant of these three polynomials is identically zero. We may however view them as three polynomials with support in the unit square $\mathcal{P} \subset \mathbb{R}^2$ and consider their homogenization relative to $\mathcal{P}$. This is equivalent to compactifying the torus $(\mathbb{C}^*)^2$ as $\mathbb{P}^1 \times \mathbb{P}^1$ and considering the natural homogenizations of our polynomials in the homogeneous coordinate ring of $\mathbb{P}^1 \times \mathbb{P}^1$, i.e. the ring of polynomials $\mathbb{C}[x_1, y_1, x_2, y_2]$ bigraded by $(\deg_{x_1, y_1}, \deg_{x_2, y_2})$. We have:

$$F_i(x_1, x_2, y_1, y_2) = a_i x_1 x_2 + b_i x_1 y_2 + c_i x_2 y_1 + d_i y_1 y_2, \quad a_i, b_i, c_i, d_i \in \mathbb{K}.$$

These polynomials have the property that

$$F_i(\lambda_1 x_1, \lambda_1 y_1, \lambda_2 x_2, \lambda_2 y_2) = \lambda_1 \lambda_2 F_i(x_1, x_2, y_1, y_2),$$

for all non zero λ_1, λ_2.

Notice that $\langle F_0, F_1, F_2 \rangle \subset \langle x_1, x_2, y_1 y_2 \rangle$ and we can take as Δ the determinant of any matrix that expresses the F_j in terms of those monomials. For example

$$\Delta = \det \begin{pmatrix} a_0 x_2 + b_0 y_2 & c_0 y_1 & d_0 \\ a_1 x_2 + b_1 y_2 & c_1 y_1 & d_1 \\ a_2 x_2 + b_2 y_2 & c_2 y_1 & d_2 \end{pmatrix}$$

We point out that in this case $\rho = (1,1) = 3(1,1) - (2,2)$, which is the bidegree of Δ. If we consider for instance the reverse lexicographic term order with $y_2 \prec y_1 \prec x_2 \prec x_1$, the least monomial of degree ρ is $y_1 y_2$. The normal form of Δ modulo a Gröbner basis of the bi-homogeneous ideal $\langle F_0, F_1, F_2 \rangle$ equals a coefficient times $y_1 y_2$. This coefficient is a rational function of $(a_0, \ldots, d_2)$ whose numerator is the $\mathbb{P}^1 \times \mathbb{P}^1$ resultant of F_0, F_1, F_2 displayed in (1.59). We invite the reader to check that its denominator equals the determinant of the 3×3 square submatrix of the matrix of coefficients of the given polynomials

$$\begin{pmatrix} a_0 & b_0 & c_0 \\ a_1 & b_1 & c_1 \\ a_2 & b_2 & c_2 \end{pmatrix}.$$

Again, this is precisely the subresultant polynomial whose vanishing is equivalent to $y_1 y_2 \in \langle F_0, F_1, F_2 \rangle$ (c.f. also [DK]).

As a final remark, we mention briefly the relation between residues, resultants and rational A-hypergeometric functions in the sense of Gel'fand, Kapranov and Zelevinsky [GZK89]. Recall that given a configuration

$$A = \{a_1, \ldots, a_n\} \subset \mathbb{Z}^p$$

or, equivalently an integral $p \times n$ matrix A, a function F, holomorphic in an open set $\mathcal{U} \subset \mathbb{C}^n$, is said to be A-hypergeometric of degree $\beta \in \mathbb{C}^p$ if and only if it satisfies the differential equations:

$$\partial^u F - \partial^v F = 0 \,,$$

for all $u, v \in \mathbb{N}^n$ such that $A \cdot u = A \cdot v$, where $\partial^u = \dfrac{\partial^{|u|}}{\partial z_1^{u_1} \dots z_n^{u_n}}$, and

$$\sum_{j=1}^{n} a_{ij} z_j \frac{\partial F}{\partial z_j} = \beta_i F$$

for all $i = 1, \dots, p$. The study of A-hypergeometric functions is a very active area of current research with many connections to computational and commutative algebra. We refer the reader to [SST00] for a comprehensive introduction and restrict ourselves to the discussion of a simple example.

Let $\Sigma(d)$ denote the set of integer points in the m-simplex

$$\{ u \in \mathbb{R}_{\geq 0}^m : \sum_{j=1}^{m} u_j \leq d \} \,.$$

Let $A \subset \mathbb{Z}^{2m+1}$ be the Cayley configuration

$$A = (\{e_0\} \times \Sigma(d)) \cup \dots \cup (\{e_m\} \times \Sigma(d)).$$

Let $f_i(t) = \sum_{\alpha \in \Sigma(d)} z_{i\alpha} t^\alpha$, $i = 0, \dots, d$ be an $m + 1$-tuple of generic polynomials supported in $\Sigma(d)$. Denote by $F_i(x_0, \dots, x_d)$ the homogenization of f_i. Given an $m + 1$-tuple of positive integers $a = (a_0, \dots, a_m)$ let $\langle F^a \rangle$ be the collection $\langle F_0^{a_0}, \dots, F_m^{a_m} \rangle$. The following result is a special case of a more general result (see [AS96, CD97, CDS01]) involving the Cayley product of a general family of configurations $A_i \subset \mathbb{Z}^m$, $i = 0, \dots, m$.

Theorem 1.7.5. *For any $b \in \mathbb{N}^{m+1}$ with $|b| = d|a| - (n+1)$, the homogeneous residue $\operatorname{res}_{\langle F^a \rangle}^{\mathbb{P}^2}(x^b)$, viewed as a function of the coefficients $x_{i\alpha}$, is a rational A-hypergeometric function of degree $\beta = (-a_0, \dots, -a_m, -b_1-1, \dots, -b_m-1)$.*

Suppose, for example, that $m = 2$ and $d = 1$. Then, we have

$$A = \begin{pmatrix} 1 & 1 & 1 & 0 & 0 & 0 & 0 & 0 & 0 \\ 0 & 0 & 0 & 1 & 1 & 1 & 0 & 0 & 0 \\ 0 & 0 & 0 & 0 & 0 & 0 & 1 & 1 & 1 \\ 0 & 1 & 0 & 0 & 1 & 0 & 0 & 1 & 0 \\ 0 & 0 & 1 & 0 & 0 & 1 & 0 & 0 & 1 \end{pmatrix}$$

and $F_i(x_0, x_1, x_2) = a_{i0} x_0 + a_{i1} x_1 + a_{i2} x_2$. Let $a = (2, 1, 1)$ and $b = (0, 1, 0)$. Then the residue $\operatorname{res}_{\langle F^a \rangle}^{\mathbb{P}^2}(x_1)$ might be computed using Algorithm 2 in Section 1.5.6 to obtain the following rational function

$$(a_{20} a_{12} - a_{10} a_{22}) / \det(a_{ij})^2.$$

Note that, according to Theorem 1.7.2 and (1.55), the denominator of the above expression is the homogeneous resultant

$$\mathrm{Res}_{2,1,1}(F_0^2, F_1, F_2) = \mathrm{Res}_{1,1,1}(F_0, F_1, F_2)^2 \,.$$

Indeed, as

$$\frac{x_1}{F_0^2 F_1 F_2} = -\frac{\partial}{\partial a_{01}} \left(\frac{1}{F_0 F_1, F_2} \right),$$

differentiation "under the integral sign" gives the equality

$$\mathrm{res}_{\langle F^a \rangle}^{\mathbb{P}^2}(x_1) = -\frac{\partial}{\partial a_{01}} \left(\frac{1}{\det(a_{ij})} \right).$$

One can also show that the determinant $\det(a_{ij})$ agrees with the discriminant of the configuration A. We should point out that Gel'fand, Kapranov and Zelevinsky have shown that the irreducible components of the singular locus of the A-hypergeometric system for any degree β have as defining equations the discriminant of A and of its facial subsets, which in this case correspond to all minors of (a_{ij}) .

In [CDS01] it is conjectured that essentially all rational A-hypergeometric functions whose denominators are a multiple of the A-discriminant arise as the toric residues of Cayley configurations. We refer to [CDS02, CD04] for further discussion of this conjecture.

Acknowledgment

We thank David Cox for his thorough reading of a preliminary version of this manuscript.

2

Solving equations via algebras

David A. Cox

Department of Mathematics and Computer Science, Amherst College,
Amherst, MA 01002 USA, dac@cs.amherst.edu

Summary. This chapter studies algebras obtained as the quotient of a polynomial ring by an ideal of finite codimension. These algebras have a rich supply of interesting linear maps whose eigenvalues, eigenvectors, and characteristic polynomials can be used to solve systems of polynomial equations. We will also discuss applications to resultants, factorization, primary decomposition, and Galois theory.

2.0 Introduction

This chapter will consider the quotient ring

$$\mathcal{A} = \mathbb{K}[x_1, \ldots, x_n] / \langle f_1, \ldots, f_s \rangle$$

where $\mathbb{K}$ is a field and $f_1, \ldots, f_s$ are polynomials in $x_1, \ldots, x_n$ with coefficients in $\mathbb{K}$. The ring $\mathcal{A}$ is also a vector space over $\mathbb{K}$ in a compatible way, so that $\mathcal{A}$ is an *algebra* over $\mathbb{K}$. These are the "algebras" in the title of the chapter. For us, the most interesting case is when $\mathcal{A}$ has finite dimension as a vector space over $\mathbb{K}$. We call $\mathcal{A}$ a *finite commutative algebra* when this happens.

What's Covered. We will first use the algebra $\mathcal{A}$ to determine the solutions of the polynomial system

$$f_1(x_1, \ldots, x_n) = \cdots = f_s(x_1, \ldots, x_n) = 0.$$

We will then use the dual space of $\mathcal{A}$ to give an interesting description of the ideal $\langle f_1, \ldots, f_s \rangle$. In the remaining sections of the chapter, we will learn that finite commutative algebras can be used in a variety of other contexts, including the following:

- Resultants.
- Factoring over number fields and finite fields.
- Primary decomposition.
- Galois theory.

In all of these applications, *multiplication maps* play a central role. Given a finite commutative algebra $\mathcal{A}$, an element $a \in \mathcal{A}$ gives a multiplication map

$$M_a : \mathcal{A} \longrightarrow \mathcal{A}$$

defined by $M_a(b) = ab$ for $b \in \mathcal{A}$. This is a linear map from a finite-dimensional vector space to itself, which means the many tools of linear algebra can be brought to bear to study M_a. Furthermore, since $\mathcal{A}$ is commutative, the linear maps M_a all commute as we vary $a \in \mathcal{A}$.

What's Omitted. We will not discuss everything of interest connected with finite commutative algebras. The three main topics not covered are:

- Gorenstein duality.
- Real solutions.
- Border bases.

Duality is covered in Chapter 3 and also in [EM96, EM98], and an introduction to real solutions of polynomial equations appears in [CLO98]. Border bases are discussed in Chapters 3 (briefly) and 4 (in more detail).

Notation. Given $\mathcal{A} = \mathbb{K}[x_1, \ldots, x_n]/\langle f_1, \ldots, f_s \rangle$ and $f \in \mathbb{K}[x_1, \ldots, x_n]$, we will use the following notation:

- $[f] \in \mathcal{A}$ is the coset of f in the quotient algebra $\mathcal{A}$.
- M_f is the multiplication map $M_{[f]}$. Thus $M_f([g]) = [fg]$ for all $[g] \in \mathcal{A}$.
- M_f is the matrix of M_f relative to a chosen basis of $\mathcal{A}$ over $\mathbb{K}$.

Other notation will be introduced as needed.

2.1 Solving equations

This section will cover basic material on solving equations using eigenvalues and eigenvectors of multiplication maps on finite dimensional algebras.

2.1.1 The finiteness theorem and Gröbner bases

Consider a system of polynomial equations

$$
\begin{aligned}
f_1(x_1, \ldots, x_n) &= 0 \\
f_2(x_1, \ldots, x_n) &= 0 \\
&\;\;\vdots \\
f_s(x_1, \ldots, x_n) &= 0
\end{aligned}
\tag{2.1}
$$

in n variables $x_1, \ldots, x_n$ with coefficients in a field $\mathbb{K}$. Here is an example from [MS95] that we will use throughout this section and the next.

Example 2.1.1. Consider the equations

$$f_1 = x^2 + 2y^2 - 2y = 0$$
$$f_2 = xy^2 - xy = 0 \qquad\qquad (2.2)$$
$$f_3 = y^3 - 2y^2 + y = 0$$

over the complex numbers $\mathbb{C}$. If we write the the first and third equations as

$$f_1 = x^2 + 2y(y - 1) = 0 \quad \text{and} \quad f_3 = y(y - 1)^2 = 0,$$

then it follows easily that the only solutions are the points $(0, 0)$ and $(0, 1)$. (*Exercise: Prove this.*) However, this ignores multiplicities, which as we will see are perfectly captured by the algebra $\mathcal{A} = \mathbb{C}[x, y]/\langle f_1, f_2, f_3 \rangle$.

Our first major result is the *Finiteness Theorem*, which gives a necessary and sufficient condition for the algebra corresponding to the equations (2.1) to be finite-dimensional over $\mathbb{K}$.

Theorem 2.1.2. *The algebra*

$$\mathcal{A} = \mathbb{K}[x_1, \ldots, x_n]/\langle f_1, \ldots, f_s \rangle$$

is finite-dimensional over $\mathbb{K}$ if and only if the equations (2.1) have only finitely many solutions over the algebraic closure $\overline{\mathbb{K}}$.

Proof. We will sketch the main ideas since this result is so important. A complete proof can be found in Chapter 5, §3 of [CLO97].

First suppose that $\mathcal{A}$ is finite-dimensional over $\mathbb{K}$. Then, for each i, the set $\{[1], [x_i], [x_i^2], \ldots\} \subset \mathcal{A}$ must be linearly dependent, so that there is a nonzero polynomial $p_i(x_i)$ such that $[p_i(x_i)] = [0]$ in $\mathcal{A}$. This means that

$$p_i(x_i) \in \langle f_1, \ldots, f_s \rangle,$$

which easily implies that p_i vanishes at all common solutions of (2.1). It follows that for each i, the solutions have only finitely many distinct ith coordinates. Hence the number of solutions is finite.

Going the other way, suppose that there are only finitely many solutions over $\overline{\mathbb{K}}$. Then in particular there are only finitely many ith coordinates, so that we can find a nonzero polynomial $q_i(x_i)$ which vanishes on all solutions of (2.1) over $\overline{\mathbb{K}}$. In this situation, *Hilbert's Nullstellensatz* (see Chapter 4, §1 of [CLO97] for a proof) asserts that

$$p_i(x_i) = q_i^N(x_i) \in \langle f_1, \ldots, f_s \rangle$$

for some sufficiently large integer N.

Now consider the lexicographic order $>_{\text{lex}}$ on monomials $x^\alpha = x_1^{a_1} \cdots x_n^{a_n}$. Recall that $x^\alpha > x^\beta$ if $a_1 > b_1$, or $a_1 = b_1$ and $a_2 > b_2$, or ... (in other words, the left-most nonzero entry of $\alpha - \beta \in \mathbb{Z}^n$ is positive). This allows us to define the *leading term* of any nonzero polynomial in $\mathbb{K}[x_1, \ldots, x_n]$.

The theory of Gröbner bases (explained in Chapters 2 and 5 of [CLO97]) implies that $\langle f_1, \ldots, f_s \rangle$ has a *Gröbner basis* $g_1, \ldots, g_t$ with the following properties:

- $g_1, \ldots, g_t$ form a basis of $\langle f_1, \ldots, f_s \rangle$.
- The leading term of every nonzero element of $\langle f_1, \ldots, f_s \rangle$ is divisible by the leading term of one of the g_j.
- The set of *remainder monomials*

$$B = \{x^\alpha \mid x^\alpha \text{ is not divisible by the leading term of any } g_j\}$$

gives cosets $[x^\alpha]$, $x^\alpha \in B$, that form a basis of $\mathcal{A}$ over $\mathbb{K}$.

Since the leading term of $p_i(x_i)$ is a power of x_i, the second bullet implies that the leading term of some g_j is a power of x_i. It follows that in any $x^\alpha \in B$, x_i must appear to strictly less than this power. Since this is true for all i, it follows that B is finite, so that $\mathcal{A}$ is finite-dimensional by the third bullet.

More details about monomial orders, leading terms, and Gröbner bases can also be found in Chapter 2 of [CLO97].

Let's apply Theorem 2.1.2 to our example.

Example 2.1.3. For the equations $f_1 = f_2 = f_3 = 0$ of Example 2.1.1, one can show that f_1, f_2, f_3 form a Gröbner basis for lexicographic order with $x > y$. Thus the leading terms of the polynomials in the Gröbner basis are

$$x^2, xy^2, y^3,$$

so that the remainder monomials (= monomials not divisible by any of these leading terms) are

$$B = \{1, y, y^2, x, xy\}.$$

(*Exercise: Verify this.*) Hence $\mathcal{A}$ has dimension 5 over $\mathbb{C}$ in this case.

2.1.2 Eigenvalues of multiplication maps

For the remainder of this section, we will assume that

$$\mathcal{A} = \mathbb{K}[x_1, \ldots, x_n]/\langle f_1, \ldots, f_s \rangle$$

is finite-dimensional over $\mathbb{K}$. For simplicity of exposition, we will also assume that

$$\mathbb{K} = \overline{\mathbb{K}}.$$

Thus $\mathbb{K}$ will always be algebraically closed.

As in the introduction, $f \in \mathbb{K}[x_1, \ldots, x_n]$ gives a multiplication map

$$M_f : \mathcal{A} \longrightarrow \mathcal{A}.$$

Our main result is the following *Eigenvalue Theorem* first noticed by Lazard in 1981 (see [Laz81]).

Theorem 2.1.4. *Assume that* (2.1) *has a finite positive number of solutions. The eigenvalues of M_f are the values of f at the solutions of* (2.1) *over* $\mathbb{K}$.

Proof. We will sketch the proof and refer to Theorem 4.5 of Chapter 2 of [CLO98] for the details.

First suppose $\lambda \in \mathbb{K}$ is not a value of f at a solution of (2.1). Then the equations

$$f - \lambda = f_1 = \cdots = f_s = 0$$

have no solutions over $\mathbb{K} = \overline{\mathbb{K}}$, so that by the Nullstellensatz, we can write

$$1 = h \cdot (f - \lambda) + \sum_{i=1}^{s} h_i f_i$$

for some polynomials $h, h_1, \ldots, h_s \in \mathbb{K}[x_1, \ldots, x_n]$. Since the multiplication map M_1 is the identity $1_{\mathcal{A}}$ and each M_{f_i} is the zero map, it follows that

$$M_f - \lambda 1_{\mathcal{A}} = M_{f-\lambda} : \mathcal{A} \longrightarrow \mathcal{A}$$

is an isomorphism with inverse M_h. Thus λ is not an eigenvalue of M_f.

Going the other way, let $p \in \mathbb{K}^n$ be a solution of (2.1). As in the proof of Theorem 2.1.2, the remainder monomials $B = \{x^{\alpha(1)}, \ldots, x^{\alpha(m)}\}$ give the basis $[x^{\alpha(1)}], \ldots, [x^{\alpha(m)}]$ of $\mathcal{A}$. The matrix of M_f relative to this basis is denoted M_f. For $j = 1, \ldots, m$, let $p^{\alpha(j)}$ be the element of $\mathbb{K}$ obtained by evaluating $x^{\alpha(j)}$ at p. Then we claim that

$$\mathsf{M}_f^{\mathsf{t}}(p^{\alpha(1)}, \ldots, p^{\alpha(m)})^{\mathsf{t}} = f(p)(p^{\alpha(1)}, \ldots, p^{\alpha(m)})^{\mathsf{t}}, \tag{2.3}$$

where t denotes transpose. Since $1 \in B$ (otherwise there are no solutions), the vector $(p^{\alpha(1)}, \ldots, p^{\alpha(m)})^{\mathsf{t}}$ is nonzero. Thus (2.3) implies that $f(p)$ is an eigenvalue of $\mathsf{M}_f^{\mathsf{t}}$ and hence also of M_f and M_f.

To prove (2.3), suppose that $\mathsf{M}_f = (m_{ij})$. This means that

$$[x^{\alpha(j)} f] = \sum_{i=1}^{m} m_{ij} [x^{\alpha(i)}]$$

for $j = 1, \ldots, m$. Then $x^{\alpha(j)} f \equiv \sum_{i=1}^{m} m_{ij} x^{\alpha(i)} \bmod \langle f_1, \ldots, f_s \rangle$. Since $f_1, \ldots, f_s$ all vanish at p, evaluating this congruence at p implies that

$$p^{\alpha(j)} f(p) = \sum_{i=1}^{m} m_{ij} p^{\alpha(i)}$$

for $j = 1, \ldots, m$. This easily implies (2.3). $\quad\blacksquare$

Example 2.1.5. For the polynomials of Examples 2.1.1 and 2.1.3, the set $B = \{1, y, y^2, x, xy\}$ gives a basis of $\mathcal{A}$. One computes that the matrix of M_x is

$$M_x = \begin{pmatrix} 0 & 0 & 0 & 0 & 0 \\ 0 & 0 & 0 & 2 & 2 \\ 0 & 0 & 0 & -2 & -2 \\ 1 & 0 & 0 & 0 & 0 \\ 0 & 1 & 1 & 0 & 0 \end{pmatrix}.$$

The first and second columns are especially easy to see since here M_x maps basis elements to basis elements. For the third column, one uses $f_2 = xy^2 - xy$ to show that

$$M_x([y^2]) = [xy^2] = [xy].$$

The fourth and fifth columns are obtained similarly. (*Exercise: Do this.*) Using Maple or Mathematica, one finds that the characteristic polynomial of M_x is $\mathrm{CharPoly}_{M_x}(u) = u^5$. By Theorem 2.1.4, it follows that all solutions of the equations (2.2) have x-coordinate equal to 0.

In a similar way, one finds that M_y has matrix

$$M_y = \begin{pmatrix} 0 & 0 & 0 & 0 & 0 \\ 1 & 0 & -1 & 0 & 0 \\ 0 & 1 & 2 & 0 & 0 \\ 0 & 0 & 0 & 0 & 0 \\ 0 & 0 & 0 & 1 & 1 \end{pmatrix}.$$

with characteristic polynomial $\mathrm{CharPoly}_{M_y}(u) = u^2(u-1)^3$. Thus the y-coordinate of a solution of (2.2) is 0 or 1. For later purposes we also note that

$$M_x \text{ has minimal polynomial } \mathrm{MinPoly}_{M_x}(u) = u^3$$
$$M_y \text{ has minimal polynomial } \mathrm{MinPoly}_{M_y}(u) = u(u-1)^2.$$

We will see that later that since y takes distinct values 0 and 1 at the solutions, the characteristic polynomial $u^2(u-1)^3$ of M_y tells us the *multiplicities* of the solutions of (2.2).

In general, the matrix of $M_f : \mathcal{A} \to \mathcal{A}$ is easy to compute once we have a Gröbner basis G of $\langle f_1, \ldots, f_s \rangle$. This is true because of the following:

- As we saw in the proof of Theorems 2.1.2 and 2.1.4, G determines the remainder monomials B that give a basis of $\mathcal{A}$.
- Given $g \in \mathbb{K}[x_1, \ldots, x_n]$, the division algorithm from Chapter 2, Section 3 of [CLO97] constructs a *normal form*

$$N(g) \in \mathrm{Span}(B)$$

with the property that $g \equiv N(g) \bmod \langle f_1, \ldots, f_s \rangle$.

This gives an easy algorithm for computing M_f with respect to the basis of $\mathcal{A}$ given by B: For each $x^\alpha \in B$, simply compute $N(x^\alpha f)$ using the division algorithm. This is what we did in Example 2.1.5.

If we compute the matrix M_{x_i} for each i, then then Theorem 2.1.4 implies that the x_i-coordinates of the solutions are given by the eigenvalues of M_{x_i}. But how do we put these coordinates together to figure out the actual solutions? This was trivial to do in Example 2.1.5. In the general case, one could simply try all possible combinations of the coordinates to find the solutions. This is very inefficient. We will learn a better method in Section 2.1.3.

Minimal Polynomials of Multiplication Maps. The minimal polynomial of a multiplication map $M_f : \mathcal{A} \to \mathcal{A}$ has an interesting interpretation. Given $f \in \mathbb{K}[x_1, \ldots, x_n]$, note that

$$M_f \text{ is the zero map} \iff f \in \langle f_1, \ldots, f_s \rangle.$$

Furthermore, given any polynomial $P(u) \in \mathbb{K}[u]$, we have

$$P(M_f) = M_{P(f)}.$$

(*Exercise: Prove these two facts.*)

As defined in linear algebra, the *minimal polynomial* P of M_f is the monic polynomial of minimal degree such that $P(M_f)$ is the zero map. Using the above two facts, it follows that P is the monic polynomial of minimal degree such that $P(f) \in \langle f_1, \ldots, f_s \rangle$.

In particular, the minimal polynomial of M_{x_i} is the monic polynomial of minimal degree such that $P(x_i) \in \langle f_1, \ldots, f_s \rangle$. In other words, $P(x_i)$ is the generator of the elimination ideal

$$\langle f_1, \ldots, f_s \rangle \cap \mathbb{K}[x_i].$$

(*Exercise: Prove this.*) Thus $P(x_i) = 0$ is the equation obtained by eliminating all variables but x_i from our original systems of equations (2.1). This gives a relation between multiplication maps and elimination theory.

2.1.3 Eigenvectors of multiplication maps

A better method for solving equations, first described in [AS88], is to use the *eigenvectors* of $\mathsf{M}_f^{\mathsf{t}}$ given by (2.3), namely

$$\mathsf{M}_f^{\mathsf{t}}(p^{\alpha(1)}, \ldots, p^{\alpha(m)})^{\mathsf{t}} = f(p)(p^{\alpha(1)}, \ldots, p^{\alpha(m)})^{\mathsf{t}}.$$

In this equation, p is a solution of (2.1), $B = \{x^{\alpha(1)}, \ldots, x^{\alpha(m)}\}$, and $p^{\alpha(j)}$ is the element of $\mathbb{K}$ obtained by evaluating $x^{\alpha(j)}$ at p. As we noted in the proof of Theorem 2.1.4, (2.3) implies that $(p^{\alpha(1)}, \ldots, p^{\alpha(m)})^{\mathsf{t}}$ is an eigenvector of $\mathsf{M}_f^{\mathsf{t}}$ for the eigenvalue $f(p)$.

This allows us to use eigenvalues to find solutions as follows. Suppose that all eigenspaces of $\mathsf{M}_f^{\mathsf{t}}$ have dimension 1 (we say that $\mathsf{M}_f^{\mathsf{t}}$ is *non-derogatory* in this case). Then suppose that λ is an eigenvalue of $\mathsf{M}_f^{\mathsf{t}}$ with eigenvector

$$\mathbf{v} = (u_1, \ldots, u_m)^{\mathbf{t}}.$$

By assumption, we know that $\mathbf{v}$ is unique up to a scalar. At this point, we also know that $\lambda = f(p)$ for some solution p, but we don't know what p is.

To determine p, observe that $(p^{\alpha(1)}, \ldots, p^{\alpha(m)})^{\mathbf{t}}$ is also an eigenvalue of $\mathtt{M}_f^{\mathbf{t}}$ for λ. Since we may assume that $x^{\alpha(1)} = 1$, the first coordinate of this eigenvector is 1. Since λ has a 1-dimensional eigenspace, our computed eigenvector $\mathbf{v}$ is a scalar multiple of $(p^{\alpha(1)}, \ldots, p^{\alpha(m)})^{\mathbf{t}}$. Hence, if we rescale $\mathbf{v}$ so that its first coordinate is 1, then

$$\mathbf{v} = (1, u_2, \ldots, u_m)^{\mathbf{t}} = (1, p^{\alpha(2)}, \ldots, p^{\alpha(m)})^{\mathbf{t}}. \tag{2.4}$$

The key point is that the monomials $x^{\alpha(j)} \in B$ include some (and often all) of the variables $x_1, \ldots, x_n$. This means that we can read off the corresponding coordinates of p from $\mathbf{v}$. Here is an example of how this works.

Example 2.1.6. Consider the matrices $\mathtt{M}_x^{\mathbf{t}}$ and $\mathtt{M}_y^{\mathbf{t}}$ from Example 2.1.5. Neither is non-derogatory since Maple or Mathematica shows that their eigenspaces all have dimension 2. However, if we set $f = 2x + 3y$, then

$$\mathtt{M}_f^{\mathbf{t}} = 2\mathtt{M}_x^{\mathbf{t}} + 3\mathtt{M}_y^{\mathbf{t}}$$

is non-derogatory, where

the eigenvalue 0 has eigenbasis $\mathbf{v} = (1, 0, 0, 0, 0)^{\mathbf{t}}$

the eigenvalue 3 has eigenbasis $\mathbf{v} = (1, 1, 1, 0, 0)^{\mathbf{t}}$.

(*Exercise: Check this.*) Since $B = \{1, y, y^2, x, xy\}$ has the variables x and y in the fourth and second positions respectively, it follows from (2.4) that the x- and y-coordinates of the solutions are the fourth and second entries of the eigenvectors. This gives the solutions

$$(0, 0) \quad \text{and} \quad (0, 1)$$

found in Example 2.1.1.

Before using this method in general, we need to answer some questions:

- What happens when some variables are missing from B?
- What does it mean for a matrix to be non-derogatory?
- Can we find $f \in \mathbb{K}[x_1, \ldots, x_n]$ such that $\mathtt{M}_f^{\mathbf{t}}$ is non-derogatory? What happens if we can't?

The remainder of Section 2.1.3 will be devoted to answering these questions.

Missing Variables. For a fixed monomial ordering, the ideal $\langle f_1, \ldots, f_s \rangle$ has a Gröbner basis G. We assume that G is *reduced*, which means the following:

- The leading coefficient of every $g \in G$ is 1.

- For any $g \in G$, its non-leading terms are not divisible by the leading terms of the remaining polynomials in G.

As we've already explained, G then determines the remainder monomials

$$B = \{x^\alpha \mid x^\alpha \text{ is not divisible by the leading term of any } g \in G\}.$$

We will assume that $G \neq \{1\}$, which implies that $1 \in B$ and that (2.1) has solutions in $\mathbb{K}$ (the latter is true by the Consistency Algorithm described in Chapter 4, §1 of [CLO97] since $\mathbb{K} = \overline{\mathbb{K}}$).

We need to understand which variables lie in B. We say that x_i is *known* if $x_i \in B$ and *missing* otherwise (this is not standard terminology). As explained above, if M_f^t is non-derogatory, then the eigenvectors determine the known coordinates of all solutions. It remains to find the missing coordinates. We will analyze this using the arguments of [MS95].

A variable x_i is missing if it is divisible by the leading term of some element of G. Since $G \neq \{1\}$ and is reduced, it follows that there is some $g_i \in G$ such that

$$g_i = x_i + \text{terms strictly smaller according to the term order.}$$

Furthermore, since this is true for every missing variable and G is reduced, it follows that other terms in the above formula for g_i involve only known variables (if a missing variable appeared in some term, it would be a missing variable $x_j \neq x_i$, so that the term would be divisible by the leading term of $g_j = x_j + \cdots \in G$). Thus

$$g_i = x_i + \text{terms involving only known variables.}$$

Now let p be a solution of (2.1). Then $g_i(p) = 0$, so that the above analysis implies that

$$0 = p_i + \text{terms involving only known coordinates.}$$

Hence the $g_i \in G$ tell us how to find the missing coordinates in terms of the known ones.

Non-Derogatory Matrices. If M is square matrix, then:

$$\text{M is non-derogatory} \iff \text{the Jordan canonical form of M has}$$
$$\text{one Jordan block per eigenvalue}$$
$$\iff \text{MinPoly}_M = \text{CharPoly}_M.$$

(*Exercise: Prove this.*) This will have a nice consequence in Section 2.1.4 below. Note also that M is nonderogatory if and only if M^t is.

Existence of Non-Derogatory Multiplication Matrices. Our first observation is that there are systems of equations such that M_f^t is derogatory for *all* polynomials $f \in \mathbb{K}[x_1, \ldots, x_n]$. Here is a simple example.

Example 2.1.7. Consider the equations

$$x^2 = y^2 = 0.$$

The only solution is $p = (0,0)$ and $B = \{1, x, y, xy\}$. Given $f = a + bx + cy + dx^2 + exy + \cdots$ in $\mathbb{K}[x, y]$, we have $f(p) = a$. One computes that

$$M_f^T = \begin{pmatrix} a & b & c & e \\ 0 & a & 0 & c \\ 0 & 0 & a & b \\ 0 & 0 & 0 & a \end{pmatrix},$$

so that $M_f^t - aI_4$ has rank ≤ 2 for all f. (*Exercise: Prove these assertions.*) It follows that the eigenspace of M_f^t for the eigenvalue $f(p) = a$ has dimension at least 2. Thus M_f^t is always derogatory.

To describe what happens in general, we need to discuss the local structure of solutions. A basic result from commutative algebra states that the ideal $\langle f_1, \ldots, f_s \rangle$ has a *primary decomposition*. Since we are over algebraically closed field and the equations (2.1) have only finitely many solutions, the primary decomposition can be written

$$\langle f_1, \ldots, f_s \rangle = \bigcap_p I_p \tag{2.5}$$

where the intersection is over all solutions p of (2.1) and each I_p is defined by

$$I_p = \{ f \in \mathbb{K}[x_1, \ldots, x_n] \mid gf \in \langle f_1, \ldots, f_s \rangle \text{ for some } g \text{ with } g(p) \neq 0 \}. \tag{2.6}$$

One can show that I_p is a primary ideal, which in this case means that $\sqrt{I_p}$ is the maximal ideal $\langle x_1 - p_1, \ldots, x_n - p_n \rangle$, $p = (p_1, \ldots, p_n)$. We will explain how to compute primary decompositions in Section 2.4.3. See also Chapter 5.

Example 2.1.8. The ideal of Example 2.1.1 has the primary decomposition

$$\begin{aligned} &\langle x^2 + 2y^2 - 2y, xy^2 - xy, y^3 - 2y^2 + y \rangle \\ &= \langle x^2, y \rangle \cap \langle x^2 + 2(y - 1), x(y - 1), (y - 1)^2 \rangle \\ &= I_{(0,0)} \cap I_{(0,1)}. \end{aligned}$$

We will prove this in Section 2.4.3.

Given the primary decomposition $\langle f_1, \ldots, f_s \rangle = \bigcap_p I_p$, we set

$$\mathcal{A}_p = \mathbb{K}[x_1, \ldots, x_n]/I_p.$$

Then (2.5) and the Chinese Remainder Theorem (see p. 245 of [KR00]) give an algebra isomorphism

$$\mathcal{A} = \mathbb{K}[x_1, \ldots, x_n]/\langle f_1, \ldots, f_s \rangle \simeq \prod_p \mathbb{K}[x_1, \ldots, x_n]/I_p = \prod_p \mathcal{A}_p. \qquad (2.7)$$

We call $\mathcal{A}_p$ the *local ring* of the solution p. This ring reflects the local structure of a solution p of (2.1). The *multiplicity* of a solution p is defined to be

$$\mathrm{mult}(p) = \dim_{\mathbb{K}} \mathcal{A}_p.$$

This definition and (2.7) imply that

$$\dim_{\mathbb{K}} \mathcal{A} = \#\,\text{solutions counted with multiplicity}.$$

We now define a special kind of solution.

Definition 2.1.9. *A solution p of* (2.1) *is* **curvilinear** *if $\mathcal{A}_p \simeq \mathbb{K}[x]/\langle x^k \rangle$ for some integer $k \geq 1$.*

Over the complex numbers, p is curvilinear if and only if we can find local analytic coordinates $u_1, \ldots, u_n$ at p and an integer $k \geq 1$ such that the equations are equivalent to

$$u_1 = u_2 = \cdots = u_{n-1} = u_n^k = 0.$$

Alternatively, let $\mathfrak{m}_p$ be the maximal ideal of $\mathcal{A}_p$. The integer

$$e_p = \dim_{\mathbb{K}} \mathfrak{m}_p/\mathfrak{m}_p^2 = \#\,\text{minimal generators of } \mathfrak{m}_p \qquad (2.8)$$

is called the *embedding dimension* of $\mathcal{A}_p$. Then one can prove that a solution p is curvilinear if and only if $\mathcal{A}_p$ has embedding dimension $e_p \leq 1$.

Example 2.1.10. For the solutions $(0,0)$ and $(0,1)$ of the polynomials given in Example 2.1.8, we compute their multiplicity and embedding dimension as follows. For $(0,0)$, we have

$$\mathcal{A}_{(0,0)} = \mathbb{K}[x,y]/\langle x^2, y \rangle \simeq \mathbb{K}[x]/\langle x^2 \rangle.$$

This shows that $(0,0)$ has multiplicity 2 and embedding dimension 1 (and hence is curvilinear). As for $(0,1)$, a Gröbner basis computation shows that

$$\mathcal{A}_{(0,1)} = \mathbb{K}[x,y]/\langle x^2 + 2(y-1), x(y-1), (y-1)^2 \rangle$$

has dimension 3, so that the multiplicity is 3. (*Exercise: Do this.*) The embedding dimension is less obvious and will be computed in Section 2.2.2.

Using curvilinear solutions, we can characterize those systems for which M_f (or $\mathtt{M}_f$ or $\mathtt{M}_f^{\mathtt{t}}$) is non-derogatory for some $f \in \mathbb{K}[x_1, \ldots, x_n]$.

Theorem 2.1.11. *There exists $f \in \mathbb{K}[x_1, \ldots, x_n]$ such that M_f is non-derogatory if and only if every solution of* (2.1) *is curvilinear. Furthermore, when the solutions are all curvilinear, then M_f is non-derogatory when f is a generic linear combination of $x_1, \ldots, x_n$.*

Proof. First observe that since there are only finitely many solutions and $\mathbb{K}$ is infinite (being algebraically closed), a generic choice of $a_1, \ldots, a_n$ guarantees that $f = a_1 x_1 + \cdots + a_n x_n$ takes distinct values at the solutions p.

Next observe that M_f is compatible with the algebra isomorphism (2.7). If we also assume that f takes distinct values at the solutions, then it follows that M_f is non-derogatory if and only if

$$M_f : \mathcal{A}_p \to \mathcal{A}_p$$

is non-derogatory for every p.

To prove the theorem, first suppose that $M_f : \mathcal{A}_p \to \mathcal{A}_p$ is non-derogatory and let

$$u = [f - f(p)] \in \mathcal{A}_p$$

be the element of $\mathcal{A}_p$ determined by $f - f(p)$. Then the kernel of $M_{f-f(p)}$ has dimension 1, which implies the following:

- u lies in the maximal ideal $\mathfrak{m}_p$ since elements in $\mathcal{A}_p \setminus \mathfrak{m}_p$ are invertible.
- The image of $M_{f-f(p)}$ has codimension 1.

Since the image is $\langle u \rangle \subset \mathfrak{m}_p$ and $\mathfrak{m}_p$ also has codimension 1 in $\mathcal{A}_p$, it follows that

$$\langle u \rangle = \mathfrak{m}_p$$

Thus p is curvilinear. (*Exercise: Supply the details.*)

Conversely, if every p is curvilinear, then it is easy to see that M_f is non-derogatory when f is a generic linear combinations of the variables. (*Exercise: Complete the proof.*)

Applying the Eigenvector Method. In order to apply the method described here, we need to find f such that M_f^t is non-derogatory. Typically, one uses $f = a_1 x_1 + \cdots + a_n x_n$. The idea is that when the solutions are curvilinear, Theorem 2.1.11 implies that f will work for most choices of the a_i.

We implement this as follows. Given a system with finitely many solutions, make a random choice of the a_i and compute the corresponding M_f^t. Then:

- Test if M_f^t is non-derogatory by computing whether $\mathrm{CharPoly}_{\mathsf{M}_f}(u)$ equals $\mathrm{MinPoly}_{\mathsf{M}_f}(u)$.
- Once a non-derogatory M_f^t is found, use the eigenvector method to find the solutions.

If our system has only curvilinear solutions, then this procedure will probably succeed after a small number of attempts. On the other hand, if we make a large number of choices of a_i, all of which give a non-derogatory M_f^t, then we are either very unlucky or our system has some non-curvilinear solutions, but we don't know which.

To overcome this problem, there are two ways to proceed:

- First, one can compute the *radical*

$$\sqrt{\langle f_1, \ldots, f_s \rangle} = \{f \in \mathbb{K}[x_1, \ldots, x_n] \mid f^k \in \langle f_1, \ldots, f_s \rangle \text{ for some } k \geq 1\}.$$

The radical gives a system of equations with the same solutions as (2.1), except that all solutions how have multiplicity 1 and hence are curvilinear. Thus Theorem 2.1.11 applies to the radical system. Furthermore, Proposition 2.7 of Chapter 2 of [CLO98] states that

$$\sqrt{\langle f_1, \ldots, f_s \rangle} = \langle f_1, \ldots, f_s, (p_1(x_1))_{\mathrm{red}}, \ldots, (p_n(x_n))_{\mathrm{red}} \rangle,$$

where $p_i(x_i)$ is the minimal polynomial of M_{x_i} written as a polynomial in x_i and $(p_i(x_i))_{\mathrm{red}}$ is the squarefree polynomial with the same roots as $p_i(x_i)$. See [KR00, Sec. 3.7B] for the squarefree part of a polynomial.

- Second, one can intersect eigenspaces of the $\mathsf{M}_{x_i}^{\mathsf{t}}$. Let p_1 be an eigenvalue of $\mathsf{M}_{x_1}^{\mathsf{t}}$, so that p_1 is the first coordinate of a solution of (2.1). Then, since $\mathsf{M}_{x_1}^{\mathsf{t}}$ and $\mathsf{M}_{x_2}^{\mathsf{t}}$ commute, $\mathsf{M}_{x_2}^{\mathsf{t}}$ induces a linear map

$$\mathsf{M}_{x_2}^{\mathsf{t}} : E_A(p_1, \mathsf{M}_{x_1}^{\mathsf{t}}) \to E_A(p_1, \mathsf{M}_{x_1}^{\mathsf{t}}).$$

where $E_A(p_1, \mathsf{M}_{x_1}^{\mathsf{t}})$ is the eigenspace of $\mathsf{M}_{x_1}^{\mathsf{t}}$ for the eigenvalue p_1. The eigenvalues of this map give the second coordinates of all solutions which have p_1 as their first coordinate. Continuing in this way gives all solutions since the intersection $\bigcap_{i=1}^{n} E_A(p_i, \mathsf{M}_{x_i}^{\mathsf{t}})$ is one-dimensional (see Theorem 3.2.3 of Chapter 3). This method is analyzed carefully in [MT01].

2.1.4 Single-Variable representation

One nice property of the non-derogatory case is that when M_f is non-derogatory, we can represent the algebra $\mathcal{A}$ using one variable. Here is the precise result.

Proposition 2.1.12. *Assume that $f \in \mathbb{K}[x_1, \ldots, x_n]$ and that M_f is non-derogatory. Then there is an algebra isomorphism*

$$\mathbb{K}[u]/\langle \mathrm{CharPoly}_{M_f}(u) \rangle \simeq \mathcal{A}.$$

Proof. Consider the map $\mathbb{K}[u] \to \mathcal{A}$ defined by $P(u) \mapsto [P(f)]$. Then $P(u)$ is in the kernel if and only if $[P(f)] = [0]$, i.e., if and only if $P(f) \in \langle f_1, \ldots, f_s \rangle$. In the discussion of minimal polynomials in Section 2.1.2, we showed that the minimal polynomial of M_f is the nonzero polynomial of smallest degree with this property. It follows easily that the kernel of this map is generated by the the minimal polynomial of M_f. Thus we get an injective algebra homomorphism

$$\mathbb{K}[u]/\langle \mathrm{MinPoly}_{M_f}(u) \rangle \longrightarrow \mathcal{A}.$$

But $\mathrm{MinPoly}_{M_f}(u) = \mathrm{CharPoly}_{M_f}(u)$ since M_f is non-derogatory, and we also know that

$$\dim_{\mathbb{K}} \mathbb{K}[u]/\langle \text{CharPoly}_{M_f}(u) \rangle = \deg \text{CharPoly}_{M_f}(u) = \dim_{\mathbb{K}} \mathcal{A}.$$

It follows that the above injection is the desired isomorphism.

Notice that this proof applies over an arbitrary field $\mathbb{K}$. Thus, when $\mathbb{K}$ is infinite and all of the solutions are curvilinear (e.g., all have multiplicity 1), then Proposition 2.1.12 applies when f is a generic linear combination of the variables.

We can use the single-variable representation to give an alternate method for finding solutions. The idea is that the isomorphism

$$\mathbb{K}[u]/\langle \text{CharPoly}_{M_f}(u) \rangle \simeq \mathcal{A}$$

enables us to express the coset $[x_i] \in \mathcal{A}$ as a polynomial in $[f]$, say

$$[x_i] = P_i([f]),$$

where $\deg(P_i) < \dim_{\mathbb{K}} \mathcal{A}$. (*Exercise: Show that P_i can be explicitly computed using a Gröbner basis for $\langle f_1, \ldots, f_s \rangle$.*) Now we get all solutions as follows.

Proposition 2.1.13. *Assume that M_f be non-derogatory and let $P_1, \ldots, P_n$ be constructed as above. Then:*

1. $\langle f_1, \ldots, f_s \rangle = \langle \text{CharPoly}_{M_f}(f), x_1 - P_1(f), \ldots, x_n - P_n(f) \rangle$.
2. *For any root λ of $\text{CharPoly}_{M_f}(u)$, the n-tuple*

$$(P_1(\lambda), \ldots, P_n(\lambda))$$

   is a solution of (2.1), and all solutions of (2.1) arise this way.
3. *If $f = \sum_{i=1}^{n} a_i x_i$, then $\sum_{i=1}^{n} a_i(x_i - P_i(f)) = 0$, so that I has n generators.*

Proof. For part 1, it is easy to see that $x_i - P_i(f) \in \langle f_1, \ldots, f_s \rangle$, and $\text{CharPoly}_{M_f}(f) \in \langle f_1, \ldots, f_s \rangle$ by the Cayley-Hamilton theorem. For the other inclusion, one uses $x_i = x_i - P_i(f) + P_i(f)$ to express an element of $\langle f_1, \ldots, f_s \rangle$ as an element of $\langle x_1 - P_1(f), \ldots, x_n - P_n(f) \rangle$ plus a polynomial in f. Then Proposition 2.1.12 implies that this polynomial is divisible by $\text{CharPoly}_{M_f}(f)$. From here, the other parts of the proposition follow easily. (*Exercise: Complete the proof.*)

The third part of Proposition 2.1.13 implies that when all solutions are curvilinear, we can rewrite the system as n equations in n unknowns. In this case, the corresponding ideal is called a *complete intersection*.

The single-variable representation will have some unexpected consequences later in the chapter.

2.1.5 Generalized eigenspaces and multiplicities

The final observation of Section 2.1 relates multiplicities and generalized eigenspaces. Given an eigenvalue λ of a linear map $T : V \to V$, recall that its *generalized eigenspace* is

$$G_V(\lambda, T) = \{v \in V \mid (T - \lambda I)^N(v) = 0 \text{ for some } N \geq 1\}.$$

It is well-known that the dimension of $G_V(\lambda, T)$ is the multiplicity of λ as a root of the characteristic polynomial of T.

Proposition 2.1.14. *The characteristic polynomial of $M_f : \mathcal{A} \to \mathcal{A}$ is*

$$\mathrm{CharPoly}_{M_f}(u) = \prod_p (u - f(p))^{\mathrm{mult}(p)}.$$

Furthermore, if $f \in \mathbb{K}[x_1, \ldots, x_n]$ takes distinct values at the solutions of (2.1) and p is one of the solutions, then the generalized eigenspace $G_{\mathcal{A}}(f(p), M_f)$ is naturally isomorphic to the local ring $\mathcal{A}_p$.

Proof. First observe that M_f is compatible with the isomorphism

$$\mathcal{A} \simeq \prod_p \mathcal{A}_p = \prod_p \mathbb{K}[x_1, \ldots, x_n]/I_p$$

where $\langle f_1, \ldots, f_s \rangle = \bigcap_p I_p$ is the primary decomposition and the product and intersection are over all solutions of (2.1).

Now fix one solution p and note that $f(p)$ is the only eigenvalue of M_f on $\mathcal{A}_p$ since $\mathcal{A}_p = \mathbb{K}[x_1, \ldots, x_n]/I_p$ and p is the only common solution of the polynomials in I_p. This easily leads to the desired formula for $\mathrm{CharPoly}_{M_f}(u)$ since $\mathrm{mult}(p) = \dim_{\mathbb{K}} \mathcal{A}_p$.

The previous paragraph also implies that $M_{f-f(p)}$ is nilpotent on $\mathcal{A}_p$, so that $\mathcal{A}_p$ is contained in the generalized eigenspace of M_f for the eigenvalue $f(p)$. If we further assume that $f(q) \neq f(p)$ for $q \neq p$, then $M_{f-f(p)}$ is invertible on $\mathcal{A}_q$. Since $\mathcal{A} \simeq \prod_p \mathcal{A}_p$, it follows that we can identify $\mathcal{A}_p$ with the generalized eigenspace $G_{\mathcal{A}}(f(p), M_f)$. $\qquad\square$

Here is a familiar example.

Example 2.1.15. For the equations $f_1 = f_2 = f_3 = 0$ of Example 2.1.5, the solutions are $(0, 0)$ and $(0, 1)$, and $\mathrm{CharPoly}_{M_y}(u) = u^2(u - 1)^3$. Since y separates the solutions, Proposition 2.1.14 implies that $(0, 0)$ has multiplicity 2 and $(0, 1)$ has multiplicity 3.

Genericity. Proposition 2.1.14 shows that we can compute multiplicities by factoring $\mathrm{CharPoly}_{M_f}(u)$, provided f takes distinct values at the solutions. Furthermore, if we let $f = a_1 x_1 + \cdots + a_n x_n$, then this is true generically. Thus, if we make a random choice of the a_i, then with high probability, the resulting f will take distinct values at the solutions. Thus, given a system of equations (2.1), we have a probabilistic algorithm for finding both the number of solutions and their respective multiplicities.

But sometimes (in a deterministic algorithm, for instance), one needs a certificate that f takes distinct values at the solutions. Here are two ways to achieve this:

- First, if $\langle f_1, \ldots, f_s \rangle$ is radical, then f takes distinct values at the solutions if and only if $\Phi = \mathrm{CharPoly}_{M_f}$ has distinct roots. (*Exercise: Prove this.*) Hence one need only compute $\gcd(\Phi, \Phi')$.
- In general, one can compute $\sqrt{\langle f_1, \ldots, f_s \rangle}$ as described in Section 2.1.3 and then proceed as in the previous bullet.

Numerical Issues. A serious numerical issue is that it is sometimes hard to distinguish between a single solution of multiplicity $k > 1$ and a cluster of k very close solutions of multiplicity 1. Several people, including Hans Stetter, are trying to come up with numerically stable methods for understanding such clusters. For example:

- While the individual points in a cluster are not stable, their center of gravity is.
- When the cluster consists of two points, the slope of the line connecting them is numerically stable.

More details can be found in [HS97a]. From a sophisticated point of view, this is equivalent to studying the numerical stability points in a Hilbert scheme of subschemes of affine space of fixed finite length supported at a fixed point.

Other Notions of Multiplicity. The multiplicity $\mathrm{mult}(p)$ defined in this section is sometimes called the *geometric multiplicity*. There is also a more subtle version of multiplicity called the *algebraic multiplicity* or *Hilbert-Samuel multiplicity* $e(p)$. A discussion of multiplicity can be found in [Cox].

2.2 Ideals defined by linear conditions

As in the previous section, we will assume that

$$\mathcal{A} = \mathbb{K}[x_1, \ldots, x_n]/\langle f_1, \ldots, f_s \rangle$$

is a finite-dimensional algebra over an algebraically closed field $\mathbb{K}$. In this section, we will continue the theme of linear algebra, focusing now on the role of linear functionals on $\mathcal{A}$ and $\mathbb{K}[x_1, \ldots, x_n]$.

2.2.1 Duals and dualizing modules

Given $\mathcal{A}$ as above, its dual space is

$$\widehat{\mathcal{A}} = \mathrm{Hom}_{\mathbb{K}}(\mathcal{A}, \mathbb{K}).$$

If $\{\ell_1, \ldots, \ell_m\}$ is a basis of $\widehat{\mathcal{A}}$, then composing with the quotient map $\mathbb{K}[x_1, \ldots, x_n] \to \mathcal{A}$ gives linear functionals

$$L_1, \ldots, L_m : \mathbb{K}[x_1, \ldots, x_n] \longrightarrow \mathbb{K}$$

with the property that

$$\langle f_1, \ldots, f_s \rangle = \{ f \in \mathbb{K}[x_1, \ldots, x_n] \mid L_i(f) = 0, i = 1, \ldots, m \}$$

Thus the ideal $\langle f_1, \ldots, f_s \rangle$ is defined by the linear conditions given by the L_i. In this section, we will explore some interesting ways of doing this.

Recall from Section 2.1 that we have the product decomposition

$$\mathcal{A} \simeq \prod_p \mathcal{A}_p$$

induced by the primary decomposition $\langle f_1, \ldots, f_s \rangle = \bigcap_p I_p$, where the product and intersection are over all solutions in $\mathbb{K}^n$ of the equations

$$f_1 = f_2 = \cdots = f_s = 0.$$

The product induces a natural isomorphism

$$\widehat{\mathcal{A}} \simeq \prod_p \widehat{\mathcal{A}}_p. \tag{2.9}$$

One feature of (2.9) is the following. For each solution p, let $\{\ell_{p,i}\}_{i=1}^{\mathrm{mult}(p)}$ be a basis of $\widehat{\mathcal{A}}_p$. As above, every $\ell_{p,i}$ gives a linear functional

$$L_{p,i} : \mathbb{K}[x_1, \ldots, x_n] \longrightarrow \mathbb{K}.$$

Then:

- If we fix p, then

$$I_p = \{ f \in \mathbb{K}[x_1, \ldots, x_n] \mid L_{p,i}(f) = 0 \text{ for } i = 1, \ldots, \mathrm{mult}(p) \}$$

  is the primary ideal such that $\mathcal{A}_p = \mathbb{K}[x_1, \ldots, x_n]/I_p$.
- If we vary over all p and i, then the $L_{p,i}$ define the ideal

$$\langle f_1, \ldots, f_s \rangle = \bigcap_p I_p.$$

This way of thinking of the linear conditions gives not only the ideal but also its primary decomposition.

The dual space $\widehat{\mathcal{A}}$ also relates to the matrices M_f and M_f^{t} appearing in Section 2.1. Since M_f is the matrix of the multiplication map $M_f : \mathcal{A} \to \mathcal{A}$ relative to the basis of $\mathcal{A}$ coming from the remainder monomials B, linear algebra implies that M_f^{t} is the matrix of the dual map $M_f^{\mathrm{t}} : \widehat{\mathcal{A}} \to \widehat{\mathcal{A}}$ relative to the dual basis of $\widehat{\mathcal{A}}$. This has the following nice application.

Example 2.2.1. Let p be a solution and let $1_p : \mathcal{A} \to \mathbb{K}$ be the linear functional defined by $1_p([f]) = f(p)$. (*Exercise: Explain why this is well-defined.*) Thus $1_p \in \widehat{\mathcal{A}}$. In Section 2.1.2, we used equation (2.3) to prove that $f(p)$ is an eigenvalue of M_f^t. In terms of M_f^t, this equation can be written

$$M_f^t(1_p) = f(p)\, 1_p.$$

(*Exercise: Prove this.*) See Theorem 3.2.3 of Chapter 3 for more details.

We also need to explain how $\widehat{\mathcal{A}}$ relates to commutative algebra. The key point is that $\widehat{\mathcal{A}} = \mathrm{Hom}_{\mathbb{K}}(\mathcal{A}, \mathbb{K})$ becomes an $\mathcal{A}$ module via

$$(a\ell)(b) = \ell(ab)$$

for $a, b \in \mathcal{A}$ and $\ell \in \widehat{\mathcal{A}}$. Similarly, in the decomposition (2.9), each $\widehat{\mathcal{A}_p}$ is a module over $\mathcal{A}_p$. We call $\widehat{\mathcal{A}}$ and $\widehat{\mathcal{A}_p}$ the *dualizing modules* of $\mathcal{A}$ and $\mathcal{A}_p$ respectively. Dualizing modules are discussed in Chapter 21 of [Eis95].

Finally, we say that $\mathcal{A}$ is *Gorenstein* if there is an $\mathcal{A}$-module isomorphism

$$\widehat{\mathcal{A}} \simeq \mathcal{A}.$$

Being Gorenstein is equivalent to the existence of a nondegenerate bilinear form

$$\langle \cdot, \cdot \rangle : \mathcal{A} \times \mathcal{A} \to \mathbb{K}$$

with the property that

$$\langle ab, c \rangle = \langle a, bc \rangle$$

for $a, b, c \in \mathcal{A}$. (*Exercise: Prove this.*) Here is an example.

Example 2.2.2. Let $P \in \mathbb{C}[z]$ be a polynomial of degree d with simple zeros. Consider the *global residue*

$$\mathrm{res}_P : \mathbb{C}[z]/\langle P \rangle \to \mathbb{C}$$

introduced in Chapter 1. By Theorem th:dual of Chapter 1, the bilinear form

$$\mathbb{C}[z]/\langle P \rangle \times \mathbb{C}[z]/\langle P \rangle \to \mathbb{C}$$

defined by $(g_1 \cdot g_2) = \mathrm{res}_P(g_1 g_2)$ is nondegenerate. Since

$$(g_1 g_2 \cdot g_3) = \mathrm{res}_P((g_1 g_2)g_3) = \mathrm{res}_P(g_1(g_2 g_3)) = (g_1 \cdot g_2 g_3),$$

we see that $\mathcal{A} = \mathbb{C}[z]/\langle P \rangle$ is Gorenstein.

See Chapter 21 of [Eis95] for more on Gorenstein duality.

2.2.2 Differential conditions defining ideals

So far, we've seen that each primary ideal I_p can be described using $\mathrm{mult}(p)$ linear conditions. We will now explain how to represent these linear conditions using constant coefficient differential operators evaluated at p. We will assume that $\mathbb{K}$ has characteristic 0.

Let's begin with some examples.

Example 2.2.3. The equation

$$x^2(x-1)^3 = 0$$

has the solutions 0 of multiplicity 2 and 1 of multiplicity 3. In terms of derivatives, we have the ideal

$$\langle x^2(x-1)^3 \rangle = \{f \in \mathbb{K}[x] \mid f(0) = f'(0) = 0, \ f(1) = f'(1) = f''(1) = 0\}.$$

Notice that the multiplicities correspond to the number of conditions defining the ideal at 0 and 1 respectively.

Example 2.2.4. Consider the three sets of equations

$$(a): \ x^2 = xy = y^2 = 0$$
$$(b): \ x^2 = y^2 = 0$$
$$(c): \ x = y^3 = 0.$$

One easily sees that $(0,0)$ is the only solution, with multiplicity 3 in case (a), multiplicity 4 in case (b), and multiplicity 3 in case (c). In terms of partial derivatives, the corresponding ideals are given by

$(a): \ \langle x^2, xy, y^2 \rangle = \{f \in \mathbb{K}[x,y] \mid f(0,0) = f_x(0,0) = f_y(0,0) = 0\}$
$(b): \ \langle x^2, y^2 \rangle = \{f \in \mathbb{K}[x,y] \mid f(0,0) = f_x(0,0) = f_y(0,0) = f_{xy}(0,0) = 0\}$
$(c): \ \langle x, y^3 \rangle = \{f \in \mathbb{K}[x,y] \mid f(0,0) = f_y(0,0) = f_{yy}(0,0) = 0\}.$

In each case, the multiplicity is equal to the number of conditions defining the ideal.

We will now generalize this description and use it to obtain interesting information about the local rings. For instance, in the above example, we will see that the descriptions of the ideals in terms of partial derivatives imply the following:

- In cases (b) and (c), the ring is Gorenstein but not in case (a).
- In case (c) the ring is curvilinear but not in cases (a) and (b).

We begin by setting up some notation. Consider the polynomial ring $\mathbb{K}[\partial_1, \ldots, \partial_n]$. Then an exponent vector $\alpha = (a_1, \ldots, a_n)$ gives the monomial ∂^α, which we regard as the partial derivative

$$\partial^\alpha = \frac{\partial^{a_1 + \cdots + a_n}}{\partial x_1^{a_1} \cdots \partial x_n^{a_n}}.$$

Thus elements of $\mathbb{K}[\partial_1, \ldots, \partial_n]$ become constant coefficient differential operators on $\mathbb{K}[x_1, \ldots, x_n]$.

In examples, we sometimes write ∂^α as ∂_{x^α}. Thus

$$\partial_{xy^2} = \partial^{(1,2)} = \frac{\partial^3}{\partial x \partial y^2}$$

when operating on $\mathbb{K}[x, y]$. Also note that Example 2.2.4 involves the operators

$$
\begin{aligned}
&\text{(a)} : 1, \partial_x, \partial_y \\
&\text{(b)} : 1, \partial_x, \partial_y, \partial_{xy} \\
&\text{(c)} : 1, \partial_y, \partial_{y^2}.
\end{aligned}
\tag{2.10}
$$

applied to polynomials in $\mathbb{K}[x, y]$ and evaluated at $(0, 0)$. Here, 1 is the identity operator on $\mathbb{K}[x, y]$.

We next define the *deflation* or *shift* of $D = \sum_\alpha c_\alpha \partial^\alpha \in \mathbb{K}[\partial_1, \ldots, \partial_n]$ by an exponent vector β to be the operator

$$\sigma_\beta D = \sum_\alpha c_\alpha \binom{\alpha}{\beta} \partial^{\alpha - \beta},$$

where $\binom{\alpha}{\beta} = \binom{a_1}{b_1} \cdots \binom{a_n}{b_n}$ and $\partial^{\alpha - \beta} = 0$ whenever $\alpha - \beta$ has a negative coordinate. The reason for the binomial coefficients in the formula for $\sigma_\beta D$ is that they give the Leibniz formula

$$D(fg) = \sum_\beta \partial^\beta(f) \sigma_\beta D(g)$$

for $f, g \in \mathbb{K}[x_1, \ldots, x_n]$. Here are some simple examples of deflations.

Example 2.2.5. Observe that

$$\partial_{xy} \text{ has nonzero deflations } \partial_{xy}, \partial_x, \partial_y, 1$$
$$\partial_{y^2} \text{ has nonzero deflations } \partial_{y^2}, 2\partial_y, 1.$$

These correspond to cases (b) and (c) of Example 2.2.4. On the other hand, the operators of case (a) are not deflations of a single operator. As we will see, this is why case (a) is not Gorenstein.

Definition 2.2.6. *A subspace $L \subset \mathbb{K}[\partial_1, \ldots, \partial_n]$ is **closed** if it has finite dimension over $\mathbb{K}$ and closed under deflation, i.e., $\sigma_\beta(L) \subset L$ for all β.*

The reader can easily check that the differential operators in cases (a), (b) and (c) of (2.10) span closed subspaces. Here is the main result of Section 2.2.2 (see [MMM93] for a proof).

Theorem 2.2.7. *For every solution p of (2.1), there is a unique closed subspace $L_p \subset \mathbb{K}[\partial_1, \ldots, \partial_n]$ of dimension $\mathrm{mult}(p)$ such that*

$$\langle f_1, \ldots, f_s \rangle = \{f \in \mathbb{K}[x_1, \ldots, x_n] \mid D(f)(p) = 0 \ \forall \text{ solution } p \text{ and } D \in L_p\},$$

where $D(f)(p)$ means the evaluation of the polynomial $D(f)$ at the point p. Furthermore, the primary component of $\langle f_1, \ldots, f_s \rangle$ corresponding to a solution p is

$$I_p = \{f \in \mathbb{K}[x_1, \ldots, x_n] \mid D(f)(p) = 0 \text{ for all } D \in L_p\},$$

and conversely,

$$L_p = \{D \in \mathbb{K}[\partial_1, \ldots, \partial_n] \mid D(f)(p) = 0 \text{ for all } f \in I_p\}.$$

It should not be surprising that Examples 2.2.3 and 2.2.4 are examples of this theorem. Here is a more substantial example.

Example 2.2.8. Consider the equations

$$\begin{aligned} f_1 &= x^2 + 2y^2 - 2y = 0 \\ f_2 &= xy^2 - xy = 0 \\ f_3 &= y^3 - 2y^2 + y = 0 \end{aligned}$$

from Example 2.1.1. There, we saw that the only solutions were $(0,0)$ and $(0,1)$. In [MS95], it is shown that

$$\begin{aligned} L_{(0,0)} &= \mathrm{Span}(1, \partial_x) \\ L_{(0,1)} &= \mathrm{Span}(1, \partial_x, \partial_{x^2} - \partial_y). \end{aligned} \tag{2.11}$$

Thus $\langle f_1, f_2, f_3 \rangle$ consists of all $f \in \mathbb{K}[x, y]$ such that

$$f(0,0) = f_x(0,0) = f(0,1) = f_x(0,1) = 0, f_{xx}(0,1) = f_y(0,1),$$

and looking at the conditions for $(0,0)$ and $(0,1)$ separately gives the primary decomposition of $\langle f_1, f_2, f_3 \rangle$. In Section 2.2.3 we will describe how (2.11) was computed.

Gorenstein and Curvilinear Points. We conclude Section 2.2.2 by explaining how special properties of the local ring $\mathcal{A}_p$ can be determined from the representation given in Theorem 2.2.7.

Theorem 2.2.9. $\mathcal{A}_p$ *is Gorenstein if and only if there is $D \in L_p$ whose deflations span L_p.*

Proof. Let L_p^{ev} denote the linear forms $\mathbb{K}[x_1, \ldots, x_n] \to \mathbb{K}$ obtained by composing elements of L_p with evaluation at p. Each such map vanishes on I_p and thus gives an element of $\widehat{\mathcal{A}_p}$. Hence

$$L_p \simeq L_p^{\mathrm{ev}} \simeq \widehat{\mathcal{A}_p}.$$

Furthermore, if $D \in L_p$ maps to $\widetilde{D} \in \widehat{\mathcal{A}_p}$, then the Leibniz formula makes it easy to see that the deflation $\sigma_\beta D$ maps to $(x - p)^\beta \widetilde{D}$, where

$$(x - p)^\beta = (x_1 - p_1)^{b_1} \cdots (x_n - p_n)^{b_n}$$

for $p = (p_1, \ldots, p_n)$ and $\beta = (b_1, \ldots, b_n)$. (*Exercise: Prove this.*) Hence these deflations span L_p if and only if $\widehat{\mathcal{A}_p}$ is generated by a single element as an $\mathcal{A}_p$-module. In the latter case, we have a surjective $\mathcal{A}_p$-module homomorphism $\mathcal{A}_p \to \widehat{\mathcal{A}_p}$ which is an isomorphism since $\mathcal{A}_p$ and $\widehat{\mathcal{A}_p}$ have the same dimension over $\mathbb{K}$. We are done by the definition of Gorenstein from Section 2.2.1.

Before stating our next result, we need some definitions from [MS95].

Definition 2.2.10. The **order** of $D = \sum_\alpha c_\alpha \partial^\alpha$ is the degree of D as a polynomial in $\mathbb{K}[\partial_1, \ldots, \partial_n]$. A basis $D_1, \ldots, D_{\mathrm{mult}(p)}$ of L_p is **consistently ordered** if for every $r \geq 1$, there is $j \geq 1$ such that

$$\mathrm{Span}(D \in L_p \mid D \text{ has order } \leq r) = \mathrm{Span}(D_1, \ldots, D_j).$$

Note that every consistently ordered basis has $D_1 = 1$. Also observe that the bases listed in (2.10) and (2.11) are consistently ordered.

We can now characterize when $\mathcal{A}_p$ is curvilinear.

Theorem 2.2.11. The embedding dimension e_p of $\mathcal{A}_p$ is the number of operators of order 1 in a consistently ordered basis of L_p. In particular, $\mathcal{A}_p$ is curvilinear if and only if any such basis has at most one operator of order 1.

Proof. Let $\mathfrak{m}_p$ be the maximal ideal of $\mathcal{A}_p$. Recall from equation (2.8) that

$$e_p = \dim_{\mathbb{K}} \mathfrak{m}_p / \mathfrak{m}_p^2$$
$$= \# \text{ minimal generators of } \mathfrak{m}_p.$$

Also let $L_p^r = \mathrm{Span}(D \in L_p \mid D \text{ has order } \leq r)$. Then $L_p^0 \subset L_p^1 \subset \cdots$ and, for $r \geq 0$, we have

$$\dim_{\mathbb{K}} L_p^r / L_p^{r-1} = \# \text{operators of order } r \text{ in a} \tag{2.12}$$
$$\text{consistently ordered basis of } L_p.$$

We claim that there is a natural isomorphism

$$L_p^1 / L_p^0 \simeq \mathrm{Hom}_{\mathbb{K}}(\mathfrak{m}_p / \mathfrak{m}_p^2, \mathbb{K}). \tag{2.13}$$

Assuming this for the moment, the first assertion of the theorem follows immediately from (2.12) for $r = 1$ and the above formula for e_p. Then the final assertion follows since by definition $\mathcal{A}_p$ is curvilinear if and only if it has embedding dimension $e_p \leq 1$.

To prove (2.13), let $\mathfrak{M}_p = \langle x_1 - p_1, \ldots, x_n - p_n \rangle \subset \mathbb{K}[x_1, \ldots, x_n]$ be the maximal ideal of p. Then any operator $D = \sum_{i=1}^{n} a_i \partial_i$ induces the linear map

$$\mathfrak{M}_p \longrightarrow \mathbb{K}$$

that sends $f \in \mathfrak{M}_p$ to $D(f)(p)$. By the product rule, this vanishes if $f \in \mathfrak{M}_p^2$, so that we get an element of the dual space

$$\mathrm{Hom}_{\mathbb{K}}(\mathfrak{M}_p/\mathfrak{M}_p^2, \mathbb{K}).$$

Furthermore, it is easy to see that every element of the dual space arises in this way. (*Exercise: Prove these assertions.*)

The isomorphism $\mathbb{K}[x_1, \ldots, x_n]/I_p \simeq \mathcal{A}_p$ induces exact sequences

$$0 \to I_p \to \mathfrak{M}_p \to \mathfrak{m}_p \to 0$$

and

$$0 \to \mathrm{Hom}_{\mathbb{K}}(\mathfrak{m}_p/\mathfrak{m}_p^2, \mathbb{K}) \to \mathrm{Hom}_{\mathbb{K}}(\mathfrak{M}_p/\mathfrak{M}_p^2, \mathbb{K}) \to \mathrm{Hom}_{\mathbb{K}}(I_p/I_p \cap \mathfrak{M}_p^2, \mathbb{K}) \to 0.$$

It follows that $D = \sum_{i=1}^{n} a_i \partial_i$ gives an element of $\mathrm{Hom}_{\mathbb{K}}(\mathfrak{m}_p/\mathfrak{m}_p^2, \mathbb{K})$ if and only if D vanishes on I_p, which is equivalent to $D \in L_p$. Since these operators represent L_p^1/L_p^0, we are done. (*Exercise: Fill in the details.*)

We get the following corollary of this result and Theorem 2.1.11.

Corollary 2.2.12. *The multiplication map M_f is non-derogatory when f is a generic linear combination of $x_1, \ldots, x_n$ if and only if for every solution p, a consistently ordered basis of L_p has at most one operator of order 1.*

Since the bases in Example 2.2.8 are consistently ordered, (2.11) shows that the solutions of the corresponding equations have embedding dimension 1 and hence are curvilinear. (This is the computation promised in Example 2.1.10.) Thus M_f is non-derogatory when f is a generic linear combination of x, y. Of course, we computed a specific instance of this in Example 2.1.6, but now we know the systematic reason for our success.

Note also that if we apply Theorems 2.2.9 and 2.2.11 to Example 2.2.4, then we see that the ring is Gorenstein in cases (b) and (c) (but not (a)) and curvilinear in case (c) (but not (a) and (b)). This proves the claims made in the two bullets from the discussion following Example 2.2.4

2.2.3 Two algorithms

We've seen that the ideal $\langle f_1, \ldots, f_s \rangle$ can be described using Gröbner bases and using conditions on partial derivatives. As we will now explain, going from one description to the other is a simple matter of linear algebra.

Gröbner Bases to Partial Derivatives. If we have a Gröbner basis for $\langle f_1, \ldots, f_s \rangle$, then we obtain the required closed subspaces L_p in a three-step process. The first step is to compute the primary decomposition

$$\langle f_1, \ldots, f_s \rangle = \bigcap_p I_p.$$

In particular, this means knowing a Gröbner basis for each I_p. We will explain how to compute such a primary decomposition in Section 2.4.3.

Given this, we fix a primary ideal I_p. We next recall a useful fact which relates I_p to the maximal ideal $\mathfrak{M}_p = \langle x_1 - p_1, \ldots, x_n - p_n \rangle$ of p in $\mathbb{K}[x_1, \ldots, x_n]$.

Lemma 2.2.13. *If* $m = \mathrm{mult}(p)$, *then* $\mathfrak{M}_p^m \subset I_p$.

Proof. It suffices to prove that $\mathfrak{m}_p^m = \{0\}$, where $\mathfrak{m}_p$ is the maximal ideal of $\mathcal{A}_p$. By Nakayama's lemma, we know that that $\mathfrak{m}_p^k \neq \mathfrak{m}_p^{k+1}$ whenever $\mathfrak{m}_p^k \neq \{0\}$. Using

$$\mathcal{A}_p \supset \mathfrak{m}_p \supset \mathfrak{m}_p^2 \supset \cdots \supset \mathfrak{m}_p^k \supset \{0\},$$

it follows that $\dim_\mathbb{K} \mathcal{A}_p \geq k + 1$ whenever $\mathfrak{m}_p^k \neq \{0\}$. The lemma follows.

This lemma will enable us to describe I_p in terms of differential operators of order at most m. However, this description works best when $p = 0$. So the second step is to translate so that $p = 0$. Hence for the rest of our discussion, we will assume that $p = 0$. Thus Lemma 2.2.13 tells us that

$$\mathfrak{M}_0^m \subset I_0, \quad m = \mathrm{mult}(0).$$

The third step is to write down the differential operators in L_0 as follows. Let B_0 be the set of remainder monomials for the Gröbner basis of I_0 and set

$$\mathrm{Mon}_m = \{x^\alpha \mid x^\alpha \notin B_0, \ \deg(x^\alpha) < m\}.$$

For each $x^\alpha \in \mathrm{Mon}_m$, let

$$x^\alpha \equiv \sum_{x^\beta \in B_0} c_{\alpha\beta} x^\beta \bmod I_0 \tag{2.14}$$

In other words, $\sum_{x^\beta \in B_0} c_{\alpha\beta} x^\beta$ is the remainder of x^α on division by the Gröbner basis of I_0. Then, for each $x^\beta \in B_0$, define

$$D_\beta = \partial^\beta + \sum_{x^\alpha \in \mathrm{Mon}_m} c_{\alpha\beta} \frac{\beta!}{\alpha!} \partial^\alpha.$$

where $\alpha! = a_1! \cdots a_n!$ for $\alpha = (a_1, \ldots, a_n)$ and similarly for $\beta!$.

Proposition 2.2.14. $f \in \mathbb{K}[x_1, \ldots, x_n]$ *lies in* I_p *if and only if* $D_\beta(f)(0) = 0$ *for all* $x^\beta \in B_0$.

Proof. Let $f = \sum_\alpha a_\alpha x^\alpha$. Since $\mathfrak{M}_0^m \subset I_0$, we can assume that $f = \sum_{\deg(\alpha) < m} a_\alpha x^\alpha$. Using (2.14), it is straightforward to show that

$$f \in I_p \iff a_\beta + \sum_{x^\alpha \in \mathrm{Mon}_m} c_{\alpha\beta} a_\alpha = 0 \text{ for all } x^\beta \in B_0.$$

(*Exercise: Prove this equivalence.*) However, since

$$\partial^\gamma(x^\delta)(0) = \begin{cases} \gamma! & \text{if } \gamma = \delta \\ 0 & \text{otherwise,} \end{cases} \tag{2.15}$$

one easily sees that for $x^\beta \in B_0$,

$$D_\beta(f)(0) = \left(\partial^\beta + \sum_{x^\alpha \in \mathrm{Mon}_m} c_{\alpha\beta} \frac{\beta!}{\alpha!} \partial^\alpha \right) \left(\sum_{\deg(\gamma) < m} a_\gamma x^\gamma \right)(0)$$
$$= \beta! \left(a_\beta + \sum_{x^\alpha \in \mathrm{Mon}_m} c_{\alpha\beta} a_\alpha \right).$$

The proposition now follows immediately since $\mathbb{K}$ has characteristic 0.

Here is an example of this result.

Example 2.2.15. For the polynomials from Example 2.2.8, we will show in Section 2.4.3 that the primary decomposition is

$$\langle x^2 + 2y^2 - 2y, xy^2 - xy, y^3 - 2y^2 + y \rangle$$
$$= \langle x^2, y \rangle \cap \langle x^2 + 2(y - 1), x(y - 1), (y - 1)^2 \rangle$$
$$= I_{(0,0)} \cap I_{(0,1)}.$$

Let's focus on $I_{(0,1)}$. If we translate this to the origin, then we get the ideal

$$I_0 = \langle x^2 + 2y, xy, y^2 \rangle.$$

The generators are a Gröbner basis for lex order with $x > y$, the remainder monomials are $B_0 = \{1, x, y\}$, and the multiplicity is $m = 3$. Thus

$$\mathrm{Mon}_3 = \{x^\alpha \mid x^\alpha \notin \{1, x, y\}, \ \deg(x^\alpha) < 3\} = \{x^2, xy, y^2\}.$$

The coefficients $c_{\alpha\beta}$ are given by

$$x^2 \equiv 0 \cdot 1 + 0 \cdot x + (-2) \cdot y \mod I_0$$
$$xy \equiv 0 \cdot 1 + 0 \cdot x + 0 \cdot y \mod I_0$$
$$y^2 \equiv 0 \cdot 1 + 0 \cdot x + 0 \cdot y \mod I_0.$$

so that

$$D_1 = 1, \quad D_x = \partial_x, \quad D_y = \partial_y + (-2)\frac{1!}{2!}\partial_{x^2} = \partial_y - \partial_{x^2}.$$

(*Exercise: Check this!*) Up to a sign, this is the basis of $L_{(0,1)}$ that appeared in Example 2.2.8. The treatment for $L_{(0,0)}$ is similar. (*Exercise: Do it.*)

We also want to remark on an alternate way to view the construction $L_0 = \mathrm{Span}(D_\beta \mid x^\beta \in B_0)$. Here, we are in the situation where $p = 0$, so that by Theorem 2.2.7, we have

$$I_0 = \{f \in \mathbb{K}[x_1, \ldots, x_n] \mid D_\beta(f)(0) = 0 \text{ for all } x^\beta \in B_0\}$$

and

$$L_0 = \{D \in \mathbb{K}[\partial_1, \ldots, \partial_n] \mid D_\beta(f)(0) = 0 \text{ for all } f \in I_0\}.$$

Now we will do something audacious: switch x_i with ∂_i. This means that I_0 becomes an ideal

$$\hat{I}_0 \subset \mathbb{K}[\partial_1, \ldots, \partial_n]$$

and L_0 becomes a subspace

$$\hat{L}_0 \subset \mathbb{K}[x_1, \ldots, x_n].$$

Observe that the pairing (2.15) is unchanged under $x_i \leftrightarrow \partial_i$. Thus:

$$\hat{L}_0 = \{\text{the polynomial solutions of the infinitely}$$
$$\text{many differential operators in } \hat{I}_0\}.$$

This is the point of view taken in Chapter 10 of [Stu02]. Here is an example.

Example 2.2.16. In Example 2.2.15, we showed that

$$I_0 = \langle x^2 + 2y, xy, y^2 \rangle \implies L_0 = \mathrm{Span}(1, \partial_x, \partial_y - \partial_{x^2}).$$

This means that under the switch $x \leftrightarrow \partial_x, y \leftrightarrow \partial_y$, the subspace

$$\hat{L}_0 = \mathrm{Span}(1, x, y - x^2) \subset \mathbb{K}[x, y]$$

is the space of all polynomial solutions of the infinitely many operators in the ideal

$$\hat{I}_0 = \langle \partial_x^2 + 2\partial_y, \partial_x\partial_y, \partial_y^2 \rangle \subset \mathbb{K}[\partial_x, \partial_y].$$

Other examples can be found in [Stu02].

We should also note that the description of I_p given by differential conditions can require a lot of space. Examples and more efficient methods can be found in Section 3.3 of [MMM96].

Partial Derivatives to Gröbner Bases. Now suppose that conversely, we are given the data of Theorem 2.2.7. This means that for each solution p we have a closed subspace L_p of dimension $\mathrm{mult}(p)$ such that

$$\langle f_1, \ldots, f_s \rangle = \{f \in \mathbb{K}[x_1, \ldots, x_n] \mid D(f)(p) = 0 \text{ for all } p \text{ and } D \in L_p\}.$$

If we pick a basis $D_{p,i}$ of each L_p, then the linear forms $f \mapsto D_{p,i}(f)(p)$ give a linear map

$$L : \mathbb{K}[x_1, \ldots, x_n] \longrightarrow \mathbb{K}^m \tag{2.16}$$

where $m = \dim_{\mathbb{K}} \mathcal{A}$. This map is surjective and its kernel is $\langle f_1, \ldots, f_s \rangle$. Given an order $>$ (with some restrictions to be noted below), our goal is to find a Gröbner basis of $\langle f_1, \ldots, f_s \rangle$ with respect to $>$ using the linear map (2.16).

The idea is to simultaneously build up the Gröbner basis G and the set of remainder monomials B. We begin with both lists being empty and feed in monomials one at a time, beginning with 1. The main loop of the algorithm is described as follows.

Main Loop: Given a monomial x^α, compute $L(x^\alpha)$ together with $L(x^\beta)$ for all $x^\beta \in B$.

- If $L(x^\alpha)$ is *linearly dependent* on the $L(x^\beta)$, then compute a linear relation

$$L(x^\alpha) = \sum_{x^\beta \in B} a_\beta L(x^\beta), \quad a_\beta \in F$$

(hence $x^\alpha - \sum_{x^\beta \in B} a_\beta x^\beta \in \langle f_1, \ldots, f_s \rangle$) and add $x^\alpha - \sum_{x^\beta \in B} a_\beta x^\beta$ to G.
- If $L(x^\alpha)$ is *linearly independent* from the $L(x^\beta)$, then add x^α to B.

Once this loop is done for x^α, we feed in the next monomial, which is the minimal element (with respect to $>$) of the set

$$N(x^\alpha, G) = \{ \text{monomials } x^\gamma > x^\alpha \text{ such that } x^\gamma \text{ is not} \\ \text{divisible by the leading term of any } g \in G \}. \tag{2.17}$$

Hence we need to find the minimal element of $N(x^\alpha, G)$. As explained in [BW93], this is easy to do whenever $>$ is a lex or total degree order. The algorithm terminates when (2.17) becomes empty.

In [MMM93], it shown that this algorithm always terminates and that when this happens, G is the desired Gröbner basis and B is the corresponding set of remainder monomials. Here is an example.

Example 2.2.17. In the notation of Example 2.2.15, let $p = (0,0)$ and $L_0 = \mathrm{Span}(1, \partial_x, \partial_y - \partial_{x^2})$. It follows that I_0 is the kernel of the map

$$L : \mathbb{K}[x, y] \longrightarrow \mathbb{K}^3$$

defined by $L(f) = (f(0,0), f_x(0,0), f_y(0,0) - f_{xx}(0,0))$. If we use lex order with $x > y$, then the above algorithm starts with $B = G = \emptyset$ and proceeds as follows:

x^α	$L(x^\alpha)$	B	G	$\min(N(x^\alpha, G))$
1^*	$(1,0,0)$	$\{1\}$	$\emptyset$	y
y^*	$(0,0,1)$	$\{1,y\}$	$\emptyset$	y^2
y^2	$(0,0,0)$	$\{1,y\}$	$\{y^2\}$	x
x^*	$(0,1,0)$	$\{1,y,x\}$	$\{y^2\}$	xy
xy	$(0,0,0)$	$\{1,y,x\}$	$\{y^2, xy\}$	x^2
x^2	$(0,0,-2)$	$\{1,y,x\}$	$\{y^2, xy, x^2 + 2y\}$	none!

$$\tag{2.18}$$

In this table, an asterisk denotes those monomials which become remainder monomials. The other monomials are leading terms of the Gröbner basis. (*Exercise: Check the steps of the algorithm to see how it works.*)

A complexity analysis of this algorithm can be found in [MMM93].

2.2.4 Ideals of points and basis conversion

We conclude by observing that the algorithm illustrated in Example 2.2.17 applies to many situations besides partial derivatives. The key point is that if

$$L : \mathbb{K}[x_1, \ldots, x_n] \longrightarrow \mathbb{K}^m$$

is *any* surjective linear map whose kernel is an ideal I, then the algorithm described in the discussion following (2.16) gives a Gröbner basis for I. Here are two situations where this is useful.

Ideals of Points. Suppose we have a finite list of points $p_1, \ldots, p_m \in \mathbb{K}^n$. Then we want to compute a Gröbner basis of the ideal

$$I = \{f \in \mathbb{K}[x_1, \ldots, x_n] \mid f(p_1) = \cdots = f(p_m) = 0\}$$

consisting of all polynomials which vanish at $p_1, \ldots, p_m$. This is now easy, for the points give the linear map $L : \mathbb{K}[x_1, \ldots, x_n] \to \mathbb{K}^m$ defined by

$$L(f) = (f(p_1), \ldots, f(p_m))$$

whose kernel is the ideal I. Furthermore, it is easy to see that L is surjective (see the proof of Theorem 2.10 of Chapter 2 of [CLO98]). Thus we can find a Gröbner basis of I using the above algorithm.

Example 2.2.18. Consider the points $(0,0), (1,0), (0,1) \in \mathbb{K}^2$. This gives the linear map $L : \mathbb{K}[x, y] \to \mathbb{K}^3$ defined by

$$L(f) = (f(0,0), f(1,0), f(0,1)).$$

If you apply the algorithm for lex order with $x > y$ as in Example 2.2.17, you will obtain a table remarkably similar to (2.18), except that the Gröbner basis will be $\{y^2 - y, xy, x^2 - x\}$. (*Exercise: Do this computation.*)

A more complete treatment appears in [MMM93]. This is also related to the *Buchberger-Möller algorithm* introduced in [BM82]. The harder problem of computing the homogeneous ideal of a finite set of points in projective space is discussed in [ABKR00].

Basis Conversion. Suppose that we have a Gröbner basis G' for $\langle f_1, \ldots, f_s \rangle$ with respect to one order $>'$ and want to find a Gröbner basis G with respect to a second order $>$.

We can do this as follows. Let B' be the set of remainder monomials with respect to G'. Then taking the remainder on division by G' gives a linear map

$$L : \mathbb{K}[x_1, \ldots, x_n] \longrightarrow \mathrm{Span}(B') \simeq \mathbb{K}^m.$$

The kernel is $\langle f_1, \ldots, f_s \rangle$ and the map is surjective since $L(x^\beta) = x^\beta$ for $x^\beta \in B'$. Then we can apply the above method to find the desired Gröbner basis G. This is the FGLM basis conversion algorithm of [FGLM93].

Example 2.2.19. By Example 2.2.17, $\{y^2, xy, x^2 + 2y\}$ is a Gröbner basis with respect to lex order with $x > y$. To convert to a lex order Gröbner basis with $y > x$, we apply the above method. After constructing a a table similar to (2.18), we obtain the Gröbner basis $\{x^3, y + \frac{1}{2}x^2\}$. (*Exercise: Do this.*)

Besides ideals of points and basis conversion, this algorithm has other interesting applications. See [MMM93] for details.

2.3 Resultants

The method for solving equations discussed in Section 2.1 assumed that we had a Gröbner basis available. In this section, we will see that when our equations have more structure, we can often compute the multiplication matrices directly, without using a Gröbner basis. This will lead to a method for solving equations closely related to the theory of resultants.

2.3.1 Solving equations

We will work in $\mathbb{K}[x_1, \ldots, x_n]$, $\mathbb{K}$ algebraically closed, but we will now assume that we have n equations in n unknowns, i.e.,

$$f_1(x_1, \ldots, x_n) = \cdots = f_n(x_1, \ldots, x_n) = 0. \tag{2.19}$$

Solutions to such a system are described by Bézout's theorem.

Theorem 2.3.1. *Consider a system of equations* (2.19) *as above and let d_i be the degree of f_i. Then:*

1. *If the system has only finitely many solutions, then the total number of solutions (counted with multiplicity) is at most $\mu = d_1 \cdots d_n$.*
2. *If $f_1, \ldots, f_n$ are generic, then there are precisely $\mu = d_1 \cdots d_n$ solutions, all of multiplicity 1.*

To find the solutions, we will use a method first described by Auzinger and Stetter [AS88]. The idea is to construct a $\mu \times \mu$ matrix whose eigenvectors will determine the solutions. For this purpose, let

$$d = d_1 + \cdots + d_n - n + 1 \tag{2.20}$$

and divide the monomials of degree $\leq d$ into $n + 1$ disjoint sets as follows:

$$S_n = \{x^\gamma : \deg(x^\gamma) \leq d, \ x_n^{d_n} \text{ divides } x^\gamma\}$$

$$S_{n-1} = \{x^\gamma : \deg(x^\gamma) \leq d, \ x_n^{d_n} \text{ doesn't divide } x^\gamma \text{ but } x_{n-1}^{d_{n-1}} \text{ does}\}$$

$$\vdots$$

$$S_0 = \{x^\gamma : \deg(x^\gamma) \leq d, \ x_n^{d_n}, \ldots, x_1^{d_1} \text{ don't divide } x^\gamma\}.$$

Note that

$$S_0 = \{x_1^{b_1} \cdots x_n^{b_n} \mid 0 \leq b_i \leq d_i - 1 \text{ for all } i\}, \text{ so that } \# S_0 = \mu. \qquad (2.21)$$

(*Exercise: Prove this.*) Since S_0 plays a special role, we will use x^α to denote elements of S_0 and x^β to denote elements of $S_1 \cup \cdots \cup S_n$. Now observe that

$$\text{if } x^\alpha \in S_0, \text{ then } x^\alpha \text{ has degree } \leq d - 1,$$

$$\text{if } x^\beta \in S_i, \ i > 0, \text{ then } x^\beta/x_i^{d_i} \text{ has degree } \leq d - d_i,$$

where the first assertion uses $d - 1 = d_1 + \cdots + d_n - n = \sum_{i=1}^n (d_i - 1)$.

Now let $f_0 = a_1 x_1 + \cdots + a_n x_n$, $a_i \in \mathbb{K}$, and consider the equations:

$$x^\alpha f_0 = 0 \quad \text{for all } x^\alpha \in S_0$$

$$(x^\beta/x_1^{d_1}) f_1 = 0 \quad \text{for all } x^\beta \in S_1$$

$$\vdots$$

$$(x^\beta/x_n^{d_n}) f_n = 0 \quad \text{for all } x^\beta \in S_n.$$

Since the $x^\alpha f_0$ and $x^\beta/x_i^{d_i} f_i$ have degree $\leq d$, we can write these polynomials as linear combinations of the x^α and x^β. We will order these monomials so that the elements $x^\alpha \in S_0$ come first, followed by the elements $x^\beta \in S_1 \cup \cdots \cup S_n$. This gives a square matrix M_0 such that

$$M_0 \begin{pmatrix} x^{\alpha_1} \\ x^{\alpha_2} \\ \vdots \\ x^{\beta_1} \\ x^{\beta_2} \\ \vdots \end{pmatrix} = \begin{pmatrix} x^{\alpha_1} f_0 \\ x^{\alpha_2} f_0 \\ \vdots \\ x^{\beta_1}/x_1^{d_1} f_1 \\ x^{\beta_2}/x_1^{d_1} f_1 \\ \vdots \end{pmatrix}, \qquad (2.22)$$

where, in the column on the left, the first two elements of S_0 and the first two elements of S_1 are listed explicitly. The situation is similar for the column on the right. Each entry of M_0 is either 0 or a coefficient of some f_i. In the literature, M_0 is called a *Sylvester-type matrix*.

We next partition M_0 so that the rows and columns of M_0 corresponding to elements of S_0 lie in the upper left hand corner. This gives

$$M_0 = \begin{pmatrix} M_{00} & M_{01} \\ M_{10} & M_{11} \end{pmatrix}, \tag{2.23}$$

where M_{00} is a $\mu \times \mu$ matrix for $\mu = d_1 \cdots d_n$, and M_{11} is also a square matrix. One can show that M_{11} is invertible for a generic choice of $f_1, \ldots, f_n$. Hence we can define the $\mu \times \mu$ matrix

$$\widetilde{M}_{f_0} = M_{00} - M_{01} M_{11}^{-1} M_{10}. \tag{2.24}$$

Also, given a point $p \in \mathbb{K}^n$, let $\mathbf{p}^\alpha$ be the column vector $(p^{\alpha_1}, p^{\alpha_2}, \ldots)^{\mathbf{t}}$ obtained by evaluating all monomials in S_0 at p (where $\mathbf{t}$ means transpose).

Theorem 2.3.2. *Let $f_1, \ldots, f_n$ be generic polynomials, where f_i has total degree d_i, and construct $\widetilde{M}_{f_0}$ as in (2.24) with $f_0 = a_1 x_1 + \cdots + a_n x_n$. Then $\mathbf{p}^\alpha$ is an eigenvector of $\widetilde{M}_{f_0}$ with eigenvalue $f_0(p)$ whenever p is a solution of (2.19). Furthermore, the vectors $\mathbf{p}^\alpha$ are linearly independent as p ranges over all solutions of (2.19).*

Proof. Let $\mathbf{p}^\beta$ be the column vector $(p^{\beta_1}, p^{\beta_2}, \ldots)^{\mathbf{t}}$ given by evaluating all monomials in $S_1 \cup \cdots \cup S_n$ at p. Then evaluating (2.22) at a solution p of (2.19) gives

$$M_0 \begin{pmatrix} \mathbf{p}^\alpha \\ \mathbf{p}^\beta \end{pmatrix} = \begin{pmatrix} f_0(p)\, \mathbf{p}^\alpha \\ 0 \end{pmatrix}.$$

In terms of (2.23), this becomes

$$\begin{pmatrix} M_{00} & M_{01} \\ M_{10} & M_{11} \end{pmatrix} \begin{pmatrix} \mathbf{p}^\alpha \\ \mathbf{p}^\beta \end{pmatrix} = \begin{pmatrix} f_0(p)\, \mathbf{p}^\alpha \\ 0 \end{pmatrix},$$

and it follows that

$$\widetilde{M}_{f_0}\, \mathbf{p}^\alpha = f_0(p)\, \mathbf{p}^\alpha. \tag{2.25}$$

(*Exercise: Prove this.*) Hence, for a solution p, $f_0(p)$ is an eigenvalue of $\widetilde{M}_{f_0}$ with eigenvector $\mathbf{p}^\alpha$. For generic $a_1, \ldots, a_n$, $f_0 = a_1 x_1 + \cdots + a_n x_n$ takes distinct values at the solutions, i.e., the eigenvalues $f_0(p)$ are distinct. This shows that the corresponding eigenvectors $\mathbf{p}^\alpha$ are linearly independent.

We can now solve (2.19) by the method of Section 2.1.3. We know that there are $\mu = d_1 \cdots d_n$ solutions p. Furthermore, the values $f_0(p)$ are distinct for a generic choice of $f_0 = a_1 x_1 + \cdots + a_n x_n$. Then Theorem 2.3.2 implies that the $\mu \times \mu$ matrix $\widetilde{M}_{f_0}$ has μ eigenvectors $\mathbf{p}^\alpha$. Hence all of the eigenspaces must have dimension 1, i.e., $\widetilde{M}_{f_0}$ is non-derogatory.

Also notice that $1 \in S_0$ by (2.21). It follows that we can assume that every $\mathbf{p}^\alpha$ is of the form

$$\mathbf{p}^\alpha = (1, p^{\alpha(2)}, \ldots, p^{\alpha(\mu)})^{\mathbf{t}}.$$

Thus, once we compute an eigenvector $\mathbf{v}$ of $\widetilde{M}_{f_0}$ for the eigenvalue $f_0(p)$, we know how to rescale $\mathbf{v}$ so that $\mathbf{v} = \mathbf{p}^\alpha$.

As in Section 2.1.3, the idea is to read off the solution p from the entries of the eigenvector $\mathbf{p}^\alpha$. If f_i has degree $d_i > 1$, then $x_i \in S_0$, so that p_i appears as a coordinate of $\mathbf{p}^\alpha$. Hence we can recover all coordinates of p_i except for those corresponding to equations with $d_i = 1$. These were called the "missing variables" in Section 2.1.3. In this situation, the missing variables correspond to linear equations. Since we can find the coordinates of the solution p for all of the other variables, we simply substitute these known values into the linear equations corresponding to the missing variables. Hence we find all coordinates of the solution by linear algebra. Details of this procedure are described in Exercise 5 of Section 3.6 of [CLO98].

This is all very nice but seems to ignore the quotient algebra

$$\mathcal{A} = \mathbb{K}[x_1, \ldots, x_n]/\langle f_1, \ldots, f_n \rangle.$$

In fact, what we did above has a deep relation to $\mathcal{A}$ as follows.

Theorem 2.3.3. *If $f_1, \ldots, f_n$ are generic polynomials, where f_i has total degree d_i, then the cosets of the monomials*

$$S_0 = \{x_1^{b_1} \cdots x_n^{b_n} \mid 0 \le b_i \le d_i - 1 \text{ for all } i\}$$

form a basis of the quotient algebra $\mathcal{A}$. Furthermore, if $f_0 = a_1 x_1 + \cdots + a_n x_n$ and $\widetilde{\mathsf{M}}_{f_0}$ is the matrix constructed in (2.24) using $f_0, f_1, \ldots, f_n$, then

$$\widetilde{\mathsf{M}}_{f_0} = \mathsf{M}_{f_0}^{\mathrm{t}},$$

where M_{f_0} is the matrix of the multiplication map $M_{f_0} : \mathcal{A} \to \mathcal{A}$ relative to the basis given by S_0.

Proof. Recall from Bézout's theorem that when $f_1, \ldots, f_n$ are generic, the equations (2.19) have $\mu = d_1 \cdots d_n$ solutions of multiplicity 1 in $\mathbb{K}^n$. It follows that $\mathcal{A}$ has dimension μ over $\mathbb{K}$. Since this is also the cardinality of S_0, the first part of the theorem will follow once we show that the cosets of the monomials in S_0 are linearly independent.

Write the elements of S_0 as $x^{\alpha(1)}, \ldots, x^{\alpha(\mu)}$ and suppose we have a linear relation among the cosets $[x^{\alpha(j)}]$, say

$$c_1[x^{\alpha(1)}] + \cdots + c_\mu[x^{\alpha(\mu)}] = 0.$$

Evaluating this equation at a solution p makes sense and implies that

$$c_1 p^{\alpha(1)} + \cdots + c_\mu p^{\alpha(\mu)} = 0. \tag{2.26}$$

In the generic case, our equations have $\mu = d_1 \cdots d_n$ solutions, so that (2.26) gives μ equations in μ unknowns $c_1, \ldots, c_\mu$. But the coefficients of the rows give the transposes of the vectors $\mathbf{p}^\alpha$, which are linearly independent by Theorem 2.3.2 It follows that $c_1 = \cdots = c_\mu = 0$. This proves that the cosets

$[x^{\alpha(1)}], \ldots, [x^{\alpha(\mu)}]$ are linearly independent. Thus S_0 gives a basis of $\mathcal{A}$ as claimed.

For the second assertion, observe from equation (2.3) that

$$M_{f_0}^t \, \mathbf{p}^\alpha = f_0(p) \, \mathbf{p}^\alpha$$

for each solution p. Comparing this to (2.25), we get

$$M_{f_0}^t \, \mathbf{p}^\alpha = \widetilde{M}_{f_0} \, \mathbf{p}^\alpha$$

for all solutions p. Since $f_1, \ldots, f_n$ are generic, we have μ solutions p, and the corresponding eigenvectors $\mathbf{p}^\alpha$ are linearly independent by Theorem 2.3.2. This implies $M_{f_0}^t = \widetilde{M}_{f_0}$.

The above proof of $M_{f_0}^t = \widetilde{M}_{f_0}$ requires that $\det(M_{11}) \neq 0$ and that all solutions have multiplicity 1. In Chapter 3, Theorem 3.5.1 will show that $M_{f_0}^t = \widetilde{M}_{f_0}$ holds under the weaker hypothesis that $\det(M_{11}) \neq 0$. (Note that the matrix M_0 defined in Section 3.5.1 is the transpose of our M_0. Thus the "Schur complement" defined in Theorem 3.5.1 is what we call $\widetilde{M}_{f_0}^t$.)

It is satisfying to see how the method described in this section relates to what we did in Section 2.1. However, there is a *lot* more going on here. Here are some items of interest.

Multiplication Matrices. By setting $f_0 = x_i$ in Theorem 2.3.3, we can construct the matrix of multiplication by x_i as $M_{x_i} = \widetilde{M}_{x_i}^t$. However, it is possible to compute all of these maps simultaneously by using $f_0 = u_1 x_1 + \cdots + u_n x_n$, where $u_1, \ldots, u_n$ are variables. In the decomposition (2.23), the matrices M_{10} and M_{11} don't involve the coefficients of f_0. Thus, we can still form the matrix $\widetilde{M}_{f_0}$ from (2.24), and it is easy to see that

$$\widetilde{M}_{f_0}^t = u_1 M_{x_1} + \cdots + u_n M_{x_n}.$$

Thus one computation gives all of the multiplication matrices M_{x_i}. See the discussion following Theorem 3.5.1 in Chapter 3 for an example.

Solving via Multivariate Factorization. As above, suppose that $f_0 = u_1 x_1 + \cdots + u_n x_n$, where $u_1, \ldots, u_n$ are variables. In this case, $\det(\widetilde{M}_{f_0})$ becomes a polynomial in $F[u_1, \ldots, u_n]$. The results of this section imply that for $f_1, \ldots, f_n$ generic, the eigenvalues of $\widetilde{M}_{f_0}$ are $f_0(p)$ as p ranges over all solutions of (2.19). Since all of the eigenspaces have dimension 1, we obtain

$$\det(\widetilde{M}_{f_0}) = \prod_p (u_1 p_1 + \cdots + u_n p_n). \tag{2.27}$$

It follows that if we can factor $\det(\widetilde{M}_{f_0})$ into irreducibles in $F[u_1, \ldots, u_n]$, then we get all solutions of (2.19). The general problem of multivariate factorization over an algebraically closed field will be discussed in Chapter 9. We will see in Section 2.3.2 that (2.27) is closely related to resultants.

Ideal Membership. Given $f \in \mathbb{K}[x_1,\ldots,x_n]$, how do we tell if $f \in \langle f_1,\ldots,f_n \rangle$? This is the *Ideal Membership Problem*. How do we do this without a Gröbner basis? One method (probably not very efficient) uses the above matrices M_{x_i} as follows:

$$f \in \langle f_1,\ldots,f_n \rangle \iff f(\mathsf{M}_{x_1},\ldots,\mathsf{M}_{x_n}) \text{ is the zero matrix.}$$

To prove this criterion, note that $f(\mathsf{M}_{x_1},\ldots,\mathsf{M}_{x_n}) = \mathsf{M}_f$ since the M_{x_i} commute. Then we are done since M_f is the zero matrix if and only if f is in the ideal. (*Exercise: Supply the details.*)

Sparse Polynomials. It is also possible to develop a sparse version of the solution method described in this section. The idea is that one fixes in advance the terms which appear in each f_i and then considers what happens when f_i is generic relative to these terms. One gets results similar to Theorems 2.3.2 and 2.3.3, and there are also nice relations to polyhedral geometry. This material is discussed in Chapter 7. See also [CLO98, Chapter 7].

Duality. The assumption that $f_1,\ldots,f_n$ have only finitely many solutions in $\mathbb{K}^n$ implies that these polynomials form a *regular sequence*. This allows us to apply the duality theory of complete intersections. There are also interesting relations with the multidimensional residues discussed in Chapters 1 and 3. This material is also covered in greater detail in [EM96] and [EM98].

2.3.2 Multivariate resultants

The classical multivariable resultant $\mathrm{Res}_{d_0,\ldots,d_n}$ in an irreducible polynomial in the coefficients of $n+1$ homogeneous polynomials

$$F_0,\ldots,F_n \in \mathbb{K}[x_0,\ldots,x_n]$$

of degrees $d_0,\ldots,d_n$ with the property that

$$\mathrm{Res}_{d_0,\ldots,d_n}(F_0,\ldots,F_n) = 0$$

if and only if the F_i have a common solution in the projective space $\mathbb{P}^n(\mathbb{K})$ (as usual, $\mathbb{K} = \overline{\mathbb{K}}$).

   This resultant has an affine version as follows. If we dehomogenize F_i by setting $x_0 = 1$, then we get polynomials $f_i \in \mathbb{K}[x_1,\ldots,x_n]$ of degree at most d_i. Since F_i and f_i have the same coefficients, we can write the resultant as

$$\mathrm{Res}_{d_0,\ldots,d_n}(f_0,\ldots,f_n).$$

Then the vanishing of this resultant means that the system of equations $f_0 = \cdots = f_n = 0$ has a solution either in $\mathbb{K}^n$ or "at ∞," i.e., a projective solution with $x_0 = 0$.

In the situation of Section 2.3.1, we have n polynomials $f_1, \ldots, f_n$ of degrees $d_1, \ldots, d_n$ in $x_1, \ldots, x_n$. To compute a resultant, we need one more polynomial. Not surprisingly, we will use

$$f_0 = a_1 x_1 + \cdots + a_n x_n.$$

We will usually assume $a_i \in \mathbb{K}$, though (as illustrated at the end of Section 2.3.1) it is sometimes useful to replace a_i with a variable u_i.

In order to compute the resultant $\mathrm{Res}_{1,d_1,\ldots,d_n}(f_0, f_1, \ldots, f_n)$, we need to study the behavior of the system $f_1 = \cdots = f_n = 0$ at ∞. Write

$$f_i = \sum_{j=0}^{d_i} f_{i,j}$$

where $f_{i,j}$ is homogeneous of degree j in $x_1, \ldots, x_n$. Then f_i homogenizes to

$$F_i = \sum_{j=0}^{d_i} f_{i,j} x_0^{d_i - j}$$

of degree d_i in $x_0, x_1, \ldots, x_n$. Then (2.19) has a *solution at* ∞ when the homogenized system

$$F_1 = \cdots = F_n = 0$$

has a nontrivial solution with $x_0 = 0$.

The following result relates the algebra $\mathcal{A} = \mathbb{K}[x_1, \ldots, x_n]/\langle f_1, \ldots, f_n \rangle$ to solutions at ∞.

Lemma 2.3.4. *The following are equivalent:*

$$f_1 = \cdots = f_n = 0 \text{ has no solutions at } \infty$$
$$\Longleftrightarrow \mathrm{Res}_{d_1,\ldots,d_n}(f_{1,d_1}, \ldots, f_{n,d_n}) \neq 0$$
$$\Longleftrightarrow \mathcal{A} \text{ has dimension } \mu = d_1 \cdots d_n \text{ over } \mathbb{K}.$$

Proof. Note that F_i reduces to f_{i,d_i} when $x_0 = 0$. Thus the f_i have a solution at ∞ if and only if the system of homogeneous equations

$$f_{1,d_1} = \cdots = f_{n,d_n} = 0$$

has a nontrivial solution. This gives the first equivalence. The second uses Bézout's theorem and some facts from algebraic geometry. See Section 3 of Chapter 3 of [CLO98] for the details.

When there are no solutions at ∞, it follows that we get our algebra $\mathcal{A}$ of dimension $d_1 \cdots d_n$ over $\mathbb{K}$. But unlike Section 2.3.1, the solutions may have multiplicities > 1. In this case, we can relate resultants and multiplication maps as follows.

Theorem 2.3.5. *If* $f_0 = u_1 x_1 + \cdots + u_n x_n$ *and* (2.19) *has no solutions at* ∞, *then*

$$
\begin{aligned}
&\mathrm{Res}_{1,d_1,\ldots,d_n}(f_0, f_1, \ldots, f_n) \\
&= \mathrm{Res}_{d_1,\ldots,d_n}(f_{1,d_1}, \ldots, f_{n,d_n}) \det(M_{f_0}) \\
&= \mathrm{Res}_{d_1,\ldots,d_n}(f_{1,d_1}, \ldots, f_{n,d_n}) \prod_p (u_1 p_1 + \cdots + u_n p_n)^{\mathrm{mult}(p)}.
\end{aligned}
$$

and

$$
\begin{aligned}
&\mathrm{Res}_{1,d_1,\ldots,d_n}(u - f_0, f_1, \ldots, f_n) \\
&= \mathrm{Res}_{d_1,\ldots,d_n}(f_{1,d_1}, \ldots, f_{n,d_n}) \, \mathrm{CharPoly}_{m_{f_0}}(u) \\
&= \mathrm{Res}_{d_1,\ldots,d_n}(f_{1,d_1}, \ldots, f_{n,d_n}) \prod_p \big(u - (u_1 p_1 + \cdots + u_n p_n)\big)^{\mathrm{mult}(p)}.
\end{aligned}
$$

Proof. In each case, the first equality uses Theorem 3.4 of [CLO98, Ch. 3] and the second uses Proposition 2.1.14 of this chapter.

While this is nice (and will have some unexpected consequences when we discuss Galois theory in Section 2.5), the relation between resultants and what we did in Section 2.3.1 goes much deeper. Here are some details.

Computing Resultants. First recall that the method given in Section 2.3.1 for computing the matrix $\mathtt{M}_{f_0}$ of the multiplication map M_{f_0} used the equality

$$
\widetilde{\mathtt{M}}_{f_0} = \mathtt{M}_{f_0}^{\mathrm{t}}
$$

from Theorem 2.3.3. As we noted after the proof, this equality requires that $\det(M_{11}) \neq 0$ (since $\det(M_{11})^{-1}$ was used in the formula for $\widetilde{\mathtt{M}}_{f_0}$ given in (2.24)). This relates to resultants as follows.

A standard method for computing $\mathrm{Res}_{1,d_1,\ldots,d_n}(f_0, f_1, \ldots, f_n)$ involves the quotient of two determinants. In our situation, the relevant formula is

$$
\det(M_0) = \mathrm{Res}_{1,d_1,\ldots,d_n}(f_0, f_1, \ldots, f_n) \det(M_0'), \tag{2.28}
$$

where M_0 is *precisely* the matrix appearing in (2.22) and M_0' is the submatrix described in Section 4 of Chapter 3 of [CLO98]. It follows that

$$
\mathrm{Res}_{1,d_1,\ldots,d_n}(f_0, f_1, \ldots, f_n) = \frac{\det(M_0)}{\det(M_0')}
$$

whenever $\det(M_0') \neq 0$. The subtle point is that $\det(M_0)$ and $\det(M_0')$ can both vanish even though $\mathrm{Res}_{1,d_1,\ldots,d_n}(f_0, f_1, \ldots, f_n)$ is nonzero. So to calculate the resultant using M_0, we definitely need $\det(M_0') \neq 0$. Yet for $\widetilde{\mathtt{M}}_{f_0}$, we need $\det(M_{11}) \neq 0$. Here is the nice relation between these determinants.

Proposition 2.3.6. $\det(M_{11}) = \mathrm{Res}_{d_1,\ldots,d_n}(f_{1,d_1}, \ldots, f_{n,d_n}) \det(M_0')$.

Proof. First observe that by (2.22) and the definition of $\widetilde{\mathsf{M}}_{f_0}$, we have

$$\det(M_0) = \det \begin{pmatrix} I & -M_{01}M_{11}^{-1} \\ 0 & I \end{pmatrix} \det \begin{pmatrix} M_{00} & M_{01} \\ M_{10} & M_{11} \end{pmatrix}$$

$$= \det \begin{pmatrix} \widetilde{\mathsf{M}}_{f_0} & 0 \\ M_{10} & M_{11} \end{pmatrix} = \det(\widetilde{\mathsf{M}}_{f_0}) \det(M_{11}).$$

whenever $\det(M_{11}) \neq 0$. Using this with (2.28) and Theorems 2.3.5 and 2.3.3, we obtain

$$\begin{aligned}
\det(\widetilde{\mathsf{M}}_{f_0}) \det(M_{11}) &= \det(M_0) \\
&= \mathrm{Res}_{1,d_1,\ldots,d_n}(f_0, f_1, \ldots, f_n) \det(M_0') \\
&= \mathrm{Res}_{d_1,\ldots,d_n}(f_{1,d_1}, \ldots, f_{n,d_n}) \det(\mathsf{M}_{f_0}) \det(M_0') \\
&= \mathrm{Res}_{d_1,\ldots,d_n}(f_{1,d_1}, \ldots, f_{n,d_n}) \det(\widetilde{\mathsf{M}}_{f_0}) \det(M_0')
\end{aligned}$$

when $f_1, \ldots, f_n$ are sufficiently generic. Cancelling $\det(\widetilde{\mathsf{M}}_{f_0})$ (which is nonzero generically) shows that the equality

$$\det(M_{11}) = \mathrm{Res}_{d_1,\ldots,d_n}(f_{1,d_1}, \ldots, f_{n,d_n}) \det(M_0')$$

holds generically. Since each side is a polynomial in the coefficients of the f_i, this equality must hold unconditionally.

We noted earlier that $\det(M_{11}) \neq 0$ implies that $\widetilde{\mathsf{M}}_{f_0}$ is defined and satisfies $\widetilde{\mathsf{M}}_{f_0} = \mathsf{M}_{f_0}^{\mathrm{t}}$. Then Proposition 2.3.6 shows that $\det(M_{11}) \neq 0$ also guarantees the following additional facts:

- $\mathrm{Res}_{d_1,\ldots,d_n}(f_{1,d_1}, \ldots, f_{n,d_n}) \neq 0$, so that (2.19) has no solutions at ∞.
- $\det(M_0') \neq 0$, so that $\mathrm{Res}_{1,d_1,\ldots,d_n}(f_0, f_1, \ldots, f_n)$ can be computed using M_0 and M_0'.

Hence the link between Section 2.3.1 and resultants is very strong.

For experts, we observe that (2.28) and Proposition 2.3.6 imply that if $\det(M_0') \neq 0$, then we have

$$\mathrm{Res}_{1,d_1,\ldots,d_n}(f_0, f_1, \ldots, f_n) = \frac{\det M_0}{\det M_0'}$$

$$\mathrm{Res}_{d_1,\ldots,d_n}(f_{1,d_1}, \ldots, f_{n,d_n}) = \frac{\det M_{11}}{\det M_0'},$$

where we observe that M_0' is a submatrix of both M_{11} and M_0. So M_0 allows us to compute not one but two resultants. Has this been noticed before?

Genericity. In Section 2.3.1, we required that $f_1, \ldots, f_n$ be "generic", which upon careful reading means first, that the system (2.19) has $d_1 \cdots d_n$ solutions of multiplicity 1, and second, that $\det(M_{11}) \neq 0$. In terms of resultants, this means the following:

- $\mathrm{Res}_{d_1,\ldots,d_n}(f_{1,d_1},\ldots,f_{n,d_n}) \neq 0.$
- $\mathrm{Res}_{d-1,d_1,\ldots,d_n}(\det\left(\frac{\partial f_i}{\partial x_j}\right), f_1,\ldots,f_n) \neq 0,$ where d is defined in (2.20).
- $\det(M_0') \neq 0.$

The first item guarantees that $\mathcal{A}$ has the correct dimension by Lemma 2.3.4 and the second guarantees that the Jacobian is nonvanishing at all solutions, so that every solution has multiplicity 1 by the implicit function theorem. Finally, the first and third conditions are equivalent to $\det(M_{11}) \neq 0$ by Proposition 2.3.6.

One historical remark is that while the formula (2.28) is due to Macaulay in 1902, many ideas of Section 2.3.2 are present in the work of Kronecker in 1882. For example, Kronecker defines

$$\mathrm{Res}_{1,d_1,\ldots,d_n}(u - f_0, f_1, \ldots, f_n)$$

and shows that as a polynomial in u, its roots are $f_0(p)$ for p a solution of (2.19). He also notes that the discriminant condition of the second bullet is needed to get solutions of multiplicity 1 and that when this is true, the ideal $\langle f_1, \ldots, f_n \rangle$ is radical (see pp. 276 and 330 of [Kro31, Vol. II]).

2.4 Factoring

In earlier sections of this chapter, we studied quotient algebras

$$\mathcal{A} = \mathbb{K}[x_1, \ldots, x_n]/\langle f_1, \ldots, f_s \rangle$$

where $\mathbb{K}$ was usually algebraically closed. Here, we will work over more general fields and consider two factorization problems: factoring polynomials into a product of irreducible polynomials and factoring an ideal into an intersection of primary ideals (primary decomposition). We will also study the Theorem of the Primitive Element.

2.4.1 Factoring over number fields

Near the end of Section 2.3.1, we gave the formula (2.27)

$$\det(\widetilde{\mathrm{M}}_{f_0}) = \prod_p (u_1 p_1 + \cdots + u_n p_n),$$

where $f_0 = u_1 x_1 + \cdots + u_n x_n$ and the product is over all solutions p. The point was that we could compute the left-hand side, so that *if* we knew how to factor multivariable polynomials over an algebraically closed field, *then* we could find all of the solutions. We will now turn the tables and use finite commutative algebras and their multiplication maps to do factoring over number fields. We begin with a lovely result of Dedekind.

Dedekind Reciprocity. Suppose that $f(x), g(x) \in \mathbb{Q}[x]$ are irreducible with roots $\alpha, \beta \in \mathbb{C}$ such that $f(\alpha) = g(\beta) = 0$. Then factor $f(x)$ into irreducibles over $\mathbb{Q}(\beta)$, say

$$f(x) = f_1(x) \cdots f_r(x), \quad f_i(x) \in \mathbb{Q}(\beta)[x]. \tag{2.29}$$

The $f_i(x)$ are distinct (i.e., none is a constant multiple of any of the others) since f is separable. Then the *Dedekind Reciprocity Theorem* describes the factorization of $g(x)$ over $\mathbb{Q}(\alpha)$ as follows.

Theorem 2.4.1. *Given the above factorization of $f(x)$ into irreducibles over $\mathbb{Q}(\beta)$, the factorization of $g(x)$ into irreducibles over $\mathbb{Q}(\alpha)$ can be written as*

$$g(x) = g_1(x) \cdots g_r(x), \quad g_i(x) \in \mathbb{Q}(\alpha)[x]$$

where

$$\frac{\deg(f_1)}{\deg(g_1)} = \frac{\deg(f_2)}{\deg(g_2)} = \cdots = \frac{\deg(f_r)}{\deg(g_r)} = \frac{\deg(f)}{\deg(g)}.$$

Proof. Consider the $\mathbb{Q}$-algebra

$$\mathcal{A} = \mathbb{Q}[x, y]/\langle f(x), g(y) \rangle.$$

Since $y \mapsto \beta$ induces $\mathbb{Q}[y]/\langle g(y) \rangle \simeq \mathbb{Q}(\beta)$ and the $f_i(x)$ are distinct irreducibles (because $f(x)$ is separable), we get algebra isomorphisms

$$\begin{aligned}
\mathcal{A} &\simeq \mathbb{Q}(\beta)[x]/\langle f(x) \rangle \\
&\simeq \mathbb{Q}(\beta)[x]/\langle f_1(x) \cdots f_r(x) \rangle \\
&\simeq \prod_{i=1}^{r} \underbrace{\mathbb{Q}(\beta)[x]/\langle f_i(x) \rangle}_{\mathbb{K}_i},
\end{aligned} \tag{2.30}$$

where $\mathbb{K}_i$ is a field since $f_i(x)$ is irreducible over $\mathbb{Q}(\beta)$. Thus $\mathcal{A}$ is isomorphic to the product of the fields $\mathbb{K}_1, \ldots, \mathbb{K}_r$. Furthermore, since $[\mathbb{K}_i : \mathbb{Q}(\beta)] = \deg(f_i)$, the degree of $\mathbb{K}_i$ over $\mathbb{Q}$ is

$$[\mathbb{K}_i : \mathbb{Q}] = [\mathbb{K}_i : \mathbb{Q}(\beta)][\mathbb{Q}(\beta) : \mathbb{Q}] = \deg(f_i) \deg(g). \tag{2.31}$$

Now interchange the roles of f and g. The factorization of $g(y)$ into s irreducibles $g_i(y)$ over $\mathbb{Q}(\alpha) \simeq \mathbb{Q}[x]/\langle f(x) \rangle$ gives an isomorphism between $\mathcal{A}$ and a product of s fields $\mathcal{A} \simeq \prod_{i=1}^{s} \mathbb{K}_i'$ such that

$$[\mathbb{K}_i' : \mathbb{Q}] = \deg(g_i) \deg(f). \tag{2.32}$$

However, the decomposition of $\mathcal{A}$ into a product of fields is unique up to isomorphism (this is proved by studying the idempotents of $\mathcal{A}$). Hence we must have $r = s$ and $\mathbb{K}_i \simeq \mathbb{K}_i'$ after a suitable permutation of indices. It follows that (2.31) and (2.32) must equal for all i, and the result follows.

According to [Edw04], Dedekind discovered this result in 1855, though his version wasn't published until 1982. Kronecker found this theorem independently and stated it in his university lectures. Theorem 2.4.1 was first published by Kneser in 1887.

A Factorization Algorithm. The algebra $\mathcal{A}$ that we used in the proof of Theorem 2.4.1 can also be used to construct the factorization of $f(x)$ over $\mathbb{Q}(\beta)$. The idea is to compute

$$\Phi(u) = \mathrm{CharPoly}_{M_{f_0}}(u), \quad f_0 = x + ty,$$

for an appropriately chosen $t \in \mathbb{Q}$. This polynomial in $\mathbb{Q}[u]$ can be computed using the methods of Section 2.1 and factored using known algorithms for factoring polynomials in $\mathbb{Q}[u]$. Kronecker observed that these factors determine the factorization of $f(x)$ over $\mathbb{Q}(\beta)$. Here is his result.

Theorem 2.4.2. *Assume that $f_0 = x + ty$ takes distinct values at the solutions of $f(x) = g(y) = 0$ and let*

$$\Phi(u) = \prod_{i=1}^{r} \Phi_i(u)$$

be the irreducible factorization of $\Phi(u)$ in $\mathbb{Q}[u]$. Then the irreducible factorization of $f(x)$ over $\mathbb{Q}(\beta)$ is

$$f(x) = c\, f_1(x) \cdots f_r(x),$$

where $c \in K^*$ and

$$f_i(x) = \gcd(\Phi_i(x + t\beta), f(x)).$$

(Note that the gcd is computed in $\mathbb{Q}(\beta)[x]$.)

Proof. If $n = \deg(f(x))$ and $m = \deg(g(y))$, then Bézout's theorem implies that the equations $f(x) = g(y) = 0$ have at most nm solutions in x and y counted with multiplicity. But in fact there are exactly nm solutions since $f(x)$ and $g(y)$ are separable. It follows that all of the multiplicities are 1. We now apply the methods of Section 2.1.

By assumption, $f_0 = x + ty$ takes distinct values at all solutions of $f(x) = g(y) = 0$. Since they have multiplicity 1, the multiplication map $M_{f_0} : \mathcal{A} \to \mathcal{A}$ is non-derogatory, where $\mathcal{A} = \mathbb{Q}[x, y]/\langle f(x), g(y) \rangle$. Then the single-variable representation (Proposition 2.1.12) implies that $u \mapsto [x + ty] \in \mathcal{A}$ induces

$$\mathbb{Q}[u]/\langle \Phi(u) \rangle \simeq \mathcal{A}$$

since $\Phi(u)$ is the characteristic polynomial of multiplication by $f_0 = x + ty$ on the algebra $\mathcal{A}$. Notice also that

$$\mathrm{Disc}(\Phi(u)) \neq 0$$

since the eigenvalues all have multiplicity 1. This implies that the above factorization of $\Phi(u)$ is a product of distinct irreducibles.

By the Chinese Remainder Theorem, such a factorization gives a decomposition

$$\mathcal{A} \simeq \mathbb{Q}[u]/\langle \Phi(u) \rangle \simeq \prod_{i=1}^{r} \mathbb{Q}[u]/\langle \Phi_i(u) \rangle$$

into a product of fields. Using the definition of $\mathcal{A}$, this transforms into the product of fields given by

$$\mathcal{A} = \mathbb{Q}[x,y]/\langle g(y), f(x) \rangle \simeq \prod_{i=1}^{r} \mathbb{Q}[x,y]/\langle g(y), f(x), \Phi_i(x+ty) \rangle. \qquad (2.33)$$

(*Exercise: Prove this.*) Since $y \mapsto \beta$ induces $\mathbb{Q}[y]/\langle g(y) \rangle \simeq \mathbb{Q}(\beta)$, we can rewrite (2.33) as a product of fields

$$\mathcal{A} \simeq \prod_{i=1}^{r} \mathbb{Q}(\beta)[x]/\langle f(x), \Phi_i(x+t\beta) \rangle. \qquad (2.34)$$

However, $\mathbb{Q}(\beta)[x]$ is a PID, so that $\langle f(x), \Phi_i(x+t\beta) \rangle$ is the principal ideal generated by $f_i(x) = \gcd(f(x), \Phi_i(x+t\beta))$. Since each factor in the product in (2.34) is a field, we see that $f_i(x)$ is irreducible over $\mathbb{Q}(\beta)$. Furthermore, each $f_i(x)$ divides $f(x)$. It remains to show that $f(x)$ equals $f_1(x) \cdots f_r(x)$ up to a constant.

We proved above that $\Phi(u)$ has distinct roots, so that the same is true for $\Phi(x+t\beta)$. It follows that in the factorization $\Phi(x+t\beta) = \prod_{i=1}^{r} \Phi_i(x+t\beta)$, the factors $\Phi_i(x+t\beta)$ have distinct roots as we vary i. Hence the same is true for the $f_i(x)$. Hence their product divides $f(x)$ since each one does. However, using (2.34) and degree calculations to those of Theorem 2.4.1, we see that

$$\dim_{\mathbb{K}} \mathcal{A} = \sum_{i=1}^{r} \deg(f_i) \deg(g).$$

(*Exercise: Supply the details.*) Since $\dim_{\mathbb{K}} \mathcal{A} = nm = \deg(f) \deg(g)$, we see that $\deg(f) = \deg(f_1 \cdots f_r)$, and the theorem follows.

This theorem leads to the following algorithm for factoring $f(x)$ over $\mathbb{Q}(\beta)$:

- Pick a random $t \in \mathbb{Q}$ and compute $\Phi(u) = \text{CharPoly}_{M_{f_0}}(u)$ for $f_0 = x + ty$. Also compute $\text{Disc}(\Phi(u))$.
- If $\text{Disc}(\Phi(u)) \neq 0$, then factor $\Phi(u) = \prod_{i=1}^{r} \Phi_i(u)$ into irreducibles in $\mathbb{Q}[u]$ and for each i compute $\gcd(\Phi_i(x+t\beta), f(x))$ in $\mathbb{Q}(\beta)[x]$. This gives the desired factorization.
- If $\text{Disc}(\Phi(u)) = 0$, then pick a new $t \in \mathbb{Q}$ and return to the first bullet.

Since $\mathrm{Disc}(\Phi(u)) \neq 0$ if and only if $x + ty$ takes distinct values at the solutions of $f(x) = g(y) = 0$, Theorem 2.4.2 implies that the second bullet correctly computes the required factorization when the discriminant is nonzero. Notice that the second bullet uses the Euclidean algorithm in $\mathbb{Q}(\beta)[x]$, which can be done constructively using the representation $\mathbb{Q}(\beta) \simeq \mathbb{Q}[y]/\langle g(y) \rangle$.

As for the third bullet, the number of $t \in \mathbb{Q}$ that satisfy the equation $\mathrm{Disc}(\Phi(u)) = 0$ is bounded above by $\frac{1}{2}nm(nm-1)$, where $n = \deg(f(x))$, $m = \deg(g(y))$. (*Exercise: Prove this.*) Thus the third bullet can occur at most $\frac{1}{2}nm(nm-1)$ times. It follows that the above algorithm is deterministic.

An alternate approach would be to follow what Kronecker does on pages 258–259 of [Kro31, Vol. II] and regard t as a variable in $f_0 = x + ty$. Then $\Phi(u)$ becomes a polynomial $\Phi(u,t) \in \mathbb{Q}[x,t]$. If one can factors $\Phi(u,t)$ into irreducibles $\mathbb{Q}[u,t]$, say $\Phi(u,t) = \prod_{i=1}^{r} \Phi_i(u,t)$, then it is straightforward to recover $f_i(x,\beta)$ from $\Phi_i(x + t\beta, t)$. A rigorously constructive version of this is described in [Edw04].

2.4.2 Finite fields and primitive elements

Here, we will give two further applications of finite commutative algebras.

Factoring over Finite Fields. We begin with a brief description for factoring a polynomial $f(x) \in \mathbb{F}_q[x]$, where $\mathbb{F}_q$ is a finite field with $q = p^\ell$ elements. We will use the algebra

$$\mathcal{A} = \mathbb{F}_q[x]/\langle f(x) \rangle$$

and the Frobenius map

$$\mathrm{Frob} : \mathcal{A} \to \mathcal{A}, \quad \mathrm{Frob}(a) = a^q.$$

This map is linear over $\mathbb{F}_q$ and has 1 as an eigenvalue since $1^q = 1$. As in Section 2.1, $E_{\mathcal{A}}(\mathrm{Frob}, 1)$ denotes the corresponding eigenspace. This eigenspace determines whether or not $f(x)$ is irreducible as follows.

Proposition 2.4.3. *If $f(x)$ has no multiple roots, i.e., $\gcd(f(x), f'(x)) = 1$, then the dimension of the eigenspace $E_{\mathcal{A}}(\mathrm{Frob}, 1)$ is the number of irreducible factors of $f(x)$.*

Proof. Since $f(x)$ has no multiple roots, a factorization $f(x) = f_1(x) \cdots f_r(x)$ into irreducible polynomials in $\mathbb{F}_q[x]$ gives an algebra isomorphism

$$\mathcal{A} \simeq \prod_{i=1}^{r} \mathbb{K}_i, \quad \mathbb{K}_i = \mathbb{F}_q[x]/\langle f_i(x) \rangle,$$

compatible with the Frobenius map Frob. If $a \in \mathbb{K}_i$, then since $\mathbb{K}_i$ is a field, we have the equivalences

$$\mathrm{Frob}(a) = a \iff a^q = a \iff a \in \mathbb{F}_q.$$

It follows that on $\mathbb{K}_i$, the eigenvalue 1 has a 1-dimensional eigenspace $E_{\mathbb{K}_i}(\text{Frob}, 1)$. Since the eigenspace $E_{\mathcal{A}}(\text{Frob}, 1)$ is the direct sum of the $E_{\mathbb{K}_i}(\text{Frob}, 1)$, the result follows.

Here is a simple example of this result.

Example 2.4.4. Let $f(x) = x^5 + x^4 + 1 \in \mathbb{F}_2[x]$. One easily sees that $f(x)$ is separable. Then $\mathcal{A} = \mathbb{F}_2[x]/\langle f(x)\rangle$ is a vector space over $\mathbb{F}_2$ of dimension 5 with basis $[1], [x], [x^2], [x^3], [x^4]$, which for simplicity we write as $1, x, x^2, x^3, x^4$.

Note that Frob : $\mathcal{A} \to \mathcal{A}$ is the squaring map since $q = 2$. To compute the matrix of Frob, we apply Frob to each basis element and represent the result in terms of the basis:

$$1 \mapsto 1$$
$$x \mapsto x^2$$
$$x^2 \mapsto x^4$$
$$x^3 \mapsto x^6 = 1 + x + x^4$$
$$x^4 \mapsto x^8 = 1 + x + x^2 + x^3 + x^4.$$

Here, $x^6 = 1 + x + x^4$ means that $1 + x + x^4$ is the remainder of x^6 on division by $f(x) = x^5 + x^4 + 1$, and similarly for the last line. Hence the matrix of Frob $- 1_{\mathcal{A}}$ is

$$\begin{pmatrix} 1&0&0&1&1 \\ 0&0&0&1&1 \\ 0&1&0&0&1 \\ 0&0&0&0&1 \\ 0&0&1&1&1 \end{pmatrix} - \begin{pmatrix} 1&0&0&0&0 \\ 0&1&0&0&0 \\ 0&0&1&0&0 \\ 0&0&0&1&0 \\ 0&0&0&0&1 \end{pmatrix} = \begin{pmatrix} 0&0&0&1&1 \\ 0&1&0&1&1 \\ 0&1&1&0&1 \\ 0&0&0&1&1 \\ 0&0&1&1&0 \end{pmatrix}$$

(remember that we are in characteristic 2). This matrix has rank 3 since the first column is zero and the sum of the last three columns is zero. (*Exercise: Check the rank carefully.*) Hence we have two linearly independent eigenvectors for the eigenvalue 1. By Proposition 2.4.3, $f(x)$ is not irreducible over $\mathbb{F}_2$.

Besides giving the number of irreducible factors of $f(x)$, one can also use the eigenspace $E_{\mathcal{A}}(\text{Frob}, 1)$ to construct the irreducible factorization of $f(x)$. The rough idea is that if $[h(x)] \in \mathcal{A}$ is a nonzero element of $E_{\mathcal{A}}(\text{Frob}, 1)$, then $\gcd(h(x), f(x))$ is a factor of f and (if $h(x)$ chosen correctly) is actually one of the irreducible factors of $f(x)$. This is Berlekamp's algorithm, which is described in Section 4.1 of [LN83].

Theorem of the Primitive Element. The single-variable representation used in the proof of Theorem 2.4.2 may remind the reader of the Theorem of the Primitive Element. As we will now show, this is no accident.

Theorem 2.4.5. *Let $\mathbb{K} \subset \mathbb{L} = \mathbb{K}(\alpha_1, \ldots, \alpha_n)$ be an extension such that $\mathbb{K}$ is infinite and each α_i is separable over $\mathbb{K}$. Then there are $t_1 \ldots, t_n \in \mathbb{K}$ such that*

$$\mathbb{L} = \mathbb{K}(\alpha), \quad \alpha = t_1\alpha_1 + \cdots + t_n\alpha_n.$$

Proof. Let f_i be the minimal polynomial of α_i over $\mathbb{K}$ and let

$$\mathcal{A} = \mathbb{K}[x_1, \ldots, x_n]/\langle f_1(x_1), \ldots, f_n(x_n)\rangle.$$

Note that we use a separate variable x_i for each polynomial f_i. Then arguing as in the proof of Theorem 2.4.2, one easily sees that by Bézout's theorem, all solutions of

$$f_1(x_1) = f_2(x_2) = \cdots = f_n(x_n) = 0 \tag{2.35}$$

have multiplicity 1. Since $\mathbb{K}$ is infinite, we can pick $t_1, \ldots, t_n \in \mathbb{K}$ such that $f_0 = t_1x_1 + \cdots + t_nx_n$ takes distinct values at all solutions of (2.35). It follows that $M_{f_0} : \mathcal{A} \to \mathcal{A}$ is non-derogatory, so that by Proposition 2.1.12, the map $u \mapsto [t_1x_1 + \cdots + t_nx_n] \in \mathcal{A}$ induces a surjection

$$\mathbb{K}[u] \longrightarrow \mathcal{A}.$$

Furthermore, the map $\mathcal{A} \to \mathbb{L}$ induced by $x_i \mapsto \alpha_i$ is surjective since $\mathbb{L} = \mathbb{K}(\alpha_1, \ldots, \alpha_n)$. The theorem follows since the composition $\mathbb{K}[u] \to \mathcal{A} \to \mathbb{L}$ is surjective and maps u to $\alpha = t_1\alpha_1 + \cdots + t_n\alpha_n$.

Here is an example to illustrate the role of separability.

Example 2.4.6. Let $\mathbb{K} = \mathbb{F}_p(t, u)$, where t and u are variables, and let $\mathbb{K} \subset \mathbb{L}$ be the field obtained by adjoining the the pth roots of t and u. This extension is purely inseparable of degree p^2 and $\mathbb{L} \neq \mathbb{K}(\alpha)$ for all $\alpha \in \mathbb{L}$ since $\alpha \in \mathbb{L}$ implies that $\alpha^p \in \mathbb{K}$. (*Exercise: Prove these assertions carefully.*)

Hence the single-variable representation of Proposition 2.1.12 must fail. To see the underlying geometric reason for this failure, first observe that

$$\mathbb{L} \simeq \mathbb{K}[x, y]/\langle x^p - t, y^p - u\rangle$$

is the algebra from the proof of Theorem 2.4.5. In the algebraic closure $\overline{\mathbb{K}}$ of $\mathbb{K}$, the only solution of

$$x^p - t = y^p - u = 0$$

is given by $x = \sqrt[p]{t}$ and $y = \sqrt[p]{u}$. The local ring at this point is

$$\overline{\mathbb{K}}[x, y]/\langle x^p - t, y^p - u\rangle = \overline{\mathbb{K}}[x, y]/\langle (x - \sqrt[p]{t})^p, (y - \sqrt[p]{u})^p\rangle \simeq \overline{\mathbb{K}}[x, y]/\langle x^p, y^p\rangle,$$

which clearly has embedding dimension 2 and hence is not curvilinear. It follows that M_f is derogatory for *all* $f \in \mathbb{K}[x, y]$. Since the single-variable representation requires that M_f be non-derogatory, we can see why the Theorem of the Primitive Element fails in this case.

2.4.3 Primary decomposition

The final task of Section 2.4 is to extend the factorizations introduced in Section 2.4.1 to the realm of ideals. Suppose that

$$f_1(x_1, \ldots, x_n) = \cdots = f_s(x_1, \ldots, x_n) = 0 \qquad (2.36)$$

is a system of equations with coefficients in a field $\mathbb{K}$ and only finitely many solutions over the algebraic closure $\overline{\mathbb{K}}$. (Thus we are back in the situation where the number of equations need not equal the number of variables.) We say that $\langle f_1, \ldots, f_s \rangle$ is *zero-dimensional* since a finite set of points has dimension 0. Our goal is to give an algorithm for computing the primary decomposition of a zero-dimensional ideal.

Theoretical Results. An ideal $I \subset \mathbb{K}[x_1, \ldots, x_n]$ is *primary* if $fg \in I$ always implies that either $f \in I$ or $g^N \in I$ for some $N \geq 1$. It is easy to see that the radical $\sqrt{I}$ of a primary ideal is prime. By Chapter 4, §7 of [CLO97], every ideal $I \subset \mathbb{K}[x_1, \ldots, x_n]$ has a *primary decomposition*

$$I = I_1 \cap \cdots \cap I_r \qquad (2.37)$$

into an intersection of primary ideals. We say that (2.37) is *minimal* when r is as small as possible.

In the zero-dimensional case, the primary components I_i of $\langle f_1, \ldots, f_s \rangle$ can be obtained from the given ideal by adding one more carefully chosen polynomial u_i. Here is the precise result.

Lemma 2.4.7. *A zero-dimensional ideal* $\langle f_1, \ldots, f_s \rangle$ *has a minimal primary decomposition*

$$\langle f_1, \ldots, f_s \rangle = I_1 \cap \cdots \cap I_r$$

such that $\sqrt{I_1}, \ldots, \sqrt{I_r}$ *are distinct maximal ideals. Furthermore, for each i,*

$$I_i \not\subset \bigcup_{j \neq i} \sqrt{I_j},$$

and any $u_i \in I_i \setminus \bigcup_{j \neq i} \sqrt{I_j}$ *has the property that*

$$I_i = \langle f_1, \ldots, f_s, u_i \rangle.$$

Proof. Let $\langle f_1, \ldots, f_s \rangle = I_1 \cap \cdots \cap I_r$ be a minimal primary decomposition. Note that I_i and hence $\sqrt{I_i}$ are zero-dimensional since $\langle f_1, \ldots, f_s \rangle$ is. We also know that $\sqrt{I_i}$ is prime. But zero-dimensional prime ideals are maximal. (*Exercise: Prove this.*) Hence the $\sqrt{I_i}$ are maximal. Furthermore, if $\sqrt{I_i} = \sqrt{I_j}$ for some $i \neq j$, then $I_i \cap I_j$ is primary (*Exercise: Supply a proof.*) This contradicts the minimality of our representation. Hence the $\sqrt{I_i}$ are distinct.

If $I_i \subset \bigcup_{j \neq i} \sqrt{I_j}$, then $I_i \subset \sqrt{I_j}$ for some $j \neq i$ by the Prime Avoidance Theorem ([Sha90, Th. 3.61]). This implies that $\sqrt{I_i} \subset \sqrt{I_j}$ and hence

$\sqrt{I_i} = \sqrt{I_j}$ since the radicals are maximal. This contradiction proves that $I_i \not\subset \bigcup_{j \neq i} \sqrt{I_j}$.

Now let $u_i \in I_i \setminus \bigcup_{j \neq i} \sqrt{I_j}$. Then we certainly have $\langle f_1, \ldots, f_s, u_i \rangle \subset I_i$. For the opposite inclusion, take $j \neq i$ and note that $u_i \notin \sqrt{I_j}$ implies that $1 + u_i r_j \in \sqrt{I_j}$ for some r_j since $\sqrt{I_j}$ is maximal. (*Exercise: Prove this.*) Thus $(1 + u_i r_j)^{N_j} \in I_j$ for some $N_j \geq 1$. Expanding the product

$$\prod_{j \neq i} (1 + u_i r_j)^{N_j} \in \prod_{j \neq i} I_j \subset \bigcap_{j \neq i} I_j,$$

we see that $1 + u_i r \in \bigcap_{j \neq i} I_j$ for some r. Now take $a \in I_i$. Then

$$a(1 + u_i r) \in I_i \cap \bigcap_{j \neq i} I_j = \langle f_1, \ldots, f_s \rangle.$$

Hence $a = a(1 + u_i r) + u_i(-ar) \in \langle f_1, \ldots, f_s \rangle + \langle u_i \rangle = \langle f_1, \ldots, f_s, u_i \rangle$, as desired.

In the zero-dimensional case, one can also prove that the ideals I_i in the primary decomposition are unique. For general ideals, uniqueness need not hold (see Exercise 6 of Chapter 4, §7 of [CLO97] for an example) due to the phenomenon of *embedded components*.

The most commonly used algorithm for computing the primary decomposition of a zero-dimensional ideal is described in [GTZ88] and uses Gröbner bases plus a change of coordinates to find the u_i of Lemma 2.4.7. However, the recent paper [Mon02] of C. Monico shows how to find the u_i using the quotient algebra

$$\mathcal{A} = \mathbb{K}[x_1, \ldots, x_n]/\langle f_1, \ldots, f_s \rangle.$$

We will describe Monico's method, beginning with the following special case.

The Rational Case. The solutions of (2.36) are *rational over* $\mathbb{K}$ if all solutions in $\overline{\mathbb{K}}^n$ actually lie in $\mathbb{K}^n$. In this situation, it is easy to see that the primary decomposition is

$$\langle f_1, \ldots, f_s \rangle = \bigcap_p I_p,$$

where the intersection is over all solutions p of (2.36). Furthermore, as we noted in (2.6), the primary component I_p is

$$I_p = \{ f \in \mathbb{K}[x_1, \ldots, x_n] \mid gf \in \langle f_1, \ldots, f_s \rangle \ \exists g \in \mathbb{K}[x_1, \ldots, x_n] \text{ with } g(p) \neq 0 \},$$

and $\sqrt{I_p}$ is the maximal ideal $\langle x_1 - p_1, \ldots, x_n - p_n \rangle$ when $p = (p_1, \ldots, p_n)$. Unfortunately, this elegant description of I_p is not useful for computational purposes. But we can use the methods of Section 2.1 to find the polynomials u_i of Lemma 2.4.7 as follows.

Proposition 2.4.8. *Suppose that $\langle f_1, \ldots, f_s \rangle$ is zero-dimensional and all solutions of (2.36) are rational over $\mathbb{K}$. If $f \in \mathbb{K}[x_1, \ldots, x_n]$ takes distinct values at the solutions of (2.36), then for each solution p, the corresponding primary component is*

$$I_p = \langle f_1, \ldots, f_s, (f - f(p))^{\mathrm{mult}(p)} \rangle.$$

Proof. Let $u_p = (f - f(p))^{\mathrm{mult}(p)}$. By Lemma 2.4.7, it suffices to show that $u_p \in I_p$ and $u_p \notin \sqrt{I_q}$ for all solutions $q \neq p$. Since $\sqrt{I_q}$ is the maximal ideal of q, the latter condition is equivalent to the non-vanishing of u_p at q, which follows since f takes distinct values at the solutions.

To prove that $u_p \in I_p$, let $v_p = \prod_{q \neq p} (f - f(q))^{\mathrm{mult}(q)}$. By Proposition 2.1.14,

$$u_p v_p = \mathrm{CharPoly}_{M_f}(f). \tag{2.38}$$

However, the Cayley-Hamilton theorem tells us that $\mathrm{CharPoly}_{M_f}(M_f)$ is the zero operator on $\mathcal{A}$. Applied to $[1] \in \mathcal{A}$, we obtain

$$[0] = \mathrm{CharPoly}_{M_f}(M_f)[1] = \mathrm{CharPoly}_{M_f}([f]) = [\mathrm{CharPoly}_{M_f}(f)].$$

Combined with (2.38), this implies

$$u_p v_p = \mathrm{CharPoly}_{M_f}(f) \in \langle f_1, \ldots, f_s \rangle \subset I_p.$$

Since I_p is primary, either u_p or some power of v_p lies in I_p. But

$$v_p(p) = \prod_{q \neq p} (f(p) - f(q))^{\mathrm{mult}(q)} \neq 0$$

since f takes distinct values at the solutions. Hence no power of v_p lies in I_p, so that $u_p \in I_p$. ∎

Here is an example of this proposition.

Example 2.4.9. Consider the ideal $\langle x^2 + 2y^2 - 2y, xy^2 - xy, y^3 - 2y^2 + y \rangle \subset \mathbb{Q}[x, y]$. We saw in Example 2.1.1 that the corresponding equations have solutions $(0, 0)$ and $(0, 1)$, which are rational over $\mathbb{Q}$. Since y takes distinct values at the solutions, we can use $f = y$ in Proposition 2.4.8 to compute the primary decomposition.

By Example 2.1.5, the characteristic polynomial of m_y is $u^2(u - 1)^3$. It follows that the primary components are

$$I_{(0,0)} = \langle x^2 + 2y^2 - 2y, xy^2 - xy, y^3 - 2y^2 + y, y^2 \rangle = \langle x^2, y \rangle$$
$$I_{(0,1)} = \langle x^2 + 2y^2 - 2y, xy^2 - xy, y^3 - 2y^2 + y, (y-1)^3 \rangle$$
$$= \langle x^2 + 2(y-1), x(y-1), (y-1)^2 \rangle.$$

(Exercise: Verify the final equality using the congruences

$$y(y-1)^2 \equiv (y-1)^2 \bmod (y-1)^3 \quad and \quad y(y-1) \equiv y - 1 \bmod (y-1)^2.)$$

Putting these together, we obtain the primary decomposition

$$\langle x^2 + 2y^2 - 2y, xy^2 - xy, y^3 - 2y^2 + y\rangle$$
$$= \langle x^2, y\rangle \bigcap \langle x^2 + 2(y-1), x(y-1), (y-1)^2\rangle$$
$$= I_{(0,0)} \bigcap I_{(0,1)}$$

given in Example 2.2.15.

We note that in Proposition 2.4.8, one can replace the characteristic polynomial with the minimal polynomial. Here is the precise result.

Proposition 2.4.10. *Suppose that $\langle f_1, \ldots, f_s\rangle$ is zero-dimensional and all solutions of (2.36) are rational over $\mathbb{K}$. If $f \in \mathbb{K}[x_1, \ldots, x_n]$ takes distinct values at the solutions of (2.36), then for each solution p, the corresponding primary component is*

$$I_p = \langle f_1, \ldots, f_s, (f - f(p))^{n(p)}\rangle,$$

where $\mathrm{MinPoly}_{M_f}(u) = \prod_p (u - f(p))^{n(p)}$.

(*Exercise: Prove this proposition.*) Here is an example.

Example 2.4.11. For the ideal of Example 2.4.9, recall from Example 2.1.5 that the minimal polynomial of y is $u(u-1)^2$. Thus

$$I_{(0,0)} = \langle x^2 + 2y^2 - 2y, xy^2 - xy, y^3 - 2y^2 + y, y\rangle = \langle x^2, y\rangle$$
$$I_{(0,1)} = \langle x^2 + 2y^2 - 2y, xy^2 - xy, y^3 - 2y^2 + y, (y-1)^2\rangle$$
$$= \langle x^2 + 2(y-1), x(y-1), (y-1)^2\rangle.$$

This gives the same primary decomposition as Example 2.4.9, though the initial description of the primary components is simpler because the minimal polynomial has smaller exponents than the characteristic polynomial.

The General Case. Now suppose that $\mathbb{K}$ is a field and that the equations (2.36) have solutions whose coordinates may lie in a strictly larger field. This means that in the primary decomposition over $\mathbb{K}$, the number of primary components no longer equals the number of solutions. Here is an example taken from [Mon02].

Example 2.4.12. The equations $x^2 - 2 = y^2 - 2 = 0$ have four solutions $(\pm\sqrt{2}, \pm\sqrt{2})$, none of which is rational over $\mathbb{Q}$. We will see below that the primary decomposition of $\langle x^2 - 2, y^2 - 2\rangle \subset \mathbb{Q}[x, y]$ is

$$\langle x^2 - 2, y^2 - 2\rangle = I_1 \cap I_2 = \langle x^2 - 2, x - y\rangle \cap \langle x^2 - 2, x + y\rangle.$$

Note that the ideal I_1 corresponds to $\pm(\sqrt{2}, \sqrt{2})$ while I_2 corresponds to $\pm(\sqrt{2}, -\sqrt{2})$.

Here is a description of the primary decomposition of an arbitrary zero-dimensional ideal.

Proposition 2.4.13. *Suppose that* $\langle f_1, \ldots, f_s \rangle$ *is zero-dimensional and* $f \in \mathbb{K}[x_1, \ldots, x_n]$ *takes distinct values at the solutions of* $f_1 = \cdots = f_s = 0$. *If the irreducible factorization of* $\mathrm{CharPoly}_{M_f}(u)$ *is*

$$\mathrm{CharPoly}_{M_f}(u) = \prod_{i=1}^{r} p_i(u)^{m_i},$$

where $p_1(u), \ldots, p_r(u)$ *are distinct monic irreducible polynomials, then the primary decomposition of* $\langle f_1, \ldots, f_s \rangle$ *is given by*

$$\langle f_1, \ldots, f_s \rangle = I_1 \cap \cdots \cap I_r,$$

where

$$I_i = \langle f_1, \ldots, f_s, p_i(f)^{m_i} \rangle.$$

Proof. We will use Galois theory to prove the proposition in the special case when $\mathbb{K}$ is *perfect* (see [Mon02] for the general case). This means that either $\mathbb{K}$ has characteristic zero, or $\mathbb{K}$ has characteristic $p > 0$ and every element of $\mathbb{K}$ is a pth power. Every finite extension of a perfect field is separable.

If I_i is a primary component of $\langle f_1, \ldots, f_s \rangle$, then its radical $\sqrt{I_i}$ is prime in $\mathbb{K}[x_1, \ldots, x_n]$. Then the following are true:

- The variety $\mathbf{V}(I_i) = \mathbf{V}(\sqrt{I_i}) \subset \overline{\mathbb{K}}^n$ is irreducible over $\mathbb{K}$.
- The Galois group $\mathrm{Gal}(\overline{\mathbb{K}}/\mathbb{K})$ acts on $\mathbf{V}(I_i)$.

These bullets imply that the action of $\mathrm{Gal}(\overline{\mathbb{K}}/\mathbb{K})$ on each $\mathbf{V}(I_i)$ is transitive. (*Exercise: Prove this.*) Hence all $p \in \mathbf{V}(I_i)$ have the same multiplicity, denoted m_i. Also note that $\mathbf{V}(I_i) \cap \mathbf{V}(I_j) = \emptyset$ for $i \neq j$. (*Exercise: Prove this.*)

By Proposition 2.1.14, we see that

$$\mathrm{CharPoly}_{M_f}(u) = \prod_{i=1}^{r} \prod_{p \in \mathbf{V}(I_i)} (u - f(p))^{m_i}.$$

Since f has coefficients in $\mathbb{K}$, we see that $\sigma(f(p)) = f(q)$ whenever $\sigma \in \mathrm{Gal}(\overline{\mathbb{K}}/\mathbb{K})$ takes p to q. But we also know that the $f(p)$ are all distinct and $\mathbb{K}$ is perfect. Thus standard arguments from Galois theory imply that $p_i(u) = \prod_{p \in \mathbf{V}(I_i)} (u - f(p))$ is irreducible over $\mathbb{K}$. (*Exercise: Supply the details.*) It follows that the above factorization coincides with the one in the statement of the proposition.

From here, the rest of the proof is similar to what we did in the proof of Proposition 2.4.8. The key point as always is that f takes distinct values at the solutions. (*Exercise: Complete the proof.*)

The above proof shows that when $\mathbb{K}$ is perfect, the m_i's compute the multiplicities of the corresponding points. However, this can fail when $\mathbb{K}$ is not perfect. We should also mention that one can weaken the hypothesis that f takes distinct values at the solutions: an analysis of the proof in [Mon02] reveals that it is sufficient to assume that $f(p) \neq f(q)$ whenever p and q are solutions of (2.36) lying in different orbits of the $\mathrm{Gal}(\overline{\mathbb{K}}/\mathbb{K})$-action. When this happens, however, the exponent m_i may fail to equal the multiplicity.

Here is an example of Proposition 2.4.13.

Example 2.4.14. For the ideal $\langle x^2 - 2, y^2 - 2 \rangle \subset \mathbb{Q}[x,y]$ of Example 2.4.12, one easily sees that $f = x + 2y$ takes distinct values at the solutions and has characteristic polynomial

$$\mathrm{CharPoly}_{M_f}(u) = (u^2 - 18)(u^2 - 2),$$

where $u^2 - 18$ and $u^2 - 2$ are irreducible over $\mathbb{Q}$. By Proposition 2.4.13, we get the primary decomposition $\langle x^2 - 2, y^2 - 2 \rangle = I_1 \cap I_2$, where

$$I_1 = \langle x^2 - 2, y^2 - 2, (x + 2y)^2 - 18 \rangle = \langle x^2 - 2, x - y \rangle$$
$$I_2 = \langle x^2 - 2, y^2 - 2, (x + 2y)^2 - 2 \rangle = \langle x^2 - 2, x + y \rangle.$$

This is the primary decomposition of Example 2.4.12. (*Exercise: Verify this.*)

We could instead have used $f = x+y$, which has characteristic polynomial $u^2(u^2 - 8)$. The function f does not take distinct values on the roots but does separate orbits of the Galois action. As noted above, the conclusion of Proposition 2.4.13 still holds for such an f. (*Exercise: Check that $u^2(u^2 - 8)$ leads to the above primary decomposition.*)

We next relate primary decomposition to the factorizations discussed in Section 2.4.1.

Example 2.4.15. As in Section 2.4.1, suppose that $f(x), g(x) \in \mathbb{Q}[x]$ are irreducible and $\alpha, \beta \in \mathbb{C}$ satisfy $f(\alpha) = g(\beta) = 0$. Also suppose that we have the irreducible factorization

$$f(x) = f_1(x) \cdots f_r(x) \text{ over } \mathbb{Q}(\beta).$$

We can relate this to primary decomposition as follows. Pick $t \in \mathbb{Q}$ such that $f = x + ty$ takes distinct values at the solutions of $f(x) = g(y) = 0$. In the proof of Theorem 2.4.2, we showed that all solutions have multiplicity 1, so that we have a factorization

$$\mathrm{CharPoly}_{M_f}(u) = \prod_{i=1}^{r} \Phi_i(u),$$

where the $\Phi_i(u) \in \mathbb{Q}[u]$ are distinct irreducibles. Proposition 2.4.13 implies that the primary decomposition of $\langle f(x), g(y) \rangle \subset \mathbb{Q}[x,y]$ is

$$\langle f(x), g(y) \rangle = \bigcap_{i=1}^{r} \langle f(x), g(y), \Phi_i(x + ty) \rangle,$$

and Theorem 2.4.2 asserts that the irreducible factors of $f(x)$ in $\mathbb{Q}(\beta)[x]$ are

$$f_i(x) = \gcd(\Phi_i(x + t\beta), f(x)).$$

Since $\mathbb{Q}(\beta)[x] \simeq \mathbb{Q}[x, y]/\langle g(y) \rangle$, there is a polynomial $f_i(x, y) \in \mathbb{Q}[x, y]$ such that $f_i(x, \beta) = f_i(x)$ in $\mathbb{Q}(\beta)[x]$. Then

$$\langle f(x), g(y), \Phi_i(x + ty) \rangle = \langle g(y), f_i(x, y) \rangle.$$

(*Exercise: Prove this.*) Hence the above primary decomposition can be written

$$\langle f(x), g(y) \rangle = \bigcap_{i=1}^{r} \langle g(y), f_i(x, y) \rangle.$$

This shows that there is a close relation between primary decomposition and factorization.

There is also a version of Proposition 2.4.13 that uses minimal polynomials instead of characteristic polynomials.

Proposition 2.4.16. *Suppose that* $\langle f_1, \ldots, f_s \rangle$ *is zero-dimensional and* $f \in \mathbb{K}[x_1, \ldots, x_n]$ *takes distinct values at the solutions of* (2.36). *If the irreducible factorization of* $\mathrm{MinPoly}_{M_f}(u)$ *is*

$$\mathrm{MinPoly}_{M_f}(u) = \prod_{i=1}^{r} p_i(u)^{n_i},$$

where $p_1(u), \ldots, p_r(u)$ *are distinct monic irreducible polynomials, then the primary decomposition of* $\langle f_1, \ldots, f_s \rangle$ *is given by*

$$\langle f_1, \ldots, f_s \rangle = I_1 \cap \cdots \cap I_r,$$

where

$$I_i = \langle f_1, \ldots, f_s, p_i(f)^{n_i} \rangle.$$

Proof. See [ABRW96] or [YNT92].

Algorithmic Aspects. From the point of view of doing primary decomposition algorithmically, one weakness of Proposition 2.4.13 is that f needs to take distinct values at the solutions. How do we do this without knowing the solutions? This problem was discussed at the end of Section 2.1.5. Another weakness of this method is that computing the characteristic polynomial of a large matrix can be time-consuming. The timings reported in [Mon02] indicate that as the number of solutions increases, methods based on [GTZ88] outperform the algorithm using Proposition 2.4.13.

Other approaches to primary decomposition are given in [EHV92] and [MMM96]. See also Chapter 5.

2.5 Galois theory

Solving equations has been our main topic of discussion. Since Galois theory is also concerned with the solutions of equations, it makes sense that there should be some link. As we will see, turning a polynomial equation $f(x) = 0$ of degree n into n equations in n unknowns is a very useful thing to do.

To illustrate our approach, consider the splitting field of $x^2 - x - 1 \in \mathbb{Q}[x]$. Two simple description of this field are

$$\mathbb{Q}(\sqrt{5}) \quad \text{and} \quad \mathbb{Q}[y]/\langle y^2 - 5 \rangle.$$

However, we will see that the splitting field can also be expressed as

$$\mathbb{Q}[x_1, x_2]/\langle x_1 + x_2 - 1, x_1 x_2 + 1 \rangle. \tag{2.39}$$

Although this may seem more complicated, it has the advantage of giving explicit descriptions of the roots (the cosets of x_1 and x_2) and the Galois action (permute these two cosets). Note also that the generators of ideal appearing in (2.39) make perfect sense since they give the sum and product of the roots of $x^2 - x - 1$.

The quotient (2.39) is an example of a *splitting algebra*. We will see that our methods, when applied to general splitting algebras, lead to some standard results in Galois theory. We will also show that primary decomposition gives an algorithm for computing Galois groups.

2.5.1 Splitting algebras

Let $\mathbb{K}$ be an infinite field and $f(x) \in \mathbb{K}[x]$ be a monic polynomial of degree n with distinct roots. We will write $f(x)$ as

$$f(x) = x^n - c_1 x^{n-1} + \cdots + (-1)^n c_n, \quad c_i \in \mathbb{K}.$$

The elementary symmetric polynomials $\sigma_1, \ldots, \sigma_n \in \mathbb{K}[x_1, \ldots, x_n]$ are defined by the identity

$$(x - x_1) \cdots (x - x_n) = x^n - \sigma_1 x^{n-1} + \cdots + (-1)^n \sigma_n. \tag{2.40}$$

Consider the system of n equations in $x_1, \ldots, x_n$ given by

$$\sigma_1(x_1, \ldots, x_n) - c_1 = 0$$
$$\sigma_2(x_1, \ldots, x_n) - c_2 = 0$$
$$\vdots \tag{2.41}$$
$$\sigma_n(x_1, \ldots, x_n) - c_n = 0.$$

The associated algebra is

$$\mathcal{A} = \mathbb{K}[x_1, \ldots, x_n]/\langle \sigma_1 - c_1, \ldots, \sigma_n - c_n \rangle.$$

This is the *splitting algebra* of f over $\mathbb{K}$. The system (2.41) and the algebra $\mathcal{A}$ were first written down by Kronecker in 1882 and 1887 respectively (see page 282 of [Kro31, Vol. II] for the equations and page 213 of [Kro31, Vol. III] for the algebra). A very nice modern treatment of the splitting algebra appears in the recent preprint [EL02].

The Universal Property. We first explain why the splitting algebra deserves its name. The natural map $\mathbb{K}[x_1, \ldots, x_n] \to \mathcal{A}$ takes σ_i to c_i, so that by (2.40), the cosets $[x_i] \in \mathcal{A}$ become roots of $f(x)$. It follows that

$$f(x) \text{ splits completely over } \mathcal{A}.$$

But more is true, for the factorization of $f(x)$ over $\mathcal{A}$ controls *all possible ways* in which $f(x)$ splits. Here is the precise statement.

Proposition 2.5.1. *Suppose that R is a $\mathbb{K}$-algebra such that $f(x)$ splits completely over R via*

$$f(x) = (x - \alpha_1) \cdots (x - \alpha_n), \quad \alpha_1, \ldots, \alpha_n \in R.$$

Then there is a $\mathbb{K}$-algebra homomorphism $\varphi : \mathcal{A} \to R$ such that this splitting is the image under φ of the splitting of $f(x)$ over $\mathcal{A}$.

Proof. Consider the $\mathbb{K}$-algebra homomorphism $\Phi : \mathbb{K}[x_1, \ldots, x_n] \to R$ determined by $x_i \mapsto \alpha_i$. This maps (2.40) to the splitting in the statement of the proposition, so that Φ maps σ_i to c_i. Hence $\Phi(\sigma_i - c_i) = 0$ for all i, which implies that Φ induces a $\mathbb{K}$-algebra homomorphism $\varphi : \mathcal{A} \to R$. It follows easily that φ has the desired property.

The splitting of $f(x)$ over $\mathcal{A}$ is thus "universal" in the sense that any other splitting is a homomorphic image of this one.

The Dimension of $\mathcal{A}$. Our next task is to compute $\dim_\mathbb{K} \mathcal{A}$. By Section 2.1.3, the dimension is the number of solutions, counted with multiplicity. Let $\overline{\mathbb{K}}$ an the algebraic closure of $\mathbb{K}$ and fix a splitting

$$f(x) = (x - \alpha_1) \cdots (x - \alpha_n) \in \overline{\mathbb{K}}[x].$$

Using this, we can describe the solutions of (2.41) as follows. If $(\beta_1, \ldots, \beta_n) \in \overline{\mathbb{K}}^n$ is a solution, then the substitutions $x_i \mapsto \beta_i$ take (2.40) to

$$f(x) = (x - \beta_1) \cdots (x - \beta_n) \in \overline{\mathbb{K}}[x].$$

Thus the β_i's are some permutation of the α_i. Since $f(x)$ has distinct roots by hypothesis, there is a unique $\sigma \in S_n$ such that $\beta_i = \alpha_{\sigma(i)}$ for all i. It follows easily that (2.41) has precisely $n!$ solutions given by

$$(\alpha_{\sigma(1)}, \ldots, \alpha_{\sigma(n)}), \quad \sigma \in S_n.$$

We can determine the multiplicities of these solutions as follows. Since $\sigma_i - c_i$ has degree i as a polynomial in $x_1, \ldots, x_n$, Bézout's theorem tells us that (2.41) has at most $1 \cdot 2 \cdot 3 \cdots n = n!$ solutions, counting multiplicity. Since we have $n!$ solutions, the multiplicities must all be 1. It follows that

$$\dim_{\mathbb{K}} \mathcal{A} = n!.$$

The Action of S_n. The symmetric group S_n acts on $\mathbb{K}[x_1, \ldots, x_n]$ by permuting the variables. Since $\sigma_i - c_i$ is invariant under this action, the action descends to an action of S_n on the splitting algebra $\mathcal{A}$.

The Emergence of Splitting Fields. Although f splits over $\mathcal{A}$, this algebra need not be a field. So how does $\mathcal{A}$ relate to the splitting fields of f over $\mathbb{K}$? We will analyze this following Kronecker's approach.

Since $\mathbb{K}$ is infinite, $f_0 = t_1 x_1 + \cdots + t_n x_n$ takes distinct values at the solutions of (2.41) for most choices of $t_1, \ldots, t_n \in \mathbb{K}$. Thus, as σ varies over the elements of S_n,

$$f_0(\alpha_{\sigma(1)}, \ldots, \alpha_{\sigma(n)}) = t_1 \alpha_{\sigma(1)} + \cdots + t_n \alpha_{\sigma(n)} \tag{2.42}$$

gives $n!$ distinct elements of $\overline{\mathbb{K}}$. Since all solutions of (2.41) have multiplicity 1, the characteristic polynomial of M_{f_0} on $\mathcal{A}$ is

$$\text{CharPoly}_{M_{f_0}}(u) = \prod_{\sigma \in S_n} \left(u - (t_1 \alpha_{\sigma(1)} + \cdots + t_n \alpha_{\sigma(n)}) \right) \tag{2.43}$$

and the linear map M_{f_0} is non-derogatory. By Proposition 2.1.12, it follows that the map sending u to $[t_1 x_1 + \cdots t_n x_n] \in \mathcal{A}$ induces a $\mathbb{K}$-algebra isomorphism

$$\mathbb{K}[u]/\langle \text{CharPoly}_{M_{f_0}}(u) \rangle \simeq \mathcal{A}.$$

Now factor $\text{CharPoly}_{M_{f_0}}(u)$ into a product of monic irreducible polynomials in $\mathbb{K}[u]$, say

$$\text{CharPoly}_{M_{f_0}}(u) = \prod_{i=1}^{r} G_i(u).$$

Since $\text{CharPoly}_{M_{f_0}}(u)$ has distinct roots, the $G_i(u)$ are distinct. Hence we get $\mathbb{K}$-algebra isomorphisms

$$\mathcal{A} \simeq \mathbb{K}[u]/\langle \text{CharPoly}_{M_{f_0}}(u) \rangle \simeq \prod_{i=1}^{r} \underbrace{\mathbb{K}[u]/\langle G_i(u) \rangle}_{\mathbb{K}_i}. \tag{2.44}$$

Each $\mathbb{K}_i$ is a field, and since the projection map $\mathcal{A} \to \mathbb{K}_i$ is surjective, each $\mathbb{K}_i$ is a splitting field of f over $\mathbb{K}$. Thus the factorization of the characteristic polynomial of M_{f_0} shows that the splitting algebra $\mathcal{A}$ is isomorphic to a product of fields, each of which is a splitting field of f over $\mathbb{K}$.

While the decomposition $\mathcal{A} \simeq \prod_{i=1}^{r} \mathbb{K}_i$ is nice, there are still some unanswered questions:

- Are the fields $\mathbb{K}_i$ isomorphic?
- When $r > 1$, $\mathcal{A}$ involves several splitting fields. Why?

We will answer these questions in Sections 2.5.2 and 2.5.3.

History. The methods described here date back to Galois and Kronecker. For example, in 1830 Galois chose $t_1, \ldots, t_n$ such that the $n!$ values (2.42) are distinct and showed that

$$V = t_1 \alpha_1 + \cdots + t_n \alpha_n$$

is a primitive element of the splitting field. He also used the polynomial on the right-hand side of (2.43). In all of this, Galois simply assumed the existence of the roots.

In 1887, Kronecker gave the first rigorous construction of splitting fields. His method was to prove the existence of $t_1, \ldots, t_n$ as above and then factor $\mathrm{CharPoly}_{M_{f_0}}(u)$ into irreducibles. Letting $G_i(u)$ be one of the factors of $\mathrm{CharPoly}_{M_{f_0}}(u)$, he showed that $\mathbb{K}[u]/\langle G_i(u) \rangle$ is a splitting field of f over $\mathbb{K}$.

2.5.2 Some Galois theory

We now use the above description of the splitting algebra $\mathcal{A}$ to prove some standard results in Galois theory. We begin by observing that $\mathcal{A}$ has two structures: an action of S_n and a product decomposition

$$\mathcal{A} \simeq \prod_{i=1}^{r} \mathbb{K}_i,$$

where $\mathbb{K}_i$ is a splitting field of f over $\mathbb{K}$. As we will see, the Galois group arises naturally from the interaction between these structures.

Since the decomposition $\mathcal{A} \simeq \prod_{i=1}^{r} \mathbb{K}_i$ is unique up to isomorphism, it follows that for $1 \le i \le r$ and $\sigma \in S_n$, we have $\sigma(\mathbb{K}_i) = \mathbb{K}_j$ for some j. Then we get the following result.

Proposition 2.5.2. S_n *acts transitively on the set of fields* $\{\mathbb{K}_1, \ldots, \mathbb{K}_r\}$ *and for each* $i = 1, \ldots, r$, *there is a natural isomorphism*

$$\mathrm{Gal}(\mathbb{K}_i/\mathbb{K}) \simeq \{\sigma \in S_n \mid \sigma(\mathbb{K}_i) = \mathbb{K}_i\}.$$

Proof. Under the isomorphism

$$\mathbb{K}[u]/\langle \mathrm{CharPoly}_{M_{f_0}}(u) \rangle \simeq \mathcal{A},$$

S_n permutes the factors of $\mathrm{CharPoly}_{M_{f_0}}(u) = \prod_{i=1}^{r} G_i(u)$. Over $\overline{\mathbb{K}}$, the factorization becomes

$$\mathrm{CharPoly}_{M_{f_0}}(u) = \prod_{\sigma \in S_n} \left(u - (t_1 \alpha_{\sigma(1)} + \cdots + t_n \alpha_{\sigma(n)}) \right).$$

This shows that S_n must permute the $G_i(u)$ transitively. By (2.44), we conclude that S_n permutes the $\mathbb{K}_i$ transitively.

For the second assertion, let $\text{Gal}_i = \{\sigma \in S_n \mid \sigma(\mathbb{K}_i) = \mathbb{K}_i\}$. Since every σ induces an automorphism of $\mathbb{K}_i$, we get an injective group homomorphism $\text{Gal}_i \to \text{Gal}(\mathbb{K}_i/\mathbb{K})$. To show that this map is surjective, take $\gamma \in \text{Gal}(\mathbb{K}_i/\mathbb{K})$. Under the projection $\mathcal{A} \to \mathbb{K}_i$, the cosets $[x_i]$ map to roots of $f(x)$ lying in $\mathbb{K}_i$. Then γ must permute these according to some $\sigma \in S_n$. Since the roots generate $\mathbb{K}_i$ over $\mathbb{K}$ and σ permutes the roots, we have $\sigma(\mathbb{K}_i) = \mathbb{K}_i$. It follows that $\sigma \in \text{Gal}_i$ maps to γ. This gives the desired isomorphism.

We can use Proposition 2.5.2 to prove some classic results of Galois theory as follows. We begin with the uniqueness of splitting fields.

Theorem 2.5.3. *All splitting fields of f over $\mathbb{K}$ are isomorphic via an isomorphism that is the identity on $\mathbb{K}$.*

Proof. Let $\mathbb{L}$ be an arbitrary splitting field of f over $\mathbb{K}$. Then splitting of f over $\mathbb{L}$ must come from the universal splitting via a $\mathbb{K}$-algebra homomorphism $\varphi : \mathcal{A} \to \mathbb{L}$. Furthermore, φ is onto since the roots of f generate $\mathbb{L}$ over $\mathbb{K}$. Using the decomposition $\mathcal{A} \simeq \prod_{i=1}^{r} \mathbb{K}_i$, we obtain a surjection

$$\prod_{i=1}^{r} \mathbb{K}_i \longrightarrow \mathbb{L}.$$

It is now easy to see that $\mathbb{L} \simeq \mathbb{K}_i$ for some i. (*Exercise: Prove this.*) Then we are done since this is an isomorphism of $\mathbb{K}$-algebras and the $\mathbb{K}_i$ are mutually isomorphic $\mathbb{K}$-algebras by the transitivity proved in Proposition 2.5.2.

Theorem 2.5.4. $\# \text{Gal}(\mathbb{K}_i/\mathbb{K}) = [\mathbb{K}_i : \mathbb{K}]$.

Proof. As above, let $\text{Gal}_i = \{\sigma \in S_n \mid \sigma(\mathbb{K}_i) = \mathbb{K}_i\}$. Thus Gal_i is the isotropy subgroup of $\mathbb{K}_i$ under the action of S_n on $\{\mathbb{K}_1, \ldots, \mathbb{K}_r\}$. Since this action is transitive by Proposition 2.5.2, we see that

$$\# \text{Gal}_i = \frac{n!}{r}.$$

However, we know that the $\mathbb{K}_1, \ldots, \mathbb{K}_r$ are mutually isomorphic. Thus

$$n! = \dim_{\mathbb{K}} \mathcal{A} = [\mathbb{K}_1 : \mathbb{K}] + \cdots + [\mathbb{K}_r : \mathbb{K}] = r\,[\mathbb{K}_i : \mathbb{K}].$$

Combining this with the previous equation gives $\# \text{Gal}_i = [\mathbb{K}_i : \mathbb{K}]$. Then we are done since $\text{Gal}_i \simeq \text{Gal}(\mathbb{K}_i/\mathbb{K})$ by Proposition 2.5.2.

Theorem 2.5.5. *If the characteristic of $\mathbb{K}$ doesn't divide $\# \text{Gal}(\mathbb{K}_i/\mathbb{K})$, then $\mathbb{K}$ is the fixed field of the action of $\text{Gal}(\mathbb{K}_i/\mathbb{K})$ on $\mathbb{K}_i$.*

Proof. Let $\alpha \in \mathbb{K}_i$ be in the fixed field and set $N = \# \operatorname{Gal}(\mathbb{K}_i/\mathbb{K})$. We may assume that $\alpha \neq 0$. Let $p \in \mathbb{K}[x_1, \ldots, x_n]$ map to $\alpha_i \in \mathbb{K}_i$ and to $0 \in \mathbb{K}_j$ for $j \neq i$. Then $P = \sum_{\sigma \in S_n} \sigma \cdot p$ is symmetric and hence is a polynomial in the σ_i by the Fundamental Theorem of Symmetric Polynomials. In $\mathcal{A}$, this means that $[P] \in \mathbb{K}$, so that P projects to an element of $\mathbb{K}$ in each of $\mathbb{K}_1, \ldots, \mathbb{K}_r$.

Now consider the projection of P onto $\mathbb{K}_i$. If

$$\sigma \in \operatorname{Gal}_i = \{\sigma \in S_n \mid \sigma(\mathbb{K}_i) = \mathbb{K}_i\} \simeq \operatorname{Gal}(\mathbb{K}_i/\mathbb{K}),$$

then $\sigma \cdot p$ projects to $\sigma(\alpha) = \alpha$. On the other hand, if $\sigma \notin \operatorname{Gal}_i$, then $\sigma \cdot p$ projects to 0 since $\mathbb{K}_i \cap \sigma(\mathbb{K}_i) = \{0\}$ for such σ. (*Exercise: Prove this.*) It follows that the projection of P onto $\mathbb{K}_i$ is $N\alpha$. Thus $N\alpha \in \mathbb{K}$, and then $\alpha \in \mathbb{K}$ follows by hypothesis.

A more general version of Theorem 2.5.5 is proved in [EL02].

Resultants. The characteristic polynomial $\operatorname{CharPoly}_{M_{f_0}}(u)$ is a resultant in disguise. More precisely, we claim that

$$\operatorname{Res}_{1,1,2,\ldots,n}(u - f_0, \sigma_1 - c_1, \ldots, \sigma_n - c_n) = -\operatorname{CharPoly}_{M_{f_0}}(u). \qquad (2.45)$$

To prove this, recall from Theorem 2.3.5 of Section 2.3 that this resultant equals the characteristic polynomial multiplied by

$$\operatorname{Res}_{1,2,\ldots,n}\big((\sigma_1 - c_1)_1, \ldots, (\sigma_n - c_n)_n\big),$$

where $(\sigma_i - c_i)_i$ consists of the terms of $\sigma_i - c_i$ of degree i. This is obviously just σ_i, so that this multiplier reduces to

$$\operatorname{Res}_{1,2,\ldots,n}(\sigma_1, \ldots, \sigma_n).$$

This resultant equals -1 by Exercise 11 of Section 3 of Chapter 3 of [CLO98]. Hence we obtain (2.45) as claimed.

Action of the Symmetric Group. Finally, we will describe the action of S_n on the product decomposition $\mathcal{A} = \mathbb{K}_1 \times \cdots \times \mathbb{K}_r$. If we let $\alpha_{ij} \in \mathbb{K}_j$ be the projection of $[x_i] \in \mathcal{A}$ onto the jth factor, then $\alpha_{1j}, \ldots, \alpha_{nj}$ are the roots of $f(x)$ in $\mathbb{K}_j$. So we have r isomorphic copies of the splitting field together with an ordered list of the roots in each field. Let

$$\mathbf{e}_j = (0, \ldots, 0, 1, 0, \ldots, 0) \in \mathbb{K}_1 \times \cdots \times \mathbb{K}_r,$$

where the 1 is in the jth position. Then by abuse of notation we can write $\alpha_{ij}\, \mathbf{e}_j \in \mathcal{A}$. Now take $\sigma \in S_n$ and suppose that $\sigma(\mathbf{e}_j) = \mathbf{e}_\ell$ (this is a precise way of saying that $\sigma(\mathbb{K}_j) = \mathbb{K}_\ell$). Then one can show without difficulty that $\sigma([x_i]) = [x_{\sigma(i)}]$ implies that

$$\sigma(\alpha_{ij}\, \mathbf{e}_j) = \alpha_{\sigma(i)\ell}\, \mathbf{e}_\ell. \qquad (2.46)$$

(*Exercise: Prove this.*) In the special case when σ comes from an element of $\operatorname{Gal}(\mathbb{K}_j/\mathbb{K})$, (2.46) gives the action of the Galois group on the roots. The nice thing about (2.46) is that it tell us what happens when we apply an arbitrary permutation, not just those coming from $\operatorname{Gal}(\mathbb{K}_j/\mathbb{K})$.

2.5.3 Primary decomposition

The Galois group of f consists of all permutations in S_n that preserve the algebraic structure of the roots. In this section, we will use primary decomposition to describe "the algebraic structure of the roots" and see how the Galois group "preserves" this structure. We will work over $\mathbb{Q}$ for simplicity.

Given $f = x^n - c_1 x^{n-1} + \cdots + (-1)^n c_n \in \mathbb{Q}[x]$ as in Section 2.5.1, the splitting algebra is

$$\mathcal{A} = \mathbb{Q}[x_1, \ldots, x_n]/\langle \sigma_1 - c_1, \ldots, \sigma_n - c_n \rangle,$$

where as usual σ_i is the ith elementary symmetric polynomial. We've seen that $\mathcal{A}$ is a product

$$\mathcal{A} = \prod_{i=1}^{r} \mathbb{K}_i,$$

where each $\mathbb{K}_i$ is a splitting field of f over $\mathbb{Q}$. But we also have the primary decomposition

$$\langle \sigma_1 - c_1, \ldots, \sigma_n - c_n \rangle = \bigcap_{i=1}^{r} I_i,$$

where I_i is maximal in $\mathbb{Q}[x_1, \ldots, x_n]$. These decompositions are related by

$$\mathbb{K}_i = \mathbb{Q}[x_1, \ldots, x_n]/I_i, \quad i = 1, \ldots, r.$$

Each I_i is larger than $\langle \sigma_1 - c_1, \ldots, \sigma_n - c_n \rangle$. The ideal $\langle \sigma_1 - c_1, \ldots, \sigma_n - c_n \rangle$ encodes the obvious relations among the roots, and the polynomials we add to get from I_i to $\langle \sigma_1 - c_1, \ldots, \sigma_n - c_n \rangle$ reflect the extra algebraic relations between the roots that hold in the splitting field $\mathbb{K}_i$. Having more relations among the roots means that I_i is larger and hence $\mathbb{K}_i$ and the Galois group are smaller.

For instance, if the Galois group of f is S_n, then $\langle \sigma_1 - c_1, \ldots, \sigma_n - c_n \rangle$ is a maximal ideal and the splitting algebra is the splitting field. This means that the *only* relations among the roots are the obvious ones relating the coefficients to the roots via the elementary symmetric polynomials.

Let's see what happens when the Galois group is smaller than S_n.

Example 2.5.6. Let $f = x^3 - c_1 x^2 + c_2 x - c_3 \in \mathbb{Q}[x]$ be an irreducible cubic. The splitting algebra of f is $\mathcal{A} = \mathbb{Q}[x_1, x_2, x_3]/\langle \sigma_1 - c_1, \sigma_2 - c_2, \sigma_3 - c_3 \rangle$. It is well-known that

$$\text{the Galois group of } f \text{ is isomorphic to } \begin{cases} S_3 & \text{if } \Delta(f) \notin \mathbb{Q}^2 \\ \mathbb{Z}/3\mathbb{Z} & \text{if } \Delta(f) \in \mathbb{Q}^2, \end{cases}$$

where $\Delta(f) \in \mathbb{Q}$ is the discriminant of f. By the above analysis, it follows that $\mathcal{A}$ is the splitting field of f when $\Delta(f) \notin \mathbb{Q}^2$.

Now suppose that $\Delta(f) = a^2$ for some $a \in \mathbb{Q}$. In this case, the splitting algebra is a product of two copies of the splitting field, i.e., $\mathcal{A} = \mathbb{K}_1 \times \mathbb{K}_2$. Let

$$\sqrt{\Delta} = (x_1 - x_2)(x_1 - x_3)(x_2 - x_3) \in \mathbb{Q}[x_1, x_2, x_3].$$

In the splitting algebra $\mathcal{A}$, we have $[\sqrt{\Delta}]^2 = [\Delta(f)]$, so that

$$[\sqrt{\Delta}]^2 = [a]^2.$$

Since $\mathcal{A}$ is not an integral domain, this does not imply $[\sqrt{\Delta}] = \pm[a]$. In fact, $[\sqrt{\Delta}] \in \mathcal{A}$ cannot have a numerical value since $[\sqrt{\Delta}]$ is not invariant under S_3. Yet once we map to a field, the value must be $\pm a$. But which sign do we choose? The answer is *both*, which explains why we need two fields in the splitting algebra.

In this case, we have the primary decomposition

$$\langle \sigma_1 - c_1, \sigma_2 - c_2, \sigma_3 - c_3 \rangle = I_1 \cap I_2,$$

where

$$I_1 = \langle \sigma_1 - c_1, \sigma_2 - c_2, \sigma_3 - c_3, \sqrt{\Delta} - a \rangle$$
$$I_2 = \langle \sigma_1 - c_1, \sigma_2 - c_2, \sigma_3 - c_3, \sqrt{\Delta} + a \rangle.$$

(*Exercise: Prove this.*) Note also that this is compatible with the action of S_3. For example, $(12) \in S_3$ maps I_1 to I_2 since $(12) \cdot \sqrt{\Delta} = -\sqrt{\Delta}$. It follows that (12) maps $\mathbb{K}_1$ to $\mathbb{K}_2$ in the decomposition $\mathcal{A} = \mathbb{K}_1 \times \mathbb{K}_2$. This is consistent with the description of the S_n action given at the end of Section 2.5.2.

Example 2.5.6 is analogous to what happens in quantum mechanics when an observation forces a mixed state (e.g. a superposition of pure states with different energy levels) to become a pure state (with a fixed energy level). In Example 2.5.6, the idea is that $[\sqrt{\Delta}]^2 = [D(f)]^2 = [a]^2$ means that $[\sqrt{\Delta}]$ is somehow a "mixed state" which becomes a "pure state" (i.e., $\pm a \in \mathbb{Q}$) when "observed" (i.e., when mapped to a field).

The quartic is more complicated since there are five possibilities for the Galois group of an irreducible quartic. We will discuss the following case.

Example 2.5.7. Let $f = x^4 - c_1 x^3 + c_2 x^2 - c_3 x + c_4 \in \mathbb{Q}[x]$ be an irreducible quartic with splitting algebra

$$\mathcal{A} = \mathbb{Q}[x_1, x_2, x_3, x_4]/\langle \sigma_1 - c_1, \sigma_2 - c_2, \sigma_3 - c_3, \sigma_4 - c_4 \rangle.$$

One of the tools used in solving the quartic is the *Ferrari resolvent*

$$x^3 - c_2 x^2 + (c_1 c_3 - 4 c_4) x - c_3^2 - c_1^2 c_4 + 4 c_2 c_4. \tag{2.47}$$

Euler showed that if $\beta_1, \beta_2, \beta_3$ are the roots of (2.47), then the roots of f are

$$\frac{1}{4}\left(c_1 \pm \sqrt{\beta_1 + c_1^2 - 4c_2} \pm \sqrt{\beta_2 + c_1^2 - 4c_2} \pm \sqrt{\beta_3 + c_1^2 - 4c_2}\right),$$

provided the signs are chosen so that the product of the square roots is $c_1^3 - 4c_1c_2 + 8c_3$. Also, as shown by Lagrange, the roots of the resolvent (2.47) are

$$\alpha_1\alpha_2 + \alpha_3\alpha_4, \ \alpha_1\alpha_3 + \alpha_2\alpha_4, \ \alpha_1\alpha_4 + \alpha_2\alpha_3. \tag{2.48}$$

The Galois group G of f over $\mathbb{Q}$ is isomorphic to one of the groups

$$S_4, \ A_4, \ D_8, \ \mathbb{Z}/4\mathbb{Z}, \ \mathbb{Z}/2\mathbb{Z} \times \mathbb{Z}/2\mathbb{Z},$$

where D_8 is the dihedral group of order 8. Three cases are easy to distinguish:

$$G \simeq \begin{cases} S_4 & \text{if } \Delta(f) \notin \mathbb{Q}^2 \text{ and (2.47) is irreducible over } \mathbb{Q} \\ A_4 & \text{if } \Delta(f) \in \mathbb{Q}^2 \text{ and (2.47) is irreducible over } \mathbb{Q} \\ \mathbb{Z}/2\mathbb{Z} \times \mathbb{Z}/2\mathbb{Z} & \text{if } \Delta(f) \in \mathbb{Q}^2 \text{ and (2.47) is reducible over } \mathbb{Q}. \end{cases}$$

The remaining case is when $\Delta(f) \notin \mathbb{Q}^2$ and (2.47) has a root in $\mathbb{Q}$. Here, the Galois group is D_8 or $\mathbb{Z}/4\mathbb{Z}$. We state without proof the following nice fact:

$$G \simeq D_8 \iff \Delta(f) \notin \mathbb{Q}^2, \text{ (2.47) has a root } b \in \mathbb{Q} \text{ and}$$
$$\langle \sigma_1 - c_1, \sigma_2 - c_2, \sigma_3 - c_3, \sigma_4 - c_4 \rangle = I_1 \cap I_2 \cap I_3,$$
is the primary decomposition, where
$$I_1 = \langle \sigma_1 - c_1, \sigma_2 - c_2, \sigma_3 - c_3, \sigma_4 - c_4, x_1x_2 + x_3x_4 - b \rangle$$
$$I_2 = \langle \sigma_1 - c_1, \sigma_2 - c_2, \sigma_3 - c_3, \sigma_4 - c_4, x_1x_3 + x_2x_4 - b \rangle$$
$$I_3 = \langle \sigma_1 - c_1, \sigma_2 - c_2, \sigma_3 - c_3, \sigma_4 - c_4, x_1x_4 + x_2x_3 - b \rangle.$$

The reason for three ideals is that b is one of the three combinations of roots given in (2.48). To get a field out of the ideal $\langle \sigma_1 - c_1, \sigma_2 - c_2, \sigma_3 - c_3, \sigma_4 - c_4 \rangle$, we must commit to which combination gives b. This gives the ideals I_1, I_2, I_3 as above.

The Galois group. We also observe that primary decomposition gives an algorithm for computing the Galois group of $f = x^n - c_1x^{n-1} + \cdots + (-1)^n c_n$. To do this, pick $f_0 = t_1x_1 + \cdots + t_nx_n$ such that $M_{f_0} : \mathcal{A} \to \mathcal{A}$ is non-derogatory and let

$$\text{CharPoly}_{M_{f_0}}(u) = \prod_{i=1}^{r} G_i(u)$$

be the irreducible factorization of the characteristic polynomial in $\mathbb{Q}[u]$. Then Proposition 2.4.13 implies that we have the primary decomposition

$$\langle \sigma_1 - c_1, \ldots, \sigma_n - c_n \rangle = \bigcap_{i=1}^{r} I_i$$

where
$$I_i = \langle \sigma_1 - c_1, \ldots, \sigma_n - c_n, G_i(f_0) \rangle.$$

Furthermore, S_n permutes the I_i since $\langle \sigma_1 - c_1, \ldots, \sigma_n - c_n \rangle$ is invariant, and

$$\mathrm{Gal}(\mathbb{K}_i/\mathbb{Q}) \simeq \{\sigma \in S_n \mid \sigma(I_i) = I_i\}.$$

Using a Gröbner basis of I_i, we can determine whether $\sigma(I_i)$ equals I_i for any given $\sigma \in S_n$. Hence, by going through the elements of S_n one-by-one, we get a (horribly inefficient) algorithm for computing the Galois group. However, in simple examples like Examples 2.5.6 or 2.5.7, the Galois group is easy to determine from the primary decomposition. (*Exercise: Do this computation.*)

Finally, we note that many of the ideas in Section 2.5 are well-known to researchers in computational Galois theory. See, for example, [AV00] and [PZ89].

Acknowledgments

I would like to thank Hal Schenck for inviting me to lecture on the preliminary version of this chapter at Texas A&M, and Alicia Dickenstein and Ioannis Emiris for inviting me to use an expanded version for my lectures at the CIMPA Summer School in Argentina.

I am grateful to Michael Möller for piquing my interest in eigenvector methods and to Alicia Dickenstein and Hal Schenck for useful comments on earlier drafts of the chapter.

3

Symbolic-numeric methods for solving polynomial equations and applications

Mohamed Elkadi[1] and Bernard Mourrain[2]

[1] UMR 6623, UNSA, B.P. 71, Parc Valrose, 06108 Nice, France,
elkadi@math.unice.fr
[2] INRIA, GALAAD, B.P. 93, Sophia-Antipolis, 06902 France,
mourrain@sophia.inria.fr

Summary. This tutorial gives an introductory presentation of algebraic and geometric methods to solve a polynomial system $f_1 = \cdots = f_m = 0$. The algebraic methods are based on the study of the quotient algebra $\mathcal{A}$ of the polynomial ring modulo the ideal $I = (f_1, \ldots, f_m)$. We show how to deduce the geometry of solutions from the structure of $\mathcal{A}$ and in particular, how solving polynomial equations reduces to eigenvalue and eigenvector computations of multiplication operators in $\mathcal{A}$. We give two approaches for computing the normal form of elements in $\mathcal{A}$, used to obtain a representation of multiplication operators. We also present the duality theory and its application to solving systems of algebraic equations. The geometric methods are based on projection operations which are closely related to resultant theory. We present different constructions of resultants and different methods for solving systems of polynomial equations based on these formulations. Finally, we illustrate these tools on problems coming from applications in computer-aided geometric design, computer vision, robotics, computational biology and signal processing.

3.0 Introduction

Polynomial system solving is ubiquitous in many applications such as computer geometric design, geometric modelling, robotics, computer vision, computational biology, signal processing, ... Specific methods like minimization, Newton iterations, ... are often used, but do not always offer guarantees on the result. In this paper, we give an introductory presentation of algebraic methods for solving a polynomial system $f_1 = \cdots = f_m = 0$. By a reformulation of the problem in terms of matrix manipulations, we obtain a better control of the structure and the accuracy of computations. The tools that we introduce are illustrated by explicit computations. A MAPLE package implements the algorithms described hereafter and is publicly available on the Internet[3]. We encourage the reader to use it for his own experimentation on

---

[3] http://www.inria.fr/galaad/logiciels/multires/

the examples illustrating the presentation. For more advanced computations described in the last section, we use the C++ library SYNAPS available on the Internet[4]. Our approach is based on the study of the quotient algebra $\mathcal{A}$ of the polynomial ring by the ideal $(f_1, \ldots, f_m)$. We describe, in the first part, the well known method of Gröbner basis to compute the normal form of elements in $\mathcal{A}$ which yields the algebraic structure of this quotient. We also mention a recent generalization of this approach which allows to combine, more safely, symbolic and numeric computations.

In the second part, we show how to deduce the geometry of solutions from the structure of $\mathcal{A}$. In particular, we show how solving polynomial systems reduces to the computation of eigenvalues or eigenvectors of operators of multiplication in $\mathcal{A}$. In the real case, we also show how to recover information on the real roots from this algebra.

We also study duality theory and show how to use it for solving polynomial systems.

Another major operation in effective algebraic geometry is projection. It is related to resultant theory. We present different notions and constructions of resultants and we derive methods to solve systems of polynomial equations. In practice, according to the class of systems that we want to solve, we will have to choose the resultant construction adapted to the geometry of the problem. Finally, we illustrate these tools on problems coming from several areas of applications.

For more details on the material presented here, see [EM].

3.1 Solving polynomial systems

The problem of solving polynomial equations goes back to the ancient Greeks and Chinese. It is not surprising that a large number of methods exists to handle this problem. We divide them into the following families and we will focus essentially on the last two classes.

3.1.1 Classes of solvers

Analytic solvers

The analytic solvers exploit the value of the functional $f = (f_1, \ldots, f_m)$ and its derivatives in order to converge to a solution or all the solutions of $f = 0$. Typical examples are Newton-like methods, Minimization methods, Weierstrass' method [Dem87, SS93, Bin96, MR02].

---

[4] http://www.inria.fr/galaad/logiciels/synaps/

Homotopic solvers

The idea behind the homotopic approaches is to deform a system with known roots into the system $f = 0$ that we want to solve. Examples of such continuation methods are based on projective [MS87b], toric [Li97, VVC94] or generally flat deformations of $f = 0$. See Chapter 8 and [AG90b] for more details.

Subdivision solvers

The subdivision methods use an exclusion criterion to remove a domain if it does not contain a root of $f = 0$. These solvers are often used to isolate the real roots, if possible. Exclusion criteria are based on Taylor's exclusion function [DY93], interval arithmetic [Kea90], the Turan test [Pan96], Sturm's method [BR90, Roy96], or Descartes' rule [Usp48, RZ03, MVY02].

Algebraic solvers

This class of methods exploits the known relations between the unknowns. They are based on normal form computations in the quotient algebra [CLO97, MT00, MT02] and reduce to a univariate or eigenvalue problem [Mou98].

Geometric solvers

These solvers project the problem onto a smaller subspace and exploit geometric properties of the set of solutions. Tools such as resultant constructions [GKZ94, EM99b, BEM00, BEM01, Bus01a] are used to reduce the solutions of the polynomial system to a univariate or eigenvalue problem. This reduction to univariate polynomials is also an important ingredient of triangular set methods [Tsü94, Wan95, ALMM99].

3.1.2 Notation

We fix the notation that will be used hereafter. Let $\mathbb{K}$ be a field, $\overline{\mathbb{K}}$ be its algebraic closure, $R = \mathbb{K}[x_1, \ldots, x_n] = \mathbb{K}[\mathbf{x}]$ be the algebra of polynomials in the variables $\mathbf{x} = (x_1, \ldots, x_n)$ with coefficients in $\mathbb{K}$. For the sake of simplicity, we will assume that $\mathbb{K}$ is of characteristic 0.

Let $f_1, \ldots, f_m \in R$ be m polynomials. Our objective is to solve the system $f_1 = 0, \ldots, f_m = 0$, also denoted by $f = 0$. If $\alpha = (\alpha_1, \ldots, \alpha_n) \in \mathbb{N}^n, |\alpha| = \alpha_1 + \cdots + \alpha_n, \mathbf{x}^\alpha = x_1^{\alpha_1} \ldots x_n^{\alpha_n}$.

Let I be the ideal generated by $f_1, \ldots, f_m$ in R and $\mathcal{Z}(I)$ be the affine variety $\{\zeta \in \overline{\mathbb{K}}^n : f_1(\zeta) = \cdots = f_m(\zeta) = 0\}$. We will assume that $\mathcal{Z}(I) = \{\zeta_1, \ldots, \zeta_d\}$ is a non-empty and finite set. The algebraic approach to solve the system $f = 0$ is based on the study of the $\mathbb{K}$-algebra $\mathcal{A} = R/I$. The hypothesis that $\mathcal{Z}(I)$ is finite implies that the $\mathbb{K}$-vector space $\mathcal{A}$ is of finite dimension over $\mathbb{K}$, see Theorem 2.1.2 in Chapter 2. We denote by $\widehat{R}$ (resp. $\widehat{\mathcal{A}}$) the dual of the vector space R (resp. $\mathcal{A}$).

Algebraic solvers exploit the properties of $\mathcal{A}$, which means that they must be able to compute effectively in this algebra. This can be performed by a so-called *normal form* algorithm. We are going to describe now two approaches to compute normal forms.

3.1.3 Gröbner bases

Gröbner bases are a major tool in effective algebraic geometry, which yields algorithmic answers to many question in this domain [CLO97, BW93, AL94, Eis95]. It is related to the use of a monomial ordering.

Definition 3.1.1. *A monomial ordering is a total order $>$ on the set of monomials of $\mathbb{K}[\mathbf{x}]$ such that*

i) $\forall \alpha \neq 0,\ 1 < \mathbf{x}^{\alpha}$,
ii) $\forall (\alpha, \beta, \gamma) \in (\mathbb{N}^n)^3$, *if* $\mathbf{x}^{\alpha} < \mathbf{x}^{\beta}$ *then* $\mathbf{x}^{\alpha+\gamma} < \mathbf{x}^{\beta+\gamma}$.

Some well known monomial orderings are defined as follows:

Let $\alpha = (\alpha_1, \ldots, \alpha_n) \in \mathbb{N}^n$ and $\beta = (\beta_1, \ldots, \beta_n) \in \mathbb{N}^n$.

– The lexicographic ordering with $x_1 > \cdots > x_n$: $\mathbf{x}^{\alpha} <_l \mathbf{x}^{\beta}$ iff there exists i such that $\alpha_1 = \beta_1, \ldots, \alpha_i = \beta_i, \alpha_{i+1} < \beta_{i+1}$.

– The graded lexicographic ordering with $x_1 > \cdots > x_n$: $\mathbf{x}^{\alpha} <_{gl} \mathbf{x}^{\beta}$ iff $|\alpha| < |\beta|$ or ($|\alpha| = |\beta|$ and $\mathbf{x}^{\alpha} <_l \mathbf{x}^{\beta}$).

Given a monomial ordering $>$, we define as in the univariate case, the leading term of $p \in R$ as the term (the coefficient times its monomial) of p whose monomial is maximal for $>$. We denote it by $\mathcal{L}_>(p)$ (or simply $\mathcal{L}(p)$). We write every $p \in R$ as $p = a_0 \mathbf{x}^{\alpha_0} + \cdots + a_l \mathbf{x}^{\alpha_l}$, with $a_i \neq 0$ and $\alpha_0 > \cdots > \alpha_l$.

Let $f, f_1, \ldots, f_m \in R$. As in the Euclidean division there are polynomials $q_1, \ldots, q_m, r$ such that $f = q_1 f_1 + \cdots + q_m f_m + r$, where no term of r divides any of $\mathcal{L}(f_1), \ldots, \mathcal{L}(f_m)$ (in this case we say that r is reduced with respect to $f_1, \ldots, f_m$). This is the multivariate division of f by $f_1, \ldots, f_m$. The polynomials $q_1, \ldots, q_m$ are the quotients and r the remainder of this division.

If I is an ideal of $R = \mathbb{K}[\mathbf{x}]$, we define $\mathcal{L}_>(I)$ (or simply $\mathcal{L}(I)$) to be the ideal generated by the set of leading terms of elements of I.

By Dickson's lemma [CLO97] or by Noetherianity of $\mathbb{K}[\mathbf{x}]$, this ideal $\mathcal{L}_>(I)$ is generated by a finite set of monomials. This leads to the definition of Gröbner bases:

Definition 3.1.2. *A finite subset $G = \{g_1, \ldots, g_t\}$ of the ideal I is a Gröbner basis of I for a given monomial order $>$ iff $\mathcal{L}_>(I) = (\mathcal{L}_>(g_1), \ldots, \mathcal{L}_>(g_t))$.*

Some interesting properties of a Gröbner basis G are:

– For any $p \in R$, the remainder of the multivariate division of p by G is unique. It is called the *normal form* of p modulo the ideal I and is denoted by $N(p)$ (see [CLO97]).

– The polynomial $p \in I$ iff its normal form $N(p) = 0$.

– A basis B of the $\mathbb{K}$-vector space $\mathcal{A} = R/I$ is given by the set of monomials *which are not in* $\mathcal{L}_>(I)$. This allows us to define the multiplication table by an element $a \in \mathcal{A}$: We multiply first the elements of B by a as usual polynomials and then normalize the products by reduction by G.

The ideal I can have several Gröbner bases but only one which is reduced (i.e. the leading coefficients of elements of G are equal to 1, and every $g \in G$ is reduced with respect to $G \setminus \{g\}$). Efficient algorithms and software have been developed over the past decades to compute reduced Gröbner bases. We mention in particular [Fau99], [GS], [GPS01], [Roba].

Example 3.1.3. Let I be the ideal of $R = \mathbb{Q}[x_1, x_2]$ generated by

$$f_1 := 13\,x_1^2 + 8\,x_1\,x_2 + 4\,x_2^2 - 8\,x_1 - 8\,x_2 + 2 \quad \text{and} \quad f_2 := x_1^2 + x_1\,x_2 - x_1 - \tfrac{1}{6}.$$

The reduced Gröbner basis G of I for the graded lexicographic ordering with $x_1 > x_2$ is (on Maple):

```
> with(Groebner); G:= gbasis([f1,f2],tdeg(x[1],x[2]));
```

$$(30\,x_1 x_2 - 30\,x_1 - 25 - 24\,x_2^{\,2} + 48\,x_2, 15\,x_1^{\,2} + 12\,x_2^{\,2} - 24\,x_2 + 10,$$
$$216\,x_2^{\,3} - 648\,x_2^{\,2} + 5\,x_1 + 632\,x_2 - 200).$$

The leading monomials of elements of G are $x_1\,x_2, x_1^{\,2}, x_2^{\,3}$. Then a basis of $\mathcal{A}$ is $\{1, x_1, x_2, x_2^2\}$. Using the reduction by G, the matrix of multiplication by x_1 in this basis is:

```
> L:= map(u->normalf(u,G,tdeg(x[1],x[2])),
>          [x[1],x[1]^2,x[1]*x[2],x[1]*x[2]^2]);
```

$$(x_1, -4/5\,x_2^{\,2}+8/5\,x_2-2/3, x_1+5/6+4/5\,x_2^{\,2}-8/5\,x_2, -\frac{839}{270}\,x_2+8/5\,x_2^{\,2}+\frac{53}{54}\,x_1+\frac{85}{54})$$

```
> matrixof(L,[[1,x[1],x[2],x[2]^2]]);
```

$$\begin{pmatrix} 0 & -2/3 & 5/6 & \frac{85}{54} \\ 1 & 0 & 1 & \frac{53}{54} \\ 0 & 8/5 & -8/5 & -\frac{839}{270} \\ 0 & -4/5 & 4/5 & 8/5 \end{pmatrix}.$$

This is the matrix of coefficients of elements of the monomial basis multiplied by x_1, expressed in this basis.

Since the variety $\mathcal{Z}(I)$ is finite, a lexicographic Gröbner basis with $x_n > \cdots > x_1$ contains elements $g_1, \ldots, g_n$ such that $g_i \in \mathbb{K}[x_1, \ldots, x_i]$ and $\mathcal{L}(g_i)$ depends only on x_i. This reduces the problem of solving $f = 0$ to solving a triangular system, hence to the problem of finding the roots of a univariate polynomial. Unfortunately the lexicographic Gröbner bases are not used in practice because of their high complexity of computation. We proceed as follows: First we compute a Gröbner basis for another monomial ordering and then we use a conversion procedure to obtain a lexicographic one. For more details see for instance [FGLM93].

3.1.4 General normal form

The construction of Gröbner bases may not be numerically stable, as shown in the following example:

Example 3.1.4. Let

```
> f1:= x[1]^2+x[2]^2-x[1]+x[2]-2; f2:= x[1]^2-x[2]^2+2*x[2]-3;
```

The Gröbner basis of (f_1, f_2) for the graded lexicographic ordering with $x_1 > x_2$ is:

```
> G:=gbasis([f1,f2],tdeg(x[1],x[2]));
```

$$(2\,x_2{}^2 - x_1 - x_2 + 1, 2\,x_1{}^2 - x_1 + 3\,x_2 - 5).$$

The leading monomials of elements of G are x_1^2 and x_2^2. A monomial basis of $\mathcal{A}$ is $\{1, x_1, x_2, x_1 x_2\}$. Consider now a small perturbation of the system $f_1 = f_2 = 0$ and compute its Gröbner basis for the same monomial ordering:

```
> gbasis([f1,f2+1.0/10000000*x[1]*x[2]],tdeg(x[1],x[2]));
```

$$\begin{aligned}
(&-2\,x_2{}^2 + x_1 + x_2 - 1 + 0.0000001\,x_1 x_2, \quad x_1{}^2 + x_2{}^2 - x_1 + x_2 - 2, \\
&x_2{}^3 - 10000000.99999999999999950000000000000125\,x_2{}^2 \\
&+5000000.2500000124999993749999687500015625000781250\,x_1 \\
&+5000000.7500000374999931249999062500171875002343750\,x_2 \\
&-5000000.2500000624999993749998437500015625003906250).
\end{aligned}$$

The leading monomials of this Gröbner basis are $x_1 x_2, x_1{}^2, x_2{}^3$ and the corresponding basis of the perturbed algebra is $\{1, x_1, x_2, x_2^2\}$. After a small perturbation, the basis of the quotient algebra may "jump" from one set of monomials to another one, though the two set of solutions are very close from a geometric point of view. Moreover, some polynomials of the Gröbner basis of the perturbed system have large coefficients.

Thus, Gröbner bases computations may introduce artificial discontinuities due to the choice of a monomial order. A recent generalization of this notion has been proposed in [Mou99, MT00]. It is based on a new criterion which gives a necessary and sufficient condition for a projection onto a vector subspace of R to be a normal form modulo the ideal I. More precisely we have:

Theorem 3.1.5. *Let B be a vector space in $R = \mathbb{K}[x_1, \ldots, x_n]$ connected to the constant polynomial 1^5. If B^+ is the vector subspace generated by $B \cup x_1 B \cup \ldots \cup x_n B$, $N : B^+ \to B$ is a linear map such that N is the identity on B, we define for $i = 1, \ldots, n$, the maps*

$$M_i : B \to B$$
$$b \mapsto M_i(b) := N(x_i b).$$

The two following properties are equivalent:

1. *For all $1 \leq i, j \leq n$, $M_i \circ M_j = M_j \circ M_i$.*
2. *$R = B \oplus I$, where I is the ideal generated by the kernel of N.*

If this holds, the B-reduction along $\ker(N)$ is canonical.

In Chapter 4, you will also find more material on this approach and a proof of Theorem 3.1.5, in the special case of 0-dimensional ideals.

This leads to a completion-like algorithm which starts with the linear subspace K_0 generated by the polynomials $f_1, \ldots, f_m$, which we wish to solve, and iterates the construction $K_{i+1} = K_i^+ \cap L$, where L is a fixed vector space. We stop when $K_{i+1} = K_i$. See [Mou99, MT00, Tré02] for more details. This approach allows us to fix first the set of monomials on which we want to do linear operations and thus to treat more safely polynomials with approximate coefficients. It can be adapted very naturally to Laurent polynomials, which is not the case for Gröbner bases computations. Moreover it can be specialized very efficiently to systems of equations for which the basis of $\mathcal{A}$ is known a priori, such as in the case of a complete projective intersection [MT00].

Example 3.1.6. For the perturbed system of the previous example, the normal forms for the monomials on the border of $B = \{1, x_1, x_2, x_1 x_2\}$ are:

$$x_1{}^2 = -0.00000005\, x_1 x_2 + 1/2\, x_1 - 3/2\, x_2 + 5/2,$$
$$x_2{}^2 = +0.00000005\, x_1 x_2 + 1/2\, x_1 + 1/2\, x_2 - 1/2,$$
$$x_1{}^2 x_2 = 0.49999999\, x_1 x_2 - 0.74999998\, x_1 + 1.75000003\, x_2 + 0.74999994,$$
$$x_1 x_2{}^2 = 0.49999999\, x_1 x_2 - 0.25000004\, x_1 - 0.74999991\, x_2 + 1.25000004.$$

This set of relations gives the matrices of multiplication by the variables x_1 and x_2 in $\mathcal{A}$. An implementation by Ph. Trébuchet of an algorithm computing this new type of normal form is available in the SYNAPS library (see `Solve(L,newmac<C>())`).

3.2 Structure of the quotient algebra

In this section we will see how to recover the solutions of the system $f = 0$ from the structure of the algebra $\mathcal{A}$, which we assume to be given through a normal form procedure.

---

[5] Any monomial $\mathbf{x}^\alpha \neq 1 \in B$ is of the form $x_i \mathbf{x}^\beta$ with $\mathbf{x}^\beta \in B$ and some i in $\{1, \ldots, n\}$.

3.2.1 Dual of the quotient algebra

First we consider the dual $\widehat{R}$ that is, the space of linear forms from R to $\mathbb{K}$. The *evaluation* $\mathbf{1}_\zeta$ at a *fixed point* ζ is an example of such linear forms: $p \in R \mapsto \mathbf{1}_\zeta(p) := p(\zeta) \in \mathbb{K}$. Another class of linear forms is obtained by differential operators, namely for $\alpha = (\alpha_1, \ldots, \alpha_n) \in \mathbb{N}^n$,

$$\mathbf{d}^\alpha : R \to \mathbb{K}$$
$$p \mapsto \frac{1}{\prod_{i=1}^n \alpha_i!} \left((\partial_1)^{\alpha_1} \cdots (\partial_n)^{\alpha_n} p \right)(0),$$

where ∂_i is the derivative with respect to the variable x_i (see also Section 2.2.2 of Chapter 2). If $\alpha = (\alpha_1, \ldots, \alpha_n) \in \mathbb{N}^n$ and $\beta = (\beta_1, \ldots, \beta_n) \in \mathbb{N}^n$,

$$\mathbf{d}^\alpha \left(\prod_{i=1}^n x_i^{\beta_i} \right) = \begin{cases} 1 \text{ if } \alpha_i = \beta_i \text{ for } i = 1, \ldots, n \\ 0 \text{ otherwise.} \end{cases}$$

It follows that $(\mathbf{d}^\alpha)_{\alpha \in \mathbb{N}^n}$ is the dual basis of the monomial basis $(\mathbf{x}^\alpha)_{\alpha \in \mathbb{N}^n}$ of R. Notice that $(\mathbf{d}^\alpha)_{\alpha \in \mathbb{N}^n}$ can be defined for every characteristic. We assume again that $\mathbb{K}$ is a field of arbitrary characteristic. We deduce that for every $\Lambda \in \widehat{R}$ we have $\Lambda = \sum_{\alpha \in \mathbb{N}^n} \Lambda(\mathbf{x}^\alpha) \mathbf{d}^\alpha$.

The vector space $\{ \sum_{\alpha \in \mathbb{N}^n} c_\alpha \mathbf{d}_1^{\alpha_1} \ldots \mathbf{d}_n^{\alpha_n} : c_\alpha \in \mathbb{K} \}$ (where $\mathbf{d}_i^{\alpha_i}$ denotes the map $p \in R \mapsto \frac{1}{\alpha_i!}(\partial_i^{\alpha_i} p)(0)$) of formal power series in $\mathbf{d}_1, \ldots, \mathbf{d}_n$ with coefficients in $\mathbb{K}$ is denoted by $\mathbb{K}[[\mathbf{d}]] = \mathbb{K}[[\mathbf{d}_1, \ldots \mathbf{d}_n]]$. The linear map

$$\Lambda \in \widehat{R} \mapsto \sum_{\alpha \in \mathbb{N}^n} \Lambda(\mathbf{x}^\alpha) \mathbf{d}^\alpha \in \mathbb{K}[[\mathbf{d}]]$$

defines a one-to-one correspondence. So we can identify $\widehat{R}$ with $\mathbb{K}[[\mathbf{d}]]$. Under this identification, the linear form evaluation at 0 corresponds to the constant power series 1; it is also denoted $\mathbf{d}^0$.

Example 3.2.1. Let $n = 3$. The value of the linear form $1 + \mathbf{d}_1 + 2\mathbf{d}_1\mathbf{d}_2 + \mathbf{d}_3^2$ on the polynomial $1 + x_1 + x_1 x_2$ is:

$$(1 + \mathbf{d}_1 + 2\mathbf{d}_1\mathbf{d}_2 + \mathbf{d}_3^2)(1 + x_1 + x_1 x_2) = 4.$$

The dual $\widehat{R}$ has a natural structure of R-module: For $(p, \Lambda) \in R \times \widehat{R}$,

$$p \cdot \Lambda : q \in R \mapsto (p \cdot \Lambda)(q) := \Lambda(pq) \in \mathbb{K}.$$

If $p \in R$ and $\alpha_i \in \mathbb{N}^*$, we check that $\mathbf{d}_i^{\alpha_i}(x_i\, p) = \frac{1}{(\alpha_i - 1)!}\left(\partial_i^{\alpha_i - 1} p\right)(0)$. Consequently, for $p \in R$ and $\alpha = (\alpha_1, \ldots, \alpha_n) \in \mathbb{N}^n$ with $\alpha_i \neq 0$ for a fixed i, we have

$$(x_i \cdot \mathbf{d}^\alpha)(p) = \mathbf{d}^\alpha(x_i\, p) = \mathbf{d}_1^{\alpha_1} \cdots \mathbf{d}_{i-1}^{\alpha_{i-1}} \mathbf{d}_i^{\alpha_i - 1} \mathbf{d}_{i+1}^{\alpha_{i+1}} \cdots \mathbf{d}_n^{\alpha_n}(p).$$

That is, x_i acts as the *inverse* of $\mathbf{d}_i$ in $\mathbb{K}[[\mathbf{d}]]$. This is the reason why in the literature such a representation is referred to as the *inverse system* (see for instance [Mac94]). If $\alpha_i = 0$, then $x_i \cdot \mathbf{d}^\alpha = 0$. Then we redefine the product $p \cdot \Lambda$ as follows:

Proposition 3.2.2. *(see also [MP00], [Fuh96]) For $p \in R$ and $\Lambda \in \mathbb{K}[[\mathbf{d}]]$,*

$$p \cdot \Lambda = \pi_+\big(p(\mathbf{d}_1^{-1}, \ldots, \mathbf{d}_n^{-1})\, \Lambda(\mathbf{d})\big),$$

where π_+ is the projection on the vector space generated by the monomials with positive exponents.

Example 3.2.1 (continued).

$$(1 + x_1 + x_1 x_2) \cdot (1 + \mathbf{d}_1 + \mathbf{d}_1 \mathbf{d}_2 + \mathbf{d}_3^2) = 3 + \mathbf{d}_1 + \mathbf{d}_1 \mathbf{d}_2 + \mathbf{d}_3^2 + \mathbf{d}_2.$$

The constant term of this expansion is the value of the linear form $1 + \mathbf{d}_1 + \mathbf{d}_1 \mathbf{d}_2 + \mathbf{d}_3^2$ at the polynomial $1 + x_1 + x_1 x_2$.

3.2.2 Multiplication operators

Since the variety $\mathcal{Z}(I)$ is finite, the $\mathbb{K}$-algebra $\mathcal{A}$ has the decomposition

$$\mathcal{A} = \mathcal{A}_1 \oplus \cdots \oplus \mathcal{A}_d, \tag{3.1}$$

where $\mathcal{A}_i$ is the local algebra associated with the root ζ_i (see also Section 2.7, Chapter 2). So there are elements $\mathbf{e}_1, \ldots, \mathbf{e}_n \in \mathcal{A}$ such that

$$\mathbf{e}_1 + \cdots + \mathbf{e}_d \equiv 1 \;,\; \mathbf{e}_i^2 \equiv \mathbf{e}_i \;,\; \mathbf{e}_i \mathbf{e}_j \equiv 0 \text{ if } i \neq j.$$

These elements are called the fundamental *idempotents* of $\mathcal{A}$, and generalize the univariate Lagrange polynomials. They satisfy $\mathcal{A}_i = \mathbf{e}_i \mathcal{A}$ and $\mathbf{e}_i(\zeta_j) = 1$ if $i = j$ and 0 otherwise. The dimension of the $\mathbb{K}$-vector space $\mathcal{A}_i$ is by definition the *multiplicity* of the root ζ_i, and it is denoted by μ_{ζ_i} .

We recall that a linear form on $\mathcal{A}$ can be identified with a linear form on R which vanishes on the ideal I. Thus the evaluation $\mathbf{1}_\zeta$, which is a linear form on R, is an element of $\widehat{\mathcal{A}}$ iff $\zeta \in \mathcal{Z}(I)$.

The first operators that come naturally in the study of $\mathcal{A}$ are the operators of multiplication by elements of $\mathcal{A}$. For any $a \in \mathcal{A}$, we define

$$\begin{aligned} M_a : \mathcal{A} &\to \mathcal{A} \\ b &\mapsto M_a(b) := a\, b. \end{aligned}$$

We also consider its transpose operator

$$\begin{aligned} M_a^{\mathrm{t}} : \widehat{\mathcal{A}} &\to \widehat{\mathcal{A}} \\ \Lambda &\mapsto M_a^{\mathrm{t}}(\Lambda) = \Lambda \circ M_a. \end{aligned}$$

The matrix of M_a^{t} in the dual basis of a basis B of $\mathcal{A}$ is the transpose of the matrix of M_a in B.

Example 3.1.3 (continued). Consider the matrix M_{x_1} of multiplication by x_1 in the basis $B = \{1, x_1, x_2, x_1 x_2\}$ of $\mathcal{A} = \mathbb{K}[x_1, x_2]/(f_1, f_2)$: We multiply the monomials of B by x_1 and reduce the products to the normal forms, so

$$1 \times x_1 \equiv x_1 \quad , \quad x_1 \times x_1 \equiv -x_1 x_2 + x_1 + \frac{1}{6} \quad , \quad x_2 \times x_1 \equiv x_1 x_2 \quad ,$$

$$x_1 x_2 \times x_1 \equiv -x_1 x_2 + \frac{55}{54} x_1 + \frac{2}{27} x_2 + \frac{5}{54}.$$

Then

$$M_{x_1} = \begin{pmatrix} 0 & \frac{1}{6} & 0 & \frac{5}{54} \\ 1 & 1 & 0 & \frac{55}{54} \\ 0 & 0 & 0 & \frac{2}{27} \\ 0 & -1 & 1 & -1 \end{pmatrix}.$$

The multiplication operators can be computed using a normal form algorithm. This can be performed, for instance by Gröbner basis computations (see Sections 3.1.3 and 3.1.4). In Section 3.5, we will describe another way to compute implicitly these operators based on resultant matrices (see also Section 2.3, Chapter 2).

Hereafter, $\mathbf{x}^E = (\mathbf{x}^\alpha)_{\alpha \in E}$ denotes a monomial basis of $\mathcal{A}$ (for instance obtained by a Gröbner basis). Then any polynomial can be reduced modulo $(f_1, \ldots, f_m)$ to a linear combination of monomials of $\mathbf{x}^E$.

The matrix approach to solve polynomial systems is based on the following fundamental theorem:

Theorem 3.2.3. *Assume that* $\mathcal{Z}(I) = \{\zeta_1, \ldots, \zeta_d\}$. *We have*

1. *Let* $a \in \mathcal{A}$. *The eigenvalues of the operator* M_a *(and its transpose* M_a^t*)* *are* $a(\zeta_1), \ldots, a(\zeta_d)$.
2. *The common eigenvectors of* $(M_a^t)_{a \in \mathcal{A}}$ *are (up to a scalar)* $\mathbf{1}_{\zeta_1}, \ldots, \mathbf{1}_{\zeta_d}$.

Proof. 1) Let $i \in \{1, \ldots, d\}$. For every $b \in \mathcal{A}$,

$$\left(M_a^t(\mathbf{1}_{\zeta_i}) \right)(b) = \mathbf{1}_{\zeta_i}(a\,b) = \left(a(\zeta_i)\,\mathbf{1}_{\zeta_i} \right)(b).$$

This shows that $a(\zeta_1), \ldots, a(\zeta_d)$ are eigenvalues of M_a and M_a^t, $\mathbf{1}_{\zeta_i}$ is an eigenvector of M_a^t associated with $a(\zeta_i)$, and $\mathbf{1}_{\zeta_1}, \ldots, \mathbf{1}_{\zeta_d}$ are common eigenvectors to $M_a^t, a \in \mathcal{A}$.

Now we will show that every eigenvalue of M_a is one $a(\zeta_i)$. For this we consider

$$p(\mathbf{x}) = \prod_{\zeta \in \mathcal{Z}(I)} \left(a(\mathbf{x}) - a(\zeta) \right) \in \mathbb{K}[\mathbf{x}].$$

This polynomial vanishes on $\mathcal{Z}(I)$. Using Hilbert's Nullstellensatz we can find an integer $m \in \mathbb{N}$ such that the operator

$$p^m(M_a) = \prod_{\zeta \in \mathcal{Z}(I)} \left(M_a - a(\zeta)\,\mathbb{I} \right)^m$$

vanishes on $\mathcal{A}$ ($\mathbb{I}$ is the identity operator). We deduce that the minimal polynomial of the operator M_a divides $\prod_{\zeta \in \mathcal{Z}(I)} \left(T - a(\zeta) \right)^m$, and that the eigenvalues of M_a belong to $\{a(\zeta) : \zeta \in \mathcal{Z}(I)\}$.

2) Let $\Lambda \in \widehat{\mathcal{A}}$ be a common eigenvector to M_a^t, $a \in \mathcal{A}$, and $\gamma = (\gamma_1, \ldots, \gamma_n)$ such that $M_{x_i}^t(\Lambda) = \gamma_i \Lambda$ for $i = 1, \ldots, n$. Then all the monomials $\mathbf{x}^\alpha$ satisfy

$$\left(M_{x_i}^t(\Lambda)\right)(\mathbf{x}^\alpha) = \Lambda(x_i \mathbf{x}^\alpha) = \gamma_i \Lambda(\mathbf{x}^\alpha).$$

From this we deduce that $\Lambda = \Lambda(1)\, \mathbf{1}_\gamma$. As $\Lambda \in \widehat{\mathcal{A}} = I^\perp$, $\Lambda(p) = \Lambda(1)p(\gamma) = 0$ for every $p \in I$, and $\mathbf{1}_\gamma \in \widehat{\mathcal{A}}$.

Since $\mathbf{x}^E = (\mathbf{x}^\alpha)_{\alpha \in E}$ is a basis of $\mathcal{A}$, the coordinates of $\mathbf{1}_{\zeta_i}$ in the dual basis of $\mathbf{x}^E$ are $(\zeta_i^\alpha)_{\alpha \in E}$. Thus if $\mathbf{x}^E$ contains $1, x_1, \ldots, x_n$ (which is often the case), we deduce the following algorithm:

Algorithm 3.2.4 SOLVING IN THE CASE OF SIMPLE ROOTS.

Let $a \in \mathcal{A}$ such that $a(\zeta_i) \neq a(\zeta_j)$ for $i \neq j$ (which is generically the case) and M_a be the matrix of multiplication by a in the basis $\mathbf{x}^E = (1, x_1, \ldots, x_n, \ldots)$ of $\mathcal{A}$.

1. *Compute the eigenvectors $\Lambda = (\Lambda_1, \Lambda_{x_1}, \ldots, \Lambda_{x_n}, \ldots)$ of M_a^t.*
2. *For each eigenvector Λ with $\Lambda_1 \neq 0$, compute and output the point $\zeta = \left(\frac{\Lambda_{x_1}}{\Lambda_1}, \ldots, \frac{\Lambda_{x_n}}{\Lambda_1}\right)$.*

The set of output points ζ contains the simple roots (i.e. roots with multiplicity 1) of $f = 0$, since for such a root the eigenspace associated to the eigenvalue $a(\zeta)$ is one-dimensional and contains $\mathbf{1}_\zeta$. But as we will see in the next example, it can also yield in some cases the multiple roots.

Example 3.1.3 (continued). The eigenvalues, their multiplicities, and the corresponding normalized eigenvectors of the transpose of the matrix of multiplication by x_1 are:

```
> neigenvects(transpose(Mx1),1);
```

$$\left\{-\frac{1}{3}, 2, V_1 = \left(1, -\frac{1}{3}, \frac{5}{6}, -\frac{5}{18}\right)\right\} \quad , \quad \left\{\frac{1}{3}, 2, V_2 = \left(1, \frac{1}{3}, \frac{7}{6}, \frac{7}{18}\right)\right\}.$$

As the basis of $\mathcal{A}$ is $(1, x_1, x_2, x_1 x_2)$, we deduce from Theorem 3.2.3 that the solutions of the system $f_1 = f_2 = 0$ can be read off from the 2nd and the 3rd coordinates of the normalized eigenvectors: So $\mathcal{Z}(I) = \{(-\frac{1}{3}, \frac{5}{6}), (\frac{1}{3}, \frac{7}{6})\}$. Moreover, the 4th coordinates of V_1 and V_2 are the products of the 2nd by the 3rd coordinates. In this example the multiplicity 2 of the two eigenvalues is exactly the multiplicity of roots ζ_1 and ζ_2 (see Chapter 2, Proposition 2.1.14).

In order to compute exactly the set of roots counted with their multiplicity, we use the following result. It is based on the fact that commuting matrices share common eigenspaces and the decomposition (3.1) of the algebra $\mathcal{A}$.

Theorem 3.2.5. *[Mou98, MP00, CGT97] There exists a basis of $\mathcal{A}$ such that for all $a \in \mathcal{A}$, the matrix of M_a in this basis is of the form*

$$M_a = \begin{pmatrix} N_a^1 & & 0 \\ & \ddots & \\ 0 & & N_a^d \end{pmatrix} \quad \text{with} \quad N_a^i = \begin{pmatrix} a(\zeta_i) & & \star \\ & \ddots & \\ 0 & & a(\zeta_i) \end{pmatrix}.$$

Proof. For every $i \in \{1, \ldots, d\}$, the multiplication operators in $\mathcal{A}_i$ by elements of $\mathcal{A}$ commute. Then using (3.1) it is possible to choose a basis of $\mathcal{A}_i$ such that the multiplication matrices N_a^i by $a \in \mathcal{A}$ in $\mathcal{A}_i$ in this basis are upper-triangular. By theorem 3.2.3, N_a^i has one eigenvalue, namely $a(\zeta_i)$.

We deduce the algorithm:

Algorithm 3.2.6 SOLVING BY SIMULTANEOUS TRIANGULATION.

INPUT: *Matrices of multiplication* $M_{x_i}, i = 1, \ldots, n$, *in a basis of* $\mathcal{A}$.

1. *Compute a (Schur) decomposition* P *such that the matrices* $T_i = PM_{x_i}P^{-1}$, $i = 1, \ldots, n$, *are upper-triangular.*
2. *Compute and output the diagonal vectors* $\mathbf{t}_i = (t_{i,i}^1, \ldots, t_{i,i}^n)$ *of triangular matrices* $T_k = (t_{i,j}^k)_{i,j}$.

OUTPUT: $\mathcal{Z}(I) = \{\mathbf{t}_i : i = 1, \ldots, \dim_{\mathbb{K}}(\mathcal{A})\}$.

The first step in this algorithm is performed by computing a Schur decomposition of M_l (where l is a generic linear form) which yields a change of basis matrix P. Then we compute the triangular matrices $T_i = PM_{x_i}P^{-1}, i = 1, \ldots, n$, since they commute with M_l.

3.2.3 Chow form and rational univariate representation

In some problems it is important to have an exact representation of the roots of the system $f = 0$. We will represent these roots in terms of solutions of a univariate polynomial. More precisely, they will be the image of these solutions by a rational map. The aim of the foregoing developments is to show how to construct explicitly such a representation.

Definition 3.2.7. *The Chow form of the ideal* I *is the homogeneous polynomial in* $\mathbf{u} = (u_0, \ldots, u_n)$ *defined by*

$$C_I(\mathbf{u}) = \det(u_0 + u_1 M_{x_1} + \cdots + u_n M_{x_n}) \in \mathbb{K}[\mathbf{u}].$$

According to Theorem 3.2.5, we have:

Proposition 3.2.8. *The Chow form*

$$C_I(\mathbf{u}) = \prod_{\zeta \in \mathcal{Z}(I)} (u_0 + u_1\zeta_1 + \cdots + u_n\zeta_n)^{\mu_\zeta}.$$

Example 3.1.3 (continued). The Chow form of $I = (f_1, f_2)$ using the matrices of multiplication by x_1 and x_2 is:

```
> factor(det(u[0]+ u[1]*Mx1+ u[2]*Mx2));
```

$$\left(u_0 + \frac{1}{3} u_1 + \frac{7}{6} u_2 \right)^2 \left(u_0 - \frac{1}{3} u_1 + \frac{5}{6} u_2 \right)^2.$$

It is a product of linear forms whose coefficients yield the roots $\zeta_1 = (-\frac{1}{3}, \frac{5}{6})$ and $\zeta_2 = (\frac{1}{3}, \frac{7}{6})$ of $f_1 = f_2 = 0$. The exponents are the multiplicities of the roots (here 2). When the points of $\mathcal{Z}(I)$ are rational (as in this example) we can easily factorize $\mathcal{C}_I(\mathbf{u})$ as a product of linear forms and get the solutions of the system $f = 0$. But usually, this factorization is possible only on an algebraic extension of the field of coefficients (see Chapter 9 for more details on this task).

From the Chow form, it is possible to deduce a rational univariate representation of $\mathcal{Z}(I)$:

Theorem 3.2.9. *(see [Ren92, ABRW96, Rou99, EM99a, Lec00]) Let $\Delta(\mathbf{u})$ be a multiple of the Chow form $\mathcal{C}_I(\mathbf{u})$. For a generic vector $\mathbf{t} \in \mathbb{K}^{n+1}$ we write*

$$\frac{\Delta}{\gcd\left(\Delta, \frac{\partial\Delta}{\partial u_0}\right)}(\mathbf{t} + \mathbf{u}) = d_0(u_0) + u_1 d_1(u_0) + \cdots + u_n d_n(u_0) + R(u),$$

where $d_i(u_0) \in \mathbb{K}[u_0], R(u) \in (u_1, \ldots, u_n)^2$, $\gcd\left(d_0(u_0), d_0'(u_0)\right) = 1$. Then for all $\zeta \in \mathcal{Z}(I)$, there exists a root ζ_0 of $d_0(u_0)$ such that

$$\zeta = \left(\frac{d_1(\zeta_0)}{d_0'(\zeta_0)}, \ldots, \frac{d_n(\zeta_0)}{d_0'(\zeta_0)} \right).$$

Proof. We decompose $\Delta(\mathbf{u})$ as

$$\Delta(\mathbf{u}) = \left(\prod_{\zeta=(\zeta_1,\ldots,\zeta_n)\in\mathcal{Z}(I)} (u_0 + \zeta_1 u_1 + \cdots + \zeta_n u_n)^{n_\zeta} \right) H(\mathbf{u}),$$

with $n_\zeta \in \mathbb{N}^*$, where $\prod_{\zeta\in\mathcal{Z}(I)}(u_0+\zeta_1 u_1+\cdots+\zeta_n u_n)^{n_\zeta}$ and $H(\mathbf{u})$ are relatively prime. Let

$$d(\mathbf{u}) = \frac{\Delta(\mathbf{u})}{\gcd\left(\Delta(\mathbf{u}), \frac{\partial\Delta}{\partial u_0}(\mathbf{u})\right)} = \left(\prod_{\zeta\in\mathcal{Z}(I)} (u_0 + \zeta_1 u_1 + \cdots + \zeta_n u_n) \right) h(\mathbf{u}),$$

where $\prod_{\zeta\in\mathcal{Z}(I)}(u_0 + \zeta_1 u_1 + \cdots + \zeta_n u_n)$ and $h(\mathbf{u})$ are relatively prime. If $t = (t_1, \ldots, t_n) \in \mathbb{K}^n$ and $\mathbf{t} = (0, t_1, \ldots, t_n) \in \mathbb{K}^{n+1}$, we have

$$d(\mathbf{t} + \mathbf{u}) = \left(\prod_{\zeta\in\mathcal{Z}(I)} ((t,\zeta) + u_0 + \zeta_1 u_1 + \cdots + \zeta_n u_n) \right) h(\mathbf{t} + \mathbf{u})$$

$$= d_0(u_0) + u_1 d_1(u_0) + \cdots + u_n d_n(u_0) + r(\mathbf{u}),$$

with $(t, \zeta) = t_1 \zeta_1 + \cdots + t_n \zeta_n$, $d_0, \ldots, d_n \in \mathbb{K}[u_0]$, $r(\mathbf{u}) \in (u_1, \ldots, u_n)^2$, and

$$h(\mathbf{t} + \mathbf{u}) = h_0(u_0) + u_1 h_1(u_0) + \cdots + u_n h_n(u_0) + s(\mathbf{u}) \ ,$$

with $h_0, \ldots, h_n \in \mathbb{K}[u_0]$ and $s(\mathbf{u}) \in (u_1, \ldots, u_n)^2$. By identification

$$d_0(u_0) = \left(\prod_{\zeta \in \mathcal{Z}(I)} ((t, \zeta) + u_0) \right) h_0(u_0) \ , \qquad \text{and for } i = 1, \ldots, n,$$

$$d_i(u_0) = \left(\sum_{\zeta \in \mathcal{Z}(I)} \zeta_i \prod_{\xi \neq \zeta} ((t, \xi) + u_0) \right) h_0(u_0) + \left(\prod_{\zeta \in \mathcal{Z}(I)} ((t, \zeta) + u_0) \right) h_i(u_0).$$

If $t \in \mathbb{K}^n$ is generic, $\prod_{\zeta \in \mathcal{Z}(I)} ((t, \zeta) + u_0)$ and $h_0(u_0)$ are relatively prime. Let $\zeta_0 = -(t, \zeta)$ be a root of $d_0(u_0)$, then $h_0(\zeta_0) \neq 0$ and

$$d_0'(\zeta_0) = \left(\prod_{\xi \neq \zeta} ((t, \xi) - (t, \zeta)) \right) h_0(\zeta_0) \ ,$$

$$d_i(\zeta_0) = \zeta_i \left(\prod_{\xi \neq \zeta} ((t, \xi) - (t, \zeta)) \right) h_0(\zeta_0) \qquad , \text{for } i = 1, \ldots, n.$$

Moreover we can assume that the generic vector t is such that $(t, \zeta) \neq (t, \xi)$ for $(\zeta, \xi) \in \mathcal{Z}(I)^2$ and $\zeta \neq \xi$. Then

$$\zeta_i = \frac{d_i(\zeta_0)}{d_0'(\zeta_0)}, \qquad \text{for } i = 1, \ldots, n.$$

This result describes the coordinates of solutions of $f = 0$ as the image by a rational map of some roots of $d_0(u_0)$. It does not imply that any root of $d_0(u_0)$ yields a point in $\mathcal{Z}(I)$, so that this representation may be redundant. However the "bad" prime factors in $d_0(u_0)$ can be removed by substituting the rational representation back into the equations $f_1, \ldots, f_m$.

In Proposition 3.5.4 we will see how to obtain a multiple of $C_I(\mathbf{u})$ without the knowledge of a basis of $\mathcal{A}$.

Algorithm 3.2.10 RATIONAL UNIVARIATE REPRESENTATION.

INPUT: *A multiple $\Delta(\mathbf{u})$ of the Chow form of the ideal $I = (f_1, \ldots, f_m)$.*

1. *Compute the square-free part $d(\mathbf{u})$ of $\Delta(\mathbf{u})$.*
2. *Choose a generic $t \in \mathbb{K}^n$ and compute the first terms of*

$$d(\mathbf{t} + \mathbf{u}) = d_0(u_0) + u_1 \, d_1(u_0) + \cdots + u_n \, d_n(u_0) + \cdots$$

3. *Compute the redundant rational representation*

$$d_0(u_0) = 0 \quad , \quad \left(\frac{d_1(u_0)}{d_0'(u_0)}, \ldots, \frac{d_n(u_0)}{d_0'(u_0)} \right).$$

4. *Factorize $d_0(u_0)$, keep the "good" prime factors and output the rational univariate representation of $\mathcal{Z}(I)$.*

Example 3.1.3 (continued). From the Chow form, we deduce the univariate representation of $\mathcal{Z}(I)$:

$$\left(u_0 + \frac{3}{2} \right) \left(u_0 + \frac{1}{2} \right) = 0 \quad , \quad \zeta(u_0) = \left(-\frac{1}{6\,(1 + u_0)}, \frac{11 + 12\,u_0}{12\,(1 + u_0)} \right).$$

This gives the solutions

$$u_0 = -\tfrac{3}{2}, \zeta_1 = \zeta(-\tfrac{3}{2}) = \left(\tfrac{1}{3}, \tfrac{7}{6} \right) \quad \text{and} \quad u_0 = -\tfrac{1}{2}, \zeta_2 = \zeta(-\tfrac{1}{2}) = \left(-\tfrac{1}{3}, \tfrac{5}{6} \right)$$

of $f_1 = f_2 = 0$.

3.2.4 Real roots

Now we assume that the polynomials $f_1, \ldots, f_m$ have real coefficients: $\mathbb{K} = \mathbb{R}$. A natural question which arises in many practical problems is *how many real solutions does the system $f = 0$ have?* We will use properties of the linear form trace to answer this question.

Definition 3.2.11. *The linear form trace, denoted by* Tr, *is defined by*

$$\mathrm{Tr} : \mathcal{A} \to \mathbb{R}$$
$$a \mapsto \mathrm{Tr}(a) := tr(M_a),$$

where $tr(M_a)$ is the trace of the linear operator M_a.

According to Theorem 3.2.5, we have

$$\mathrm{Tr} = \sum_{\zeta \in \mathcal{Z}(I)} \mu_\zeta \, \mathbf{1}_\zeta.$$

We associate to Tr and to any $h \in \mathcal{A}$ the *quadratic form*:

$$Q_h : (a, b) \in \mathcal{A} \times \mathcal{A} \mapsto Q_h(a, b) := \text{Tr}(hab) \in \mathbb{R},$$

which gives the following generalization of a result due to Hermite for counting the number of real roots.

Theorem 3.2.12. *(See [PRS93, GVRR99]) Let $h \in \mathbb{R}[\mathbf{x}]$. We have:*

1. *The rank of the quadratic form Q_h is the number of distinct complex roots ζ of $f = 0$ such that $h(\zeta) \neq 0$.*
2. *The signature of Q_h is equal to*

$$\#\{\zeta \in \mathbb{R}^n : f_1(\zeta) = \cdots = f_m(\zeta) = 0, h(\zeta) > 0\} - \#\{\zeta \in \mathbb{R}^n : f_1(\zeta) = \cdots = f_m(\zeta) = 0, h(\zeta) < 0\}, \text{ where } \# \text{ denotes the cardinality of a set.}$$

In particular, if $h = 1$, the rank of Q_1 is the number of distinct complex roots of $f = 0$ and its signature is the number of real roots of this system. This allows us to analyze the geometry of the real roots as illustrated in the following example:

Example 3.1.3 (continued). By direct computations, we have

$$\text{Tr}(1) = 4 \ , \ \text{Tr}(x_1) = 0 \ , \ \text{Tr}(x_2) = 4 \ , \ \text{Tr}(x_1 x_2) = \frac{2}{9}.$$

We deduce the value of the linear form Tr on the other interesting monomials by using the transpose operators $\text{M}^t_{x_i}$ as follows:

```
> T0   := evalm([4,0,4,2/9]):
> T1   := evalm(transpose(Mx1)&*T0):   T2:= evalm(transpose(Mx2)&*T0):
> T11  := evalm(transpose(Mx1)&*T1):   T12:= evalm(transpose(Mx2)&*T1):
> T112:= evalm(transpose(Mx2)&*T11):
> Q1   := matrix(4,4,[T0,T1,T2,T12]);
> Qx1  := matrix(4,4,[T1,T11,T12,T112]);
```

So we obtain

$$Q_1 = \begin{pmatrix} 4 & 0 & 4 & \frac{2}{9} \\ 0 & 4 & \frac{2}{9} & \frac{4}{9} \\ 4 & \frac{2}{9} & \frac{37}{9} & \frac{4}{9} \\ \frac{2}{9} & \frac{4}{9} & \frac{4}{9} & \frac{37}{81} \end{pmatrix}, \quad Q_{x_1} = \begin{pmatrix} 0 & \frac{4}{9} & \frac{2}{9} & \frac{4}{9} \\ \frac{4}{9} & 0 & \frac{4}{9} & \frac{2}{9} \\ \frac{2}{9} & \frac{4}{9} & \frac{4}{9} & \frac{37}{81} \\ \frac{4}{9} & \frac{2}{81} & \frac{37}{81} & \frac{4}{81} \end{pmatrix}.$$

The rank and the signatures of the quadratic forms Q_1 and Q_{x_1} are

```
> rank(Q1), signature(Q1), rank(Qx1), signature(Qx1);
```

$$2 \ , \ (2,0) \ , \ 2 \ , \ (1,1) \ ,$$

which tells us (without computing these roots) that there are 2 real roots, one with $x_1 < 0$ and another with $x_1 > 0$.

3.3 Duality

In this section $m = n$. Let us define the notion of Bezoutian matrix that will be useful in the following.

Definition 3.3.1. The Bezoutian $\Theta_{f_0,\ldots,f_n}$ of $f_0, \ldots, f_n \in R$ is the polynomial

$$\Theta_{f_0,\ldots,f_n}(\mathbf{x}, \mathbf{y}) = \begin{vmatrix} f_0(\mathbf{x}) & \theta_1(f_0)(\mathbf{x},\mathbf{y}) & \cdots & \theta_n(f_0)(\mathbf{x},\mathbf{y}) \\ \vdots & \vdots & \vdots & \vdots \\ f_n(\mathbf{x}) & \theta_1(f_n)(\mathbf{x},\mathbf{y}) & \cdots & \theta_n(f_n)(\mathbf{x},\mathbf{y}) \end{vmatrix} \in \mathbb{K}[\mathbf{x}, \mathbf{y}],$$

where

$$\theta_i(f_j)(\mathbf{x}, \mathbf{y}) = \frac{f_j(y_1, \ldots, y_{i-1}, x_i, \ldots, x_n) - f_j(y_1, \ldots, y_i, x_{i+1}, \ldots, x_n)}{x_i - y_i}.$$

Set $\Theta_{f_0,\ldots,f_n}(\mathbf{x}, \mathbf{y}) = \sum_{\alpha,\beta} a_{\alpha,\beta} \mathbf{x}^\alpha \mathbf{y}^\beta$ with $a_{\alpha,\beta} \in \mathbb{K}$, we order the monomials $\mathbf{x}^\alpha \mathbf{y}^\beta$, then the matrix $B_{f_0,\ldots,f_n} := (a_{\alpha,\beta})_{\alpha,\beta}$ is called the Bezoutian matrix of $f_0, \ldots, f_n$.

The Bezoutian was initially used by E. Bézout to construct the resultant of two polynomials in one variable [Béz64].

When f_0 is the constant 1 and f is the polynomial map $(f_1, \ldots, f_n)$, the Bezoutian $\Theta_{1, f_1, \ldots, f_n}$ will be denoted by Δ_f.

We will define the residue τ_f associated to $f = (f_1, \ldots, f_n)$ and we will give some of its important properties (for more details see [SS75], [Kun86], [EM96], [BCRS96], also Chapter 1 of this book).

The dual $\widehat{\mathcal{A}}$ of the vector space $\mathcal{A}$ has a natural structure of $\mathcal{A}$-module: If $(a, \Lambda) \in \mathcal{A} \times \widehat{\mathcal{A}}$, the linear form $a.\Lambda : b \in \mathcal{A} \mapsto (a.\Lambda)(b) := \Lambda(ab)$.

Definition 3.3.2. *The finite $\mathbb{K}$-algebra $\mathcal{A}$ is called Gorenstein if the $\mathcal{A}$-modules $\widehat{\mathcal{A}}$ and $\mathcal{A}$ are isomorphic.*

Set $\Delta_f = \sum_{\alpha,\beta} a_{\alpha,\beta} \, \mathbf{x}^\alpha \mathbf{y}^\beta$ with $a_{\alpha,\beta} \in \mathbb{K}$, we define the linear map

$$\Delta_f^\triangleright : \widehat{R} \to R$$

$$\Lambda \mapsto \Delta_f^\triangleright(\Lambda) := \sum_\alpha \left(\sum_\beta a_{\alpha,\beta} \, \Lambda(\mathbf{y}^\beta) \right) \mathbf{x}^\alpha.$$

This map induces naturally a linear one also denoted by $\Delta_f^\triangleright : \widehat{\mathcal{A}} \to \mathcal{A}$. Since the number of polynomials m is equal to the number n of variables and the affine variety $\mathcal{Z}(I)$ is finite, one can prove that $\Delta_f^\triangleright$ is an isomorphism of $\mathcal{A}$-modules (see [SS75], [Kun86], [EM96], [BCRS96]). Then $\mathcal{A}$ is a Gorenstein algebra. Thus we can state the following definition:

Definition 3.3.3. The residue τ_f of $f = (f_1, \ldots, f_n)$ is the linear form on R such that

1. $\tau_f(h) = 0, \forall h \in I,$
2. $\Delta_f{}^{\triangleright}(\tau_f) - 1 \in I.$

In the univariate case, let $f = f_d x^d + \cdots + f_0$ be a polynomial of degree d. For $h \in R$ let $r = r_{d-1} x^{d-1} + \cdots + r_0$ be the remainder in the Euclidean division of h by f, then

$$\tau_f(h) = \frac{r_{d-1}}{f_d}. \tag{3.2}$$

In the multivariate case, if for each $i = 1, \ldots, n$, f_i depends only on x_i, then

$$\tau_f(x_1^{\alpha_1} \ldots x_n^{\alpha_n}) = \tau_{f_1}(x_1^{\alpha_1}) \ldots \tau_{f_n}(x_n^{\alpha_n}). \tag{3.3}$$

If the roots of $f_1 = \cdots = f_n = 0$ are simple (this is equivalent to the fact that the Jacobian of f, denoted by $\mathrm{Jac}(f)$, does not vanish on $\mathcal{Z}(I)$), then $\tau_f = \sum_{\zeta \in \mathcal{Z}(I)} \frac{1_\zeta}{\mathrm{Jac}(f)(\zeta)}.$

But in the general multivariate setting the situation is more complicated. We will show how to compute effectively τ_f for an arbitrary map f.

An important tool in the duality theory is the transformation law.

Proposition 3.3.4. *(Classical transformation law)*
   Let $g = (g_1, \ldots, g_n)$ be another polynomial map such that the variety defined by $g_1, \ldots, g_n$ is finite and

$$\forall i = 1, \ldots, n \quad , \quad g_i = \sum_{j=1}^{n} a_{i,j} f_j \quad \text{with} \quad a_{i,j} \in \mathbb{K}[x].$$

Then $\tau_f = \det(a_{i,j}) \cdot \tau_g$.

Proposition 3.3.5. *(Generalized transformation law [BY99, EM96]).*
   Let $(f_0, \ldots, f_n)$ and $(g_0, \ldots, g_n)$ be two maps of $\mathbb{K}[x_0, \mathbf{x}] = \mathbb{K}[x_0, x_1, \ldots, x_n]$ which define finite affine varieties. We assume that $f_0 = g_0$ and there are positive integers m_i and polynomials $a_{i,j}$ such that

$$\forall i = 1, \ldots, n \quad , \quad f_0^{m_i} g_i = \sum_{j=1}^{n} a_{i,j} f_j.$$

Then $\tau_{(f_0, \ldots, f_n)} = \det(a_{i,j}) \cdot \tau_{(g_0^{m_1 + \cdots + m_n + 1}, g_1 \ldots, g_n)}$.

If $f_0 = x_0$ and $m_1 = \cdots = m_n = 0$, the generalized transformation law reduces to the classical one.

Another important fact in this theory is the following formula:

$$\mathrm{Jac}(f) \cdot \tau_f = \mathrm{Tr}, \tag{3.4}$$

where $\mathrm{Tr} : a \in R \mapsto \mathrm{Tr}(a) \in \mathbb{K}$ ($\mathrm{Tr}(a)$ is the trace of the endomorphism of multiplication by a in the vector space $\mathcal{A}$). If the characteristic of $\mathbb{K}$ is 0, we deduce from this formula that $\dim_{\mathbb{K}}(\mathcal{A}) = \tau_f(\mathrm{Jac}(f))$.

3.3.1 Residue calculus

The effective construction of the residue of the polynomial map $f = (f_1, \ldots, f_n)$ is based on the computation of algebraic relations between $f_1, \ldots, f_n$ and the coordinate functions x_i (see also Section 1.5.4 of Chapter 1). We give here a method using Bezoutian matrices to get them.

Let $f_0, \ldots, f_n$ be $n+1$ elements of R such that the n polynomials $f_1, \ldots, f_n$ are algebraically independent over $\mathbb{K}$. For algebraic dimension reasons there is a nonzero polynomial P such that $P(f_0, \ldots, f_n) = 0$. We will show how to find such a P by means of the Bezoutian matrix.

Proposition 3.3.6. *(see [EM00]) Let* $u = (u_0, \ldots, u_n)$ *be new parameters. Then every nonzero maximal minor* $P(u_0, \ldots, u_n)$ *of the Bezoutian matrix of the elements* $f_0 - u_0, \ldots, f_n - u_n$ *in* $\mathbb{K}[u_0, \ldots, u_n][\mathbf{x}]$ *satisfies the identity* $P(f_0, \ldots, f_n) = 0$.

This proposition comes from the fact that we can write the Bezoutian matrix of $f_0 - u_0, \ldots, f_n - u_n$ (up to invertible matrices with coefficients in $\mathbb{K}(u_1, \ldots, u_n)$) as

$$
\left(\begin{array}{c|c} \mathbf{M}_{f_0} - u_0 \mathbb{I} & \mathbf{0} \\ \hline \mathbf{0} & * \end{array} \right)
\tag{3.5}
$$

where $\mathbb{I}$ is the identity matrix, $\mathbf{M}_{f_0}$ is the matrix of multiplication by f_0 in the vector space $\mathbb{K}(u_1, \ldots, u_n)[\mathbf{x}]/(f_1 - u_1, \ldots, f_n - u_n)$. By Cayley-Hamilton's theorem every maximal minor of this Bezoutian matrix gives an algebraic relation between $f_0, \ldots, f_n$ (for more details see [EM00]).

In practice, we use a fraction free Gaussian elimination (Bareiss method) in order to find a nonzero maximal minor of the Bezoutian matrix (see the implementation of the function `melim` in the MULTIRES package).

We will see now how to compute effectively the residue τ_f.

Proposition 3.3.7. *For* $i \in \{1, \ldots, n\}$, *let*

$$
P_i(u_0, \ldots, u_n) = a_{i,0}(u_1, \ldots, u_n) u_0^{m_i} + \cdots + a_{i,m_i}(u_1, \ldots, u_n)
$$

be an algebraic relation between $x_i, f_1, \ldots, f_n$. *If for each* i *there is* $k_i \in \{0, \ldots, m_i - 1\}$ *such that* $a_{i,k_i}(0) \neq 0$, *then for* $h \in R$ *the computation of the multivariate residue* $\tau_f(h)$ *reduces to univariate residue calculus.*

Proof. If $j_i = \min\{k : a_{i,k}(0) \neq 0\}$, we have

$$
g_i(x_i) = a_{i,j_i}(0)x_i^{m_i - j_i} + \cdots + a_{i,m_i}(0) = \sum_{j=1}^{n} A_{i,j} f_j , \quad A_{i,j} \in \mathbb{K}[x].
$$

By the transformation law and (3.3) there are scalars c_α such that

$$\tau_f(h) = \tau_{(g_1,\ldots,g_n)}\big(h\det(A_{i,j})\big) = \sum_{\alpha=(\alpha_1,\ldots,\alpha_n)} c_\alpha \, \tau_{g_1}(x_1^{\alpha_1})\ldots\tau_{g_n}(x_n^{\alpha_n}).$$

If w are formal parameters, similarly for every $h \in R$, $\tau_{f-w}(h)$ is a rational function in w whose denominator is the product of powers of $a_{1,0}(w),\ldots,a_{n,0}(w)$. But it is not clear how to recover $\tau_f(h)$ from this function. For an arbitrary map f, $\tau_f(h)$ can be computed using the generalized transformation law.

For $(\alpha_1,\ldots,\alpha_n) \in \mathbb{K}^n$ and a new variable x_0, we define the multi-index $m = (m_1,\ldots,m_n)$ and the polynomials R_i, S_i as follows: If $P_i(u_0,\ldots,u_n)$ is an algebraic relation between $x_i, f_1,\ldots,f_n$, then there are $B_{i,j} \in \mathbb{K}[x_0,\ldots,x_n]$ such that

$$P_i(x_i,\alpha_1 x_0,\ldots,\alpha_n x_0) = \sum_{j=1}^n (f_j - \alpha_j x_0)B_{i,j} \tag{3.6}$$

$$= x_0^{m_i}\big(R_i(x_i) - x_0 S_i(x_i,x_0)\big). \tag{3.7}$$

From the transformation laws we deduce the following result:

Proposition 3.3.8. *If for each $i = 1\ldots n$, the univariate polynomial R_i does not vanish identically, then for $h \in R$ we have*

$$\tau_f(h) = \sum_{k\in\mathbb{N}^n:|k|\leq|m|} \tau_{(x_0^{|m|+1-|k|},R_1^{k_1+1},\ldots,R_n^{k_n+1})}\big(S_1^{k_1}\ldots S_n^{k_n} \, h\det(B_{i,j})\big).$$

Proof. From (3.6) and Proposition 3.3.5, we have

$$\tau_f(h) = \tau_{(x_0,f_1-\alpha_1 x_0,\ldots,f_n-\alpha_n x_0)}(h) = \tau_{(x_0^{|m|+1},R_1-x_0 S_1,\ldots,R_n-x_0 S_n)}\big(h\det(B_{i,j})\big).$$

Using the identities

$$R_i^{|m|+1} - (x_0 S_i)^{|m|+1} = (R_i - x_0 S_i)\sum_{k_i=0}^{|m|} R_i^{|m|-k_i}(x_0 S_i)^{k_i}, i = 1\ldots n,$$

and Proposition 3.3.4 we deduce the formula in Proposition 3.3.8.

Propositions 3.3.6 and 3.3.8 give an effective algorithm to compute the residue of a map in the multivariate setting. They reduce the multivariate residue calculus to the univariate one.

We will show how to use the residue for solving polynomial systems. Let $\zeta_1,\ldots,\zeta_D$ be the solutions of the system $f = 0$ (each solution appears as many times as its multiplicity). Let us fix $i \in \{1,\ldots,n\}$. Using formula (3.4) and Theorem 3.2.5, we can compute the Newton sums

$$S_j = \tau_f\big(x_i^j \operatorname{Jac}(f)\big) = \operatorname{Tr}(x_i^j) = \zeta_{1,i}^{\,j} + \cdots + \zeta_{D,i}^{\,j},$$

where $\zeta_{1,i}, \ldots, \zeta_{D,i}$ are the i-th coordinates of $\zeta_1, \ldots, \zeta_D$. If $\sigma_1, \ldots, \sigma_D$ are the elementary symmetric functions of $\zeta_{1,i}, \ldots, \zeta_{D,i}$ (i.e. $\sigma_j = \sum_{1 \le i_1 < \cdots < i_j \le D} \zeta_{1,i_1} \cdots \zeta_{D,i_j}$), we can obtain the univariate polynomial

$$A_i(T) = (T - \zeta_{1,i}) \ldots (T - \zeta_{D,i}) = T^D + \sigma_1 T^{D-1} + \cdots + \sigma_D$$

by means of the Newton identities:

$$k\sigma_k = -S_k - \sigma_1 S_{k-1} - \cdots - \sigma_{k-1} S_1 \, , \ 1 \le k \le D. \tag{3.8}$$

The residue τ_f allows us to find the univariate polynomials $A_i(T), 1 \le i \le n$, and then to deduce the i-th coordinates of the roots of the system $f_1 = \cdots = f_n = 0$.

For other applications of residue theory see [EM98, EM].

3.4 Resultant constructions

Projection is one of the most used operations in effective algebraic geometry [Eis95, CLO98]. It reduces the dimension of the problem that we have to solve and often simplifies it. The resultant is a tool to perform such a projection and has many applications in this domain. It leads to efficient methods for solving polynomial equations based on matrix formulations [EM99b]. We present here different notions of resultants (see also Chapter 1).

We recall that a resultant of a polynomial system $\mathbf{f_c}$ on a complete variety X is a polynomial $\operatorname{Res}_X(\mathbf{f_c})$ on the coefficients $\mathbf{c}$ of this system (considered as variables) such that the vanishing of $\operatorname{Res}_X(\mathbf{f_c})$ is a necessary and sufficient condition for $\mathbf{f_c}$ to have a solution in the variety X. The best known formulation of the resultant is in the case of two univariate polynomials. It is given by the Sylvester matrix. Another classical one is the projective resultant of n homogeneous polynomials in n variables. It can be computed using Macaulay matrices (see Chapter 2, Section 2.3, or [DD01]). Recently a refined notion of resultants (on toric varieties) has been studied. It takes into account the actual monomials appearing in the polynomials. Its construction follows the same process as in the projective case except that the notion of degree is replaced by the support of a polynomial (for more details see Chapter 7). Here we will focus on an even more recent generalization of these resultant notions.

3.4.1 Resultant over a unirational variety

A natural extension of the toric resultant is to replace the monomial parameterization by a rational one. The polynomial system $\mathbf{f_c}$ is defined on an open subset of $\mathbb{K}^n$ and is of the form

$$\mathbf{f_c} := \begin{cases} f_0(\mathbf{t}) = \sum_{j=0}^{k_0} c_{0,j}\, \kappa_{0,j}(\mathbf{t}) \\ \quad\vdots \\ f_n(\mathbf{t}) = \sum_{j=0}^{k_n} c_{n,j}\, \kappa_{n,j}(\mathbf{t}) \end{cases} \tag{3.9}$$

where $\mathbf{t} = (t_1, \ldots, t_n) \in \mathbb{K}^n$ and the $\kappa_{i,j}$ are nonzero rational functions, which we can assume to be polynomials by reduction to the same denominator.

Let $\mathcal{K}_i = (\kappa_{i,j})_{j=0,\ldots,k_i}$ and U be the open subset of $\mathbb{K}^n$ such that $\mathcal{K}_i(\mathbf{t}) \neq 0$ on U for $i = 0, \ldots, n$. Assume that there exists $\sigma_0, \ldots, \sigma_N \in R$ defining a map

$$\sigma : U \to \mathbb{P}^N$$
$$\mathbf{t} \mapsto \big(\sigma_0(\mathbf{t}) : \cdots : \sigma_N(\mathbf{t})\big),$$

and homogeneous polynomials $\psi_{i,j}(x_0, \ldots, x_N)$, $i = 0, \ldots, n$, $j = 0, \ldots, k_i$, satisfying

$$\kappa_{i,j}(\mathbf{t}) = \psi_{i,j}\big(\sigma_0(\mathbf{t}), \ldots, \sigma_N(\mathbf{t})\big) \text{ and } \deg(\psi_{i,j}) = \deg(\psi_{i,0}) \geq 1.$$

Let X^o be the image of σ and X be its closure in $\mathbb{P}^N$. In order to construct the resultant associated to the system (3.9) on the variety X we assume the following conditions $(\mathbf{D})$:

$\begin{cases} (\mathbf{D1}) \text{ The Jacobian matrix of } \sigma = (\sigma_i)_{i=0,\ldots,N} \text{ is of rank } n \text{ at one point of } U, \\ (\mathbf{D2}) \text{ For generic } \mathbf{c}, \; f_1 = \cdots = f_n = 0 \text{ has a finite number of solutions in } U. \end{cases}$

We will show that these conditions are sufficient to define the resultant. Let $U^o = \{\mathbf{t} \in U : \kappa_{i,0}(\mathbf{t}) \neq 0 \text{ for } i = 0, \ldots, n\}$ be the dense open subset of U and consider the parameterization

$$\tau : \mathbb{P}^{k_0-1} \times \cdots \times \mathbb{P}^{k_n-1} \times U^o \to \mathbb{P}^{k_0} \times \cdots \times \mathbb{P}^{k_n} \times \mathbb{P}^N$$
$$(\tilde{\mathbf{c}}_0, \ldots, \tilde{\mathbf{c}}_n, \mathbf{t}) \mapsto (\mathbf{c}_0, \ldots, \mathbf{c}_n, \sigma(\mathbf{t}))$$

with $\mathbf{c}_i = (c_{i,0}, \tilde{\mathbf{c}}_i)$ and $c_{i,0} = -\frac{1}{\kappa_{i,0}(\mathbf{t})} \sum_{j=1}^{k_i} c_{i,j}\kappa_{i,j}(\mathbf{t})$. We denote by W^o the image of τ, W its closure in $\mathbb{P}^{k_0} \times \cdots \times \mathbb{P}^{k_n} \times \mathbb{P}^N$, $\pi_1 : \mathbb{P}^{k_0} \times \cdots \times \mathbb{P}^{k_n} \times \mathbb{P}^N \to \mathbb{P}^{k_0} \times \cdots \times \mathbb{P}^{k_n}$, and $\pi_2 : \mathbb{P}^{k_0} \times \cdots \times \mathbb{P}^{k_n} \times \mathbb{P}^N \to \mathbb{P}^N$ the canonical projections.

Theorem 3.4.1. *Under the conditions* $(\mathbf{D})$, *the variety W is irreducible and projects onto a hypersurface* $Z = \pi_1(W)$. *Moreover if* $\mathrm{Res}_X(\mathbf{f_c})$ *is one equation of Z, for any specialization of the parameters* $\mathbf{c} = (c_{i,j})$, $\mathrm{Res}_X(\mathbf{f_c}) = 0$ *if and only if there exists* $(\mathbf{c}, x) \in W$ *such that* $\tilde{f}_i(x) := \sum_{j=0}^{k_i} c_{i,j}\, \psi_{i,j}(x) = 0$ *for* $i = 0, \ldots, n$.

Proof. The variety W is the closure of a parameterized variety, so it is irreducible and its projection Z is also irreducible.

According to $(\mathbf{D1})$, the Jacobian of σ is of rank n on an open subset of U. This implies that the dimension of the variety X is n. The fibers of the projection $\pi_2 : W^o \to X^o$ are linear spaces of dimension $\sum_{i=0}^n k_i - n - 1$, for we have $\mathcal{K}_i(\mathbf{t}) \neq 0$ when $\mathbf{t} \in U$. By the fiber theorem ([Sha77] or [Har95]), we deduce that W is of dimension $\sum_{i=0}^{n} k_i - 1$.

Consider now the restriction of π_1 to W^o. According to **(D2)**, there exists an open subset of $\mathbb{P}^{k_0} \times \cdots \times \mathbb{P}^{k_n}$ on which the number of solutions of the system $f_1 = \cdots = f_n = 0$ is finite. The fibers of π_1 on this open subset is therefore of dimension 0. This shows that the projection $\pi_1(W^o)$, and thus Z, is of the same dimension as W, that is a hypersurface of $\mathbb{P}^{k_0} \times \cdots \times \mathbb{P}^{k_n}$ defined (up to a scalar) by one equation $\operatorname{Res}_X(\mathbf{f_c})$.

As the fibers of π_2 above X^o are of dimension $\sum_{i=0}^{n} k_i - n - 1$ and W is of dimension $\sum_{i=0}^{n} k_i - 1$, $\pi_2(W)$ is an irreducible variety of dimension n containing X^o. This shows that $X = \pi_2(W)$. Consequently for a specialization of the coefficients $\mathbf{c}$, $\operatorname{Res}_X(\mathbf{f_c}) = 0$ iff there exists $x \in X$ such that $(\mathbf{c}, x) \in W$, i.e. $\tilde{f}_i(x) = 0$ for $i = 0, \ldots, n$.

The degree of the resultant $\operatorname{Res}_X(\mathbf{f_c})$ in the coefficients $c_{i,j}$ of f_i is bounded by (but not necessarily equal to) the generic number of points of $V_i = \mathcal{Z}(\tilde{f}_0, \ldots, \tilde{f}_{i-1}, \tilde{f}_{i+1}, \ldots, \tilde{f}_n) \cap X$. In the case where the linear forms $\tilde{f}_i(\zeta)$, $\zeta \in V_i$, in $c_{i,j}$, are all distinct, the degree of $\operatorname{Res}_X(\mathbf{f_c})$ in the coefficients of f_i is exactly the number of generic roots of V_i. This is the case when $t_1, \ldots, t_n$ appear among the $\kappa_{i,j}, j = 0, \ldots, k_i$, as it is illustrated below.

We can compute a non-trivial multiple of $\operatorname{Res}_X(\mathbf{f_c})$ using the Bezoutian matrix.

Theorem 3.4.2. *Assume that the conditions* **(D)** *are satisfied. Then any maximal minor of the Bezoutian matrix $B_{f_0, \ldots, f_n}$ is divisible by $\operatorname{Res}_X(\mathbf{f_c})$.*

This theorem is a consequence of hypotheses **(D)** and the fact that if the variety defined by $f_1, \ldots, f_n$ is finite then the Bezoutian of $f_0, \ldots, f_n$ admits a block decomposition of the form (3.5), for more details see [BEM00].

Example 3.4.3. Consider the three following polynomials:

$$\begin{cases} f_0 = c_{0,0} + c_{0,1}t_1 + c_{0,2}t_2 + c_{0,3}(t_1{}^2 + t_2{}^2) \\ f_1 = c_{1,0} + c_{1,1}t_1 + c_{1,2}t_2 + c_{1,3}(t_1{}^2 + t_2{}^2) + c_{1,4}(t_1{}^2 + t_2{}^2)^2 \\ f_2 = c_{2,0} + c_{2,1}t_1 + c_{2,2}t_2 + c_{2,3}(t_1{}^2 + t_2{}^2) + c_{2,4}(t_1{}^2 + t_2{}^2)^2. \end{cases}$$

We are looking for conditions on the coefficients $c_{i,j}$ such that these three elements have a common "root". The projective resultant of these polynomials in $\mathbb{P}^2$ is zero (for all the values of parameters $c_{i,j}$), because the corresponding homogenized polynomials vanish at the points $(0 : 1 : \mathbf{i})$ and $(0 : 1 : -\mathbf{i})$). The toric resultant also vanishes (these polynomials have common roots in the associated toric variety). Now we consider the map

$$\sigma : \mathbb{K}^2 \to \mathbb{P}^3$$
$$(t_1, t_2) \mapsto (1 : t_1 : t_2 : t_1^2 + t_2^2).$$

The rank of the Jacobian matrix of σ is 2 and

$$\psi_0 = (x_0, x_1, x_2, x_3) \quad , \quad \psi_1 = \psi_2 = (x_0^2, x_0 x_1, x_0 x_2, x_0 x_3, x_3^2) \quad ,$$

where $(x_0 : x_1 : x_2 : x_3)$ are the homogeneous coordinates in $\mathbb{P}^3$. We have $f_i = \sum c_{i,j} \psi_{i,j} \circ \sigma$ for $i = 0, 1, 2$. For generic values of the coefficients $c_{i,j}$, the system $f_1 = f_2 = 0$ has a finite number of solutions in $\mathbb{K}^2$. By Theorem 3.4.2, any nonzero maximal minor of B_{f_0, f_1, f_2} is divisible by $\mathrm{Res}_X(f_0, f_1, f_2)$.

```
> mbezout([f1,f2,f3],[t1,t2]);
```

The Bezoutian matrix of f_1, f_2, f_3 is of size 12×12 and has rank 10. A maximal minor is a huge polynomial in $(c_{i,j})$ containing 207805 monomials. It can be factored as $q_1 q_2 (q_3)^2 \rho$, with

$$q_1 = -c_{0,2} c_{1,3} c_{2,4} + c_{0,2} c_{1,4} c_{2,3} + c_{1,2} c_{0,3} c_{2,4} - c_{2,2} c_{0,3} c_{1,4}$$
$$q_2 = c_{0,1} c_{1,3} c_{2,4} - c_{0,1} c_{1,4} c_{2,3} - c_{1,1} c_{0,3} c_{2,4} + c_{2,1} c_{0,3} c_{1,4}$$
$$q_3 = c_{0,3}{}^2 c_{1,1}{}^2 c_{2,4}{}^2 - 2 c_{0,3}{}^2 c_{1,1} c_{2,1} c_{2,4} c_{1,4} + c_{0,3}{}^2 c_{2,4}{}^2 c_{1,2}{}^2 + \cdots$$
$$\rho = c_{2,0}{}^4 c_{1,4}{}^4 c_{0,2}{}^4 + c_{2,0}{}^4 c_{1,4}{}^4 c_{0,1}{}^4 + c_{1,0}{}^4 c_{2,4}{}^4 c_{0,2}{}^4 + c_{1,0}{}^4 c_{2,4}{}^4 c_{0,1}{}^4 + \cdots$$

The polynomials q_3 and ρ contain respectively 20 and 2495 monomials. As for generic equations f_0, f_1, f_2, the number of points in the varieties $\mathcal{Z}(f_0, f_1)$, $\mathcal{Z}(f_0, f_2)$, $\mathcal{Z}(f_1, f_2)$ is 4 (see for instance [Mou96]), the resultant $\mathrm{Res}_X(f_0, f_1, f_2)$ is homogeneous of degree 4 in the coefficients of each f_i. Thus, $\mathrm{Res}_X(f_0, f_1, f_2)$ is equal to the factor ρ.

3.4.2 Residual resultant

In practical situations the equations have common zeroes which are independent of the parameters of the problem. These "degenerate" zeroes are not interesting for the resolution of this problem. We present here a resultant construction which allows us to remove these degenerate solutions when they form a complete intersection [BEM01] (for more details see [BEM01], [BKM90, CU02, Bus01a]).

We denote by S (resp. S_ν for $\nu \in \mathbb{N}$) the set of homogeneous polynomials (resp. of degree ν) in the variables $x_0, \ldots, x_n$ with coefficients in $\mathbb{K}$.

Let $g_1, \ldots, g_r$ be r (with $r \leq n + 1$) homogeneous polynomials in S of degree $k_1 \geq \cdots \geq k_r$, and let $d_0 \geq \cdots \geq d_n$ be $n + 1$ integers such that $d_n \geq \max(k_1, k_r + 1)$. We assume that $G = (g_1, \ldots, g_r)$ is a complete intersection and we consider the system

$$\mathbf{f_c} := \begin{cases} f_0(\mathbf{x}) = \sum_{i=1}^{r} h_{i,0}(\mathbf{x}) \, g_i(\mathbf{x}) \\ \vdots \\ f_n(\mathbf{x}) = \sum_{i=1}^{r} h_{i,n}(\mathbf{x}) \, g_i(\mathbf{x}) \end{cases}$$

where $h_{i,j}(\mathbf{x}) = \sum_{|\alpha| = d_j - k_i} c_\alpha^{i,j} \mathbf{x}^\alpha$ is the generic homogeneous polynomial of degree $d_j - k_i$. We look for a condition on the coefficients $\mathbf{c} = (c_\alpha^{i,j})$ such that $\mathbf{f_c}$ has a solution "outside" the variety defined by G. Such a condition is given by the residual resultant defined in [BEM01]. This resultant is constructed as a resultant over the blow-up $\pi : \tilde{X} \to X = \mathbb{P}^n$ of $\mathbb{P}^n$ along the coherent sheaf of ideals $\mathcal{G}$ associated to G ([Har83]).

If $\tilde{\mathcal{G}}$ is the sheaf on $\tilde{X}$ inverse image of $\mathcal{G}$ by π and $\tilde{\mathcal{G}}_{d_i} = \tilde{\mathcal{G}} \otimes \pi^*(\mathcal{O}_X(d_i))$, the degree of the residual resultant in the coefficients of each f_i is $N_i = \int_{\tilde{X}} \prod_{j \neq i} c_1(\tilde{\mathcal{G}}_{d_j})$, with $c_1(\tilde{\mathcal{G}}_{d_j})$ is the first Chern class of $\tilde{\mathcal{G}}_{d_j}$. Using intersection theory [Ful98], we can give an explicit formula for N_i if G is a complete intersection. More precisely we have:

Theorem 3.4.4. *[BEM01] There exists an irreducible and homogeneous polynomial* $\mathrm{Res}_{G,d_0,\ldots,d_n}$ *in* $\mathbb{K}[\mathbf{c}]$ *which satisfies*

$$\mathrm{Res}_{G,d_0,\ldots,d_n}(f_0,\ldots,f_n) = 0 \Leftrightarrow \mathcal{Z}(F : G) \neq \emptyset.$$

Moreover, if for a fixed $j \in \{0,\ldots,n\}$ *we denote by* $\mathbf{d}$ *the n-tuple* $\mathbf{d} = (d_0,\ldots,d_{j-1},d_{j+1},\ldots,d_n)$, $\sigma_0(\mathbf{d}) = (-1)^n$, $\sigma_1(\mathbf{d}) = (-1)^{n-1}\sum_{l \neq j} d_l$, $\sigma_2(\mathbf{d}) = (-1)^{n-2}\sum_{j_1 \neq j, j_2 \neq j, j_1 < j_2} d_{j_1} d_{j_2}$, $\ldots$, $\sigma_n(\mathbf{d}) = \prod_{l \neq j} d_l$, $r_j(T) = \sigma_n(\mathbf{d}) + \sum_{l=r}^n \sigma_{n-l}(\mathbf{d}) T^l$, *and*

$$P_{r_j}(y_1,\ldots,y_r) = \det \begin{pmatrix} r_j(y_1) & \cdots & r_j(y_r) \\ y_1 & \cdots & y_r \\ \vdots & & \vdots \\ y_1^{r-1} & \cdots & y_r^{r-1} \end{pmatrix}.$$

The degree of $\mathrm{Res}_{G,d_0,\ldots,d_n}$ *in the coefficients of each polynomial f_j is*

$$N_j = \frac{P_{r_j}}{P_1}(k_1,\ldots,k_r).$$

The polynomial $\mathrm{Res}_{G,d_0,\ldots,d_n}$ is called the residual resultant. In order to compute it, let $\Delta_{i_1\ldots i_r}$ be the $r \times r$ minor of the matrix $(h_{i,j})_{1 \leq i \leq r, 0 \leq j \leq n}$ corresponding to the columns $i_1,\ldots,i_r$, $(e_0,\ldots,e_n)$ and $(\tilde{e}_0,\ldots,\tilde{e}_n)$ be two bases of the S-module S^{n+1}. A matrix whose determinant is a non-trivial multiple of $\mathrm{Res}_{G,d_0,\ldots,d_n}$ can be constructed using the following result:

Theorem 3.4.5. *[BEM01] For* $\nu \geq \nu_{\mathbf{d},\mathbf{k}} = \sum_{i=0}^n d_i - n - (n - r + 2)k_r$, *the map*

$$\partial_\nu : \left(\bigoplus_{0 \leq i_1 < \ldots < i_r \leq n} S_{\nu - d_{i_1} - \cdots - d_{i_r} + \sum_{i=1}^r k_i} e_{i_1} \wedge \ldots \wedge e_{i_r} \right) \oplus \left(\bigoplus_{i=0}^{i=n} S_{\nu - d_i} \tilde{e}_i \right) \longrightarrow S_\nu$$

$$e_{i_1} \wedge \ldots \wedge e_{i_r} \longrightarrow \Delta_{i_1\ldots i_r}$$

$$\tilde{e}_i \longrightarrow f_i$$

is surjective if and only if $\mathcal{Z}(F : G) = \emptyset$. *In this case, every nonzero maximal minors of size* $\dim_{\mathbb{K}}(S_\nu)$ *of the matrix of* ∂_ν *is a multiple of* $\mathrm{Res}_{G,d_0,\ldots,d_n}$, *and the gcd of all these minors is exactly the residual resultant.*

This result is based on the resolution of the ideal $((f_0, \ldots, f_n) : G)$ given in [BKM90].

Example 3.4.6. (The residual of two points in $\mathbb{P}^2$). We consider the following system in $\mathbb{P}^2$:

$$\begin{cases} f_0 = a_0 x_0^2 + a_1 x_0 x_1 + a_2 x_0 x_2 + a_3 (x_1^2 + x_2^2) \\ f_1 = b_0 x_0^2 + b_1 x_0 x_1 + b_2 x_0 x_2 + b_3 (x_1^2 + x_2^2) \\ f_2 = c_0 x_0^2 + c_1 x_0 x_1 + c_2 x_0 x_2 + c_3 (x_1^2 + x_2^2). \end{cases}$$

If $G = (x_0, x_1^2 + x_2^2)$, $\nu_{\mathbf{d,k}} = 2$ and a nonzero maximal minor of the matrix of ∂_ν is

$$\begin{vmatrix} a_0 & b_0 & c_0 & 0 & 0 & 0 \\ 0 & 0 & 0 & -b_1 c_3 + c_1 b_3 & -b_2 c_3 + c_2 b_3 & -c_1 a_3 + a_1 c_3 \\ a_1 & b_1 & c_1 & 0 & -c_3 b_0 + b_3 c_0 & 0 \\ c_2 & b_2 & c_2 & -c_3 b_0 + b_3 c_0 & 0 & a_0 c_3 - c_0 a_3 \\ a_3 & b_3 & c_3 & 0 & -b_1 c_3 + c_1 b_3 & 0 \\ a_3 & b_3 & c_3 & -b_2 c_3 + c_2 b_3 & 0 & -c_2 a_3 + a_2 c_3 \end{vmatrix}.$$

The formula for the degrees gives $N_0 = N_1 = N_2 = 2$ and we check that this minor is the residual resultant times $c_3(c_1 b_3 - c_3 b_1)$. It has the minimal degree N_0 in the coefficients of f_0. In this example the projective and toric resultants vanish identically.

Example 3.4.7. (The residual of a curve in $\mathbb{P}^3$). We consider the following system of cubics in $\mathbb{P}^3$ containing the umbilic:

$$\begin{cases} f_0 = (a_0 x_0 + a_1 x_1 + a_2 x_2 + a_3 x_3)(x_0^2 + x_1^2 + x_2^2) + (a_4 x_0^2 + a_5 x_1^2 + a_6 x_2^2 + a_7 x_3^2 + \\ \quad a_8 x_0 x_1 + a_9 x_0 x_2 + a_{10} x_0 x_3 + a_{11} x_1 x_2 + a_{12} x_1 x_3 + a_{13} x_2 x_3)x_3 \\ f_1 = (b_0 x_0 + b_1 x_1 + b_2 x_2 + b_3 x_3)(x_0^2 + x_1^2 + x_2^2) + (b_4 x_0^2 + b_5 x_1^2 + b_6 x_2^2 + b_7 x_3^2 + \\ \quad b_8 x_0 x_1 + b_9 x_0 x_2 + b_{10} x_0 x_3 + b_{11} x_1 x_2 + b_{12} x_1 x_3 + b_{13} x_2 x_3)x_3 \\ f_2 = (c_0 x_0 + c_1 x_1 + c_2 x_2 + c_3 x_3)(x_0^2 + x_1^2 + x_2^2) + (c_4 x_0^2 + c_5 x_1^2 + c_6 x_2^2 + c_7 x_3^2 + \\ \quad c_8 x_0 x_1 + c_9 x_0 x_2 + c_{10} x_0 x_3 + c_{11} x_1 x_2 + c_{12} x_1 x_3 + c_{13} x_2 x_3)x_3 \\ f_3 = (d_0 x_0 + d_1 x_1 + d_2 x_2 + d_3 x_3)(x_0^2 + x_1^2 + x_2^2) + (d_4 x_0^2 + d_5 x_1^2 + d_6 x_2^2 + d_7 x_3^2 + \\ \quad d_8 x_0 x_1 + d_9 x_0 x_2 + d_{10} x_0 x_3 + d_{11} x_1 x_2 + d_{12} x_1 x_3 + d_{13} x_2 x_3)x_3 \end{cases}$$

Let $G = (x_3, x_0^2 + x_1^2 + x_2^2)$. The previous construction gives $N_0 = N_1 = N_2 = N_3 = 15$. The size of the matrix M_ν of ∂_ν is a 84×200. A maximal minor of rank 84 whose determinant has degree 15 in the coefficients of f_0 has been constructed as follows. We extract from M_ν 69 independent columns (by considering a random specialization). We add to this submatrix the columns of M_ν depending on the coefficients of f_0 and independent of the 69 columns, in order to get a 84×84 matrix with a nonzero determinant. It yields a nonzero multiple of the residual resultant. Notice that the projective and toric resultants are identically 0 in this example.

3.5 Geometric solvers

Let us describe now how to exploit the resultant constructions to solve polynomial systems.

3.5.1 Multiplicative structure

Let $f_0, \ldots, f_n \in R$ and $\mathrm{M}_0 = \left(\begin{array}{c|c} \mathrm{M}_{00} & \mathrm{M}_{01} \\ \hline \mathrm{M}_{10} & \mathrm{M}_{11} \end{array} \right)$ be the transpose of the matrix defined in Section 2.3 of Chapter 2. Here, we use the natural convention that the columns of the resultant matrices represent multivariate polynomials.

Theorem 3.5.1. *[PS96, ER94, MP00, CLO98] For generic systems $f_1, \ldots, f_n$, the matrix of multiplication by f_0 in the basis*

$$\mathbf{x}^{E_0} = \{x_0^{\alpha_0} \ldots x_n^{\alpha_n} : 0 \le \alpha_i < \deg f_i, i = 1, \ldots, n\}$$

of $A = R/(f_1, \ldots, f_n)$ is the Schur complement of $\mathrm{M}_{1,1}$ in M_0, namely $\mathrm{M}_{f_0} = \mathrm{M}_{00} - \mathrm{M}_{01}\mathrm{M}_{11}^{-1}\mathrm{M}_{10}$.

Proof. (see also proof of Theorem 2.3.2 of Chapter 2) Since $\mathbf{x}^{E_0}$ is a basis of the quotient by the polynomials $x_1^{d_1}, \ldots, x_n^{d_n}$, it remains a basis for generic polynomials $f_1, \ldots, f_n$ of degree $d_1, \ldots, d_n$.

In order to compute the matrix of M_{f_0} in this basis, we have first to multiply the elements of the basis by f_0. This is represented in a matrix form by the block $\mathrm{C}_0 := \left(\begin{array}{c} \mathrm{M}_{00} \\ \mathrm{M}_{10} \end{array} \right)$. Then we have to reduce these polynomials in terms of the basis $\mathbf{x}^{E_0}$ by multiples of polynomials $f_1, \ldots, f_n$. The multiples that we use are represented by the coefficient matrix $\mathrm{C}_1 := \left(\begin{array}{c} \mathrm{M}_{01} \\ \mathrm{M}_{11} \end{array} \right)$. The reduction corresponds to the matrix operation $\mathrm{C}_0 - \mathrm{C}_1 \mathrm{M}_{11}^{-1}\mathrm{M}_{10}$ which yields the block

$$\mathrm{M}_{f_0} := \mathrm{M}_{00} - \mathrm{M}_{01}\mathrm{M}_{11}^{-1}\mathrm{M}_{10}.$$

Example 3.1.3 (continued). The matrix M_0 associated to the polynomials f_1, f_2 of example 3.1.3, and a generic linear form $f_0 = u_0 + u_1 x_1 + u_2 x_2$ is:

```
> M_0 := mresultant([u[0]+u[1]*x[1]+u[2]*x[2],f1,f2],[x[1],x[2]]);
```

$$\mathrm{M}_0 := \left(\begin{array}{cccc|ccc|ccc} u_0 & 0 & 0 & 0 & 0 & 0 & 2 & 0 & 0 & -\frac{1}{6} \\ ub_2 & u_0 & 0 & 0 & 2 & 0 & -8 & 0 & -\frac{1}{6} & 0 \\ ub_1 & 0 & u_0 & 0 & 0 & 2 & -8 & -\frac{1}{6} & 0 & -1 \\ 0 & u_1 & u_2 & u_0 & -8 & -8 & 8 & 0 & -1 & 1 \\ \hline 0 & 0 & u_1 & 0 & 0 & -8 & 13 & -1 & 0 & 1 \\ 0 & u_2 & 0 & 0 & -8 & 0 & 4 & 0 & 0 & 0 \\ 0 & 0 & 0 & 0 & 0 & 13 & 0 & 1 & 0 & 0 \\ 0 & 0 & 0 & 0 & 4 & 0 & 0 & 0 & 0 & 0 \\ 0 & 0 & 0 & u_1 & 13 & 8 & 0 & 1 & 1 & 0 \\ 0 & 0 & 0 & u_2 & 8 & 4 & 0 & 0 & 1 & 0 \end{array} \right).$$

In this example a basis of $\mathcal{A}$ is $S_0 = \{1, x_1, x_2, x_1 x_2\}$. The Schur complement $M_{00} - M_{01} M_{11}^{-1} M_{10}$ of M_{11} in M_0 is the 4×4 matrix:

```
> M(u):= uschur(M_0,4);
```

$$M(u) := \begin{pmatrix} u_0 & -\frac{25}{24} u_2 & \frac{1}{6} u_1 & \frac{5}{54} u_1 - \frac{5}{54} u_2 \\ u_2 \, u_0 + 2\, u_2 & 0 & \frac{2}{27} u_1 + \frac{5}{54} u_2 \\ u_1 & -\frac{5}{4} u_2 & u_0 + u_1 & \frac{55}{54} u_1 - \frac{55}{54} u_2 \\ 0 & u_1 + \frac{5}{4} u_2 & u_2 - u_1 & u_0 - u_1 + 2\, u_2 \end{pmatrix}.$$

By Theorem 3.5.1, the coefficient of u_i in $M(u)$ is the matrix of the operator M_{x_i}.

An advantage of this approach is that we have a direct matrix representation of the multiplication operator without using an algorithm to compute a normal form in $\mathcal{A}$. This formula is a continuous function of the coefficients of input polynomials in the open set of systems such that M_{11} is invertible. Thus it can be used with approximated coefficients, which is useful in many practical applications. However the main drawback is that the size of the matrix M_0 increases very quickly with the number of variables. One way to tackle this problem consists in exploiting the structure of the matrices (i.e. their sparsity and quasi-Toeplitz structure) as described in [MP00, BMP00]. Another way to handle it and to keep a continuous representation of the matrix of multiplication has been proposed in [MT00]. In some sense, it combines the previous resultant approach with the normal form method proposed in section 3.1.4, replacing the computation of a big Schur complement $M_{00} - M_{01} M_{11}^{-1} M_{10}$ by the inversion of much smaller systems.

In the next table, we compare the size of different systems to invert (first lines) with the size m of the matrix M_{11} to invert in Macaulay's formulation, in the case of projective resultants of quadrics ($d_i = 2$) in $\mathbb{P}^n$. Here D is the Bézout bound or the dimension of the $\mathbb{K}$-vector space $\mathcal{A}$.

n	5	6	7	8	9	10	11
	5	6	7	8	9	10	11
	20	30	42	56	72	90	110
	30	**60**	105	168	252	360	495
	20	**60**	**140**	**280**	504	840	1320
	5	30	105	**280**	**630**	**1260**	2310
		6	42	168	504	**1260**	**2772**
			7	56	252	840	2310
				8	72	360	1320
					9	90	495
						10	110
							11
Σ	80	192	448	1024	2304	5120	11264
m	430	1652	6307	24054	91866	351692	1350030
D	32	64	128	256	512	1024	2048

3.5.2 Solving by hiding a variable

Another approach to solve a system of polynomial equations consists in *hiding* a variable (that is, in considering one of the variables as a *parameter*), and in searching the values of this hidden variable for which the system has a solution. Typically, if we have n equations $f_1 = 0, \ldots, f_n = 0$ in n variables, we "hide" a variable, say x_n, and apply one of resultant constructions described before to the overdetermined system $f_1 = 0, \ldots, f_n = 0$ in the $n - 1$ variables $x_1, \ldots, x_{n-1}$ and a parameter x_n. This leads to a resultant matrix $\mathsf{S}(x_n)$ with polynomial entries in x_n. It can be decomposed as

$$\mathsf{S}(x_n) = \mathsf{S}_d\, x_n^d + \mathsf{S}_{d-1} x_n^{d-1} + \cdots + \mathsf{S}_0,$$

where S_i has coefficients in $\mathbb{K}$ and the same size than $\mathsf{S}(x_n)$. We look for the values ζ_n of x_n for which the system has a *solution* $\zeta' = (\zeta_1, \ldots, \zeta_{n-1})$ in the corresponding variety X' (of dimension $n - 1$) associated with the resultant formulation. This implies that

$$\mathbf{v}(\zeta')^{\mathsf{t}}\, \mathsf{S}(\zeta_n) = \mathbf{0}, \tag{3.10}$$

where $\mathbf{v}(\zeta')$ is the vector of monomials indexing the rows of S evaluated at ζ'. Conversely, for generic systems of the corresponding resultant formulation there is only one point ζ' above the value ζ_n. Thus the vectors $\mathbf{v}$ satisfying $\mathsf{S}(\zeta_n)^{\mathsf{t}}\, \mathbf{v} = \mathbf{0}$ are scalar multiples of $\mathbf{v}(\zeta')$. From the entries of these vectors, we can deduce the other coordinates of the point ζ'. This will be assumed hereafter[6].

The relation (3.10) implies that $\mathbf{v}(\zeta')$ is a generalized eigenvector of $\mathsf{S}^t(x_n)$. Computing such vectors can be transformed into the following linear generalized eigenproblem

$$\left(\begin{bmatrix} \mathbf{0} & \mathbb{I} & \cdots & \mathbf{0} \\ \vdots & \ddots & \ddots & \vdots \\ \mathbf{0} & \cdots & \mathbf{0} & \mathbb{I} \\ \mathsf{S}_0^t & \mathsf{S}_1^t & \cdots & \mathsf{S}_{d-1}^t \end{bmatrix} - \zeta_n \begin{bmatrix} \mathbb{I} & \mathbf{0} & \cdots & \mathbf{0} \\ \mathbf{0} & \ddots & \ddots & \vdots \\ \vdots & \ddots & \mathbb{I} & \mathbf{0} \\ \mathbf{0} & \cdots & \mathbf{0} & -\mathsf{S}_d^t \end{bmatrix} \right) \mathbf{w} = 0. \tag{3.11}$$

The set of eigenvalues of (3.11) contains the values of ζ_n for which (3.10) has a solution. The corresponding eigenvectors $\mathbf{w}$ are decomposed as $\mathbf{w} = (\mathbf{w}_0, \ldots, \mathbf{w}_{d-1})$ so that the solution vector $\mathbf{v}(\zeta')$ of (3.10) is

$$\mathbf{v}(\zeta') = \mathbf{w}_0 + \zeta_n \mathbf{w}_1 + \cdots + \zeta_n^{d-1} \mathbf{w}_{d-1}.$$

This yields the following algorithm:

---

[6] Notice however that this genericity condition can be relaxed by using duality, in order to compute the points ζ' above ζ_n (when they form a zero-dimensional fiber) from the eigenspace of $\mathsf{S}(\zeta_n)$.

Algorithm 3.5.2 SOLVING BY HIDING A VARIABLE.

INPUT: $f_1, \ldots, f_n \in R$.

1. *Construct the resultant matrix* $S(x_n)$ *of* $f_1, \ldots, f_n$ *(as polynomials in* $x_1, \ldots, x_{n-1}$, *with coefficients in* $\mathbb{K}[x_n]$*) adapted to the geometry of the problem.*
2. *Solve the generalized eigenproblem* $S(x_n)\, \mathbf{v} = \mathbf{0}$.
3. *Deduce the coordinates of roots* $\zeta = (\zeta_1, \ldots, \zeta_n)$ *of* $f_1 = \cdots = f_n = 0$.

OUTPUT: *The roots of* $f_1 = \cdots = f_n = 0$.

Here again, we reduce the resolution of $f_1 = 0, \ldots, f_n = 0$ to an eigenvector problem.

Example 3.5.3. We illustrate this algorithm on the system

$$\begin{cases} f_1 = x_1\, x_2 + x_3 - 2 \\ f_2 = x_1{}^2 x_3 + 2\, x_2\, x_3 - 3 \\ f_3 = x_1\, x_2 + x_2{}^2 + x_2\, x_3 - x_1\, x_3 - 2. \end{cases}$$

We hide x_3 and use the projective resultant formulation (see Section 2.3 in Chapter 2). We obtain a 15×15 matrix $S(x_3)$, and compute its determinant:

```
> S:=mresultant([f1,f2,f3],[t1,t2]):det(S);
```

$$\det(S) := x_3{}^4 \, (x_3 - 1) \left(2\, x_3{}^5 - 11\, x_3{}^4 + 20\, x_3{}^3 - 10\, x_3{}^2 + 10\, x_3 - 27 \right).$$

The root $x_3 = 0$ does not yield an affine root of the system $f_1 = f_2 = f_3 = 0$ (the corresponding point is at infinity). Substituting $x_3 = 1$ in $S(x_3)$, we get a matrix of rank 14. The kernel of $S(1)^t$ is generated by

$$(1, 1, 1, 1, 1, 1, 1, 1, 1, 1, 1, 1, 1, 1, 1).$$

This implies that the corresponding root is $(1, 1, 1)$. For the other eigenvalues (which are the roots of the last factor in $\det(S)$), we proceed similarly in order to obtain the 5 other (simple) roots of $f_1 = f_2 = f_3 = 0$. Here are numerical approximation of these roots:

$$(0.511793 - 1.27671\, \mathrm{i}, 0.037441 + 1.92488\, \mathrm{i}, -0.476671 - 0.937337\, \mathrm{i}),$$
$$(0.511793 + 1.27671\, \mathrm{i}, 0.037441 - 1.92488\, \mathrm{i}, -0.476671 + 0.937337\, \mathrm{i}),$$
$$(-1.38186 + 0.699017\, \mathrm{i}, -0.171994 + 0.704698\, \mathrm{i}, 2.25492 + 1.09402\, \mathrm{i}),$$
$$(-1.38186 - 0.699017\, \mathrm{i}, -0.171994 - 0.704698\, \mathrm{i}, 2.25492 - 1.09402\, \mathrm{i}),$$
$$(0.0734678, 0.769107, 1.9435).$$

3.5.3 Isolated points from resultant matrices

In this section, we consider n equations $f_1, \ldots, f_n$ in n unknowns, but we do not assume necessarily that they define a finite affine variety $\mathcal{Z}(f_1, \ldots, f_n)$.

We are interested in computing a rational univariate representation of the isolated points of this variety. We denote by I_0 the intersection of the primary components of $I = (f_1, \ldots, f_n)$ corresponding to isolated points of $\mathcal{Z}(I)$ and $\mathcal{Z}_0 = \mathcal{Z}(I_0)$. We denote by $\mathcal{C}_0(\mathbf{u})$ the Chow form associated to the ideal I_0 (see Section 3.2.3).

First we consider that $I = I_0$. Let $f_0 = u_0 + u_1 x_1 + \cdots + u_n x_n$ be a generic affine form (the u_i are considered as variables). We choose one of the previous resultant constructions for $f_0, \ldots, f_n$ which yields a matrix

$$M_0 = \begin{pmatrix} M_{00} & M_{01} \\ M_{10} & M_{11} \end{pmatrix}$$

such that M_{11} is invertible (if it exists). The blocks M_{00}, M_{10} depend only on the coefficients of f_0. From Section 3.5.1 and according to the relation

$$\begin{pmatrix} M_{00} & M_{01} \\ M_{10} & M_{11} \end{pmatrix} \begin{pmatrix} \mathbb{I} & \mathbf{0} \\ -M_{11}^{-1}M_{10} & \mathbb{I} \end{pmatrix} = \begin{pmatrix} M_{00} - M_{01}M_{11}^{-1}M_{10} & M_{01} \\ \mathbf{0} & M_{11} \end{pmatrix}$$

we deduce that $\det(M_0) = \det(M_{f_0}) \det(M_{11})$. This means that $\det(M_0)$ is a scalar multiple of the Chow form of the ideal I. Such a construction applies for a system which is generic for one of the mentioned resultant formulations. We can obtain a rational univariate representation of $\mathcal{Z}(I)$ applying Algorithm 3.2.10.

If the affine variety $\mathcal{Z}(I)$ is not finite, we can still deduce a rational univariate representation of the isolated points from the previous resultant construction in (at least) two ways.

When the system is not generic for a given construction, a perturbation technique can be used. Introducing a new parameter ϵ and considering a perturbed system f_ϵ (for instance $f_\epsilon = f + \epsilon f_0$), we obtain a resultant matrix $S_\epsilon(\mathbf{u})$ whose determinant is of the form

$$\Delta(\mathbf{u}, \epsilon) = \epsilon^k \Delta_k(\mathbf{u}) + \epsilon^{k+1} \Delta_{k+1}(\mathbf{u}) + \cdots \quad \text{with} \quad \Delta_k \neq 0.$$

It can be shown that $\Delta_k(\mathbf{u})$ is a multiple of the Chow form of I_0. Applying Algorithm 3.2.10 to this multiple of the Chow form yields a rational univariate representation of $\mathcal{Z}_0$ (see [Gri86, Chi86, Can90, GH91, LL91] for more details).

The use of a new parameter ϵ has a cost that we want to remove. This can be done by exploiting the properties of the Bezoutian matrix.

Proposition 3.5.4. *[EM99a, BEM00] Any nonzero maximal minor $\Delta(\mathbf{u})$ of the Bezoutian matrix of polynomials $f_0 = u_0 + u_1 x_1 + \cdots + u_n x_n, f_1, \ldots, f_n$ is divisible by the Chow form $\mathcal{C}_0(\mathbf{u})$ of the isolated points of $I = (f_1, \ldots, f_n)$.*

The interesting point here is that we get directly the Chow form of the isolated points of $\mathcal{Z}(I)$ even if this variety is not finite. In other words, we do not need to perturb the system for computing a multiple of $\mathcal{C}_0(\mathbf{u})$. Another advantage of this approach is that it yields an "explicit" formulation for $\Delta(\mathbf{u})$, and its structure can be handled more carefully (for instance, by working directly on the matrix form instead of dealing with the expansion of minors). So we have the following algorithm:

Algorithm 3.5.5 RATIONAL UNIVARIATE REPRESENTATION OF THE ISO-LATED POINTS.

INPUT : $f_1, \ldots, f_n \in \mathbb{K}[x_1, \ldots, x_n]$

1. *Compute a nonzero multiple $\Delta(\mathbf{u})$ of the Chow form of $f_1, \ldots, f_n$, from an adapted resultant formulation of $f_0 = u_0 + u_1 x_1 + \cdots + u_n x_n, f_1, \ldots, f_n$ (for instance using the Bezoutian matrix).*
2. *Get a rational univariate representation of the isolated (and maybe some embedded) roots of $f_1 = \cdots = f_n = 0$ by applying Algorithm 3.2.10.*

In practice, instead of expanding completely the polynomial $d(\mathbf{t} + \mathbf{u})$ in Algorithm 3.2.10, it would be advantageous to consider $u_1, \ldots, u_n$ as *infinitesimal numbers* (i.e. $u_i^2 = u_i u_j = 0$ for $i, j = 1, \ldots, n$) in order to get only the first terms $d_0(u_0) + u_1 d_1(u_0) + \cdots + u_n d_n(u_0)$ of the expansion of $d(\mathbf{t} + \mathbf{u})$. Moreover, we can describe these terms as sums of determinants of matrices deduced from resultant matrices. This allows us to use fast interpolation methods to compute efficiently $d_0(u_0), \ldots, d_n(u_0)$.

3.5.4 Solving overdetermined systems

In many problems (such as in reconstruction in computer vision, autocalibration in robotics, identification of sources in signal processing, ...), each observation yields an equation. Thus, we can generate as many (approximated) equations as we want but usually only one solution is of (physical) interest. Thus we are dealing with overconstrained systems which have approximate coefficients (due to measurement errors for instance).

Here again we are interested in matrix methods which allow us to handle systems with approximate coefficients. The methods of the previous sections for the construction of resultant matrices M_0 admit natural generalizations [Laz77] to overconstrained systems, that is, to systems of equations $f_1 = \ldots = f_m = 0$, with $m > n$, defining a finite number of roots. We consider a map of the form

$$\mathcal{S} : \mathcal{V}_1 \times \cdots \times \mathcal{V}_m \to \mathcal{V}$$

$$(q_1, \ldots, q_m) \mapsto \sum_{i=1}^{m} f_i \, q_i$$

where $\mathcal{V}$ and $\mathcal{V}_i$ are linear subspaces generated by monomials of R. This yields a rectangular matrix S.

A case of special interest is when this matrix is of rank $N - 1$, where N is the number of rows of S. In this case, it can be proved [EM] that $\mathcal{Z}(f_1, \ldots, f_m)$

is reduced to one point $\zeta \in \mathbb{K}^n$, and if $\mathbf{x}^F = (\mathbf{x}^\alpha)_{\alpha \in F}$ is the set of monomials indexing the rows of S that

$$(\zeta^\alpha)_{\alpha \in F} \, S = 0.$$

Using Cramer's rule, we see that $\zeta^\alpha / \zeta^\beta$ $(\alpha, \beta \in F, \zeta^\beta \neq 0)$ can be expressed as the ratio of two maximal minors of S. If $1, x_1, \ldots, x_n \in \mathbf{x}^F$ (which is the case most of the time), we obtain ζ as a rational function of maximal minors of S, and thus of input coefficients of $f_1, \ldots, f_m$.

Algorithm 3.5.6 SOLVING AN OVERCONSTRAINED SYSTEM DEFINING A SINGLE ROOT

INPUT: *A system* $f_1, \ldots, f_m \in \mathbb{K}[x_1, \ldots, x_n]$ *(with* $m > n$*) defining a single solution.*

1. *Compute the resultant matrix* S *for one of the proposed resultant formulations.*
2. *Compute the kernel of* S *and check that it is generated by one vector* $\mathbf{w} = (\mathbf{w}_1, \mathbf{w}_{x_1}, \ldots, \mathbf{w}_{x_n}, \ldots)$.

OUTPUT: $\zeta = \left(\frac{\mathbf{w}_{x_1}}{\mathbf{w}_1}, \ldots, \frac{\mathbf{w}_{x_n}}{\mathbf{w}_1}\right)$.

Let us illustrate this algorithm, with a projective resultant construction.

Example 3.5.7. We consider the case of 3 conics:

```
> f1:= x1^2-x1*x2+x2^2-3;
> f2:= x1^2-2*x1*x2+x2^2+x1-x2;
> f3:= x1*x2+x2^2-x1+2*x2-9;
> S:=mresultant([f1,f2,f3],[x1,x2]);
```

$$S := \begin{pmatrix}
-3 & 0 & 0 & 0 & 0 & 0 & 0 & 0 & 0 & 0 & 0 & -9 & 0 & 0 & 0 \\
0 & -3 & 0 & 0 & 0 & 0 & -1 & 0 & 0 & 0 & -9 & 2 & 0 & 0 & 0 \\
0 & 0 & -3 & 0 & 0 & 0 & 1 & 0 & 0 & 0 & 0 & -1 & 0 & 0 & -9 \\
-1 & 0 & 0 & -3 & 0 & -1 & -2 & 0 & 1 & 0 & -1 & 1 & -9 & 0 & 2 \\
0 & 1 & -1 & 0 & -1 & -2 & 0 & 1 & 1 & 0 & 0 & 0 & -1 & 2 & 1 \\
0 & -1 & 1 & 0 & 0 & 1 & 0 & -1 & -2 & -1 & 1 & 0 & 2 & 0 & 1 \\
0 & 0 & 0 & 1 & -2 & 0 & 0 & 1 & 0 & 0 & 0 & 0 & 0 & 1 & 0 \\
0 & 0 & 0 & -1 & 1 & 0 & 0 & -2 & 0 & 0 & 0 & 0 & 1 & 1 & 0 \\
0 & 0 & 0 & 1 & 0 & 0 & 0 & 1 & 0 & 1 & 0 & 0 & 1 & 0 & 0 \\
1 & 0 & 0 & 0 & 0 & 1 & 1 & 0 & 0 & 0 & 0 & 0 & 0 & -9 & -1 \\
1 & 0 & 0 & 0 & 0 & 0 & 1 & 0 & -1 & -9 & 2 & 1 & 0 & 0 & 0 \\
0 & 0 & 1 & 0 & 1 & 1 & 0 & 0 & 0 & 0 & 0 & 0 & 0 & -1 & 0 \\
0 & 1 & 0 & 0 & 0 & 0 & 0 & 0 & 1 & 2 & 1 & 0 & 0 & 0 & 0 \\
0 & 0 & 0 & 0 & 1 & 0 & 0 & 0 & 0 & 0 & 0 & 0 & 0 & 0 & 0 \\
0 & 0 & 0 & 0 & 0 & 0 & 0 & 0 & 0 & 1 & 0 & 0 & 0 & 0 & 0
\end{pmatrix}.$$

The rows of S are indexed by

$$(1, x_2, x_1, x_1 x_2, x_1{}^2 x_2, x_1 x_2{}^2, x_1{}^3 x_2, x_1{}^2 x_2{}^2, x_1 x_2{}^3, x_1{}^2, x_2{}^2, x_1{}^3, x_2{}^3, x_1{}^4, x_2{}^4).$$

We compute the kernel of S^t in order to check its rank and to deduce the common root ζ of the system:

```
> kernel(transpose(S));
```

$$\{(1,2,1,2,2,4,2,4,8,1,4,1,8,1,16)\}.$$

Considering the list of monomials which index the rows of S we deduce that $\zeta = (1,2)$.

In case that the overdetermined system has more than one root, we can follow the same approach. We chose a subset E of F (if possible containing the monomials $1, x_1, \ldots, x_n$) such that the rank of the matrix indexed by the monomials $\mathbf{x}^{F \setminus E}$ is the rank $r = N - D$ of S. The set $\mathbf{x}^E$ will be the basis of $\mathcal{A}$. Assuming that the monomials $x_i \mathbf{x}^E$, $i = 1, \ldots, n$, are also in $\mathbf{x}^F$, we complete the matrix S with the block of coefficients of $f_0 \mathbf{x}^{E_0}$, where $f_0 = u_0 + u_1 x_1 + \cdots + u_n x_n$. By a Schur complement computation, we deduce the matrix of multiplication by f_0 in the basis $\mathbf{x}^E$ of $\mathcal{A}$. Now, by applying the algorithms of Section 3.2.2, we deduce the roots of the overdetermined system $f_1, \ldots, f_m$ (see [EM99b] for more details on this approach).

3.6 Applications

We will use the tools and methods developed above to solve some problems coming from several areas of applications.

3.6.1 Implicitization of a rational surface

A rational surface (S) in $\mathbb{K}^3$ may be represented by a parametric representation:

$$(S) : x = \frac{f(s,t)}{d_1(s,t)} \quad , \quad y = \frac{g(s,t)}{d_2(s,t)} \quad , \quad z = \frac{h(s,t)}{d_3(s,t)} \quad ,$$

where $f, g, h, d_1, d_2, d_3 \in \mathbb{K}[s,t]$ or by an implicit equation (i.e. $F \in \mathbb{K}[x,y,z]$ of minimal degree satisfying $F(a,b,c) = 0$ for all $(a,b,c) \in (S)$). These two representations are important for different reasons. For instance, the first one is useful for drawing (S) and the second one to intersect surfaces or to decide whether a point is in (S) or not.

We will investigate the implicitization problem, that is the problem of converting a parametric representation of a rational surface into an implicit one.

These last decades have witnessed a renewal of this problem motived by applications in computer-aided geometric design and geometric modelling ([SAG84], [Buc88a], [Hof89], [Kal91], [CM92], [AGR95], [CGZ00], [AS01], [CGKW01]). Its solution is given by resultants, Gröbner bases, moving surfaces (see [SC95], [BCD03], [D'A01]). The techniques based on resultants and moving surfaces fail in the presence of base points (i.e. common roots of f, g, h, d_1, d_2, d_2). The Gröbner bases methods are fairly expensive in practice even if the dimension is small. Recently, methods using residual resultants and approximation complexes have been proposed but only under some restrictive geometric hypotheses on the zero-locus of base points which are difficult to verify ([Bus01b], [BJ03], [BC]). We propose an approach based on the residue calculus extending [GV97]. This method works in the presence of base points and no geometric hypotheses on the zero-locus of base points are needed.

In order to find an implicit equation of (S), as in Proposition 3.3.6 we can compute a nonzero maximal minor of the Bezoutian matrix of polynomials $xd_1 - f, yd_2 - g, zd_3 - h$ with respect to s, t. In general, this yields a multiple of the implicit equation as shown below.

Example 3.6.1. Let (S) be the surface parameterized by

$$x = s \quad, \quad y = \frac{t^2 s + 2t + s}{t^2} \quad, \quad z = \frac{t^2 - 2st - 1}{t^2}.$$

The Bezoutian matrix of $x - s, yt^2 - t^2 s - 2t - s, zt^2 - t^2 + 2ts + 1$ in $(\mathbb{K}[x, y, z])[s, t]$ is a 4×4 matrix.

```
> melim([x*d1-f,y*d2-g,z*d3-h],[s,t]);
```

$$(z - 1)^2 (4x^4 - 4x^3 y + x^2 z^2 - 8x^2 z + 2xyz + 4x^2 + y^2 + 4z - 4).$$

The second factor in this expression is the expected implicit equation.

The use of the Bezoutian matrix produces an extraneous term along with the implicit equation. We will see how to use the residue calculus in order to remove it from this equation.

Let us consider the polynomials in $(\mathbb{K}[x, y, z])[s, t]$

$$\begin{cases} F(s, t) = x \, d_1(s, t) - f(s, t) \\ G(s, t) = y \, d_2(s, t) - g(s, t) \\ H(s, t) = z \, d_3(s, t) - h(s, t). \end{cases}$$

Let $\mathcal{Z}_0 = \{\zeta \in \overline{\mathbb{K}(y, z)}^2 : G(\zeta) = H(\zeta) = 0\} = \mathcal{Z}_1 \cup \mathcal{Z}_2$, where $\mathcal{Z}_1$ is the algebraic variety $\mathcal{Z}_0 \cap \mathcal{Z}(d_1 d_2 d_3) = \{\zeta \in \overline{\mathbb{K}(y, z)}^2 : G(\zeta) = H(\zeta) = d_1 d_2 d_3(\zeta) = 0\}$ and $\mathcal{Z}_2 = \mathcal{Z}_0 \setminus \mathcal{Z}_1$. If $\mathcal{Z}_2$ is finite, let $Q(x, y, z)$ be the following nonzero element

$$Q(x, y, z) = \prod_{\zeta \in \mathcal{Z}_2} F(\zeta) = \left(\prod_{\zeta \in \mathcal{Z}_2} d_1(\zeta)\right)\left(x^m + \sigma_1(y, z)x^{m-1} + \cdots + \sigma_m(y, z)\right)$$

where m is the number of points (counting their multiplicities) in $\mathcal{Z}_2$ and $\sigma_i(y, z)$ is the i-th elementary symmetric function of $\{\frac{f(\zeta)}{d_1(\zeta)} : \zeta \in \mathcal{Z}_2\}$.

Theorem 3.6.2. *The implicit equation of the surface* (S) *is the square-free part of the numerator of*

$$E(x, y, z) := x^m + \sigma_1(y, z)x^{m-1} + \cdots + \sigma_m(y, z) \in \mathbb{K}(y, z)[x].$$

Proof. Let us choose a point (y_0, z_0) in the open subset U of $\overline{\mathbb{K}}^2$ such that the specialization $\tilde{\mathcal{Z}}_2$ of $\mathcal{Z}_2$ is finite in $\overline{\mathbb{K}}^2$ and the denominators of $\sigma_1, \ldots, \sigma_m$ do not vanish. Then we have $Q(x_0, y_0, z_0) = 0$ if and only if

$$x_0^m + \sigma_1(y_0, z_0)x_0^{m-1} + \cdots + \sigma_m(y_0, z_0) = 0 \,,$$

which is equivalent to the existence of an element $\zeta_0 \in \tilde{\mathcal{Z}}_2$ such that $x_0 = \frac{f(\zeta_0)}{d_1(\zeta_0)}$. In other words, the numerator of $E(x, y, z)$ vanishes on a point $(x_0, y_0, z_0) \in U$ if and only if it belongs to (S), which implies that the square-free part of the numerator of $E(x, y, z)$ is up to a scalar the implicit equation of the surface (S). $\square$

The coefficients $\sigma_i(y, z)$ in Theorem 3.6.2 can be computed using the Newton identities (3.8). So we need to compute the Newton sums $S_i(y, z) = \sum_{\zeta \in \mathcal{Z}_2}\left(\frac{f(\zeta)}{d_1(\zeta)}\right), i = 0, \ldots, m$. By adding a variable we can assume that $d_1 = 1$.

Algorithm 3.6.3 IMPLICITIZATION OF A RATIONAL SURFACE

INPUT: *Polynomials* f, g, h, d_1, d_2, d_3 *in* $\mathbb{K}[s, t]$.

1. *Compute an algebraic relation* $A_s(u_0, u_1, u_2)$ *(resp.* $A_t(u_0, u_1, u_2)$*) between* s, $G = y\, d_2 - g$, $H = z\, d_3 - h$ *(resp.* t, G, H*) in* $\mathbb{K}[y, z][s, t]$.
   - *If the univariate polynomials* $R_s = A_s(s, 0, 0)$, $R_t = A_t(t, 0, 0)$ *do not vanish identically (which is often the case), let* M *be the* 2×2 *matrix such that* $\begin{pmatrix} R_s \\ R_t \end{pmatrix} = M \begin{pmatrix} G \\ H \end{pmatrix}$.
   – *Compute the degree*

   $$m = \tau_{(G,H)}\big(\mathrm{Jac}(G, H)\big) = \tau_{(R_s, R_t)}\big(\mathrm{Jac}(G, H)\det(M)\big)$$

   in x *of the polynomial* $E(x, y, z) \in \mathbb{K}(y, z)[x]$ *in Theorem 3.6.2.*
   – *For* i *from 1 to* m*, compute*

   $$S_i(y, z) = \tau_{(G,H)}\big(\mathrm{Jac}(G, H)f^i\big) = \tau_{(R_s, R_t)}\big(\mathrm{Jac}(G, H)\det(M)f^i\big).$$

   - *If the polynomial* $R_s R_t \equiv 0$*, the power sums* $S_i(y, z)$*, for* $i = 0, \ldots, m$*, are computed using the algebraic relations* $A_s(u_0, u_1, u_2)$*,* $A_t(u_0, u_1, u_2)$ *and the formula in Proposition 3.3.8.*
2. *Use the Newton identities (3.8) to obtain the elementary symmetric functions* $\sigma_i(y, z)$ *from the Newton sums* $S_i(y, z)$*,* $i = 1, \ldots, m$*.*

OUTPUT: *The numerator of* $x^m + \sigma_1(y, z)x^{m-1} + \cdots + \sigma_m(y, z) \in \mathbb{K}(y, z)[x]$*.*

Example 3.6.1 (continued). In this case, the univariate polynomials R_s and R_t are equal to

$$R_s = -4 + 4z^3 + 4s^4 + 4s^2 + 21s^2z^2 - 16s^2z - 4s^3y - 12z^2 - 10z^3s^2$$
$$+z^4s^2 - 8zs^4 + 4z^2s^4 + 2yz^3s + 8ys^3z - 4ys^3z^2 - 4z^2ys$$
$$+2zsy + y^2 - 2y^2z + z^2y^2 + 12z,$$
$$R_t = 4z - 4 - 8t^3y + 8t^3yz + 16t^2 - 20t^2z + 4z^2t^2 + 4t^4z^2 - 8t^4z + 4t^4.$$

The computation of the Newton sums gives

$$S_0 = 4, S_1 = y, S_2 = -\frac{1}{2}z + 4z + y - 2, \quad S_3 = \frac{1}{4}y(-3z^2 + 18z - 12 + 4y^2)$$
$$S_4 = \frac{1}{8}z^4 - 2z^3 - z^2y^2 + 9z^2 - 12z + 6y^2z + y^4 - 5y^2 + 6.$$

And the implicit equation of (S) is

$$x^4 - x^3y + \frac{1}{4}x^2z^2 - 2x^2z + x^2 + \frac{1}{2}zxy + z + \frac{1}{4}y^2 - 1.$$

3.6.2 The position of a camera

We consider a camera which is observing a scene. In this scene, three points A, B, C are identified. The center of the camera is denoted by X. We assume that the camera is calibrated, that is, we know the focal distance, the projection of the center of the camera, ... Then, we easily deduce the angles between the rays XA, XB, XC from the images of the points A, B, C.

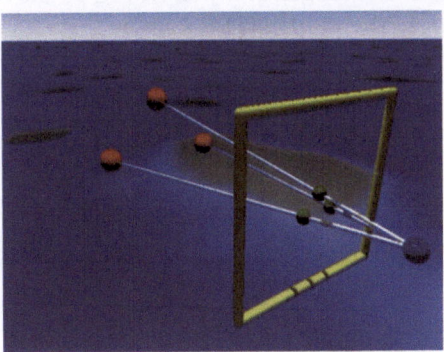

We denote by α the angle between XB and XC, β the angle between XA and XC, γ between XA and XB. These angles are deduced from the measurements in the image. We also assume that the distances a between B and C, b

between A and C, c between A and B are known. This leads to the following system of polynomial constraints:

$$\begin{cases} x_1^2 + x_2^2 - 2\cos(\gamma)x_1x_2 - c^2 = 0 \\ x_1^2 + x_3^2 - 2\cos(\beta)x_1x_3 - b^2 = 0 \\ x_2^2 + x_3^2 - 2\cos(\alpha)x_2x_3 - a^2 = 0 \end{cases} \tag{3.12}$$

where $x_1 = |XA|$, $x_2 = |XB|$, $x_3 = |XC|$. Once we know the distances x_1, x_2, x_3, the two symmetric positions of the center X are easily deduced. The system (3.12) can be solved by direct polynomial manipulations, expressing x_2 and x_3 in terms of x_1 from the two first equations and substituting in the last one. After removing the square roots, we obtain a polynomial of degree 8 in x_1, which implies at most 16 positions of the center X in this problem. Another simple way to get this equation is to eliminate the variables x_2, x_3, using the Bezoutian construction (from the MULTIRES package), and we obtain

```
> melim([f1,f2,f3], [x2,x3]);
```

$2\cos(\alpha)\left(64\cos(\beta)^2\cos(\alpha)^2\cos(\gamma)^2 - 64\cos(\beta)^3\cos(\alpha)\cos(\gamma) - 64\cos(\beta)\cos(\alpha)^3\cos(\gamma) + 16\cos(\gamma)^4\right.$
$-64\cos(\beta)\cos(\alpha)\cos(\gamma)^3 + 16\cos(\beta)^4 + 32\cos(\beta)^2\cos(\alpha)^2 + 32\cos(\beta)^2\cos(\gamma)^2 + 16\cos(\alpha)^4$
$\left.+32\cos(\alpha)^2\cos(\gamma)^2 + 64\cos(\beta)\cos(\alpha)\cos(\gamma) - 32\cos(\beta)^2 - 32\cos(\alpha)^2 - 32\cos(\gamma)^2 + 16\right)x_1^8 + \cdots$

Once this equation of degree 8 in x_1 is known, the numerical solving is easy.

3.6.3 Autocalibration of a camera

We consider here the problem of computing the intrinsic parameters of a camera from observations and measurements in 3 images of the same scene. Following the approach described in [Fau93], the camera is modeled by a pine hole projection. From the 3 images, we suppose that we are able to compute the fundamental matrices relating a pair of points in correspondence in two images. If **m, m'** are the images of a point $M \in \mathbb{R}^3$ in two photos, we have **m** F**m'**=0, where F is the fundamental matrix.

From 3 images and the 3 corresponding fundamental matrices, we deduce the so-called Kruppa equations on the 6 intrinsic parameters of the camera. See [Kru13], [Fau93] for more details. This is a system of 6 quadratic homogeneous equations in 6 variables. We solve this overdetermined system by choosing 5 equations among the six, solving the corresponding affine system and choosing the best solutions for the last equation among the 32 solutions. This took $0.38s$ on a Alpha 500Mhz workstation for the following experimentation:

Exact root	Computed root
1.049401330318981	1.049378730793354
4.884653820635368	4.884757558650871
6.011985256613766	6.011985146332036
.1726009605860270	.1725610425715577
1.727887086410446	1.727898150468536

The solver used for this computation has been developed by Ph. Trébuchet [Tré02] and is available in the library SYNAPS [DRMRT02] (see `Solve(L, newmac<C>())`).

3.6.4 Cylinders through 4 and 5 points

We consider the problem of finding cylinders through 4 or 5 points. The system that we use is described in [DMPT03].

The number of solutions for the problems that we consider are the following:

- Cylinders through 5 points: $6 = 3 \times 3 - 3$ solutions.
- Cylinders through 4 points and fixed radius: $12 = 3 \times 4$ solutions.
- Lines tangent to 4 unit balls: 12 solutions.
- Cylinders through 4 points and extremal radius: $18 = 3 \times 10 - 3 \times 4$ solutions.

Here are experimental results also performed with the solver developed by Ph. Trébuchet:

Problem	time	$max(|f_i|)$
Cylinders through 5 points	0.03s	$5 \cdot 10^{-9}$
Parallel cylinders through 2×4 points	0.03s	$5 \cdot 10^{-9}$
Cylinders through 4 points, extremal radius	2.9s	10^{-6}

The computation was performed on an Intel PII 400 128 MB of RAM. $max(|f_i|)$ is the maximum of the norm of the defining polynomials f_i evaluated at the approximated roots. The relatively huge time spent in the last problem is due to the treatment of multiple roots.

3.6.5 Position of a parallel robot

Consider a parallel robot, which is a platform controlled by 6 arms:

From the measurements of the length of the arms, we would like to know the position of the platform. This problem is a classical benchmark in polynomial system solving. We know from [RV95, Laz93, Mou93] that this problem has at most 40 solutions and that this bound is reached [Die98]. Here is the 40 degree curve that we obtain when we remove an arm of the mechanism:

The geometric constraints describing the position of the platform are transformed into a system of 6 polynomial equations:

$$\|RY_i + T - X_i\|^2 - d_i^2 = 0 \quad , \quad i = 1, \ldots, 6,$$

where R equals

$$\frac{1}{a^2 + b^2 + c^2 + d^2} \begin{pmatrix} a^2 - b^2 - c^2 + d^2 & 2\,ab - 2\,cd & 2\,ac + 2\,bd \\ 2\,ab + 2\,cd & -a^2 + b^2 - c^2 + d^2 & 2\,bc - 2\,ad \\ 2\,ac - 2\,bd & 2\,ad + 2\,bc & -a^2 - b^2 + c^2 + d^2 \end{pmatrix}$$

i.e. the rotation of the platform with respect to a reference frame, and $T = (u, v, w)$ is its translation. Using again the solver by Ph. Trébuchet and a different modelisation (with point coordinates in the first column, and quaternions in the second column), and one deduced from the residual resultant construction (in the column "redundant") as described in [Bus01a], and different numerical precision, we obtain the following results:

Direct modelisation		Quaternions		Redundant	
250 b. 3.21s	128 b. –	250 b. 8.46s	128 b. 6.25s	250 b. 1.5s	128 b. 1.2s

Here n b. denotes the number n of bits used in the computation.

3.6.6 Direct kinematic problem of a special parallel robot

Resultant constructions can also be used for some special geometry of the platform. Here is an example where two attached points of the arms on the platform are identical. We solve this problem by using the Bezoutian formulation, which yields a 20×20 matrix of polynomials in one variable. The number of complex solutions is also 40. The code for the construction of the matrix is generated in a pre-processing step and the parameters defining the geometry of the platform are instantiated at run time. This yields the following results. There are 6 real solutions, one being of multiplicity 2:

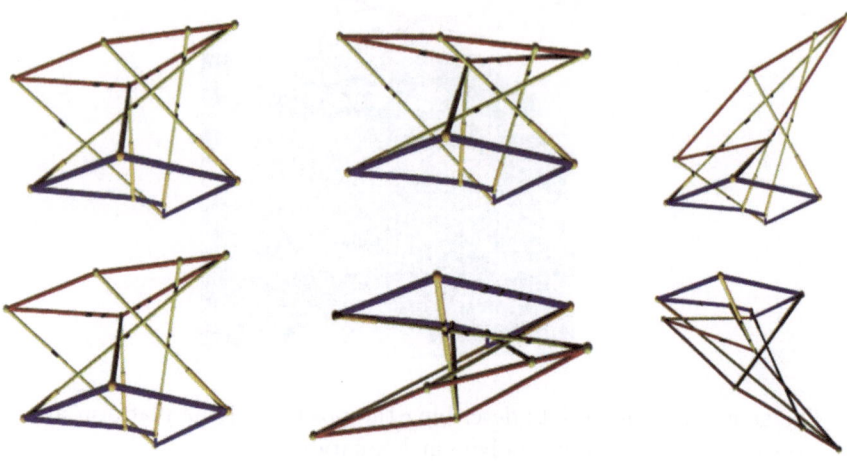

We obtain the following error $|\|RY_i + T - X_i\|^2 - d_i^2| < 10^{-6}$ and the time for solving is $0.5s$ on an Intel PII 400, 128 MB of RAM.

3.6.7 Molecular conformation

Similar resultant constructions can also be used, in order to compute the possible conformations of a molecule when the position and orientation of the links at the extremity are known. The approach is similar to the one described in [RR95]. It was developed by O. Ruatta, based on the SYNAPS library. Here also, the resultant matrix is constructed in a preprocessing step and we instantiate the parameters describing the geometry of the molecule at run-time. In this example, we obtain 6 real solutions among the 16 complex possible roots:

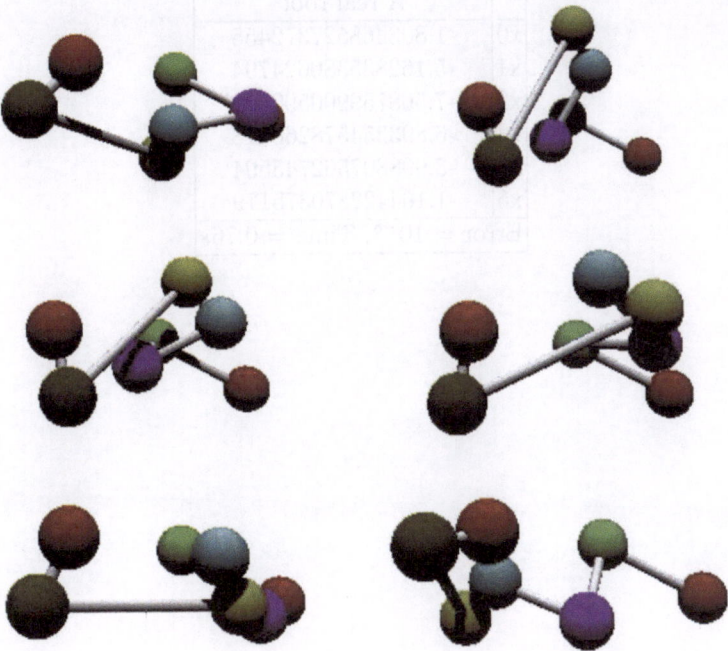

The numeric error on the solutions is bounded by 10^{-6} and the time for solving is $0.090s$, on a standard workstation.

3.6.8 Blind identification in signal processing

Finally, we consider a problem from signal processing described in detail in [GCMT02]. It is related to the transmission of an input signal $\mathbf{x}(n)$ of size p depending on the discrete time n into a convolution channel of length L. The output is $\mathbf{y}(n)$ and we want to compute the impulse response matrix $H(n)$ satisfying:

$$\mathbf{y}(n) = \sum_{m=0}^{L-1} H(m)\mathbf{x}(n-m) + \mathbf{b}(\mathbf{n}),$$

where $\mathbf{b}(n)$ is the noise. If $\mathbf{b}(n)$ is Gaussian centered, a statistic analysis of the output signal yields the equations:

$$\sum_{m=0}^{L-1}\sum_{i=1}^{p} h_{\alpha,i}(m)h_{\beta,i}(m)(-1)^{n-m} = E(y_\alpha(n)y_\beta(n-l)),$$

where $h_{\alpha,i}(m)$ are the unknowns and the $E(y_\alpha(n)y_\beta(n-l))$ are known from the output signal measurements. We solve this system of polynomial equations of degree 2 in 6 variables, which has 64 solutions for $p = 1$, with the algebraic solver of Ph. Trébuchet and we obtain the following results:

	A real root
x0	-1.803468527372455
x1	-5.162835380624794
x2	-7.568759900599482
x3	-6.893354578266418
x4	-3.998807562745594
x5	-1.164422870375179
Error $= 10^{-8}$, Time $= 0.76s$	

An algebraist's view on border bases

Achim Kehrein[1], Martin Kreuzer[1], and Lorenzo Robbiano[2]

[1] Universität Dortmund, Fachbereich Mathematik, 44221 Dortmund, Germany,
   achim.kehrein@mathematik.uni-dortmund.de
   martin.kreuzer@mathematik.uni-dortmund.de
[2] Dipartimento di Matematica, Via Dodecaneso 35, 16146 Genova, Italy,
   robbiano@dima.unige.it

Summary. This chapter is devoted to laying the algebraic foundations for border bases of ideals. Using an order ideal $\mathcal{O}$, we describe a zero-dimensional ideal *from the outside*. The first and higher borders of $\mathcal{O}$ can be used to measure the distance of a term from $\mathcal{O}$ and to define $\mathcal{O}$-border bases. We study their existence and uniqueness, their relation to Gröbner bases, and their characterization in terms of commuting matrices. Finally, we use border bases to solve a problem coming from statistics.

4.0 Introduction

El infinito tango me lleva hacia todo
[The infinite tango takes me towards everything]
(Jorge Luis Borges)

The third author was invited to teach a course at the CIMPA school in July 2003. When the time came to write a contribution to the present volume, he was still inspired by the tunes of classical tango songs which had been floating in his mind since his stay in Buenos Aires. He had the idea to create some variations on one of the themes of his lectures. Together with the first and second authors, he formed a *trio* of algebraists. They started to collect scattered phrases and tunes connected to the main theme, and to rework them into a survey on border bases. Since the idea was welcomed by the organizers, you have now the opportunity to enjoy their composition.

In the last few years it has become increasingly evident how Gröbner bases are changing the mathematical landscape. To use a lively metaphor, we can say that by considering a Gröbner basis of an ideal I in the polynomial ring $P = K[x_1, \ldots, x_n]$, we are looking at I *from the inside*, i.e. by describing a special set of generators. But a Gröbner basis grants us another perspective. We can look at I *from the outside*, i.e. by describing a set of polynomials which forms a K-vector space basis of P/I, namely the set of terms outside $\mathrm{LT}_\sigma(I)$ for some term ordering σ. However, Gröbner bases are not optimal from the latter point of view, for instance, because the bases they provide tend to be numerically ill-behaved.

This leads us to one of the main ideas behind the concept of a border basis. We want to find more "general" systems of generators of I which give rise to a K-basis of P/I. Quotation marks are in order here, since so far the generalization only works for the subclass of zero-dimensional ideals I. In the zero-dimensional case, the theory of border bases is indeed an extension of the theory of Gröbner bases, because there are border bases which cannot be associated to Gröbner bases. Moreover, border bases do not require the choice of a term ordering. Our hope is that the greater freedom they provide will make it possible to construct bases of P/I having additional good properties such as numerical stability or symmetry.

Even if these considerations convince you that studying border bases is useful, you might still ask why we want to add this survey to the current literature on that topic? Our main reason is that we believe that the algebraic foundations of border bases have not yet been laid out solidly enough. Important contributions are scattered across many publications (some in less widely distributed journals), and do not enjoy a unified terminology or a coherent set of hypotheses. We hope that this chapter can be used as a first solid foundation of a theory which will surely expand quickly.

Now let us look at the content more closely. In Section 4.1 we describe some techniques for treating pairwise commuting endomorphisms of finitely generated vector spaces. In particular, we describe a Buchberger-Möller type algorithm (see Theorem 4.1.7) for computing the defining ideal of a finite set of commuting matrices. Given pairwise commuting endomorphisms $\varphi_1, \ldots, \varphi_n$ of a finite dimensional K-vector space V, we can view V as a P-module via $f \cdot v = f(\varphi_1, \ldots, \varphi_n)(v)$ for $f \in P$ and $v \in V$. Then Theorem 4.1.9 yields an algorithm for checking whether V is a cyclic P-module, i.e. whether it is isomorphic to P/I for some zero-dimensional ideal $I \subseteq P$.

Section 4.2 is a technical interlude where order ideals, borders, indices, and marked polynomials have their solos. An order ideal is a finite set of terms which is closed under taking divisors. We use order ideals to describe a zero-dimensional ideal "from the outside". The first and higher borders of an order ideal can be used to measure the "distance" of a term from the order ideal. The main tune in Section 4.2 is played by the Border Division Algorithm 4.2.10. It imitates the division algorithm in Gröbner basis theory and allows us to divide a polynomial by a border prebasis, i.e. by a list of polynomials which are "marked" by the terms in the border of an order ideal.

And then, as true stars, border bases appear late in the show. They enter the stage in Section 4.3 and solve the task of finding a system of generators of a zero-dimensional polynomial ideal having good properties. After we discuss the existence and uniqueness of border bases (see Theorem 4.3.4), we study their relation to Gröbner bases (see for instance Propositions 4.3.6 and 4.3.9). Then we define normal forms with respect to an order ideal, and use border bases to compute them. Many useful properties of normal forms are collected in Proposition 4.3.13.

In the final part of Section 4.3, we explain the connection between border bases and commuting matrices. This variation leads to the fundamental Theorem 4.3.17 which characterizes border bases in terms of commuting matrices and opens the door for our main application. Namely, we use border bases to solve a problem coming from statistics. This application is presented in Section 4.4, where we discuss the statistical background and explain the role of border bases in this field.

Throughout the text, we have tried to provide a generous number of examples. They are intended to help the reader master the basics of the theory of border bases. Moreover, we have tried to keep this survey as self-contained and elementary as possible. When we had to quote "standard results" of computer algebra, we preferred to rely on the book by the second and third authors [KR00]. This does not mean that those results are not contained in other books on the subject; we were merely more familiar with it.

Albert Einstein is said to have remarked that the secret of creativity was to know how to hide ones sources. Since none of us is Albert Einstein, we try to mention all sources of this survey. We apologize if we are unaware of some important contribution to the topic. First and foremost, we would like to acknowledge the work of Hans J. Stetter (see [AS88], [AS89], and [Ste04]) who used border bases in connection with problems arising in numerical analysis. Later H. Michael Möller recognized the usefulness of these results for computer algebra (see [Möl93], [MS95], and [MT01]). These pioneering works triggered a flurry of further activities in the area, most notably by Bernard Mourrain (see for instance [Mou99]) from the algorithmic point of view. A good portion of the material presented here is taken from the papers [CR97], [CR01], [KK03a], [Rob98], [Robb], and [RR98]. Moreover, many results we discuss are closely related to other surveys in this volume.

Naturally, much work still has to be done; or, as we like to put it, there is still a huge TODO-list. A path which deserves further attention is the connection between border bases and numerical computation. Many ideas about the interplay of numerical and symbolic computation were proposed by Stetter, but we believe that there remains a large gap between the two areas which has to be addressed by algebraists. What about the algorithmic aspects? Almost no computer algebra system has built-in facilities for computing border bases. Naive algorithms for computing border bases, e.g. algorithms based on Gröbner basis computations, require substantial improvements in order to be practically feasible. This is an area of ongoing research. Some results in this direction are contained in Chapter 3. On the theoretical side we can ask whether the analogy between border bases and Gröbner bases can be further extended. First results in this direction are contained in [KK03a], but there appears to be ample scope for extending the algebraic theory of border bases.

Finally, wouldn't it be wonderful to remove the hypothesis that I is zero-dimensional, i.e. to develop a theory of border bases for the case when P/I is an infinite-dimensional vector space? At the moment, despite the *infinite tango*, we are unfortunately lacking the inspiration to achieve this goal. Some ideas are presented in [Ste04, Ch. 11].

As for our notation, we refer the readers to [KR00]. In particular, we let $P = K[x_1, \ldots, x_n]$ be a polynomial ring over a field K. A polynomial of the form $x_1^{\alpha_1} \cdots x_n^{\alpha_n}$, where $\alpha_1, \ldots, \alpha_n \in \mathbb{N}$, is called a **term** (or a **power product**). The monoid of all terms in P is denoted by $\mathbb{T}^n$.

4.1 Commuting endomorphisms

> *Tango has the habit of waiting*
> (Aníbal Troilo, virtuoso bandoneonist)

Every polynomial ideal I is accompanied by the quotient algebra P/I. A zero-dimensional ideal I corresponds to an algebra P/I of finite vector space dimension over K. The first part of this section reviews how the K-algebra P/I is characterized by its P-module structure and how the latter is given by n pairwise commuting multiplication endomorphisms of the K-vector space P/I. In particular, for zero-dimensional ideals these endomorphisms can be represented by pairwise commuting multiplication matrices. Then we address the converse realization problem: Which collections of n pairwise commuting matrices can be preassigned as multiplication matrices corresponding to a zero-dimensional ideal? A necessary and sufficient condition is that these matrices induce a cyclic P-module structure. Whether a P-module structure on a finite-dimensional K-vector space is cyclic can be checked effectively – an algorithm is presented in the second part.

4.1.1 Multiplication endomorphisms

Given a K-vector space V which carries a P-module structure, there exist endomorphisms of V which are associated to the multiplications by the indeterminates.

Definition 4.1.1. *For $i = 1, \ldots, n$, the P-linear map*

$$\varphi_i : V \longrightarrow V \qquad \text{defined by} \qquad v \mapsto x_i v$$

*is called the $\mathbf{i^{th}}$ **multiplication endomorphism** of V.*

The multiplication endomorphisms of V are pairwise commuting, i.e. we have $\varphi_i \circ \varphi_j = \varphi_j \circ \varphi_i$ for $i, j \in \{1, \ldots, n\}$. The prototype of such a vector space is given by the following example.

Example 4.1.2. Let $I \subseteq P$ be an ideal. The quotient algebra P/I possesses a natural P-module structure $P \times P/I \to P/I$ given by $(f, g+I) \mapsto fg+I$. Hence there are canonical multiplication endomorphisms $X_i : P/I \longrightarrow P/I$ such that $X_i(f+I) = x_i f + I$ for $f \in P$ and $i = 1, \ldots, n$. Note that P/I is a cyclic P-module with generator $1 + I$.

Remark 4.1.3. Let $\varphi_1, \ldots, \varphi_n$ be pairwise commuting endomorphisms of a vector space V. The following three constructions will be used frequently.

1. There is a natural way of equipping V with a P-module structure such that φ_i is the ith multiplication endomorphism of V, namely the structure defined by

$$P \times V \longrightarrow V \qquad \text{such that} \qquad (f, v) \mapsto f(\varphi_1, \ldots, \varphi_n)(v)$$

2. There is a ring homomorphism

$$\eta : P \longrightarrow \text{End}_K(V) \qquad \text{such that} \qquad f \mapsto f(\varphi_1, \ldots, \varphi_n)$$

3. Every ring homomorphism $\eta : P \longrightarrow \text{End}_K(V)$ induces a P-module structure on V via the rule $f \cdot v = \eta(f)(v)$.

The following result allows us to compute the **annihilator** of V, i.e. the ideal $\text{Ann}_P(V) = \{f \in P \mid f \cdot V = 0\}$.

Proposition 4.1.4. *Let V be a K-vector space equipped with a P-module structure corresponding to a ring homomorphism $\eta : P \longrightarrow \text{End}_K(V)$. Then we have $\text{Ann}_P(V) = \ker(\eta)$.*

Proof. By Remark 4.1.3, we have $f \cdot V = 0$ if and only if $\eta(f) = 0$. $\qquad \square$

Of particular interest are P-module structures on V for which V is a cyclic P-module. The following proposition shows that such structures are essentially of the type given in Example 4.1.2.

Proposition 4.1.5. *Let V be a K-vector space and a cyclic P-module. Then there exist an ideal $I \subseteq P$ and a P-linear isomorphism*

$$\Theta : P/I \longrightarrow V$$

such that the multiplication endomorphisms of V are given by the formula $\varphi_i = \Theta \circ X_i \circ \Theta^{-1}$ for $i = 1, \ldots, n$.

Proof. Let $w \in V$ be a generator of the P-module V. Then the P-linear map $\tilde{\Theta} : P \longrightarrow V$ given by $1 \mapsto w$ is surjective. Let $I = \ker \tilde{\Theta}$ be its kernel and consider the induced isomorphism of P-modules $\Theta : P/I \to V$. The P-linearity of Θ shows $\Theta(X_i(g+I)) = \varphi_i(\Theta(g+I))$ for $1 \le i \le n$ and $g + I \in P/I$. $\qquad \square$

By [KR00], Proposition 3.7.1, zero-dimensional ideals $I \subseteq P$ are characterized by $\dim_K(P/I) < \infty$. Hence, if the vector space V in this proposition is finite-dimensional, the ideal I is necessarily zero-dimensional. Now we want to answer the question, given $\varphi_1, \ldots, \varphi_n$, when is V a cyclic P-module via the structure defined in Remark 4.1.3.1? We note that if the P-module V is cyclic, then there exists an element $w \in V$ such that $\mathrm{Ann}_P(w) = \mathrm{Ann}_P(V)$.

Proposition 4.1.6 (Characterization of Cyclic P-Modules).
Let V be a K-vector space which carries the structure of a P-module.

1. *Given $w \in V$, we have $\mathrm{Ann}_P(V) \subseteq \mathrm{Ann}_P(w)$. In particular, there exists a P-linear map $\Psi_w : P/\mathrm{Ann}_P(V) \longrightarrow V$ defined by $f + \mathrm{Ann}_P(V) \mapsto f \cdot w$.*
2. *Let $w \in V$. The map Ψ_w is an isomorphism of P-modules if and only if w generates V as a P-module.*

Proof. The first claim follows from the definitions. To prove the second claim, we note that if Ψ_w is an isomorphism, then we have $V = P \cdot w$. Conversely, suppose that $V = P \cdot w$. Then the map Ψ_w is surjective. Let $f \in P$ be such that $f + \mathrm{Ann}_P(V) \in \ker(\Psi_w)$. Then $f(\varphi_1, \ldots, \varphi_n) \cdot w = 0$ implies $f(\varphi_1, \ldots, \varphi_n) = 0$ since w generates V. Hence we see that $f \in \mathrm{Ann}_P(V)$ and Ψ_w is injective. $\square$

4.1.2 Commuting matrices

In what follows, we let V be a finite-dimensional K-vector space and μ its dimension. We fix a K-basis $\mathcal{V} = (v_1, \ldots, v_\mu)$ of V. Thus every endomorphism of V can be represented by a matrix of size $\mu \times \mu$ over K. In particular, when V is a P-module, then $\mathcal{M}_1, \ldots, \mathcal{M}_n$ denote the matrices corresponding to the multiplication endomorphisms $\varphi_1, \ldots, \varphi_n$.

Using the following variant of the Buchberger-Möller algorithm, we can calculate $\mathrm{Ann}_P(V)$ as the kernel of the composite map

$$\eta : P \longrightarrow \mathrm{End}_K(V) \cong \mathrm{Mat}_\mu(K)$$

where η is the map defined in Remark 4.1.3.2. Moreover, the algorithm provides a vector space basis of $P/\mathrm{Ann}_P(V)$. To facilitate the formulation of this algorithm, we use the following convention. Given a matrix $\mathcal{A} = (a_{ij}) \in \mathrm{Mat}_\mu(K)$, we order its entries by letting $a_{ij} \prec a_{k\ell}$ if $i < k$, or if $i = k$ and $j < \ell$. In this way we "flatten" the matrix to a vector in K^{μ^2}. Then we can reduce $\mathcal{A}$ against a list of matrices by using the usual Gaußian reduction procedure.

Theorem 4.1.7 (The Buchberger-Möller Algorithm for Matrices).
Let σ be a term ordering on $\mathbb{T}^n$, and let $\mathcal{M}_1, \ldots, \mathcal{M}_n \in \mathrm{Mat}_\mu(K)$ be pairwise commuting matrices. Consider the following sequence of instructions.

M1. Start with empty lists $G = [\,], \mathcal{O} = [\,], S = [\,], N = [\,]$, and a list $L = [1]$.

M2. *If $L = [\,]$, return the pair $(G, \mathcal{O})$ and stop. Otherwise let $t = \min_\sigma(L)$ and delete it from L.*

M3. *Compute $t(\mathcal{M}_1, \ldots, \mathcal{M}_n)$ and reduce it against $N = ([\mathcal{N}_1, \ldots, \mathcal{N}_k])$ to obtain*

$$\mathcal{R} = t(\mathcal{M}_1, \ldots, \mathcal{M}_n) - \sum_{i=1}^{k} c_i \mathcal{N}_i \qquad \text{with} \qquad c_i \in K$$

M4. *If $\mathcal{R} = 0$, then append the polynomial $t - \sum_i c_i s_i$ to the list G, where s_i denotes the ith element of S. Remove from L all multiples of t. Continue with step M2.*

M5. *If $\mathcal{R} \neq 0$, then append $\mathcal{R}$ to the list N and $t - \sum_i c_i s_i$ to the list S. Append the term t to $\mathcal{O}$, and append to L those elements of $\{x_1 t, \ldots, x_n t\}$ which are neither multiples of a term in L nor in $\mathrm{LT}_\sigma(G)$. Continue with step M2.*

This is an algorithm which returns the reduced σ-Gröbner basis G of $\mathrm{Ann}_P(V)$ and a list of terms $\mathcal{O}$ whose residue classes form a K-vector space basis of $P/\mathrm{Ann}_P(V)$.

Proof. Let $I = \mathrm{Ann}_P(V)$, and let H be the reduced σ-Gröbner basis of I.

First we prove termination. In each iteration either step M4 or step M5 is performed. By its construction, the list N always contains linearly independent matrices. Hence step M5, which appends an element to N, can be performed only finitely many times. By Dickson's Lemma (see [KR00], Corollary 1.3.6), step M4 can be performed only finitely many times. Thus the algorithm terminates.

To show correctness, we prove that after a term t has been treated by the algorithm, the following holds: the list G contains all elements of H whose leading terms are less than or equal to t, and the list $\mathcal{O}$ contains all elements of $\mathbb{T}^n \setminus \mathrm{LT}_\sigma(I)$ which are less than or equal to t.

This is true after the first term $t = 1$ has been treated, i.e. appended to $\mathcal{O}$. Now suppose that the algorithm has finished an iteration. By the method used to append new terms to L in step M5, all elements of the set $(x_1\mathcal{O} \cup \cdots x_n\mathcal{O}) \setminus (\mathcal{O} \cup \mathrm{LT}_\sigma(I))$ are contained in L. From this it follows that the next term t chosen in step M2 is the smallest term in $\mathbb{T}^n \setminus (\mathcal{O} \cup \mathrm{LT}_\sigma(I))$. Furthermore, the polynomials appended to S in step M5 are supported in $\mathcal{O}$. Hence the polynomial $t - \sum_{i=1}^{k} c_i s_i$ resulting from step M3 of the next iteration has leading term t.

Now suppose that $\mathcal{R} = 0$ in step M4. By construction, the matrix of the endomorphism $\eta(s_i)$ is $\mathcal{N}_i$ for $i = 1, \ldots, k$. Therefore the polynomial $g = t - \sum_{i=1}^{k} c_i s_i$ is an element of $I = \mathrm{Ann}_P(V)$. Since the support of $\sum_{i=1}^{k} c_i s_i$ is contained in $\mathcal{O}$, the polynomial g is a new element of H.

On the other hand, if $\mathcal{R} \neq 0$ in step M5, then we claim that the term t is not contained in $\mathrm{LT}_\sigma(I)$. In view of the way we update L in step M5, the term t is not in $\mathrm{LT}_\sigma(G)$ for the current list G. By induction, the term t is not a proper multiple of a term in $\mathrm{LT}_\sigma(H)$. Furthermore, the term t is not the leading term of an element of H because such an element would be of the form $t - \sum_{i=1}^{k} c_i' s_i \in I$ with $c_i' \in K$ in contradiction to $\mathcal{R} \neq 0$. Altogether it follows that t is an element of $\mathbb{T}^n \setminus \mathrm{LT}_\sigma(I)$ and can be appended to $\mathcal{O}$.

In both cases we see that the claim continues to hold. Therefore, when the algorithm terminates, we have computed the desired lists G and $\mathcal{O}$.   □

Let us illustrate the performance of this algorithm with an example.

Example 4.1.8. Let $V = \mathbb{Q}^3$, and let $\mathcal{V} = (e_1, e_2, e_3)$ be its canonical basis. Since the two matrices

$$\mathcal{M}_1 = \begin{pmatrix} 0 & 1 & 1 \\ 0 & 2 & 1 \\ 0 & 1 & 1 \end{pmatrix} \quad \text{and} \quad \mathcal{M}_2 = \begin{pmatrix} 0 & 1 & 0 \\ 0 & 1 & 1 \\ 0 & 1 & 0 \end{pmatrix}$$

commute, they define a $\mathbb{Q}[x, y]$-module structure on V. Let us follow the iterations of the algorithm in computing the reduced σ-Gröbner basis of $\mathrm{Ann}_P(V)$, where $\sigma = \mathtt{DegLex}$.

1. $t = 1$, $L = [\ ]$, $\mathcal{R} = \begin{pmatrix} 1 & 0 & 0 \\ 0 & 1 & 0 \\ 0 & 0 & 1 \end{pmatrix} = \mathcal{I}_3$, $N = [\mathcal{I}_3]$, $S = [1]$, $\mathcal{O} = [1]$, $L = [x, y]$.

2. $t = y$, $L = [x]$, $\mathcal{R} = \begin{pmatrix} 0 & 1 & 0 \\ 0 & 1 & 1 \\ 0 & 1 & 0 \end{pmatrix} = \mathcal{M}_2$, $N = [\mathcal{I}_3, \mathcal{M}_2]$, $S = [1, y]$, $\mathcal{O} = [1, y]$,

   $L = [x, y^2]$.

3. $t = x$, $L = [y^2]$, $\mathcal{R} = \begin{pmatrix} 0 & 0 & 1 \\ 0 & 1 & 0 \\ 0 & 0 & 1 \end{pmatrix} = \mathcal{M}_1 - \mathcal{M}_2$, $N = [\mathcal{I}_2, \mathcal{M}_2, \mathcal{M}_1 - \mathcal{M}_2]$,

   $S = [1, y, x - y]$, $\mathcal{O} = [1, x, y]$, $L = [x^2, xy, y^2]$.

4. $t = y^2$, $L = [x^2, xy]$, $\mathcal{R} = \begin{pmatrix} 0 & 0 & 0 \\ 0 & 0 & 0 \\ 0 & 0 & 0 \end{pmatrix} = \mathcal{M}_2^2 - \mathcal{M}_2 - (\mathcal{M}_1 - \mathcal{M}_2)$, $G = [y^2 - x]$.

5. $t = xy$, $L = [x^2]$, $\mathcal{R} = \begin{pmatrix} 0 & 0 & 0 \\ 0 & 0 & 0 \\ 0 & 0 & 0 \end{pmatrix} = \mathcal{M}_1 \mathcal{M}_2 - 2\mathcal{M}_2 - (\mathcal{M}_1 - \mathcal{M}_2)$,

   $G = [y^2 - x, xy - x - y]$.

6. $t = x^2$, $L = [\ ]$, $\mathcal{R} = \begin{pmatrix} 0 & 0 & 0 \\ 0 & 0 & 0 \\ 0 & 0 & 0 \end{pmatrix} = \mathcal{M}_1^2 - 3\mathcal{M}_2 - 2(\mathcal{M}_1 - \mathcal{M}_2)$,

   $G = [y^2 - x, \, xy - x - y, \, x^2 - 2x - y]$.

Thus we have $\mathrm{Ann}_P(V) = (y^2 - x, \, xy - x - y, \, x^2 - 2x - y)$, and $\mathcal{O} = \{1, x, y\}$ represents a K-basis of $P / \mathrm{Ann}_P(V)$.

Now we are ready for the main algorithm of this subsection: we can check effectively whether a P-module structure given by commuting matrices defines a cyclic module.

Theorem 4.1.9 (Cyclicity Test).

Let V be a finite-dimensional K-vector space with basis $\mathcal{V} = (v_1, \dots, v_m)$, and let $\varphi_1, \dots, \varphi_n$ be pairwise commuting endomorphisms of V given by their respective matrices $\mathcal{M}_1, \dots, \mathcal{M}_n$. We equip V with the P-module structure defined by $\varphi_1, \dots, \varphi_n$. Consider the following sequence of instructions.

C1. Using Theorem 4.1.7, compute a tuple of terms $\mathcal{O} = (t_1, \dots, t_\mu)$ whose residue classes form a K-basis of $P/\operatorname{Ann}_P(V)$.

C2. If $\dim_K(V) \neq \mu$, then return "V is not cyclic" and stop.

C3. Let $z_1, \dots, z_\mu$ be further indeterminates and $\mathcal{A} \in \operatorname{Mat}_\mu(K[z_1, \dots, z_\mu])$ the matrix whose columns are $t_i(\mathcal{M}_1, \dots, \mathcal{M}_n) \cdot (z_1, \dots, z_\mu)^{\mathrm{tr}}$ for $i = 1, \dots, \mu$. Compute the determinant $d = \det(\mathcal{A}) \in K[z_1, \dots, z_\mu]$.

C4. Check if there exists a tuple $(c_1, \dots, c_\mu) \in K^\mu$ such that the polynomial value $d(c_1, \dots, c_\mu)$ is non-zero. In this case return "V is cyclic" and $w = c_1 v_1 + \cdots + c_\mu v_\mu$. Then stop.

C5. Return "V is not cyclic" and stop.

This is an algorithm which checks whether V is a cyclic P-module via $\varphi_1, \dots, \varphi_n$ and, in the affirmative case, computes a generator.

Proof. This procedure is clearly finite. Hence we only have to prove correctness. By Proposition 4.1.6, we have to check whether $\Psi_w : P/\operatorname{Ann}_P(V) \longrightarrow V$ is an isomorphism for some $w \in V$. For this it is necessary that the dimensions of the two vector spaces agree. This condition is checked in step C2. Then we use the basis elements $\{\bar{t}_1, \dots, \bar{t}_\mu\}$ and examine their images for linear independence. Since we have $\Psi_w(\bar{t}_i) = t_i(\varphi_1, \dots, \varphi_n)(w)$ for $i = 1, \dots, \mu$, the map Ψ_w is an isomorphism for some $w \in V$ if and only if the vectors $\{t_i(\mathcal{M}_1, \dots, \mathcal{M}_n)(c_1, \dots, c_\mu)^{\mathrm{tr}} \mid 1 \leq i \leq \mu\}$ are K-linearly independent for some tuple $(c_1, \dots, c_\mu) \in K^\mu$. This is checked in step C4. $\qquad\square$

If the field K is infinite, the check in step C4 can be simplified to checking $d \neq 0$. For a finite field K, we can, in principle, check all tuples in K^μ. Let us apply this algorithm by applying it in the setting of Example 4.1.8.

Example 4.1.10. Let V and $\mathcal{M}_1, \mathcal{M}_2$ be defined as in Example 4.1.8. We follow the steps of the cyclicity test.

C1. The residue classes of $\mathcal{O} = \{1, x, y\}$ form a K-basis of $P/\operatorname{Ann}_P(V)$.

C2. We have $\mu = 3 = \dim_{\mathbb{Q}}(V)$.

C3. We compute $\mathcal{I}_3 \cdot (z_1, z_2, z_3)^{\mathrm{tr}} = (z_1, z_2, z_3)^{\mathrm{tr}}$ as well as $\mathcal{M}_1 \cdot (z_1, z_2, z_3)^{\mathrm{tr}} = (z_2 + z_3, 2z_2 + z_3, z_2 + z_3)^{\mathrm{tr}}$ and $\mathcal{M}_2 \cdot (z_1, z_2, z_3)^{\mathrm{tr}} = (z_2, z_2 + z_3, z_2)^{\mathrm{tr}}$.

Thus we let $\mathcal{A} = \begin{pmatrix} z_1 & z_2 + z_3 & z_2 \\ z_2 & 2z_2 + z_3 & z_2 + z_3 \\ z_3 & z_2 + z_3 & z_2 \end{pmatrix}$ and calculate $d = \det(\mathcal{A}) = (z_1 - z_3)(z_2^2 - z_2 z_3 - z_3^2)$.

C4. Since K is infinite and $d \neq 0$, the algorithm returns "V is cyclic". For instance, since $d(1, 1, 0) = 1$, the element $w = e_1 + e_2$ generates V as a P-module.

The following example shows that V can fail to be cyclic even when the dimensions of V and $P/\operatorname{Ann}_P(V)$ agree.

Example 4.1.11. Let $V = \mathbb{Q}^3$, and let $\mathcal{V} = (e_1, e_2, e_3)$ be its canonical basis. We equip V with the $\mathbb{Q}[x,y]$-module structure defined by the commuting matrices

$$\mathcal{M}_1 = \begin{pmatrix} 0\,0\,0 \\ 1\,0\,0 \\ 0\,0\,0 \end{pmatrix} \quad \text{and} \quad \mathcal{M}_2 = \begin{pmatrix} 0\,0\,0 \\ 0\,0\,1 \\ 0\,0\,0 \end{pmatrix}$$

Let us apply the cyclicity test step-by-step.

C1. The algorithm of Theorem 4.1.7 yields $\mathcal{O} = \{1, x, y\}$.
C2. We have $\mu = 3 = \dim_{\mathbb{Q}}(V)$.
C3. We calculate $\mathcal{A} = \begin{pmatrix} z_1\,0\,\,0 \\ z_2\,z_1\,z_3 \\ z_3\,0\,\,0 \end{pmatrix}$ and $d = \det(\mathcal{A}) = 0$.
C5. The algorithm returns "V is not cyclic".

We end this section by considering the special case $n = 1$. In this univariate case some of the topics discussed in this section look very familiar.

Example 4.1.12. Suppose we are given a finitely generated K-vector space V and an endomorphism φ of V. We let $P = K[x]$ and observe that V becomes a P-module via the rule $(f, v) \mapsto f(\varphi)(v)$. When is it a cyclic P-module? Let us interpret the meaning of the steps of our cyclicity test in the univariate case. To start with, let $\mathcal{M}$ be a matrix representing φ.

C1. The algorithm of Theorem 4.1.7 applied to $\mathcal{M}$ yields a monic polynomial $f(x) = x^d + c_{d-1}x^{d-1} + \cdots + c_0$, which is the *minimal polynomial* of $\mathcal{M}$ (and of φ), and the tuple $\mathcal{O} = (1, x, x^2, \ldots, x^{d-1})$.
C2. The minimal polynomial of $\mathcal{M}$ is a divisor of the characteristic polynomial of $\mathcal{M}$, and the degree of the latter is $\dim_K(V)$. So the algorithm stops at step C2 only if the minimal polynomial and the characteristic polynomial differ.
C3. The matrix $\mathcal{A}$ can be interpreted as the matrix whose columns are the vectors $v, \varphi(v), \ldots, \varphi^{d-1}(v)$ for a generic v. If $\det(\mathcal{A}) = 0$, then the endomorphisms $1, \varphi, \ldots, \varphi^{d-1}$ are linearly dependent, a contradiction. Hence $\det(\mathcal{M})$ necessarily is non-zero and V is a cyclic P-module.

In conclusion, steps C3, C4, C5 are redundant in the univariate case. This corresponds to the well-known fact that V is a cyclic $K[x]$-module if and only if the minimal polynomial and the characteristic polynomial of φ coincide.

4.2 Border prebases

Given a zero-dimensional polynomial ideal I, we want to study the residue class ring P/I by choosing a K-basis and examining the multiplication matrices with respect to that basis. How can we find a basis having "nice" properties? One possibility is to take the residue classes of the terms in an order ideal, i.e. in a finite set of terms which is closed under forming divisors.

The choice of an order ideal $\mathcal{O}$ yields additional structure on the monoid of terms $\mathbb{T}^n$. For instance, there are terms forming the border of $\mathcal{O}$, i.e. terms t outside $\mathcal{O}$ such that there exist an indeterminate x_i and a term t' in $\mathcal{O}$ with $t = x_i t'$. Moreover, every term t has an $\mathcal{O}$-index which measures the distance from t to $\mathcal{O}$. The properties of order ideals, borders, and $\mathcal{O}$-indices are collected in the first subsection.

The second subsection deals with $\mathcal{O}$-border prebases. These are sets of polynomials each of which consists of one term in the border of $\mathcal{O}$ and a linear combination of terms in $\mathcal{O}$. Using $\mathcal{O}$-border prebases, we construct a division algorithm and define normal remainders.

4.2.1 Order ideals

Let $\mathbb{T}^n$ denote the monoid of terms in n indeterminates. Moreover, for every $d \geq 0$, we let $\mathbb{T}^n_d$ be the set of terms of degree d and $\mathbb{T}^n_{<d} = \bigcup_{i=0}^{d-1} \mathbb{T}^n_i$. The following kind of subset of $\mathbb{T}^n$ is central to this section.

Definition 4.2.1. *A non-empty, finite set of terms $\mathcal{O} \subset \mathbb{T}^n$ is called an* **order ideal** *if it is closed under forming divisors, i.e. if $t \in \mathcal{O}$ and $t' \mid t$ imply $t' \in \mathcal{O}$.*

Order ideals have many other names in the literature. For instance, statisticians sometimes call them **complete sets of estimable terms** (see Section 4.4). In Chapter 3, the more general notion of "sets of polynomials connected to 1" is used.

Definition 4.2.2. *Let $\mathcal{O} \subset \mathbb{T}^n$ be an order ideal.*

1. *The* **border** *of $\mathcal{O}$ is the set*

$$\partial \mathcal{O} = \mathbb{T}^n_1 \cdot \mathcal{O} \setminus \mathcal{O} = (x_1 \mathcal{O} \cup \cdots \cup x_n \mathcal{O}) \setminus \mathcal{O}$$

   The **first border closure** *of $\mathcal{O}$ is the set $\overline{\partial \mathcal{O}} = \mathcal{O} \cup \partial \mathcal{O}$.*
2. *For every $k \geq 1$, we inductively define the $(k+1)^{st}$* **border** *of $\mathcal{O}$ by $\partial^{k+1} \mathcal{O} = \partial(\overline{\partial^k \mathcal{O}})$ and the $(k+1)^{st}$* **border closure** *of $\mathcal{O}$ by the rule $\overline{\partial^{k+1} \mathcal{O}} = \overline{\partial^k \mathcal{O}} \cup \partial^{k+1} \mathcal{O}$. For convenience, we let $\partial^0 \mathcal{O} = \overline{\partial^0 \mathcal{O}} = \mathcal{O}$.*

The kth border closure of an order ideal $\mathcal{O}$ is an order ideal for every $k \geq 0$. In Chapter 3, the kth border of $\mathcal{O}$ is denoted by $\mathcal{O}^{[k]}$.

Example 4.2.3. Let $\mathcal{O}$ be the order ideal $\{1, x, y, x^2, xy, y^2, x^3, x^2y, y^3, x^4, x^3y\}$ in $\mathbb{T}^2$. Then we visualize $\mathcal{O}$ and its first two borders as follows.

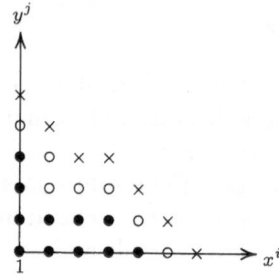

Let us collect some properties of order ideals, their borders and border closures.

Proposition 4.2.4 (Basic Properties of Borders).
Let $\mathcal{O} \subset \mathbb{T}^n$ be an order ideal.

1. *For every $k \geq 0$, we have a disjoint union $\overline{\partial^k \mathcal{O}} = \bigcup_{i=0}^{k} \partial^i \mathcal{O}$.*
2. *For every $k \geq 1$, we have $\partial^k \mathcal{O} = \mathbb{T}_k^n \cdot \mathcal{O} \setminus \mathbb{T}_{<k}^n \cdot \mathcal{O}$.*
3. *We have a disjoint union $\mathbb{T}^n = \bigcup_{i=0}^{\infty} \partial^i \mathcal{O}$.*
4. *A term $t \in \mathbb{T}^n$ is divisible by a term in $\partial \mathcal{O}$ if and only if $t \in \mathbb{T}^n \setminus \mathcal{O}$.*

Proof. The definition of the first border closure of $\mathcal{O}$ yields $\overline{\partial \mathcal{O}} = \mathcal{O} \cup \mathbb{T}_1^n \cdot \mathcal{O}$. Inductively, it follows that $\overline{\partial^{k+1} \mathcal{O}} = \overline{\partial^k \mathcal{O}} \cup \mathbb{T}_1^n \cdot \overline{\partial^k \mathcal{O}} = \overline{\partial^k \mathcal{O}} \cup \mathbb{T}_{k+1}^n \mathcal{O}$. This proves the first claim. Then the second claim is a consequence of the equality $\partial^{k+1} \mathcal{O} = \overline{\partial^{k+1} \mathcal{O}} \setminus \overline{\partial^k \mathcal{O}}$. The third claim follows from the observation that every term is in $\overline{\partial^k \mathcal{O}}$ for some $k \geq 0$.

Finally, the fourth claim holds because the second claim implies the fact that $t \in \partial^k \mathcal{O}$ for some $k \geq 1$ is equivalent to the existence of a factorization $t = t't''$ where $\deg(t') = k - 1$ and $t'' \in \partial \mathcal{O}$. $\qquad \square$

The above partition of $\mathbb{T}^n$ allows us to define a "distance" between a term and an order ideal.

Definition 4.2.5. *Let $\mathcal{O} \subset \mathbb{T}^n$ be an order ideal.*

1. *For every $t \in \mathbb{T}^n$, there exists a unique number $k \in \mathbb{N}$ such that $t \in \partial^k \mathcal{O}$. We call k the **index** of t with respect to $\mathcal{O}$ and write $\mathrm{ind}_{\mathcal{O}}(t) = k$.*
2. *For an arbitrary polynomial $f \in P \setminus \{0\}$, we define the **index** of f with respect to $\mathcal{O}$ by $\mathrm{ind}_{\mathcal{O}}(f) = \max\{\mathrm{ind}_{\mathcal{O}}(t) \mid t \in \mathrm{Supp}(f)\}$.*

By this definition, the kth border of $\mathcal{O}$ consists precisely of the terms of index k. Notice that every polynomial $f \in P \setminus \{0\}$ has a representation $f = c_1 t_1 + \cdots + c_s t_s$ where $c_1, \ldots, c_s \in K \setminus \{0\}$ and such that $t_1, \ldots, t_s \in \mathbb{T}^n$ satisfy $\mathrm{ind}_{\mathcal{O}}(t_1) \geq \cdots \geq \mathrm{ind}_{\mathcal{O}}(t_s)$. However, this representation is in general not unique since several terms in the support of f may have the same index with respect to $\mathcal{O}$.

Let us point out some of the most useful properties of the index.

Proposition 4.2.6. *Let $\mathcal{O} \subset \mathbb{T}^n$ be an order ideal.*

1. *For a term $t \in \mathbb{T}^n$, the number $k = \operatorname{ind}_\mathcal{O}(t)$ is the smallest natural number such that $t = t't''$ with a term $t' \in \mathbb{T}^n$ of degree k and with $t'' \in \mathcal{O}$.*
2. *Given two terms $t, t' \in \mathbb{T}^n$, we have $\operatorname{ind}_\mathcal{O}(t\,t') \leq \deg(t) + \operatorname{ind}_\mathcal{O}(t')$.*
3. *For $f, g \in P \setminus \{0\}$ such that $f + g \neq 0$, we have*

$$\operatorname{ind}_\mathcal{O}(f + g) \leq \max\{\operatorname{ind}_\mathcal{O}(f), \operatorname{ind}_\mathcal{O}(g)\}$$

4. *For $f, g \in P \setminus \{0\}$, we have*

$$\operatorname{ind}_\mathcal{O}(f\,g) \leq \min\{\deg(f) + \operatorname{ind}_\mathcal{O}(g), \deg(g) + \operatorname{ind}_\mathcal{O}(f)\}$$

Proof. The first claim follows from the proof of Proposition 4.2.4.4. The second claim follows from the first. The third claim is a consequence of the inclusion $\operatorname{Supp}(f+g) \subseteq \operatorname{Supp}(f) \cup \operatorname{Supp}(g)$. The last claim follows from the observation that $\operatorname{Supp}(fg) \subseteq \{t't'' \mid t' \in \operatorname{Supp}(f),\ t'' \in \operatorname{Supp}(g)\}$ and from the second claim. $\qquad\square$

Although the partial ordering on $\mathbb{T}^n$ defined by the index appears similar to a term ordering, it has a serious drawback: this ordering is incompatible with term multiplication, i.e. $\operatorname{ind}_\mathcal{O}(t) \geq \operatorname{ind}_\mathcal{O}(t')$ does not, in general, imply $\operatorname{ind}_\mathcal{O}(t\,t'') \geq \operatorname{ind}_\mathcal{O}(t'\,t'')$. Our next example is a case in point.

Example 4.2.7. Let $\mathcal{O} = \{1, x, x^2\} \subset \mathbb{T}(x, y)$. Then $\mathcal{O}$ is an order ideal with border $\partial\mathcal{O} = \{y, xy, x^2y, x^3\}$. The following sketch illustrates the situation.

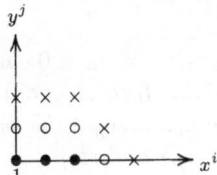

Multiplying the terms on both sides of the inequality $\operatorname{ind}_\mathcal{O}(y) > \operatorname{ind}_\mathcal{O}(x^2)$ by x^2, we get $\operatorname{ind}_\mathcal{O}(x^2 \cdot y) < \operatorname{ind}_\mathcal{O}(x^2 \cdot x^2)$. Similarly, if we multiply the terms on both sides of the equality $\operatorname{ind}_\mathcal{O}(y) = \operatorname{ind}_\mathcal{O}(x^2y)$ by x, we get the inequality $\operatorname{ind}_\mathcal{O}(x \cdot y) < \operatorname{ind}_\mathcal{O}(x \cdot x^2y)$.

4.2.2 Border division

In this subsection we introduce an important tool for dealing with zero-dimensional ideals: an $\mathcal{O}$-border prebasis, i.e. a set of polynomials of which each is a linear combination of one term in $\partial\mathcal{O}$ and terms in $\mathcal{O}$. In this way we imitate the definition of a Gröbner basis where each polynomial is a linear combination of the leading term and smaller terms. Then we present a process for dividing arbitrary polynomials by those of an $\mathcal{O}$-border prebasis. However, the remainder of this division process is not uniquely determined. This indicates that $\mathcal{O}$-border prebases are a first step in the right direction and that we must take one more step in the next section.

Definition 4.2.8. *Let* $\mathcal{O} = \{t_1, \ldots, t_\mu\}$ *be an order ideal, and let* $\partial\mathcal{O} = \{b_1, \ldots, b_\nu\}$ *be its border. A set of polynomials* $G = \{g_1, \ldots, g_\nu\}$ *is called an* $\mathcal{O}$-**border prebasis** *if the polynomials have the form* $g_j = b_j - \sum_{i=1}^{\mu} \alpha_{ij} t_i$ *such that* $\alpha_{ij} \in K$ *for* $1 \leq i \leq \mu$ *and* $1 \leq j \leq \nu$.

In particular, a border prebasis can be interpreted as a tuple of polynomials *marked* by the border terms $(b_1, \ldots, b_\nu)$ in the following sense.

Definition 4.2.9. *A pair* (g, b) *is said to be a* **marked polynomial** *if* g *is a non-zero polynomial and* $b \in \mathrm{Supp}\,(g)$ *with coefficient 1. A tuple of polynomials* $(g_1, \ldots, g_\nu)$ *is* **marked** *by a tuple of terms* $(b_1, \ldots, b_\nu)$ *if* $(g_1, b_1), \ldots, (g_\nu, b_\nu)$ *are marked polynomials.*

The definition of a border prebasis only fixes the shape of our generators. Note that this notion requires a bit more than that of marked polynomials – the unmarked terms in the polynomial's support have to be in the order ideal. Border prebases are already sufficient to perform polynomial division with remainder. The following algorithm provides a fundamental tool in working with border prebases. It is similar to the procedure called "B-reduction" in Chapter 3.

Proposition 4.2.10 (Border Division Algorithm).
Let $\mathcal{O} = \{t_1, \ldots, t_\mu\}$ *be an order ideal, let* $\partial\mathcal{O} = \{b_1, \ldots, b_\nu\}$ *be its border, and let* $\{g_1, \ldots, g_\nu\}$ *be an* $\mathcal{O}$-*border prebasis. Given a polynomial* $f \in P$, *consider the following instructions.*

D1. *Let* $f_1 = \cdots = f_\nu = 0$, $c_1 = \cdots = c_\mu = 0$, *and* $h = f$.
D2. *If* $h = 0$, *then return* $(f_1, \ldots, f_\nu, c_1, \ldots, c_\mu)$ *and stop.*
D3. *If* $\mathrm{ind}_\mathcal{O}(h) = 0$, *then find* $c_1, \ldots, c_\mu \in K$ *such that* $c_1 t_1 + \cdots + c_\mu t_\mu = h$. *Return* $(f_1, \ldots, f_\nu, c_1, \ldots, c_\mu)$ *and stop.*
D4. *If* $\mathrm{ind}_\mathcal{O}(h) > 0$, *then let* $h = a_1 h_1 + \cdots + a_s h_s$ *such that* $a_1, \ldots, a_s \in K \setminus \{0\}$ *and* $h_1, \ldots, h_s \in \mathbb{T}^n$ *satisfy* $\mathrm{ind}_\mathcal{O}(h_1) = \mathrm{ind}_\mathcal{O}(h)$. *Determine the smallest index* $i \in \{1, \ldots, \nu\}$ *such that* h_1 *factors as* $h_1 = t' b_i$ *and so that the term* $t' \in \mathbb{T}^n$ *has degree* $\mathrm{ind}_\mathcal{O}(h) - 1$. *Subtract* $a_1 t' g_i$ *from* h, *add* $a_1 t'$ *to* f_i, *and continue with step D2.*

This is an algorithm which returns a tuple $(f_1, \ldots, f_\nu, c_1, \ldots, c_\mu) \in P^\nu \times K^\mu$ such that

$$f = f_1 g_1 + \cdots + f_\nu g_\nu + c_1 t_1 + \cdots + c_\mu t_\mu$$

and $\deg(f_i) \leq \mathrm{ind}_\mathcal{O}(f) - 1$ for all $i \in \{1, \ldots, \nu\}$ with $f_i g_i \neq 0$. This representation does not depend on the choice of the term h_1 in Step D4.

For the reader's convenience we reproduce the proof from [KK03a].

Proof. First we show that Step D4 can be executed. Let $k = \mathrm{ind}_\mathcal{O}(h_1)$. By Proposition 4.2.4.2, there is a factorization $h_1 = \tilde{t} t_i$ for some term $\tilde{t}$ of degree k and some $t_i \in \mathcal{O}$, and there is no such factorization with a term $\tilde{t}$ of smaller degree. Since $k > 0$, we can write $\tilde{t} = t' x_j$ for some $t' \in \mathbb{T}^n$ and $j \in \{1, \ldots, n\}$. Then we have $\deg(t') = k - 1$, and the fact that $\tilde{t}$ has the smallest possible degree implies $x_j t_i \in \partial\mathcal{O}$. Thus we have $h_1 = t'(x_j t_i) = t' b_k$ for some $b_k \in \partial\mathcal{O}$.

Next we prove termination. We show that Step D4 is performed only finitely many times. Let us investigate the subtraction $h - a_1 t' g_i$ in Step D4. Using the representation $g_i = b_i - \sum_{k=1}^{\mu} \alpha_{ki} t_k$ given in Definition 4.2.8, this subtraction becomes

$$h - a_1 t' g_i = a_1 h_1 + \cdots + a_s h_s - a_1 t' b_i + a_1 t' \sum_{k=1}^{\mu} \alpha_{ki} t_k$$

Now $a_1 h_1 = a_1 t' b_i$ shows that a term of index $\mathrm{ind}_\mathcal{O}(h)$ is removed from h and replaced by terms of the form $t' t_l \in \overline{\partial^{k-1}\mathcal{O}}$ which have strictly smaller index. The algorithm terminates after finitely many steps because, for a given term, there are only finitely many terms of smaller or equal index.

Finally, we prove correctness. To this end, we show that the equation

$$f = h + f_1 g_1 + \cdots + f_\nu g_\nu + c_1 t_1 + \cdots + c_\mu t_\mu$$

is an invariant of the algorithm. It is satisfied at the end of Step D1. A polynomial f_i is only changed in Step D4. There the subtraction $h - a_1 t' g_i$ is compensated by the addition $(f_i + a_1 t') g_i$. The constants $c_1, \ldots, c_\mu$ are only changed in Step D3 in which h is replaced by the expression $c_1 t_1 + \cdots + c_\mu t_\mu$. When the algorithm stops, we have $h = 0$. This proves the stated representation of f. The additional claim that this representation does not depend on the choice of h_1 in Step D4 follows from the observation that h_1 is replaced by terms of strictly smaller index. Thus the different executions of Step D4 corresponding to the reduction of several terms of a given $\mathcal{O}$-index in h do not interfere with one another, and the final result – after all those terms have been rewritten – is independent of the order in which they have been taken care of. $\square$

Notice that in Step D4 the algorithm uses a representation of h which is not necessarily unique. Moreover, to make the factorization of h_1 unique, we chose the index i minimally, but this choice had not been forced upon us. Finally, the result of the division depends on the numbering of the elements of $\partial\mathcal{O}$, as our next example shows.

Example 4.2.11. Let $n = 2$, and let $\mathcal{O} = \{t_1, t_2, t_3\}$ with $t_1 = 1$, $t_2 = x$, and $t_3 = y$. Then the border of $\mathcal{O}$ is $\partial\mathcal{O} = \{b_1, b_2, b_3\}$ with $b_1 = x^2$, $b_2 = xy$, and $b_3 = y^2$. The polynomials $g_1 = x^2 + x + 1$, $g_2 = xy + y$, and $g_3 = y^2 + x + 1$ constitute an $\mathcal{O}$-border prebasis. We want to divide the polynomial $f = x^3 y^2 - xy^2 + x^2 + 2$ by this $\mathcal{O}$-border prebasis.

For easy reference, the next borders are $\partial^2 \mathcal{O} = \{x^3, x^2 y, xy^2, y^3\}$, $\partial^3 \mathcal{O} = \{x^4, x^3 y, x^2 y^2, xy^3, y^4\}$, and $\partial^4 \mathcal{O} = \{x^5, x^4 y, x^3 y^2, x^2 y^3, xy^4, y^5\}$. We apply the Border Division Algorithm and follow its steps.

D1. Let $f_1 = f_2 = f_3 = 0$, $c_1 = c_2 = c_3 = 0$, and $h = x^3y^2 - xy^2 + x^2 + 2$. The $\mathcal{O}$-indices of the terms in h are 4,2,1 and 0 respectively, so h has index 4.

D4. We have $x^3y^2 = xy^2 \cdot b_1$ with $\deg xy^2 = \text{ind}(h) - 1$. Thus we put $f_1 = xy^2$ and $h = x^3y^2 - xy^2 + x^2 + 2 - xy^2(x^2 + x + 1)$. The terms in the support of $h = -x^2y^2 - 2xy^2 + x^2 + 2$ have $\mathcal{O}$-indices 3,2,1 and 0 respectively.

D4. We have $x^2y^2 = y^2 \cdot b_1$ with $\deg y^2 = \text{ind}(h) - 1$. Add $-y^2$ to f_1 to obtain $f_1 = xy^2 - y^2$ and put $h = -x^2y^2 - 2xy^2 + x^2 + 2 + y^2(x^2 + x + 1)$. The terms in the support of $h = -xy^2 + x^2 + y^2 + 2$ have $\mathcal{O}$-indices 2,1,1 and 0 respectively.

D4. We have $xy^2 = y \cdot b_2$ with $\deg y = \text{ind}(h) - 1$. Put $f_2 = -y$ and put $h = -xy^2 + x^2 + y^2 + 2 + y(xy + y)$. The terms in the support of $h = x^2 + 2y^2 + 2$ have $\mathcal{O}$-indices 1,1 and 0 respectively.

D4. We have $x^2 = 1 \cdot b_1$ with $\deg 1 = \text{ind}(h) - 1$. Add 1 to f_1 to obtain $f_1 = xy^2 - y^2 + 1$ and put $h = x^2 + 2y^2 + 2 - 1(x^2 + x + 1)$. The terms in the support of $h = 2y^2 - x + 1$ have $\mathcal{O}$-indices 1,0 and 0 respectively.

D4. We have $y^2 = 1 \cdot b_3$ with $\deg 1 = \text{ind}(h) - 1$. Add 2 to f_3 to obtain $f_3 = 2$ and put $h = 2y^2 - x + 1 - 2(y^2 + x + 1)$. The terms in the support of $h = -3x - 1$ have $\mathcal{O}$-indices 0 and 0. Thus $\text{ind}_\mathcal{O}(h) = 0$.

D3. We have $h = -1 \cdot t_1 - 3t_2 + 0t_3$. The algorithm returns the following tuple $(xy^2 - y^2 + 1, -y, 2, 1, -3, 0)$ and stops.

Therefore we have a representation

$$f = (xy^2 - y^2 + 1)g_1 - y\, g_2 + 2\, g_3 - 1\, t_1 - 3\, t_2 + 0\, t_3$$

Second we perform the algorithm with respect to the shuffled tuple $(g_1', g_2', g_3') = (g_3, g_2, g_1)$.

D1. Let $f_1 = f_2 = f_3 = 0$, $c_1 = c_2 = c_3 = 0$, and $h = x^3y^2 - xy^2 + x^2 + 2$. The $\mathcal{O}$-indices of the terms in in the support of h are 4,2,1 and 0 respectively, so h has index 4.

D4. We have $x^3y^2 = x^3 \cdot b_1'$ with $\deg x^3 = \text{ind}(h) - 1$. Thus we put $f_1' = x^3$ and $h = x^3y^2 - xy^2 + x^2 + 2 - x^3(y^2 + x + 1)$. The terms in the support of $h = -x^4 - x^3 - xy^2 + x^2 + 2$ have $\mathcal{O}$-indices 3,2,2,1 and 0 respectively.

D4. We have $x^4 = x^2 \cdot b_3'$ with $\deg x^2 = \text{ind}(h) - 1$. Add $-x^2$ to f_3' to obtain $f_3' = x^2$ and put $h = -x^4 - x^3 - xy^2 + x^2 + 2 + x^2(x^2 + x + 1)$. The terms in the support of $h = -xy^2 + 2x^2 + 2$ have $\mathcal{O}$-indices 2,1, and 0 respectively.

D4. We have $xy^2 = x \cdot b_1'$ with $\deg x = \text{ind}(h) - 1$. Add x to f_1' to obtain $f_1' = x^3 + x$ and put $h = -xy^2 + 2x^2 + 2 + x(y^2 + y + 1)$. The terms in the support of $h = 2x^2 + xy + x + 2$ have $\mathcal{O}$-indices 1,1,0 and 0 respectively.

D4. We have $x^2 = 1 \cdot b_3'$ with $\deg 1 = \text{ind}(h) - 1$. Add 2 to f_3' to obtain $f_3' = x^2 + 2$ and put $h = 2x^2 + xy + x + 2 - 2(x^2 + x + 1)$. The terms in the support of $h = xy - x$ have $\mathcal{O}$-indices 1 and 0 respectively.

D4. We have $xy = 1 \cdot b_2'$ with $\deg 1 = \text{ind}(h) - 1$. Add 1 to f_2' to obtain $f_2' = 1$ and put $h = xy - x - 1(xy + y)$. The terms in the support of $h = x - y$ have $\mathcal{O}$-indices 0 and 0. Thus we have $\text{ind}_\mathcal{O}(h) = 0$.

D3. We write $h = 0t_1 + 1t_2 - 1t_3$. The algorithm returns the following tuple $(x^3 + x, -1, x^3 + x, 0, 1, -1)$ and stops.

Therefore we have a representation

$$f = (x^3 + x)g_1' - 1\,g_2' + (x^2 + 2)\,g_3' + 1\,t_1 - 3\,t_2 - 1\,t_3$$
$$= (x^2 + 2)g_1 - 1\,g_2 + (x^3 + x)\,g_3 + 0\,t_1 + 1\,t_2 - 1\,t_3$$

These calculations show that the order of the polynomials does affect the outcome of the Border Division Algorithm.

If we fix the tuple $(g_1, \ldots, g_\nu)$ then the result of the Border Division Algorithm is uniquely determined. The given polynomial f is represented in $P/(g_1, \ldots, g_\nu)$ by the residue class of the linear combination $c_1 t_1 + \cdots + c_\mu t_\mu$. We introduce a name for this linear combination.

Definition 4.2.12. *Let $\mathcal{O} = \{t_1, \ldots, t_\mu\}$ be an order ideal, let $G = \{g_1, \ldots, g_\nu\}$ be an $\mathcal{O}$-border prebasis, and let $\mathcal{G} = (g_1, \ldots, g_\nu)$. The **normal $\mathcal{O}$-remainder** of a polynomial f with respect to $\mathcal{G}$ is*

$$\mathrm{NR}_{\mathcal{O}, \mathcal{G}}(f) = c_1 t_1 + \cdots + c_\mu t_\mu$$

where $f = f_1 g_1 + \cdots + f_\nu g_\nu + c_1 t_1 + \cdots + c_\mu t_\mu$ is the representation computed by the Border Division Algorithm.

Example 4.2.13. Let $\mathcal{G} = (g_1, g_2, g_3)$ and $\mathcal{G}' = (g_1', g_2', g_3')$ be the tuples considered in Example 4.2.11. The above computations lead to

$$\mathrm{NR}_{\mathcal{O}, \mathcal{G}}(f) = -3x - 1 \quad \text{and} \quad \mathrm{NR}_{\mathcal{O}, \mathcal{G}'}(f) = x - y$$

So the normal $\mathcal{O}$-remainder depends on the ordering of the polynomials in $\mathcal{G}$. In the next section we shall encounter a special kind of border prebasis for which this unwanted dependence disappears.

An important consequence of the Border Division Algorithm is that the residue classes of the elements of $\mathcal{O}$ generate $P/(g_1, \ldots, g_\nu)$ as a K-vector space. But, as the above examples show, this system of generators is not necessarily a basis.

Corollary 4.2.14. *Let $\mathcal{O} = \{t_1, \ldots, t_\mu\}$ be an order ideal and $G = \{g_1, \ldots, g_\nu\}$ an $\mathcal{O}$-border prebasis. Then the residue classes of the elements of $\mathcal{O}$ generate $P/(g_1, \ldots, g_\nu)$ as a K-vector space. More precisely, the residue class of every polynomial $f \in P$ can be represented as a linear combination of the residue classes $\{\bar{t}_1, \ldots, \bar{t}_\nu\}$ by computing the normal remainder $\mathrm{NR}_{\mathcal{O}, \mathcal{G}}(f)$ for $\mathcal{G} = (g_1, \ldots, g_\nu)$.*

4.3 Border bases

After all these preparations we are ready to introduce the fundamental notion of this article: border bases. They are special systems of generators of zero-dimensional ideals which do not depend on the choice of a term ordering, but the choice of an order ideal. We discuss their existence and uniqueness and compare them to Gröbner bases of the given ideal. Then we show how one can use border bases to define normal forms, and we characterize border bases by the property that the associated multiplication matrices are pairwise commuting.

4.3.1 Existence and uniqueness of border bases

As above, let $P = K[x_1, \ldots, x_n]$ be a polynomial ring over a field K. Moreover, let I be a zero-dimensional ideal in P.

Definition 4.3.1. Let $\mathcal{O} = \{t_1, \ldots, t_\mu\}$ be an order ideal and $G = \{g_1, \ldots, g_\nu\}$ an $\mathcal{O}$-border prebasis consisting of polynomials in I. We say that the set G is an $\mathcal{O}$-**border basis** of I if the residue classes of $t_1, \ldots, t_\mu$ form a K-vector space basis of P/I.

Next we see that this definition implies that an $\mathcal{O}$-border basis of I actually generates I.

Proposition 4.3.2. Let $\mathcal{O} = \{t_1, \ldots, t_\mu\}$ be an order ideal, and let G be an $\mathcal{O}$-border basis of I. Then I is generated by G.

Proof. By definition, we have $(g_1, \ldots, g_\nu) \subseteq I$. To prove the converse inclusion, let $f \in I$. Using the Border Division Algorithm 4.2.10, the polynomial f can be expanded as $f = f_1 g_1 + \cdots + f_\nu g_\nu + c_1 t_1 + \cdots + c_\mu t_\mu$, where $f_1, \ldots, f_\nu \in P$ and $c_1, \ldots, c_\mu \in K$. This implies the equality of residue classes $0 = \bar{f} = c_1 \bar{t}_1 + \cdots + c_\mu \bar{t}_\mu$ in P/I. By assumption, the residue classes $\bar{t}_1, \ldots, \bar{t}_\mu$ form a K-vector space basis. Hence $c_1 = \cdots = c_\mu = 0$, and the expansion of f turns out to be $f = f_1 g_1 + \cdots + f_\nu g_\nu$. This completes the proof. $\qquad\square$

Remark 4.3.3. Let $\mathcal{O} = \{t_1, \ldots, t_\mu\}$ be an order ideal and G an $\mathcal{O}$-border prebasis which generates an ideal I. We let $\langle \mathcal{O} \rangle_K = Kt_1 + \cdots + Kt_\mu$ be the vector subspace of P generated by $\mathcal{O}$. Then Corollary 4.2.14 shows that the residue classes of the elements of $\mathcal{O}$ generate P/I. Since the border basis property requires that these residue classes are linearly independent, the following conditions are equivalent.

1. The set G is an $\mathcal{O}$-border basis of I.
2. We have $I \cap \langle \mathcal{O} \rangle_K = \{0\}$.
3. We have $P = I \oplus \langle \mathcal{O} \rangle_K$.

Having defined a new mathematical object, it is natural to look for its existence and possibly its uniqueness. In the following theorem, we mention the field of definition of an ideal. For a discussion on this concept, see [KR00], Section 2.4. Furthermore, given an ideal $I \subseteq P$ and a term ordering σ, we denote the order ideal $\mathbb{T}^n \setminus \mathrm{LT}_\sigma(I)$ by $\mathcal{O}_\sigma(I)$.

Theorem 4.3.4 (Existence and Uniqueness of Border Bases).
Let $\mathcal{O} = \{t_1, \ldots, t_\mu\}$ be an order ideal, let I be a zero-dimensional ideal in P, and assume that the residue classes of the elements in $\mathcal{O}$ form a K-vector space basis of P/I.

1. *There exists a unique $\mathcal{O}$-border basis G of I.*
2. *Let G' be an $\mathcal{O}$-border prebasis whose elements are in I. Then G' is the $\mathcal{O}$-border basis of I.*
3. *Let k be the field of definition of I. Then we have $G \subset k[x_1, \ldots, x_n]$.*

Proof. First we prove Claim 1. Let $\partial \mathcal{O} = \{b_1, \ldots, b_\nu\}$. For every $i \in \{1, \ldots, \nu\}$, the hypothesis implies that the residue class of b_i in P/I is linearly dependent on the residue classes of the elements of $\mathcal{O}$. Therefore I contains a polynomial of the form $b_i - \sum_{j=1}^{\mu} \alpha_{ij} t_j$ such that $\alpha_{ij} \in K$. Then $G = \{g_1, \ldots, g_\nu\}$ is an $\mathcal{O}$-border prebasis, and hence an $\mathcal{O}$-border basis of I by Definition 4.3.1. Let $G' = \{g'_1, \ldots, g'_\nu\}$ be another $\mathcal{O}$-border basis of I. If, for contradiction, there exists a term $b \in \partial \mathcal{O}$ such that the polynomials in G and G' marked by b differ, their difference is a non-zero polynomial in I whose support is contained in $\mathcal{O}$. This contradicts the hypothesis and Claim 1 is proved.

To prove the second claim, it suffices to observe that, by Definition 4.3.1, the set G' is an $\mathcal{O}$-border basis of I and to apply the first part. Finally, we prove Claim 3. Let k be the field of definition of I, let $P' = k[x_1, \ldots, x_n]$, and let $I' = I \cap P'$. Given a term ordering σ, the ideals I and I' have the same reduced σ-Gröbner basis (see [KR00], Lemma 2.4.16). Hence we have $\mathcal{O}_\sigma(I) = \mathcal{O}_\sigma(I')$, and therefore $\dim_k(P'/I') = \dim_K(P/I)$. The elements of $\mathcal{O}$ are in P' and they are linearly independent modulo I'. Hence their residue classes form a k-vector space basis of P'/I'. Let G' be the $\mathcal{O}$-border basis of I'. Then G' is an $\mathcal{O}$-border prebasis whose elements are contained in I. Thus the statement follows from Claim 2. $\qquad\qquad\square$

Given an order ideal $\mathcal{O}$ consisting of $\dim_K(P/I)$ many terms, does the $\mathcal{O}$-border basis of I always exist? The answer is negative, as our next example shows.

Example 4.3.5. Let $P = \mathbb{Q}[x, y]$, and let I be the vanishing ideal of the set of five points $X = \{(0, 0), (0, -1), (1, 0), (1, 1), (-1, 1)\}$ in the affine space $\mathbb{A}^2(\mathbb{Q})$, i.e. let $I = \{f \in P \mid f(p) = 0 \text{ for all } p \in X\}$. It is known that $\dim_K(P/I) = 5$. In $\mathbb{T}^2$, the following order ideals contain five elements:

$$\mathcal{O}_1 = \{1, x, x^2, x^3, x^4\}, \quad \mathcal{O}_2 = \{1, x, x^2, x^3, y\}, \quad \mathcal{O}_3 = \{1, x, x^2, y, y^2\},$$

$$\mathcal{O}_4 = \{1, x, x^2, y, xy\}, \quad \mathcal{O}_5 = \{1, x, y, y^2, y^3\}, \quad \mathcal{O}_6 = \{1, y, y^2, y^3, y^4\},$$

$$\mathcal{O}_7 = \{1, x, y, xy, y^2\}$$

Not all of these are suitable for border bases of I. For example, the residue classes of the elements of $\mathcal{O}_1$ cannot form a K-vector space basis of P/I since $x^3 - x \in I$. Similarly, the residue classes of the elements of $\mathcal{O}_6$ cannot form a K-vector space basis of P/I since $y^3 - y \in I$.

So, let us strive for less and ask another question. Does a given zero-dimensional ideal possess a border basis at all? Using Theorem 4.3.4, we can rephrase the question in the following way. Given a zero-dimensional ideal I, are there order ideals such that the residue classes of their elements form a K-vector space basis of P/I? This time the answer is yes, as we can show with the help of Gröbner bases.

Given an order ideal $\mathcal{O} \subset \mathbb{T}^n$, its complement $\mathbb{T}^n \setminus \mathcal{O}$ is the set of terms of a monomial ideal. Recall that every monomial ideal has a unique minimal set of generators (see [KR00], Proposition 1.3.11). The elements of the minimal set of generators of the monomial ideal corresponding to $\mathbb{T}^n \setminus \mathcal{O}$ are called the **corners** of $\mathcal{O}$. A picture illustrates the significance of this name.

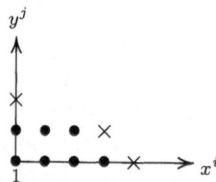

Proposition 4.3.6. *Let σ be a term ordering on $\mathbb{T}^n$. Then there exists a unique $\mathcal{O}_\sigma(I)$-border basis G of I, and the reduced σ-Gröbner basis of I is the subset of G consisting of the polynomials marked by the corners of $\mathcal{O}_\sigma(I)$.*

Proof. By Macaulay's Basis Theorem (see [KR00], Theorem 1.5.7), the residue classes of the elements in $\mathcal{O}_\sigma(I)$ form a K-vector space basis of P/I. Thus Theorem 4.3.4.1 implies the existence and uniqueness of the $\mathcal{O}_\sigma(I)$-border basis G of I.

To prove the second claim, we let $b \in \mathbb{T}^n \setminus \mathcal{O}_\sigma(I)$ be a corner of $\mathcal{O}_\sigma(I)$. The element of the minimal σ-Gröbner basis of I with leading term b has the form $b - \mathrm{NF}_{\sigma,I}(b)$, where $\mathrm{NF}_{\sigma,I}(b)$ is contained in the span of $\mathcal{O}_\sigma(I)$. Since the $\mathcal{O}_\sigma(I)$-border basis of I is unique, this Gröbner basis element agrees with the border basis element marked by b. Thus the second claim follows and the proof is complete. □

To summarize the discussion, the ideal I does not necessarily have an $\mathcal{O}$-border basis for every order ideal $\mathcal{O}$ consisting of $\dim_K(P/I)$ terms, but there always is an $\mathcal{O}$-border basis if $\mathcal{O}$ is of the form $\mathcal{O} = \mathcal{O}_\sigma(I)$ for some term ordering σ. This motivates our next question. Do all border bases belong to order ideals of the form $\mathcal{O}_\sigma(I)$? In other words, is there a bijection between the reduced Gröbner bases and the border bases of I? The answer is no, as our next example shows.

Example 4.3.7. Let $P = \mathbb{Q}[x, y]$, and let $X \subset \mathbb{A}^2(\mathbb{Q})$ be the set of points $X = \{p_1, p_2, p_3, p_4, p_5)\}$, where $p_1 = (0, 0)$, $p_2 = (0, -1)$, $p_3 = (1, 0)$, $p_4 = (1, 1)$, and $p_5 = (-1, 1)$. Furthermore, let $I \subset P$ be the vanishing ideal of X (see Example 4.3.5). The map eval : $P/I \longrightarrow \mathbb{Q}^5$ defined by $f + I \mapsto (f(p_1), \ldots, f(p_5))$ is an isomorphism of K-vector spaces.

Consider the order ideal $\mathcal{O} = \{1, x, y, x^2, y^2\}$. The matrix of size 5×5 whose columns are $(\text{eval}(1), \text{eval}(x), \ldots, \text{eval}(y^2))$ is invertible. Therefore the residue classes of the terms in $\mathcal{O}$ form a $\mathbb{Q}$-vector space basis of P/I, and I has an $\mathcal{O}$-border basis by Theorem 4.3.4.1.

The border of $\mathcal{O}$ is $\partial\mathcal{O} = \{xy, x^3, y^3, xy^2, x^2y\}$. The $\mathcal{O}$-border basis of I is $G = \{g_1, \ldots, g_5\}$ with $g_1 = x^3 - x$, $g_2 = x^2y - \frac{1}{2}y - \frac{1}{2}y^2$, $g_3 = xy - x - \frac{1}{2}y + x^2 - \frac{1}{2}y^2$, $g_4 = xy^2 - x - \frac{1}{2}y + x^2 - \frac{1}{2}y^2$, and $g_5 = y^3 - y$. To show that this border basis is not of the form $\mathcal{O}_\sigma(I)$, consider the polynomial g_3 in more detail. For any term ordering σ we have $x^2 >_\sigma x$ and $y^2 >_\sigma y$. Moreover, either $x^2 >_\sigma xy >_\sigma y^2$ or $y^2 >_\sigma xy >_\sigma x^2$. This leaves either x^2 or y^2 as the leading term of g_3. Since these terms are contained in $\mathcal{O}$, the order ideal $\mathcal{O}$ cannot be the complement of $\text{LT}_\sigma(I)$ in $\mathbb{T}^2$ for any term ordering σ.

The upshot of this example is that the set of border bases of a given zero-dimensional ideal is strictly larger than the set of its reduced Gröbner bases. Therefore there is a better chance of finding a "nice" system of generators of I among border bases than among Gröbner bases. For instance, sometimes border bases are advertised by saying that they *keep symmetry*. While this is true in many cases, the claim has to be taken with a grain of salt. Just have a look at the following example.

Example 4.3.8. Let $P = \mathbb{Q}[x, y]$ and $I = (x^2 + y^2 - 1, \; xy - 1)$. The ideal I is symmetric with respect to the indeterminates x and y. Moreover, we have $\dim_K(P/I) = 4$. The only symmetric order ideal consisting of four terms is $\mathcal{O} = \{1, x, y, xy\}$. But I does not have an $\mathcal{O}$-border basis, since we have $xy - 1 \in I$. It may be interesting to observe that the residue classes of the elements $1, x - y, x + y, x^2 - y^2$ form a K-vector space basis of P/I.

Let us investigate the relationship between Gröbner bases and border bases a little further. A list (or a set) of marked polynomials $((g_1, b_1), \ldots, (g_\nu, b_\nu))$ is said to be **marked coherently** if there exists a term ordering σ such that $\text{LT}_\sigma(g_i) = b_i$ for $i = 1, \ldots, \nu$. Furthermore, recall that an $\mathcal{O}$-border (pre)basis can be viewed as a tuple of polynomials marked by terms in the border of $\mathcal{O}$.

Proposition 4.3.9. *Let $\mathcal{O}$ be an order ideal such that the residue classes of the elements of $\mathcal{O}$ form a K-vector space basis of P/I. Let G be the $\mathcal{O}$-border basis of I, and let G' be the subset of G consisting of the elements marked by the corners of $\mathcal{O}$. Then the following conditions are equivalent.*

1. *There exists a term ordering σ such that $\mathcal{O} = \mathcal{O}_\sigma(I)$.*
2. *The elements in G' are marked coherently.*

3. *The elements in G are marked coherently.*

Moreover, if these conditions are satisfied, then G' is the reduced σ-Gröbner basis of I.

Proof. Let us prove that 1) implies both 2) and the additional claim. The fact that G' is the reduced σ-Gröbner basis of I follows from Proposition 4.3.6. Hence G' is marked coherently. Now we show that 2) implies 3). For every polynomial $g \in G \setminus G'$, there exists a polynomial $g' \in G'$ such that the marked term of g is of the form $b = t \operatorname{LT}_\sigma(g')$. Then the support of the polynomial $g - t g'$ is contained in $\mathcal{O}$, and therefore $g = t g'$. This proves that also g is marked coherently with respect to σ.

Since 3) $\Rightarrow$ 2) is obvious, only 2) $\Rightarrow$ 1) remains to be shown. Let σ be a term ordering which marks G' coherently. Denote the monomial ideal generated by the leading terms of the elements in G' by $\operatorname{LT}_\sigma(G')$. Since $\operatorname{LT}_\sigma(I) \supseteq \operatorname{LT}_\sigma(G')$, we get $\mathcal{O}_\sigma(I) = \mathbb{T}^n \setminus \operatorname{LT}_\sigma(I) \subseteq \mathbb{T}^n \setminus \operatorname{LT}_\sigma(G') = \mathcal{O}$. Also the residue classes of the elements of $\mathcal{O}_\sigma(I)$ form a K-vector space basis of P/I, and hence the inclusion is indeed an equality. $\qquad\square$

The proposition applies for instance to the monomial ideal I generated by the corners of $\mathcal{O}$. Later we shall see that the equivalent conditions of this proposition apply for a particular type of zero-dimensional ideals, namely the vanishing ideals of distracted fractions (see Example 4.4.5). The following remark will be useful in the last section.

Remark 4.3.10. Assume that there exists a term ordering σ such that every corner of $\mathcal{O}$ is σ-greater than every element in $\mathcal{O}$. Then we have $\mathcal{O} = \mathcal{O}_\sigma(I)$ for all ideals I such that the residue classes of the terms in $\mathcal{O}$ form a K-vector space basis of P/I. We do not know whether the converse holds, but we believe it does.

4.3.2 Normal forms

In Gröbner basis theory one can define a unique representative of a residue class in P/I by using the normal form of a polynomial f. The normal form is obtained by computing the normal remainder of f under the division by a Gröbner basis. It does not depend on the Gröbner basis, but only on the given term ordering and the ideal I. Hence it can be used to make the ring operations in P/I effectively computable. In this subsection we imitate this approach and generalize the normal form to border basis theory.

Let $\mathcal{O} = \{t_1, \ldots, t_\mu\}$ be an order ideal, let $G = \{g_1, \ldots, g_\nu\}$ be the $\mathcal{O}$-border basis of a zero-dimensional ideal I, and let $\mathcal{G}$ be the tuple $(g_1, \ldots, g_\nu)$. In this situation the normal $\mathcal{O}$-remainder of a polynomial does not depend on the order of the elements in $\mathcal{G}$.

Proposition 4.3.11. *Let $\pi : \{1, \ldots, \nu\} \longrightarrow \{1, \ldots, \nu\}$ be a permutation, and let $\mathcal{G}' = (g_{\pi(1)}, \ldots, g_{\pi(\nu)})$ be the corresponding permutation of the tuple $\mathcal{G}$. Then we have $\operatorname{NR}_{\mathcal{O},\mathcal{G}}(f) = \operatorname{NR}_{\mathcal{O},\mathcal{G}'}(f)$ for every polynomial $f \in P$.*

Proof. The Border Division Algorithm applied to $\mathcal{G}$ and $\mathcal{G}'$, respectively, yields representations

$$f = f_1 g_1 + \cdots + f_\nu g_\nu + \mathrm{NR}_{\mathcal{O},\mathcal{G}}(f) = f'_1 g_{\pi(1)} + \cdots + f'_\nu g_{\pi(\nu)} + \mathrm{NR}_{\mathcal{O},\mathcal{G}'}(f)$$

where $f_i, f'_j \in P$. Therefore we have $\mathrm{NR}_{\mathcal{O},\mathcal{G}}(f) - \mathrm{NR}_{\mathcal{O},\mathcal{G}'}(f) \in \langle \mathcal{O} \rangle_K \cap I$. The hypothesis that I has an $\mathcal{O}$-border basis implies $\langle \mathcal{O} \rangle_K \cap I = \{0\}$. Hence the claim follows. □

This result allows us to introduce the following definition.

Definition 4.3.12. *Let $\mathcal{O} = \{t_1, \ldots, t_\mu\}$ be an order ideal and $G = \{g_1, \ldots, g_\nu\}$ an $\mathcal{O}$-border basis of I. The **normal form** of a polynomial $f \in P$ with respect to $\mathcal{O}$ is the polynomial $\mathrm{NF}_{\mathcal{O},I}(f) = \mathrm{NR}_{\mathcal{O},\mathcal{G}}(f)$.*

The normal form $\mathrm{NF}_{\mathcal{O},I}(f)$ of $f \in P$ can be calculated by dividing f by the $\mathcal{O}$-border basis of I. It is zero if and only if $f \in I$. Further basic properties of normal forms are collected in the following proposition.

Proposition 4.3.13 (Basic Properties of Normal Forms).
Let $\mathcal{O}$ be an order ideal, and suppose that I has an $\mathcal{O}$-border basis.

1. *If there exists a term ordering σ such that $\mathcal{O} = \mathcal{O}_\sigma(I)$, then we have $\mathrm{NF}_{\mathcal{O},I}(f) = \mathrm{NF}_{\sigma,I}(f)$ for all $f \in P$.*
2. *For $f_1, f_2 \in P$, we have $\mathrm{NF}_{\mathcal{O},I}(f_1 - f_2) = \mathrm{NF}_{\mathcal{O},I}(f_1) - \mathrm{NF}_{\mathcal{O},I}(f_2)$.*
3. *For $f \in P$, we have $\mathrm{NF}_{\mathcal{O},I}(\mathrm{NF}_{\mathcal{O},I}(f)) = \mathrm{NF}_{\mathcal{O},I}(f)$.*
4. *For $f_1, f_2 \in P$, we have $\mathrm{NF}_{\mathcal{O},I}(f_1 f_2) = \mathrm{NF}_{\mathcal{O},I}(\mathrm{NF}_{\mathcal{O},I}(f_1) \, \mathrm{NF}_{\mathcal{O},I}(f_2))$.*
5. *Let $\mathcal{M}_1, \ldots, \mathcal{M}_n \in \mathrm{Mat}_n(K)$ be the matrices of the multiplication endomorphisms of P/I with respect to the basis given by the residue classes of the terms in $\mathcal{O}$. Suppose that $t_1 = 1$, and let e_1 be the first standard basis vector of K^ν. Then we have*

$$\mathrm{NF}_{\mathcal{O},I}(f) = (t_1, \ldots, t_\nu) \cdot f(\mathcal{M}_1, \ldots, \mathcal{M}_n) \cdot e_1$$

for every $f \in P$.

Proof. Claim 1) follows because both $\mathrm{NF}_{\mathcal{O},I}(f)$ and $\mathrm{NF}_{\sigma,I}(f)$ are equal to the uniquely determined polynomial in $f + I$ whose support is contained in $\mathcal{O}$. Claims 2), 3), and 4) follow from the same uniqueness. To prove the last claim, we observe that e_1 is the coordinate tuple of $1 + I$ in the basis of P/I given by the residue classes of the terms in $\mathcal{O}$. Since $\mathcal{M}_i$ is the matrix of the multiplication by x_i, the tuple $f(\mathcal{M}_1, \ldots, \mathcal{M}_n) \cdot e_1$ is the coordinate tuple of $f + I$ in this basis. From this the claim follows immediately. □

4.3.3 Border bases and commuting matrices

The purpose of this subsection is to provide the link between border bases and the theory of commuting endomorphisms discussed in the second section. More precisely, we shall characterize border bases by the property that their corresponding formal multiplication matrices commute.

Let $\mathcal{O} = \{t_1, \ldots, t_\mu\}$ be an order ideal with border $\partial\mathcal{O} = \{b_1, \ldots, b_\nu\}$, and let $G = \{g_1, \ldots, g_\nu\}$ be an $\mathcal{O}$-border prebasis. For $j = 1, \ldots, \nu$, we write $g_j = b_j - \sum_{i=1}^{\mu} \alpha_{ij} t_i$ with $\alpha_{1j}, \ldots, \alpha_{\mu j} \in K$.

In Section 4.1 we saw that a K-vector space basis of P/I allows us to describe the multiplicative structure of this algebra via a tuple of commuting matrices. If G is a border basis, we can describe these matrices as follows.

Remark 4.3.14. In the above setting, assume that G is a border basis. Then $\{\bar{t}_1, \ldots, \bar{t}_\mu\}$ is a K-vector space basis of P/I, and each multiplication endomorphism X_k of P/I corresponds to a matrix $\mathcal{X}_k = (\xi_{ij})$, i.e.,

$$X_k(\bar{t}_1) = \xi_{11}\bar{t}_1 + \cdots + \xi_{\mu 1}\bar{t}_\mu$$

$$\vdots$$

$$X_k(\bar{t}_\mu) = \xi_{1\mu}\bar{t}_1 + \cdots + \xi_{\mu\mu}\bar{t}_\mu$$

In these expansions only two cases occur. The product $x_k \, t_j$ either equals some term in the order ideal $t_r \in \mathcal{O}$ or some border term $b_s \in \partial\mathcal{O}$. In the former case we have

$$X_k(\bar{t}_j) = 0\,\bar{t}_1 + \cdots + 0\,\bar{t}_{r-1} + 1\,\bar{t}_r + 0\,\bar{t}_{r+1} + \cdots + 0\,\bar{t}_\mu$$

i.e., the jth column of $\mathcal{X}_k$ is the rth standard basis vector e_r. In the latter case we have $x_k t_j + I = b_s + I = \alpha_{1s} t_1 + \cdots + \alpha_{\mu s} t_\mu + I$, where the coefficients α_{is} are given by $g_s = b_s - \sum_i \alpha_{is} t_i$. Therefore we have

$$X_k(\bar{t}_j) = \alpha_{1s}\bar{t}_1 + \cdots + \alpha_{\mu s}\bar{t}_\mu$$

i.e., the jth column of $\mathcal{X}_k$ is $(\alpha_{1s}, \ldots, \alpha_{\mu s})^{\text{tr}}$. Observe that all matrix components ξ_{ij} are determined by the polynomials $g_1, \ldots, g_\nu$.

In view of this remark, at least formally, multiplication matrices can be defined for any border prebasis.

Definition 4.3.15. Let $\mathcal{O}=\{t_1, \ldots, t_\mu\}$ be an order ideal and $G = \{g_1, \ldots, g_\nu\}$ an $\mathcal{O}$-border prebasis. For $1 \leq k \leq n$, define the kth **formal multiplication matrix** $\mathcal{X}_k$ column-wise by

$$(\mathcal{X}_k)_{*j} = \begin{cases} e_r, & \text{if } x_k \, t_j = t_r \\ (\alpha_{1s}, \ldots, \alpha_{\mu s})^{\text{tr}}, & \text{if } x_k \, t_j = b_s \end{cases}$$

To get some insight into the meaning of this definition, let us have a look at example 4.3.7 "from the outside."

Example 4.3.16. Let $P = \mathbb{Q}[x, y]$, and let $\mathcal{O} = \{t_1, t_2, t_3, t_4, t_5\}$ be the order ideal given by $t_1 = 1$, $t_2 = x$, $t_3 = y$, $t_4 = x^2$, and $t_5 = y^2$. The border of $\mathcal{O}$ is $\partial\mathcal{O} = \{b_1, b_2, b_3, b_4, b_5\}$ where $b_1 = xy$, $b_2 = x^3$, $b_3 = y^3$, $b_4 = x^2 y$, and $b_5 = xy^2$. The polynomials $g_1 = xy - x - \frac{1}{2}y + x^2 - \frac{1}{2}y^2$, $g_2 = x^3 - x$, $g_3 = y^3 - y$, $g_4 = x^2 y - \frac{1}{2}y - \frac{1}{2}y^2$, and $g_5 = xy^2 - x - \frac{1}{2}y + x^2 - \frac{1}{2}y^2$ define a border prebasis of $I = (g_1, \ldots, g_5)$. Now we compute the formal multiplication matrices $\mathcal{X}$ and $\mathcal{Y}$.

On the one hand, we have $x\, t_1 = t_2$, $x\, t_2 = t_4$, $x\, t_3 = b_1$, $x\, t_4 = b_2$, and $x\, t_5 = b_5$. On the other hand, we have $y\, t_1 = t_3$, $y\, t_2 = b_1$, $y\, t_3 = t_5$, $y\, t_4 = b_4$, and $y\, t_5 = b_3$. Thus we obtain

$$\mathcal{X} = \begin{pmatrix} 0 & 0 & 0 & 0 & 0 \\ 1 & 0 & 1 & 1 & 1 \\ 0 & 0 & 1/2 & 0 & 1/2 \\ 0 & 1 & -1 & 0 & -1 \\ 0 & 0 & 1/2 & 0 & 1/2 \end{pmatrix} \quad \text{and} \quad \mathcal{Y} = \begin{pmatrix} 0 & 0 & 0 & 0 & 0 \\ 0 & 1 & 0 & 0 & 0 \\ 1 & 1/2 & 0 & 1/2 & 1 \\ 0 & -1 & 0 & 0 & 0 \\ 0 & 1/2 & 1 & 1/2 & 0 \end{pmatrix}$$

By Example 4.3.7, this border prebasis is even a border basis of I. Hence the formal multiplication matrices are the actual multiplication matrices. As such they commute.

The following theorem is the main result of this subsection. We characterize border bases by the property that their formal multiplication matrices commute. A more general theorem is contained in Chapter 3.

Theorem 4.3.17 (Border Bases and Commuting Matrices).

Let $\mathcal{O} = \{t_1, \ldots, t_\mu\}$ be an order ideal. An $\mathcal{O}$-border prebasis $\{g_1, \ldots, g_\nu\}$ is an $\mathcal{O}$-border basis of $I = (g_1, \ldots, g_\nu)$ if and only if its formal multiplication matrices are pairwise commuting. In that case the formal multiplication matrices represent the multiplication endomorphisms of P/I with respect to the basis $\{\bar{t}_1, \ldots, \bar{t}_\mu\}$.

Proof. Let $\mathcal{X}_1, \ldots, \mathcal{X}_n$ be the formal multiplication matrices corresponding to the given $\mathcal{O}$-border prebasis $G = \{g_1, \ldots, g_\nu\}$. If G is an $\mathcal{O}$-border basis, then Remark 4.3.14 shows that $\mathcal{X}_1, \ldots, \mathcal{X}_n$ represent the multiplication endomorphisms of P/I. Hence they are pairwise commuting.

It remains to show sufficiency. Without loss of generality, let $t_1 = 1$. The matrices $\mathcal{X}_1, \ldots, \mathcal{X}_n$ define a P-module structure on $\langle \mathcal{O} \rangle_K$ via

$$f \cdot (c_1 t_1 + \ldots c_\mu t_\mu) = (t_1, \ldots, t_\mu) f(\tilde{\mathcal{X}}_1, \ldots, \tilde{\mathcal{X}}_n)(c_1, \ldots, c_\mu)^{\mathrm{tr}}$$

First we show that this P-module is cyclic with generator t_1. To do so, we use induction on the degree to show $t_i \cdot t_1 = t_i$ for $i = 1, \ldots, \mu$. The induction starts with $t_1 = (t_1, \ldots, t_\mu)\mathcal{I}_\mu \cdot e_1$. For the induction step, let $t_i = x_j t_k$. Then we have

$$t_i \cdot t_1 = (t_1, \ldots, t_\mu) t_i(\mathcal{X}_1, \ldots, \mathcal{X}_n) e_1 = (t_1, \ldots, t_\mu) \mathcal{X}_j t_k(\mathcal{X}_1, \ldots, \mathcal{X}_n) e_1$$
$$= (t_1, \ldots, t_\mu) \mathcal{X}_j e_k = (t_1, \ldots, t_\mu) e_i = t_i$$

Thus we obtain a surjective P-linear map $\tilde{\Theta} : P \to \langle \mathcal{O} \rangle_K$ such that $f \mapsto f \cdot t_1$ and an induced isomorphism of P-modules $\Theta : P/J \to \langle \mathcal{O} \rangle_K$ with $J = \ker \tilde{\Theta}$. In particular, the residue classes $t_1 + J, \ldots, t_\mu + J$ are K-linearly independent.

Next we show $I \subseteq J$. Let $b_j = x_k t_l$. Then we have

$$g_j(\mathcal{X}_1, \ldots, \mathcal{X}_n) e_1 = b_j(\mathcal{X}_1, \ldots, \mathcal{X}_n) e_1 - \sum_{i=1}^{\mu} \alpha_{ij} t_i(\mathcal{X}_1, \ldots, \mathcal{X}_n) e_1$$

$$= \mathcal{X}_k t_l(\mathcal{X}_1, \ldots, \mathcal{X}_n) e_1 - \sum_{i=1}^{\mu} \alpha_{ij} e_i = \mathcal{X}_k e_l - \sum_{i=1}^{\mu} \alpha_{ij} e_i$$

$$= \sum_{i=1}^{\mu} \alpha_{ij} e_i - \sum_{i=1}^{\mu} \alpha_{ij} e_i = 0$$

Therefore we have $g_j \in \ker \tilde{\Theta}$ for $j = 1, \ldots, \nu$ and $I \subseteq J$, as desired.

Hence there is a natural surjective ring homomorphism $\Psi : P/I \to P/J$. Since the set $\{t_1 + I, \ldots, t_\mu + I\}$ generates the K-vector space P/I, and since the set $\{t_1 + J, \ldots, t_\mu + J\}$ is K-linearly independent, both sets must be bases and $I = J$. This shows that G is an $\mathcal{O}$-border basis of I. □

The following example shows that the formal multiplication matrices corresponding to an $\mathcal{O}$-border prebasis are not always commuting.

Example 4.3.18. Let $P = \mathbb{Q}[x, y]$ and $\mathcal{O} = \{t_1, t_2, t_3, t_4, t_5\}$ with $t_1 = 1$, $t_2 = x$, $t_3 = y$, $t_4 = x^2$, and $t_5 = y^2$. Then the border of $\mathcal{O}$ is $\partial \mathcal{O} = \{b_1, b_2, b_3, b_4, b_4\}$ with $b_1 = xy$, $b_2 = x^3$, $b_3 = y^3$, $b_4 = x^2 y$, and $b_5 = xy^2$. Consider the set of polynomials $G = \{g_1, g_2, g_3, g_4, g_5\}$ with $g_1 = xy - x^2 - y^2$, $g_2 = x^3 - x^2$, $g_3 = y^3 - y^2$, $g_4 = x^2 y - x^2$, and $g_5 = xy^2 - y^2$. It is an $\mathcal{O}$-border prebasis of the ideal $I = (g_1, \ldots, g_5)$. Its multiplication matrices

$$\mathcal{X} = \begin{pmatrix} 0 & 0 & 0 & 0 & 0 \\ 1 & 0 & 0 & 0 & 0 \\ 0 & 0 & 0 & 0 & 0 \\ 0 & 1 & 1 & 1 & 0 \\ 0 & 0 & 1 & 0 & 1 \end{pmatrix} \quad \text{and} \quad \mathcal{Y} = \begin{pmatrix} 0 & 0 & 0 & 0 & 0 \\ 0 & 0 & 0 & 0 & 0 \\ 1 & 0 & 0 & 0 & 0 \\ 0 & 1 & 0 & 1 & 0 \\ 0 & 1 & 1 & 0 & 1 \end{pmatrix}$$

do not commute:

$$\mathcal{X} \cdot \mathcal{Y} = \begin{pmatrix} 0 & 0 & 0 & 0 & 0 \\ 0 & 0 & 0 & 0 & 0 \\ 0 & 0 & 0 & 0 & 0 \\ 1 & 1 & 0 & 1 & 0 \\ 1 & 1 & 1 & 0 & 1 \end{pmatrix} \neq \mathcal{Y} \cdot \mathcal{X} = \begin{pmatrix} 0 & 0 & 0 & 0 & 0 \\ 0 & 0 & 0 & 0 & 0 \\ 0 & 0 & 0 & 0 & 0 \\ 1 & 1 & 1 & 1 & 0 \\ 1 & 0 & 1 & 0 & 1 \end{pmatrix}$$

By the theorem, the set G is not an $\mathcal{O}$-border basis of I.

The condition that the formal multiplication matrices of a border basis have to commute can also be interpreted in terms of the syzygies of that basis (see [Ste04]). Based on the results of this section one can now imitate the development of Gröbner basis theory for border bases. For instance, the border basis analogues of the conditions A – D which characterize Gröbner bases in [KR00], Chapter 2, are examined by the first two authors in [KK03a].

4.4 Application to statistics

> *Fifty percent of the citizens of this country*
> *have a below average understanding of statistics.*
> (Anonymous)

In this last section we see how to solve a problem in computational commutative algebra whose motivation comes from statistics. Does this sound strange to you? Well, come and see. Our problem comes up in the branch of statistics called *design of experiments*. If you want to get a more detailed understanding of this theory, we suggest that you start exploring it by reading [Rob98]. Or, if you prefer the statisticians' point of view, you can consult [PRW00].

To get to the heart of the problem, let us introduce some fundamental concepts of design of experiments. A **full factorial design** is a finite set of points in affine space $\mathbb{A}^n(K) \cong K^n$ of the form $D = D_1 \times \cdots \times D_n$ where D_i is a finite subset of K. Associated to it we may consider the vanishing ideal $I_D = \{f \in P \mid f(p) = 0 \text{ for all } p \in D\}$. It is a complete intersection $I_D = (f_1, \ldots, f_n)$ such that $f_i \in K[x_i]$ is a product of linear forms for $i = 1, \ldots, n$. For instance, in $\mathbb{A}^2(\mathbb{Q})$ we have the full factorial design

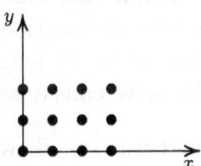

whose vanishing ideal in $\mathbb{Q}[x, y]$ is $I_D = \big(x(x-1)(x-2)(x-3),\, y(y-1)(y-2)\big)$. The particular shape of the generators of I_D implies that they are the reduced σ-Gröbner basis of I_D with respect to any term ordering σ. Hence the order ideal $\mathcal{O}_D = \mathbb{T}^n \setminus \mathrm{LT}_\sigma(I_D)$ is canonically associated to D. In the example at hand we have for instance

$$\mathcal{O}_D = \{1,\, x,\, y,\, x^2,\, xy,\, y^2,\, x^3,\, x^2y,\, xy^2,\, x^3y,\, x^2y^2,\, x^3y^2\}$$

If a particular problem depends on n parameters and each parameter can assume finitely many values $D_i \subseteq K$, the full factorial design $D = D_1 \times \cdots \times D_n$ corresponds to the set of all possible experiments. The main task in the design of experiments is to identify an unknown function

$f : D \longrightarrow K$. This function is a mathematical **model** of a quantity which has to be computed or optimized. Since it is defined on a finite set, it can be determined by performing all experiments in D and measuring the value of f each time. Notice that a function f defined on a finite set is necessarily a polynomial function.

However, in most cases it is impossible to perform all experiments corresponding to the full factorial design. The obstacles can be, for instance, lack of time, lack of money, or lack of patience. Only a subset of those experiments can be performed. The question is how many and which? In statistical jargon a subset F of a full factorial design D is called a **fraction**. Our task is to choose a fraction $F \subseteq D$ that allows us to identify the model. In particular, we need to describe the order ideals whose residue classes form a K-basis of P/I_F. Statisticians express this property by saying that such order ideals (or complete sets of estimable terms, as they call them) are *identified* by F.

Even more important is the so-called *inverse problem*. Suppose we are given an order ideal $\mathcal{O}$. We would like to determine all fractions $F \subseteq D$ such that the residue classes of the elements of $\mathcal{O}$ form a K-basis of P/I_F. The main result of [CR97] was a partial solution of this inverse problem. More precisely, all fractions $F \subseteq D$ were found such that $\mathcal{O} = \mathcal{O}_\sigma(I_F)$ for some term ordering σ. However, we have already pointed out that some order ideals $\mathcal{O}$ do not fit into this scheme (see Example 4.3.7). Later, in the paper [CR01] the full solution was presented, and the main idea was to use border bases.

Before delving into the general solution of the inverse problem following the technique employed in [CR01], let us briefly explain an example of an actual statistical problem. This example is taken from [BHH78] and adapted to our setting and terminology.

Example 4.4.1. A number of similar chemical plants had been successfully operating for several years in different locations. In a newly constructed plant the filtration cycle took almost twice as long as in the older plants. Seven possible causes of the difficulty were considered by the experts.

1. The water for the new plant was somehow different in mineral content.
2. The raw material was not identical in all respects to that used in the older plants.
3. The temperature of filtration in the new plant was slightly lower than in the older plants.
4. A new recycle device was absent in the older plants.
5. The rate of addition of caustic soda was higher in the new plant.
6. A new type of filter cloth was being used in the new plant.
7. The holdup time was lower than in the older plants.

These causes lead to seven variables $x_1, \ldots, x_7$. Each of them can assume only two values, namely *old* and *new* which we denote by 0 and 1, respectively. Our full factorial design $D \subseteq \mathbb{A}^7(\mathbb{Q})$ is therefore the set $D = \{0,1\}^7$. Its vanishing ideal is $I_D = (x_1^2 - x_1, \ x_2^2 - x_2, \ldots, \ x_7^2 - x_7)$ in the polynomial ring $P = \mathbb{Q}[x_1, x_2, \ldots, x_7]$.

The model $f : D \longrightarrow \mathbb{Q}$ is the length of a filtration cycle. In order to identify it, we would have to perform 128 cycles. This is impracticable, since it would require too much time and money. On the other hand, suppose for a moment that we conduct all experiments and the output is $f = a + b x_1 + c x_2$ for some $a, b, c \in \mathbb{Q}$. At this point it becomes clear that we wasted many resources. Had we known in advance that the polynomial has only three unknown coefficients, we could have identified them by performing only *three* suitable experiments! Namely, if we determine three values of the polynomial $a + b x_1 + c x_2$, we can find a, b, c by solving a system of three linear equations in these three indeterminates. If the matrix of coefficients is invertible, this is an easy task.

However, a priori one does not know that the answer has that shape indicated above. In practice, one has to make some guesses, perform well-chosen experiments, and possibly modify the guesses until the process yields the desired answer. In the case of the chemical plant, it turned out that only x_1 and x_5 were relevant for identifying the model.

In this example there is one point which needs additional explanation. How can we choose the fraction F such that the matrix of coefficients is invertible? In other words, given a full factorial design D and an order ideal $\mathcal{O} \subseteq \mathcal{O}_D$, which fractions $F \subseteq D$ have the property that the residue classes of the elements of $\mathcal{O}$ are a K-basis of P/I_F? This is precisely the inverse problem stated above. In order to explain its solution, we introduce the following terminology.

Definition 4.4.2. *For $i = 1, \ldots, n$, let $\ell_i \geq 1$ and $D_i = \{a_{i1}, a_{i2}, \ldots, a_{i\ell_i}\} \subseteq K$. Then we say that the full factorial design $D = D_1 \times \cdots \times D_n \subseteq \mathbb{A}^n(K)$ has* **levels** $(\ell_1, \ldots, \ell_n)$.

The polynomials $f_i = (x_i - a_{i1}) \cdots (x_i - a_{i\ell_i})$ with $i = 1, \ldots, n$ generate the vanishing ideal I_D of D. They are called the **canonical polynomials** *of D. Since $\{f_1, \ldots, f_n\}$ is a universal Gröbner basis of I_D (i.e. a Gröbner basis with respect to every term ordering), the order ideal*

$$\mathcal{O}_D = \{x_1^{\alpha_1} \cdots x_n^{\alpha_n} \mid 0 \leq \alpha_i < \ell_i \text{ for } i = 1, \ldots, n\}$$

represents a K-basis of P/I_D. We call it the **complete set of estimable terms** *of D.*

The following auxiliary result will be useful for proving the main theorem.

Lemma 4.4.3. *Let D be a full factorial design, let $\{f_1, \ldots, f_n\}$ be its canonical polynomials, let $\overline{K}$ be the algebraic closure of K, and let I be a proper ideal of $\overline{K}[x_1, \ldots, x_n]$ such that $I_D \subseteq I$.*

1. The ideal I is a radical ideal. It is the vanishing ideal of a fraction of D.
2. The ideal I is generated by elements of P, and $I \cap P$ is a radical ideal.

3. *The polynomials of every border basis of I are elements of P.*

Proof. First we prove Claim 1. Let $\mathbb{A}^n(\overline{K})$ be the affine space of dimension n over $\overline{K}$, and let $F \subset \mathbb{A}^n(\overline{K})$ be the set of zeros of I. Since $I_D \subseteq I$, we have $F \subseteq D$. By localizing the ring $A = \overline{K}[x_1, \dots, x_n]/I_D$ at the maximal ideals $\mathfrak{m}$ corresponding to the points of d, we see that either $IA_{\mathfrak{m}} = (1)$ or $IA_{\mathfrak{m}} = \mathfrak{m}A_{\mathfrak{m}}$. Therefore I is a radical ideal, and hence it is the defining ideal of F.

Since I is the defining ideal of a finite set of points with coordinates in K, it is the intersection of ideals generated by linear forms having coefficients in K. Consequently, the ideal I is defined over K which proves Claim 2. The third claim follows from Theorem 4.3.4. □

Now we are ready to state the main result of this section. Our goal is to solve the inverse problem. The idea is to proceed as follows. We are given a full factorial design D and an order ideal $\mathcal{O}$. By Theorem 4.3.4, ideals I such that $\mathcal{O}$ represents a K-basis of P/I are in 1-1 correspondence with border bases whose elements are marked by the terms in $\partial\mathcal{O}$. Except for the border basis elements which are canonical polynomials of D, we can write them down using indeterminate coefficients and require that the corresponding formal multiplication matrices are pairwise commuting. For I to be the vanishing ideal of a fraction contained in D, we have to make sure that I contains I_D. To this end, we require that the normal $\mathcal{O}$-remainders of the canonical polynomials of D are zero. By combining these requirements, we arrive at the following result.

Theorem 4.4.4 (Computing All Fractions).
Let D be a full factorial design with levels $(\ell_1, \dots, \ell_n)$, and let $\mathcal{O} = \{t_1, \dots, t_\mu\}$ be a complete set of estimable terms contained in $\mathcal{O}_D$ with $t_1 = 1$. Consider the following definitions.

1. *Let $C = \{f_1, \dots, f_n\}$ be the set of canonical polynomials of D, where f_i is marked by $x_i^{\ell_i}$ for $i = 1, \dots, n$.*
2. *Decompose $\partial\mathcal{O}$ into $\partial\mathcal{O}_1 = \{x_1^{\ell_1}, \dots, x_n^{\ell_n}\} \cap \partial\mathcal{O}$ and $\partial\mathcal{O}_2 = \partial\mathcal{O} \setminus \partial\mathcal{O}_1$.*
3. *Let C_1 be the subset of C marked by $\partial\mathcal{O}_1$, and let $C_2 = C \setminus C_1$.*
4. *Let $\eta = \#(\partial\mathcal{O}_2)$. For $i = 1, \dots, \eta$ and $j = 1, \dots, \mu$, introduce new indeterminates z_{ij}.*
5. *For every $b_k \in \partial\mathcal{O}_2$, let $g_k = b_k - \sum_{j=1}^{\mu} z_{kj}t_j \in K(z_{ij})[x_1, \dots, x_n]$.*
6. *Let $G = \{g_1, \dots, g_\eta\}$ and $H = G \cup C_1$. Let $\mathcal{M}_1, \dots, \mathcal{M}_n$ be the formal multiplication matrices associated to the $\mathcal{O}$-border prebasis H.*
7. *Let $\mathcal{I}(\mathcal{O})$ be the ideal in $K[z_{ij}]$ generated by the entries of the matrices $\mathcal{M}_i\mathcal{M}_j - \mathcal{M}_j\mathcal{M}_i$ for $1 \leq i < j \leq n$, and by the entries of the column matrices $f(\mathcal{M}_1, \dots, \mathcal{M}_n) \cdot e_1$ for all $f \in C_2$.*

Then $\mathcal{I}(\mathcal{O})$ is a zero-dimensional ideal in $K[z_{ij}]$ whose zeros are in 1-1 correspondence with the solutions of the inverse problem, i.e. with fractions $F \subseteq D$ such that $\mathcal{O}$ represents a K-basis of P/I_F.

Proof. Let $p = (\alpha_{11}, \ldots, \alpha_{\mu\eta}) \in \overline{K}^{\mu\eta}$ be a zero of $\mathcal{I}(\mathcal{O})$. When we substitute the indeterminates z_{ij} by the coordinates of p in the matrices $\mathcal{M}_1, \ldots, \mathcal{M}_n$, we obtain pairwise commuting matrices $\overline{\mathcal{M}}_1, \ldots, \overline{\mathcal{M}}_n$ which feature the additional property that $f(\overline{\mathcal{M}}_1, \ldots, \overline{\mathcal{M}}_n) \cdot e_1 = 0$ for every $f \in C_2$.

Now we substitute the coordinates of p in the polynomials of G and get polynomials $\bar{g}_k = b_k - \sum_{j=1}^{\mu} \alpha_{kj} t_j \in P$. Then we form the sets $\overline{G} = \{\bar{g}_1, \ldots, \bar{g}_\eta\}$ and $\overline{H} = \overline{G} \cup C_1$, and we let $\overline{I}$ be the ideal generated by $\overline{H}$. Since the set H is an $\mathcal{O}$-border prebasis of the ideal generated by it, the set $\overline{H}$ is an $\mathcal{O}$-border prebasis of $\overline{I}$. Moreover, the fact that $\mathcal{M}_1, \ldots, \mathcal{M}_n$ are the formal multiplication matrices of H implies that $\overline{\mathcal{M}}_1, \ldots, \overline{\mathcal{M}}_n$ are the formal multiplication matrices of $\overline{H}$. Hence we can apply Theorem 4.3.17 and conclude that $\overline{H}$ is the $\mathcal{O}$-border basis of $\overline{I}$.

By definition, we have $C_1 \subseteq \overline{I}$. Using Proposition 4.3.13.5, we see that $f(\overline{\mathcal{M}}_1, \ldots, \overline{\mathcal{M}}_n) \cdot e_1 = 0$ implies $\mathrm{NF}_{\mathcal{O}, \overline{I}}(f) = 0$, and therefore $f \in \overline{I}$ for all $f \in C_2$. Altogether, we have $C = C_1 \cup C_2 \subseteq \overline{I}$, and thus $I_D \subseteq \overline{I}$. By Lemma 4.4.3.1, it follows that $\overline{I}$ is the vanishing ideal of a fraction of D.

Conversely, let F be a fraction of D such that $\mathcal{O}$ represents a K-basis of P/I_F. Consider the $\mathcal{O}$-border basis B of I_F and write $B = B_1 \cup B_2$ such that B_1 contains the polynomials marked by $\partial\mathcal{O}_1$ and B_2 contains the polynomials marked by $\partial\mathcal{O}_2$. Since $\partial\mathcal{O}_1 \subseteq \partial\mathcal{O}_D$, the polynomials in B_1 have the shape required for $\mathcal{O}_D$-border basis elements of I_D, i.e. they agree with the polynomials in C_1. The polynomials in B_2 are of the form $\bar{g}_k = b_k - \sum_{j=1}^{\mu} \alpha_{kj} t_j$ where $b_k \in \partial\mathcal{O}_2$ and $\alpha_{kj} \in K$. Let $p = (\alpha_{ij}) \in K^{\mu\eta}$. We claim that p is a zero of $\mathcal{I}(\mathcal{O})$.

The point p is a zero of the entries of the matrices $\mathcal{M}_i\mathcal{M}_j - \mathcal{M}_j\mathcal{M}_j$ for $1 \le i < j \le n$, since the matrices $\overline{\mathcal{M}}_1, \ldots, \overline{\mathcal{M}}_n$ obtained by substituting p in $\mathcal{M}_1, \ldots, \mathcal{M}_n$ are the formal multiplication matrices of B and thus commute by Theorem 4.3.17. The point p is a zero of the entries of $f(\mathcal{M}_1, \ldots, \mathcal{M}_n) \cdot e_1$ for $f \in C_2$, since $f(\overline{\mathcal{M}}_1, \ldots, \overline{\mathcal{M}}_n) \cdot e_1$ equals $\mathrm{NF}_{\mathcal{O}, I_F}(f)$ by Proposition 4.3.13.5, and this normal form is zero because $f \in C_2 \subseteq I_D \subseteq I_F$. Altogether, we have shown that p is a zero of $\mathcal{I}(\mathcal{O})$, as claimed. $\qquad\square$

Using *distracted fractions* (see [RR98]), one can show that there always exists at least one solution of the inverse problem. Let us look at an example to illustrate the method.

Example 4.4.5. Let D be the full factorial design $D = \{0, 1, 2, 3\} \times \{0, 1, 2\}$ contained in $\mathbb{A}^2(\mathbb{Q})$, and let $\mathcal{O} = \{1, x, y, x^2, xy, y^2, x^3, x^2y\} \subset \mathcal{O}_D$. The order ideal $\mathcal{O}$ can be visualized as follows.

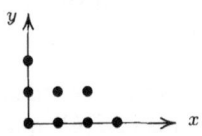

We want to find a fraction $F \subseteq D$ such that $\mathcal{O}$ represents a K-basis of P/I_F. One solution is to use the distracted fraction whose points are exactly the points marked by bullets in the above sketch, i.e. the following set $F = \{(0,0), (0,1), (0,2), (1,0), (1,1), (2,0), (2,1), (3,0)\}$. An easy computation shows that the vanishing ideal of F is

$$I_F = \big(x(x-1)(x-2)(x-3), \; x(x-1)(x-2)y, \; xy(y-1), \; y(y-1)(y-2)\big)$$

Moreover, these three generators are a universal Gröbner basis of I_F and $\mathcal{O}_\sigma(I_F) = \mathcal{O}$ for every term ordering σ.

We end this section with two examples intended to explain how Theorem 4.4.4 solves the inverse problem.

Example 4.4.6. Let D be the full factorial design $D = \{-1, 0, 1\} \times \{-1, 1\}$ with levels $(3, 2)$ contained in $\mathbb{A}^2(\mathbb{Q})$. The complete set of estimable terms of D is $\mathcal{O}_D = \{1, x, y, x^2, xy, x^2 y\}$. We want to solve the inverse problem for the order ideal $\mathcal{O} = \{1, x, y\}$ and follow the steps of Theorem 4.4.4.

1. The set of canonical polynomials of D is $C = \{f_1, f_2\}$, where $f_1 = x^3 - x$ and $f_2 = y^2 - 1$.
2. We decompose $\partial \mathcal{O} = \{x^2, xy, y^2\}$ into $\partial \mathcal{O}_1 = \{y^2\}$ and $\partial \mathcal{O}_2 = \{x^2, xy\}$.
3. Let $C_1 = \{f_2\}$ and $C_2 = \{f_1\}$.
4. Let $\eta = 2$. Choose six new indeterminates $z_{11}, z_{12}, z_{13}, z_{21}, z_{22}, z_{23}$.
5. Define $g_1 = x^2 - (z_{11} + z_{12}x + z_{13}y)$ and $g_2 = xy - (z_{21} + z_{22}x + z_{23}y)$.
6. Let $G = \{g_1, g_2\}$ and $H = \{g_1, g_2, f_2\}$. The formal multiplication matrices associated to H are

$$\mathcal{M}_1 = \begin{pmatrix} 0 & z_{11} & z_{21} \\ 1 & z_{12} & z_{22} \\ 0 & z_{13} & z_{23} \end{pmatrix} \quad \text{and} \quad \mathcal{M}_2 = \begin{pmatrix} 0 & z_{21} & 1 \\ 0 & z_{22} & 0 \\ 1 & z_{23} & 0 \end{pmatrix}$$

7. Let $\mathcal{I}(\mathcal{O}) \subseteq \mathbb{Q}[z_{11}, \ldots, z_{23}]$ be the ideal generated by the entries of the matrices $\mathcal{M}_1 \mathcal{M}_2 - \mathcal{M}_2 \mathcal{M}_1$ and $f_1(\mathcal{M}_1, \mathcal{M}_2) \cdot e_1 = (\mathcal{M}_1^3 - \mathcal{M}_1) \cdot e_1$. We obtain $\mathcal{I}(\mathcal{O}) = (z_{12} z_{21} - z_{11} z_{22} - z_{21} z_{23} + z_{13}, \; z_{21} z_{22} + z_{23}, \; z_{22} z_{23} + z_{21}, \; z_{22}^2 - 1, \; z_{13} z_{22} - z_{12} z_{23} + z_{23}^2 - z_{11}, \; z_{22} z_{23} + z_{21}, \; z_{11} z_{12} + z_{13} z_{21}, \; z_{12}^2 + z_{13} z_{22} + z_{11} - 1, \; z_{12} z_{13} + z_{13} z_{23})$.

Using a computer algebra system, for instance CoCoA, we can check that $\mathcal{I}(O)$ is a zero-dimensional, radical ideal of multiplicity 18. This means that among the $20 = \binom{6}{3}$ triples of points of D, there are 18 triples which solve the inverse problem. The two missing fractions are $\{(0,0), (0,1), (0,2)\}$ and $\{(1,0), (1,1), (1,2)\}$.

When we apply the theorem to larger full factorial designs, the calculations involved in determining the zeros of $\mathcal{I}(\mathcal{O})$ quickly become voluminous.

Example 4.4.7. Let D be the full factorial design $D = \{-1, 0, 1\} \times \{-1, 0, 1\}$ with levels $(3, 3)$ contained in $\mathbb{A}^2(\mathbb{Q})$. The complete set of estimable terms of D is $\mathcal{O}_D = \{1, x, y, x^2, xy, y^2, x^2y, xy^2, x^2y^2\}$. We want to solve the inverse problem for the order ideal $\mathcal{O} = \{1, x, y, x^2, y^2\}$ and follow the steps of Theorem 4.4.4.

1. The set of canonical polynomials of D is $C = \{f_1, f_2\}$, where $f_1 = x_1^3 - x_1$ and $f_2 = x_2^3 - x_2$.
2. We decompose $\partial\mathcal{O} = \{x^3, x^2y, xy, xy^2, y^3\}$ into $\partial\mathcal{O}_1 = \{x^3, y^3\}$ and $\partial\mathcal{O}_2 = \{x^2y, xy, xy^2\}$.
3. Let $C_1 = \{f_1, f_2\}$ and $C_2 = \emptyset$.
4. Let $\eta = 3$. Choose 15 new indeterminates $z_{11}, z_{12}, \ldots, z_{35}$.
5. Define $g_1 = x^2y - (z_{11} + z_{12}x + z_{13}y + z_{14}x^2 + z_{15}y^2)$ and $g_2 = xy - (z_{21} + z_{22}x + z_{23}y + z_{24}x^2 + z_{25}y^2)$ and $g_3 = xy^2 - (z_{31} + z_{32}x + z_{33}y + z_{34}x^2 + z_{35}y^2)$.
6. Let $G = \{g_1, g_2, g_3\}$ and $H = \{g_1, g_2, g_3, f_1, f_2\}$. The formal multiplication matrices associated to H are

$$\mathcal{M}_1 = \begin{pmatrix} 0 & 0 & z_{21} & 0 & z_{31} \\ 1 & 0 & z_{22} & 1 & z_{32} \\ 0 & 0 & z_{23} & 0 & z_{33} \\ 0 & 1 & z_{24} & 0 & z_{34} \\ 0 & 0 & z_{25} & 0 & z_{35} \end{pmatrix} \qquad \mathcal{M}_2 = \begin{pmatrix} 0 & z_{21} & 0 & z_{11} & 0 \\ 0 & z_{22} & 0 & z_{12} & 0 \\ 1 & z_{23} & 0 & z_{13} & 1 \\ 0 & z_{24} & 0 & z_{14} & 0 \\ 0 & z_{25} & 1 & z_{15} & 0 \end{pmatrix}$$

7. Let $\mathcal{I}(\mathcal{O})$ be the ideal in $\mathbb{Q}[z_{11}, \ldots, z_{35}]$ generated by the entries of the matrix $\mathcal{M}_1\mathcal{M}_2 - \mathcal{M}_2\mathcal{M}_1$. Thus $\mathcal{I}(\mathcal{O})$ is the ideal generated by the following 20 polynomials:

$$z_{21}z_{23} + z_{25}z_{31} - z_{11} \qquad z_{21}z_{22} + z_{11}z_{24} - z_{31}$$
$$z_{13}z_{21} + z_{15}z_{31} - z_{21} \qquad z_{21}z_{32} + z_{11}z_{34} - z_{21}$$
$$z_{22}z_{23} + z_{25}z_{32} - z_{12} + z_{21} + z_{24} \qquad z_{22}^2 + z_{12}z_{24} - z_{32}$$
$$z_{13}z_{22} + z_{15}z_{32} + z_{11} + z_{14} - z_{22} \qquad z_{22}z_{32} + z_{12}z_{34} - z_{22}$$
$$z_{23}^2 + z_{25}z_{33} - z_{13} \qquad z_{22}z_{23} + z_{13}z_{24} + z_{21} + z_{25} - z_{33}$$
$$z_{13}z_{23} + z_{15}z_{33} - z_{23} \qquad z_{23}z_{32} + z_{13}z_{34} - z_{23} + z_{31} + z_{35}$$
$$z_{23}z_{24} + z_{25}z_{34} - z_{14} + z_{22} \qquad z_{14}z_{24} + z_{22}z_{24} - z_{34}$$
$$z_{13}z_{24} + z_{15}z_{34} + z_{12} - z_{24} \qquad z_{24}z_{32} + z_{14}z_{34} - z_{24}$$
$$z_{23}z_{25} + z_{25}z_{35} - z_{15} \qquad z_{15}z_{24} + z_{22}z_{25} + z_{23} - z_{35}$$
$$z_{13}z_{25} + z_{15}z_{35} - z_{25} \qquad z_{25}z_{32} + z_{15}z_{34} - z_{25} + z_{33}$$

Again we can use a computer algebra system and check that $\mathcal{I}(O)$ is a zero-dimensional, radical ideal of multiplicity 81. This means that among the $126 = \binom{9}{5}$ five-tuples of points in D there are 81 five-tuple which solve the inverse problem.

One of the zeros of $\mathcal{I}(\mathcal{O})$ is the point $p \in \mathbb{Q}^{15}$ whose coordinates are

$$z_{11} = 0 \quad z_{12} = 0 \quad z_{13} = -\tfrac{1}{2} \quad z_{14} = 0 \quad z_{15} = -\tfrac{1}{2}$$
$$z_{21} = 0 \quad z_{22} = -1 \quad z_{23} = -\tfrac{1}{2} \quad z_{24} = 1 \quad z_{25} = -\tfrac{1}{2}$$
$$z_{31} = 0 \quad z_{32} = -1 \quad z_{33} = -\tfrac{1}{2} \quad z_{34} = 1 \quad z_{35} = -\tfrac{1}{2}$$

The corresponding $\mathcal{O}$-border basis is $\{x^3 - x,\ x^2y - \frac{1}{2}y - \frac{1}{2}y^2,\ xy - x - \frac{1}{2}y + x^2 - \frac{1}{2}y^2,\ xy^2 - x - \frac{1}{2}y + x^2 - \frac{1}{2}y^2,\ y^3 - y\}$. The fraction defined by this basis is

$$F_0 = \{(0,0),\ (0,-1),\ (1,0),\ (1,1),\ (-1,1)\}$$

This is our old friend of Example 4.3.7!

In view of our discussion in Section 4.3.1 it is natural to ask how many of the 81 fractions F found above have the property that $\mathcal{O}$ is not of the form $\mathcal{O}_\sigma(I_F)$ for any term ordering σ. We have seen in Example 4.3.7 that at least the fraction F_0 is of that type. By combining Theorem 4.4.4 and some techniques discussed in [CR97], one can show that 36 of those 81 fractions are of that type. This is a surprisingly high number which shows that border bases provide sometimes a much more flexible environment for working with zero-dimensional ideals than Gröbner bases do.

There will never be a last tango
(Brad Hooper)

5

Tools for computing primary decompositions and applications to ideals associated to Bayesian networks

Michael Stillman*

Department of Mathematics, Cornell University, Ithaca, NY 14853 USA,
mike@math.cornell.edu

5.0 Introduction

These lectures, prepared for the CIMPA School on "Systems of polynomial equations" (Argentina, 2003), have two goals: to present the underlying ideas and tools for computing primary decompositions of ideals, and to apply these techniques to a recent interesting class of ideals related to statistics. Primary decompositions generalize the notion of solving systems of polynomial equations, to the cases where there are infinitely many solutions, or to the case when the multiplicity of solutions is important.

Primary decompositions are an important notion both in algebraic geometry and for applications. There are several algorithms available (the two closest to what we present are [GTZ88] and [SY96]). A good overview of the state of the art is the paper [DGP99]. Primary decompositions, and related computations, such as finding minimal and associated primes, the radical of an ideal, and the equidimensional decomposition of an ideal, are all implemented in most specialized computer algebra systems, such as CoCoA [Roba], *Macaulay 2* [GS], and Singular [GPS01]. Several years ago, these algorithms and their implementations could handle only very small examples. Now, with improved implementations, and more efficient computers, larger ideals can be handled.

However, if the number of indeterminates is large, the implemented algorithms often are unable to find a primary decomposition, or even to find the minimal primes. This is the case for many of the ideals associated to Bayesian networks that we consider here.

Our first goal in these lectures is to describe some basic methods for manipulating components of an ideal. We put these together into an algorithm for primary decomposition. We challenge our students to combine these techniques in novel ways to obtain more efficient useful algorithms.

---

* The author would like to acknowledge partial financial support by the National Science Foundation through grant DMS-9970348.

Our second goal is to define some interesting ideals, called Markov ideals, associated to a Bayesian network. In applications, Bayesian networks have been used in many ways, e.g. in machine learning, in vision and speech recognition, in attempting to reconstruct gene regulatory networks, and in the analysis of DNA micro-array data. These Markov ideals provide a very interesting relationship between multivariate statistics and algebra and geometry. In these lectures, we do little more than provide a glimpse into this potentially powerful relationship. Here is one short glimpse: hidden variables in some Markov models correspond to secant loci of Segre embeddings of products of projective spaces (see [GSS] for details).

These Markov ideals often have many components, and can have relatively complicated primary decompositions. We apply the techniques that we have learned to compute some of these primary decompositions. Instead of giving canned algorithms for computing primary decompositions, we will describe several tricks of the trade that can be used on a given ideal, to help find the primary decomposition "by hand" (although with the help of a computer algebra system!). It is likely that superior algorithms exist. Again, we challenge our students to find one!

Lecture 1: We set up the situation, and describe the first two tools of computing primary decompositions: *ideal quotients*, and *splitting principles*. As an example, we find fixed points of some finite dynamical systems.

Lecture 2: We define Bayesian networks and consider independence statements between a set of discrete random variables. Given a Bayesian network, we can associate an ideal, whose primary decomposition is often hard to compute, yet very likely carries interesting information. These ideals provide a striking new link between algebra/geometry and statistics.

Lecture 3: We describe several more tools for computing primary decompositions. We ask several questions: (1) How do we find zero divisors to use with our splitting principles? (2) How do we detect if an ideal is prime, or primary? The tools we develop include birational maps, and the flattener: a polynomial obtained by analyzing the fibers of a projection map. Both of these techniques rely heavily on a Gröbner basis using product orders. We use *Macaulay 2* to investigate these methods on a simple example.

Lecture 4: In the final lecture, we put all of these techniques together and write relatively complete algorithms. A final technique that we address is removing redundancy in the computation as soon as possible. We also present some open problems related to the primary decompositions of Markov ideals.

Throughout, we provide both straightforward and challenging exercises. It is worthwhile to do these! One important exercise is to prove each of the lemmas and propositions which are presented without proof. During the lectures, we spend more time using these results than proving them, although we do include some proofs.

A good elementary introduction to get ready for these lectures is the book by Cox, Little, and O'Shea [CLO97]. The first chapter of the recent book by Hal Schenck [Sch03a] introduces ideal quotients and primary decompositions in a very nice way. His book also has Macaulay 2 examples throughout. A good overview of the known algorithms for primary decomposition is presented in [DGP99]. For delving more deeply into the Bayesian network material, look at [GSS], and the references contained in there.

Example computer sessions are included for *Macaulay 2* [GS]. This is a system that Dan Grayson and I have been working on for the last ten years. The system is freely available, and easy to install on most computers. The web page can be found on the Internet[2].

5.1 Lecture 1: Algebraic varieties and components

Throughout these lectures, let k be a field, and let $R = k[x_1, \ldots, x_n]$. If $J = (f_1, \ldots, f_r) \subset R$ is an ideal, we let

$$V(J) = \{p \in k^n \mid f_1(p) = \ldots = f_r(p) = 0\}.$$

If the base field k is algebraically closed, then there is a beautiful dictionary which relates the geometry of $X = V(J)$ to algebraic properties of the ideal J. We refer the reader to Cox-Little-O'Shea [CLO97] or Schenck [Sch03a] for the details. (If the field is not algebraically closed, the dictionary still exists, but relates the algebra of J to properties of the *scheme* corresponding to J).

For example, if the base field is algebraically closed, and if $J \subset R$ is a *prime* ideal (that is, $fg \in J$ implies $f \in J$ or $g \in J$), then $V(J)$ is irreducible (that is, cannot be written as a union $V(I_1) \cup V(I_2)$ of zero sets which are properly contained in $V(J)$).

Every ideal J in R has a *primary decomposition*, that is, a decomposition

$$J = Q_1 \cap \ldots \cap Q_r,$$

where each Q_i is primary (i.e. if $fg \in Q_i$, then $f \in Q_i$ or $g^N \in Q_i$, for some integer N.) The radical

$$P = \sqrt{Q} = \{g \in R \mid g^N \in Q, \text{ for some } N\}$$

is a prime ideal, and Q is called P-primary.

The primary decomposition is called *irredundant* if each $P_i := \sqrt{Q_i}$ is distinct, and if removing any one term Q_i breaks the equality. Every primary decomposition can be pruned to obtain an irredundant primary decomposition.

If the primary decomposition is irredundant, then the $P_1, \ldots, P_r$ are called the *associated primes* of J. This set is independent of the particular (irredundant) primary decomposition. The minimal elements of this set of primes (with respect to inclusion) are called the *minimal primes* of J. The *radical*

---

[2] http://www.math.uiuc.edu/Macaulay2

of J is the intersection of these minimal primes. If P is a minimal prime, the corresponding primary ideal is unique (i.e. doesn't depend on the specific irredundant primary decomposition). If P is an associated prime, but not minimal, then P is called an *embedded* prime. The primary ideal of an embedded prime is *not* unique.

Example 5.1.1. Let $J = (xy) \subset k[x, y]$. Geometrically, the zero set $xy = 0$ is the union of the two coordinate axes. Algebraically, this is seen in the primary decomposition $J = (x) \cap (y)$. Both (x) and (y) are minimal primes.

If $J = (x^3 y) \subset k[x, y]$, then geometrically, the zero set is the union of the x-axis and a "triple line" $x = 0$. The primary decomposition is $J = (x^3) \cap (y)$. Both associated primes are minimal, but this time the ideal (x^3) is primary, but not prime.

Example 5.1.2. Let $J = (xy, xz) \subset k[x, y, z]$. Geometrically, the zero set $xy = xz = 0$ is the union of the plane $x = 0$ and the line $y = z = 0$. The primary decomposition of J is $J = (x) \cap (y, z)$.

Example 5.1.3. Let $J = (x^2, xy) \subset k[x, y]$. For each $N \geq 1$, we obtain a different primary decomposition of J:

$$J = (x) \cap (x^2, y) = (x) \cap (x^2, xy, y^N).$$

The associated primes are $P_1 = (x)$ and $P_2 = (x, y)$, where P_1 is the only minimal prime, and P_2 is embedded. The primary ideal $Q_1 = (x)$ is the same no matter which primary decomposition we use, but the primary ideal Q_2 of P_2 depends on the decomposition. Geometrically, $V(J)$ is simply the line $x = 0$. Thinking algebraically (or, using schemes), the zero set should really be considered as the union of this line, and a "fat" embedded point at the origin.

Exercise 5.1.4. Find (by hand) a primary decomposition of the ideal $J = (x^3, xy^2 z, y^2 z^3) \subset k[x, y, z]$.

In these lectures, what computations concern us? Given J, we would like to be able to compute (in roughly increasing order of difficulty):

- The radical of J.
- The set of minimal primes of J.
- P-primary component Q of J, where P is a minimal prime,
- The set of associated primes of J.
- An irredundant primary decomposition of J.

5.1.1 Tool #1: Ideal quotients

One of the most important constructions in ideal theory is the operation of ideal quotient.

Definition 5.1.5 (Ideal quotient and saturation). *If $I \subset R$ is an ideal, and $f \in R$, then define the ideal quotient*

$$(I : f) := \{g \in R \mid gf \in I\},$$

and the saturation of I by f:

$$(I : f^\infty) := \{g \in R \mid gf^N \in I, \text{ for some } N\},$$

This somewhat opaque definition gives little clue of their importance.

Lemma 5.1.6. *Let Q be a P-primary ideal, and let $f \in R$. Then*
(a) If $f \notin P$, then $(Q : f) = Q$.
(b) If $f \in P$, but $f \notin Q$, then $(Q : f)$ is P-primary.
(c) If $f \in Q$, then $(Q : f) = (1)$.

An elementary fact, which follows directly from the definition, is that

$$(I_1 \cap I_2) : f = (I_1 : f) \cap (I_2 : f),$$

and so

Lemma 5.1.7. *If $J = Q_1 \cap \cdots \cap Q_r$ is an irredundant primary decomposition of J, where Q_i is P_i-primary, and if $f \in Q_j$ only if $j \geq s + 1$, then*

$$J : f = (Q_1 : f) \cap \cdots \cap (Q_s : f)$$

is an irredundant primary decomposition of $J : f$.

Saturations have even simpler behavior.

Lemma 5.1.8. *Let Q be a P-primary ideal, and let $f \in R$. Then*
(a) If $f \notin P$, then $(Q : f^\infty) = Q$.
(b) If $f \in P$, then $(Q : f^\infty) = (1)$.

Lemma 5.1.9. *If $J = Q_1 \cap \cdots \cap Q_r$ is an irredundant primary decomposition of J, where Q_i is P_i-primary, and if $f \in P_j$ if and only if $j \geq s + 1$, then*

$$J : f^\infty = Q_1 \cap \cdots \cap Q_s$$

is an irredundant primary decomposition of $(J : f^\infty)$.

This says that, geometrically, the components of $V(J : f^\infty)$ are precisely the components of $V(J)$ which do not lie on the hypersurface $f = 0$.

What makes ideal quotients so useful is that they may be computed using Gröbner bases.

Proposition 5.1.10. *Let $J \subset R = k[x_1, \ldots, x_n]$ be an ideal, where k is a ring (e.g. a field, or a PID), and let $f \in R$. If $L = J + (tf - 1) \subset k[t, x_1, \ldots, x_n]$, then*

$$(J : f^\infty) = L \cap k[x_1, \ldots, x_n].$$

This is not always the most efficient method to compute saturations. It also doesn't allow one to compute ideal quotients easily. There are (at least) two further ways to compute ideal quotients which are often used: the reverse lexicographic order, and syzygies. We'll describe the method using the reverse lexicographic order, but we'll leave out the syzygy method.

If $f = x_n$ is a variable, and if J is homogeneous, then $(J : x_n)$ and $(J : x_n^\infty)$ may be computed using a single reverse lexicographic Gröbner basis. The key insight is that if $>$ is the term order (or ordering) in the following proposition and g is a homogeneous polynomial, then $x_n | g$ if and only if $x_n | in(g)$.

Proposition 5.1.11 (Bayer). *Let $J \subset k[x_1, \ldots, x_n]$ be a homogeneous ideal, and let $>$ be the graded reverse lexicographic order (GrevLex) with $x_1 > \ldots > x_n$. If the Gröbner basis of J is*

$$\{g_1, \ldots, g_r, h_{r+1}, \ldots, h_s\},$$

where $g_i = x_n^{a_i} h_i$, each $a_i > 1$, and x_n does not divide the h_i, then
(a) $\{x_n^{a_1-1} h_1, \ldots, x_n^{a_r-1} h_r, h_{r+1}, \ldots, h_s$ is a Gröbner basis of $(I : x_n)$, and
(b) $\{h_1, \ldots, h_s\}$ is a Gröbner basis of $J : f^\infty$.

Exercise 5.1.12. This idea can be used to compute $J : f$ and $J : f^\infty$ when f is not an indeterminate.

(a) Show that if J is homogeneous, and f is homogeneous of degree d, then Bayer's method applied to the homogeneous ideal $J + (f - z)$, where z is a new variable having degree d, can be used to compute $J : f$ and $J : f^\infty$.

(b) Show how to compute the *homogenization* of an ideal by using saturation.

(c) Show how to use homogenization and the trick in (a), to compute $J : f$ and $J : f^\infty$ when J and f arc not necessarily homogeneous.

Example 5.1.13. Consider the ideal $J = (c^2 - bd, bc - ad) \subset \mathbb{Q}[a, b, c, d]$. Notice that the plane $c = d = 0$ is contained in the zero set of J. Let's look at this ideal in *Macaulay 2*.

```
i1 : R = QQ[a..d];

i2 : J = ideal(c^2-b*d, b*c-a*d)

          2
o2 = ideal (c  - b*d, b*c - a*d)

o2 : Ideal of R
```

First, here is the primary decomposition of J:

```
i3 : primaryDecomposition J

                           2                        2
o3 = {ideal (d, c), ideal (c  - b*d, b*c - a*d, b  - a*c)}

o3 : List
```

The reverse lexicographic order is the default in *Macaulay 2*:

```
i4 : gens gb J

o4 = | c2-bd bc-ad b2d-acd |

              1       3
o4 : Matrix R  <--- R

i5 : J : d

          2                    2
o5 = ideal (c  - b*d, b*c - a*d, b  - a*c)

o5 : Ideal of R

i6 : saturate(J,d)

          2                    2
o6 = ideal (c  - b*d, b*c - a*d, b  - a*c)

o6 : Ideal of R

i7 : J == intersect(ideal(c,d),J:d)

o7 = true
```

5.1.2 Tool #2: Splitting principles

The key technique on which almost all algorithms for primary decomposition are based is the following very simple lemma.

Proposition 5.1.14. *If* $(J : f^\infty) = (J : f^\ell)$, *then*

$$J = (J : f^\infty) \cap (J, f^\ell).$$

Proof. Suppose that $g \in (J : f^\infty)$ and also that $g \in (J, f^\ell)$. We want to show that $g \in J$. So $g = a + bf^\ell$, for some $a \in J$ and $b \in R$. However, $gf^\ell \in J$, so $bf^{2\ell} \in J$. Therefore $b \in (J : f^\infty) = (J : f^\ell)$, and so $g \in J$. $\square$

If a polynomial f satisfies $(J : f) \neq J$ and $f^\ell \notin J$, for any ℓ, we'll call f a *splitting polynomial* for J. As a simple exercise, show that there is no splitting polynomial for J if and only if J is a primary ideal.

If we are only interested in finding the set of minimal primes, we may take the radicals of both sides to obtain: for any $f \in R$,

$$\sqrt{J} = \sqrt{J : f^\infty} \cap \sqrt{J, f}.$$

Another useful splitting formula is: if $f_1 f_2 \dots f_r \in J$, then

$$\sqrt{J} = \sqrt{J, f_1} \cap \dots \cap \sqrt{J, f_r}.$$

If we have a way of finding, given an ideal J, a splitting polynomial for J, then we may build a recursive algorithm to compute a decomposition of J.

5.1.3 An example: Finite dynamical systems

As an example, let's consider finite dynamical systems: given a prime number p, let $\mathbb{F}_p$ be the finite field with p elements, let $R = \mathbb{F}_p[x_1, \ldots, x_n]$, and let $F : \mathbb{F}_p^n \longrightarrow \mathbb{F}_p^n$ be defined by

$$a = (a_1, \ldots, a_n) \mapsto (f_1(a), \ldots, f_n(a)),$$

where $f_i \in R$.

All finite dynamical systems can be written in this form:

Exercise 5.1.15. Show that, for any natural number $n > 0$ and any function $f : \mathbb{F}_p^n \longrightarrow \mathbb{F}_p^n$ there are polynomials $g_j \in \mathbb{F}_p[x_1, .., x_n]$ such that $f(a) = (g_1(a), \ldots, g_n(a))$ for all $a \in \mathbb{F}_p^n$.

By iterating F, we obtain a directed graph whose vertices are the p^n points of $\mathbb{F}_p^n$, and there are directed edges from a to $F(a)$.

In this example, we are interested in finding the fixed points of F, or more generally, of F^r (apply F r times) for some integer r. The fixed points of F are the zeros of the ideal $J = (x_1 - f_1, \ldots, x_n - f_n)$ which have all coordinates in $\mathbb{F}_p$. The problem is that there may be solutions over an extension field of $\mathbb{F}_p$ and we are not particularly interested in these solutions. Notice that if $x \in \overline{\mathbb{F}_p}$, then $x \in \mathbb{F}_p$ if and only if $x^p - x = 0$. So, if we include these polynomials, then our zero set will only contain elements of the field we are interested in.

These ideals are always equal to their own radical, and so we need not worry about embedded components:

Lemma 5.1.16. *Let* $J = (g_1, \ldots, g_s, x_1^p - x_1, \ldots, x_n^p - x_n) \subset k[x_1, \ldots, x_n]$. *For any choice of* g_i*'s,* $J = \sqrt{J}$.

Exercise 5.1.17. Prove this lemma. Use (or prove!) the fact that if $J \subset k[x_1, \ldots, x_n]$ is a zero dimensional ideal, then the radical of J is

$$\sqrt{J} = J + (h_1, \ldots, h_n),$$

where h_i is the squarefree part of the generator of the ideal $J \cap k[x_i]$.

See for example Chapter 2, Section 2.1.2.

We may use any of these splitting principles to compute the minimal primes (and therefore the primary decomposition) of J, since we have many zero-divisors around: each x_i is (potentially) a zero-divisor!

Example 5.1.18. Let $R = k[x_1, \ldots, x_4]$, where $k = \mathbb{F}_2$. Let $F : k^4 \longrightarrow k^4$.

The associated directed graph has $2^4 = 16$ nodes. Let's find the fixed points of one such finite dynamical system, with the aid of *Macaulay 2*. In such a small example, we can compute the fixed points by hand. For larger examples, e.g. $p = 3$, $n = 20$, this is not so easy!

```
i8 : R = ZZ/2[x_1 .. x_4];
```

```
i9 : L = ideal(x_1^2 + x_1, x_2^2 + x_2, x_3^2 + x_3, x_4^2 + x_4);

o9 : Ideal of R
```

Our sample finite dynamical system:

```
i10 : F = matrix {{x_1*x_2*x_4+x_1+x_4,
              x_1*x_3*x_4+x_2*x_4+x_2,
              x_1*x_3+x_3*x_4+x_3,
              x_1*x_3*x_4+x_1+x_4}}

o10 = | x_1x_2x_4+x_1+x_4 x_1x_3x_4+x_2x_4+x_2 x_1x_3+x_3x_4+x_3 x_1x_ ···

            1       4
o10 : Matrix R  <--- R
```

Fixed points of F are precisely the zeros of the following ideal.

```
i11 : J = L + ideal (vars R - F);

o11 : Ideal of R

i12 : transpose gens gb J

o12 = {-1} | x_1+x_4      |
      {-2} | x_4^2+x_4    |
      {-2} | x_3x_4+x_1   |
      {-2} | x_2x_4+x_3x_4 |
      {-2} | x_3^2+x_3    |
      {-2} | x_2^2+x_2    |

            6       1
o12 : Matrix R  <--- R
```

Although we could solve these equations by hand, we instead blindly follow the recursion using indeterminates as (potential) zero divisors. We start with x_1.

```
i13 : J1 = J : x_1

o13 = ideal (x   + 1, x   + 1, x   + 1, x   + 1)
              4        3        2        1

o13 : Ideal of R

i14 : J2 = ideal gens gb(J + ideal(x_1))

                 2           2
o14 = ideal (x , x , x   + x , x   + x )
              4   1   3     3   2     2

o14 : Ideal of R
```

The intersection of these ideals is J.

```
i15 : J == intersect(J1,J2)

o15 = true
```

The first ideal is already linear, so its zero set is a point. From the description of J_2 we could write down the rest of the solutions, but let's continue. Split using x_3:

```
i16 : J21 = J2 : x_3
```

```
                            2
o16 = ideal (x , x  + 1, x , x  + x )
             4   3        1   2    2

o16 : Ideal of R

i17 : J22 = ideal gens gb(J2 + ideal(x_3))

                          2
o17 = ideal (x , x , x , x  + x )
             4   3   1   2    2

o17 : Ideal of R
```

Now we can split each of these using x_2, obtaining 5 solutions total. Already, one can imagine ways to improve the efficiency of even this small example. For larger problems, these improvements can make the difference between obtaining an answer and waiting forever!

We could have computed this directly in *Macaulay 2*. The `decompose` routine provides the list of minimal primes. The `primaryDecomposition` routine provides an irredundant primary decomposition.

```
i18 : C = decompose J;
```

Display these ideals:

```
i19 : C/(I -> (<< toString I << endl));
ideal(x_2,x_3,x_4,x_1)
ideal(x_2+1,x_3,x_4,x_1)
ideal(x_2+1,x_3+1,x_4,x_1)
ideal(x_2,x_3+1,x_4,x_1)
ideal(x_2+1,x_3+1,x_4+1,x_1+1)
```

Exercise 5.1.19. Let $R = \mathbb{F}_3[x_1, \ldots, x_{20}]$. Choose $F = (f_1, \ldots, f_{20})$ such that each f_ℓ is a sum of two randomly chosen quadratic monomials $x_i x_j$. Find the fixed points of this finite dynamical system. Also find the points of order 2, i.e. those points a such that $a = F(F(a))$.

Here is an open question: can you characterize the graphs (of 3^{20} vertices) which arise from F in this way?

5.2 Lecture 2: Bayesian networks and Markov ideals

The emerging field of *algebraic statistics* [PRW00] advocates polynomial algebra as a tool in the statistical analysis of experiments and discrete data. Statistics textbooks define a *statistical model* as a family of probability distributions, and a closer look reveals that these families are often algebraic varieties: they are the zeros of some polynomials in the probability simplex [GHKM01], [SS00].

We begin by reviewing the general algebraic framework for independence models presented in [Stu02, §8]. Let $X_1, \ldots, X_n$ be discrete random variables where X_i takes values in the finite set $[d_i] = \{1, 2, \ldots, d_i\}$. We write $D = [d_1] \times [d_2] \times \cdots \times [d_n]$ so that $\mathbb{C}^D$ denotes the complex vector space of n-dimensional tables of format $d_1 \times \cdots \times d_n$. We introduce an indeterminate $p_{u_1 u_2 \cdots u_n}$ which represents the probability of the event $X_1 = u_1$, $X_2 = u_2, \ldots, X_n = u_n$. These indeterminates generate the ring $\mathbb{C}[D]$ of polynomial functions on the space of tables $\mathbb{C}^D$. We could also use the field $\mathbb{R}$. The points of interest from statistics are those in the *probability simplex* Δ: the set of points whose coordinates are in the interval $[0, 1]$, and whose coordinates sum to 1.

A *conditional independence statement* has the form

$$A \text{ is independent of } B \text{ given } C \quad (\text{ in symbols: } A \perp\!\!\!\perp B \mid C) \quad (5.1)$$

where A, B and C are pairwise disjoint subsets of $\{X_1, \ldots, X_n\}$. If C is empty then (5.1) means that A is independent of B.

Example 5.2.1. Let X_1 be the statement: it will rain today. Let X_2 be the statement: a puddle will form next to my car door. Let X_3 be the statement: I will get wet when I step out of the car. These are all binary random variables. Given that the puddle has formed, the other two are independent statements: $X_1 \perp\!\!\!\perp X_3 \mid X_2$.

By [Stu02, Proposition 8.1], the statement (5.1) translates into a set of homogeneous quadratic polynomials in $\mathbb{C}[D]$, and we write $I_{A \perp\!\!\!\perp B \mid C}$ for the ideal generated by these polynomials. The following example gives the basic idea and method for finding these ideals.

Example 5.2.2. Let X_1, X_2, X_3 be three random variables, with $d_1 = d_3 = 2$ and $d_2 = 3$. Let's write down the ideal in $k[p_{u_1 u_2 u_3}]$ (12 variables) which defines the set of probability distributions which satisfy $X_1 \perp\!\!\!\perp X_2 \mid X_3$.

A probability distribution satisfies this independence condition if

$$Pr(X_1 = u_1, X_2 = u_2 \mid X_3 = u_3) =$$
$$Pr(X_1 = u_1 \mid X_3 = u_3)Pr(X_2 = u_2 \mid X_3 = u_3),$$

for all choices of $u_i \in [d_i]$. By removing the conditional probabilities, and multiplying by $Pr(X_3 = u_3)$, we obtain

$$p_{++u_3} p_{u_1 u_2 u_3} = p_{u_1 + u_3} p_{+u_2 u_3},$$

where we have replaced Pr by p, and a "+" means sum over all possible values in that variable (i.e. marginalize over that variable). For example,

$$p_{1+2} = p_{112} + p_{122} + p_{132}.$$

It is a simple exercise in determinants to show that the ideal generated by

$$\{p_{++u_3} p_{u_1 u_2 u_3} - p_{u_1 + u_3} p_{+u_2 u_3} \mid \text{all } u_1, u_2, u_3\},$$

is the same as the ideal generated by the six 2 by 2 minors of the matrices M_1 and M_2, where

$$M_i = \begin{pmatrix} p_{11i} & p_{12i} & p_{13i} \\ p_{21i} & p_{22i} & p_{23i} \end{pmatrix}$$

Note that all 12 indeterminates appear, and each matrix has 6 of them.

The general case goes the same way: The ideal $I_{A \perp\!\!\!\perp B|C}$ is generated by the 2 by 2 minors of matrices M_i, for $i = 1..c$, where c is the number of possible values of C. Each matrix is obtained by making an $a \times b$ matrix where the (j, k)th entry is the linear polynomial in the $p_{u_1 \ldots u_n}$ which represents $Pr(A = j, B = k, C = i)$.

Since the ideal generated by the 2 by 2 minors of a generic matrix of indeterminates is prime, we have the following fact (see [Stu02]).

Proposition 5.2.3. *For any choice of A, B, and C, the ideal $I_{A \perp\!\!\!\perp B|C}$ is prime.*

The interesting part begins when we have more than one independence statement.

Definition 5.2.4. *If $\mathcal{M} = \{A_1, A_2, \ldots, A_r\}$ is a set of independence statements, define*

$$I_{\mathcal{M}} = I_{A_1} + \cdots + I_{A_r}.$$

Example 5.2.5 (The contraction lemma). In statistics, there is a lemma that says that any probability distribution which satisfies the two independence statements $X_1 \perp\!\!\!\perp X_2 \mid X_3$, and $X_2 \perp\!\!\!\perp X_3$, also satisfies $X_2 \perp\!\!\!\perp \{X_1, X_3\}$.

In this example, we investigate the algebraic analog of this statement. Let

$$\mathcal{M} = \{X_1 \perp\!\!\!\perp X_2 \mid X_3, \ X_2 \perp\!\!\!\perp X_3\}.$$

Let's suppose for now that $d_1 = d_2 = d_3 = 2$, i.e. we have three binary random variables. The first independence statement translates into two quadratics:

$$\phi_1 = \det \begin{pmatrix} p_{111} & p_{121} \\ p_{211} & p_{221} \end{pmatrix}, \quad \phi_2 = \det \begin{pmatrix} p_{112} & p_{122} \\ p_{212} & p_{222.} \end{pmatrix}$$

The second statement translates into a single determinant:

$$\phi = \det \begin{pmatrix} p_{+11} & p_{+12} \\ p_{+21} & p_{+22} \end{pmatrix},$$

where for example $p_{+11} = p_{111} + p_{211}$.

So $I_{\mathcal{M}} = (\phi_1, \phi_2, \phi)$.

If we consider the indeterminates of our polynomial ring to be p_{+jk} and p_{ijk}, for $i \geq 2$ (instead of the p_{ijk}), the ideal $I_{\mathcal{M}}$ is a *binomial* ideal in $\mathbb{C}[D]$, i.e. generated by polynomials which are differences of two monomials. Binomial ideals enjoy many nice properties. For instance, a reduced Gröbner basis, in any term order, consist of binomials, and they have primary decompositions where each associated prime and primary ideal is binomial. For more details, see [ES96].

The algebraic analog of the contraction lemma is the primary decomposition of this ideal. The ideal $I_{\mathcal{M}}$ has 3 components in its primary decomposition (all prime).

$$I_{\mathcal{M}} = P_1 \cap P_2 \cap I_{X_2 \perp\!\!\!\perp \{X_1, X_3\}},$$

where $P_1 = (p_{+11}, p_{+21}, \phi_2)$, and $P_2 = (p_{+12}, p_{+22}, \phi_1)$. This implies that any probability distribution which satisfies the two independence statements M also satisfies the statement: $X_2 \perp\!\!\!\perp \{X_1, X_3\}$. The algebraic picture is more complicated: outside of the probability simplex Δ, these two zero sets differ.

As a warmup for computing primary decompositions later, try

Exercise 5.2.6. (a) Show, using Macaulay 2, that this is a primary decomposition of $I_{\mathcal{M}}$.

(b) Consider the same $\mathcal{M}$, but now suppose that $d_1 = d_2 = 2$ and $d_3 = 3$. Write down the ideal $I_{\mathcal{M}}$ and find a primary decomposition for $I_{\mathcal{M}}$. Is this ideal radical? What if $d_3 \geq 4$?

5.2.1 Bayesian networks and associated ideals

A *Bayesian network* is an acyclic directed graph G with vertices $X_1, \ldots, X_n$.

For a given node X_i, let $pa(X_i)$ denote the set of parents of vertex X_i in G (a node X_j is a parent of X_i if there is a directed edge from X_j to X_i, and let $nd(X_i)$ be the set of non-descendants of X_i, *excluding* the parents of X_i. (A non-descendant of X_i is a vertex X_j such that there is no directed path from X_i to X_j. Since the graph is not acyclic, parents are non-descendants). The *local Markov property* on G is the set of independence statements

$$\text{local}(G) \quad = \quad \{X_i \perp\!\!\!\perp \text{nd}(X_i) \mid \text{pa}(X_i) : \ i = 1, 2, \ldots, n\},$$

The *global Markov property* , global(G), is the set of independence statements $A \perp\!\!\!\perp B \mid C$, for any triple A, B, C of subsets of pairwise disjoint vertices of G such that A and B are *d-separated* by C.

The notion of $d - separated$ ("directed separated") is a bit technical. The intuition is that the nodes of C block directed paths from nodes of A to nodes of B, but the notion is slightly more subtle. Since we don't really need the definition for these lectures, we refer to [GSS] or to [Lau96] for the definition.

For any Bayesian network G, we have $local(G) \subset global(G)$. Therefore we have inclusions $I_{\text{local}(G)} \subset I_{\text{global}(G)}$, and $V_{\text{global}(G)} \subset V_{\text{local}(G)}$.

Example 5.2.7. Let G be the network on four binary random variables shown in 5.1. Download the file `markov.m2` from the website[3]. This file contains code for displaying a directed acyclic graph, computing independence conditions (given a graph), and for computing the ideals corresponding to these independence conditions. The documentation for the code is contained in the file.

```
i20 : load "markov.m2"
```

---
[3] http://www.math.cornell.edu/~mike/bayes/

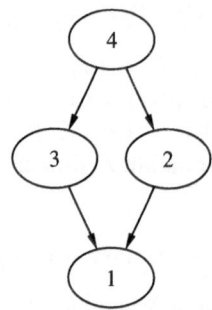

Fig. 5.1. A Bayesian network on 4 vertices

The function `makeGraph` takes as input a list of lists: the ith list is the list of direct descendants of the ith node.

```
i21 : G = makeGraph {{},{1},{1},{2,3}};
```

The Markov conditions come as a list of triples of sets of integers. Each triple represents a single independence statement.

```
i22 : LM = localMarkovStmts G;

i23 : LM/print;
{Set {1}, Set {4}, Set {2, 3}}
{Set {3}, Set {2}, Set {4}}

i24 : GM = globalMarkovStmts G;

i25 : GM/print;
{Set {1}, Set {4}, Set {2, 3}}
{Set {3}, Set {2}, Set {4}}
```

Note that for this example, local(G) and global(G) are both the same set:

$$\{1 \perp\!\!\!\perp 4 \mid \{2,3\}, 2 \perp\!\!\!\perp 3 \mid 4\}.$$

A polynomial ring with the indeterminates $p_{u_1 u_2 \ldots u_n}$ is created via:

```
i26 : R = markovRing(2,2,2,2);

i27 : numgens R

o27 = 16

i28 : gens R

o28 = {p        , p        , p        , p        , p        , p        , p   ···
         1,1,1,1   1,1,1,2   1,1,2,1   1,1,2,2   1,2,1,1   1,2,1,2   1, ···

o28 : List
```

Our two independence statements translate to the 2 by 2 minors of the following six matrices (and, since each is only 2 by 2, the ideal is generated by six quadrics).

```
i29 : M = markovMatrices(R, LM);
```

```
i30 : M/(m -> (<< m << endl << endl));
| p_(1,1,1,1) p_(1,1,1,2) |
| p_(2,1,1,1) p_(2,1,1,2) |

| p_(1,1,2,1) p_(1,1,2,2) |
| p_(2,1,2,1) p_(2,1,2,2) |

| p_(1,2,1,1) p_(1,2,1,2) |
| p_(2,2,1,1) p_(2,2,1,2) |

| p_(1,2,2,1) p_(1,2,2,2) |
| p_(2,2,2,1) p_(2,2,2,2) |

| p_(1,1,1,1)+p_(2,1,1,1) p_(1,2,1,1)+p_(2,2,1,1) |
| p_(1,1,2,1)+p_(2,1,2,1) p_(1,2,2,1)+p_(2,2,2,1) |

| p_(1,1,1,2)+p_(2,1,1,2) p_(1,2,1,2)+p_(2,2,1,2) |
| p_(1,1,2,2)+p_(2,1,2,2) p_(1,2,2,2)+p_(2,2,2,2) |
```

By changing coordinates as discussed above so that e.g. p_{1111} refers to p_{+111}, and p_{2111} still refers to p_{2111}), the ideal will be binomial in the new coordinates. The function `marginMap` makes a ring map which will make this change of coordinates.

```
i31 : F = marginMap(1,R);

o31 : RingMap R <--- R

i32 : F p_(1,1,1,1)

o32 = p        - p
       1,1,1,1    2,1,1,1

o32 : R
```

The routine `markovIdeal` yields the ideal generated by the 2 by 2 minors of the above matrices. After changing coordinates, the ideal is binomial:

```
i33 : J = F markovIdeal(R,LM);

o33 : Ideal of R

i34 : transpose generators J

o34 = {-2} | -p_(1,1,1,2)p_(2,1,1,1)+p_(1,1,1,1)p_(2,1,1,2) |
       {-2} | -p_(1,1,2,2)p_(2,1,2,1)+p_(1,1,2,1)p_(2,1,2,2) |
       {-2} | -p_(1,2,1,2)p_(2,2,1,1)+p_(1,2,1,1)p_(2,2,1,2) |
       {-2} | -p_(1,2,2,2)p_(2,2,2,1)+p_(1,2,2,1)p_(2,2,2,2) |
       {-2} | -p_(1,1,2,1)p_(1,2,1,1)+p_(1,1,1,1)p_(1,2,2,1) |
       {-2} | -p_(1,1,2,2)p_(1,2,1,2)+p_(1,1,1,2)p_(1,2,2,2) |

                 6        1
o34 : Matrix R  <--- R
```

The ideal J is minimally generated by 6 binomial quadrics.

One of the most useful aspects of Bayesian networks is that they provide a factorization of the joint probability distribution of the n random variables. In this example, note that

$$Pr(X_1 = u_1, \ldots, X_4 = u_4) =$$
$$Pr(X_4 = u_4) \times Pr(X_3 = u_3 \mid X_4 = u_4) \times Pr(X_2 = u_2 \mid X_3 = u_3, X_4 = u_4)$$
$$\times Pr(X_1 = u_1 \mid X_2 = u_2, X_3 = u_3, X_4 = u_4)$$
$$= Pr(X_4 = u_4) \times Pr(X_3 = u_3 \mid X_4 = u_4) \times Pr(X_2 = u_2 \mid X_4 = u_4)$$
$$\times Pr(X_1 = u_1 \mid X_2 = u_2, X_3 = u_3)$$

If we set $Pr(X_4 = 1) := a$ and $Pr(X_4 = 2) := 1 - a$, and similarly let $Pr(X_3 = 1 \mid X_4 = k) := b_k$, let $Pr(X_2 = 1 \mid X_4 = k) := c_k$, and $Pr(X_1 = 1 \mid X_2 = j, X_3 = k) := d_{jk}$, then the joint probabilities factor. For example, $p_{1111} = ab_1c_1d_{11}$, $p_{1112} = (1-a)b_2c_2d_{11}$, $p_{1121} = a(1-b_1)c_1d_{12}$, and so on. Instead of requiring 15 parameters, such a probability distribution may be specified using 10 numbers. This is a small example; when the number of vertices is large and the graph is sparse, the savings is dramatic.

If we denote $\mathbb{C}[E] := \mathbb{C}[a, b_1, b_2, c_1, c_2, d_{11}, \ldots, d_{22}]$, we may define a ring map

$$\Phi : \mathbb{C}[D] \longrightarrow \mathbb{C}[E].$$

In what follows we shall assume that every edge (i, j) of the Bayesian network G satisfies $i > j$. In particular, the node 1 is always a sink and the node n is always a source.

For any integer $r \in [n]$ and $u_i \in [d_i]$ as before, we abbreviate the *marginalization* over the first r random variables as follows:

$$p_{++\cdots+u_{r+1}\cdots u_n} \quad := \quad \sum_{i_1=1}^{d_1} \sum_{i_2=1}^{d_2} \cdots \sum_{i_r=1}^{d_r} p_{i_1 i_2 \cdots i_r u_{r+1} \cdots u_n}.$$

This is a linear form in our polynomial ring $\mathbb{C}[D]$. We denote by **p** the product of all of these linear forms.

As in the example, given a Bayesian network G, we obtain a factorization map:

$$\Phi : \mathbb{C}[D] \longrightarrow \mathbb{C}[E].$$

The main theorem, which is the algebraic analog of the factorization for the joint probabilities for a Bayesian network is the following:

Theorem 5.2.8. *The prime ideal* $\ker(\Phi)$ *is a minimal primary component of both of the ideals* $I_{\mathrm{local}(G)}$ *and* $I_{\mathrm{global}(G)}$. *More precisely,*

$$\left(I_{\mathrm{local}(G)} : \mathbf{p}^\infty\right) \;=\; \left(I_{\mathrm{global}(G)} : \mathbf{p}^\infty\right) \;=\; \ker(\Phi). \qquad (5.2)$$

For the precise definition of Φ and a proof, see [GSS].

This result suggests many questions, most of them unsolved. For example:

Problem 5.2.9. Find conditions on G so that $I_{\mathrm{global}(G)}$ is a prime ideal (and therefore equal to $\ker \Phi$).

Problem 5.2.10. Find the dictionary relating basic operations on directed acyclic graphs (e.g. deletion of an edge, or of a node, or contraction of an edge) with properties of the primary decomposition of the corresponding ideals.

Problem 5.2.11. Find the primary decomposition of $I_{\text{local}(G)}$ or of $I_{\text{global}(G)}$.

Perhaps more manageable is to determine certain features of the primary decomposition (e.g. the ideal being radical, or having no embedded components) in terms of the data G, and $(d_1, \ldots, d_n)$.

In the remainder of these lectures, we will develop the tools needed to answer these questions computationally, for small networks G.

5.3 Lecture 3: Tools for computing primary decompositions

In order to use the techniques we have already considered to make an algorithm for computing a primary decomposition, we must answer the following questions.

- Question #1: How do we find splitting polynomials or zero divisors to use with one of our splitting principles?
- Question #2: How can we detect that an ideal is prime or primary?
- Question #3: Practice shows that the splitting tree is highly redundant. How should we fight this problem?

We will provide answers to these questions. But: keep your mind open. We challenge you to find better methods yourself!

Example 5.3.1. As a running example throughout this lecture, let's consider the simple example which occurred in the contraction lemma in the second lecture. This is an ideal generated by 3 quadrics, in 8 indeterminates. Let's rename the indeterminates so that we can avoid indices.

```
i35 : R = QQ[a..h];

i36 : J = ideal(a*d-b*c, e*h-f*g, a*f-b*e);

o36 : Ideal of R
```

Just so we know the answer ahead of time, here is the primary decomposition:

```
i37 : (primaryDecomposition J)/print;
ideal (b, a, f*g - e*h)
ideal (f, e, b*c - a*d)
ideal (f*g - e*h, d*g - c*h, b*g - a*h, d*e - c*f, b*e - a*f, b*c - a*d)
```

There are three primary components. In many ways, this is too simple an example: all of the components have the same dimension (five), and all of the primary components are prime, so this is a radical ideal. The example still provides a good picture of the different tools and also some of the problems which occur.

5.3.1 Finding splitting polynomials and zero divisors

Given an ideal $J = (f_1, \ldots, f_r) \subset k[x_1, \ldots, x_n]$, how can we find a zero divisor $g \mod J$ (i.e. an element g for which $J : g \neq J$)? One method that often works is to examine the generators f_i and see if they factor. If so, use a factor as the zero divisor g. Often no f_i will factor. In this case, one may start computing a Gröbner basis, and examine each new Gröbner basis element g_i. If g_i factors, use this factorization to split the ideal (This is the basic description of what is known as the factorizing Gröbner basis algorithm.) The exact details of how best to use this are not clear, and vary with the problem domain. There is definitely room for improvement here in existing algorithms!

Suppose that you cannot find a factor with one of these methods, or, perhaps, are unwilling or unable to look there for zero divisors? What then? Our answer is obtained by analyzing projection maps.

5.3.2 Projections and elimination of variables

Let $R = k[x] = k[x_1, \ldots, x_n]$, where k is a field. Choose a subset of variables

$$t = \{t_1, \ldots, t_d\} \subset x = \{x_1, \ldots, x_n\},$$

and let $u = x \setminus t$. The inclusion $k[t] \subset k[u, t] = k[x]$ corresponds geometrically to the projection map $k^n \longrightarrow k^d$ defined by sending a point $(u, t) = (u_1, \ldots, u_{n-d}, t_1, \ldots, t_d)$ to $t \in k^d$. The map of rings $\phi : k[t] \longrightarrow k[u, t]/J$ corresponds to the projection map $\pi : V(J) \subset k^n \longrightarrow k^d$, and the map of rings $k[t]/J_1 \hookrightarrow k[u, t]/J$ corresponds to the projection map $\pi : V(J) \longrightarrow V(J_1) = \pi(V(J))$, where $J_1 = \ker(\phi)$. If J is not a radical ideal, or k is not an algebraically closed field such as $\mathbb{C}$, then this correspondence between the algebra and geometry needs to be defined more carefully: this is where schemes enter the algebraic geometry picture. For us though, we will think geometrically, but work algebraically, and so we won't be concerned with these subtleties.

Recall that we can compute $L = \ker(\phi)$ by using Gröbner bases. A term order on $k[x] = k[u, t]$ is called an *elimination* order (eliminating u) if $in(f) \in k[t]$ implies that $f \in k[t]$.

Proposition 5.3.2. *If $>$ is an elimination order eliminating u, and $J \subset k[u, t]$ is an ideal, with Gröbner basis $\{f_1, \ldots, f_r, h_1, \ldots, h_s\}$, where $h_i \in k[t]$, but each $f_i \notin k[t]$, then $\{h_1, \ldots, h_s\}$ is a Gröbner basis (and therefore a generating set) of $J_1 = J \cap k[t]$.*

For the purpose of analyzing projection maps, the *product order* $u \gg t$ is a good choice (this is sometimes called a *block order*): $u^a t^b > u^c t^d$ if $u^a >_{grevlex} u^c$, or $u^a = u^c$ and $t^b >_{grevlex} t^d$.

Example 5.3.3. Continuing Example 5.3.1, suppose that $t = \{a, b, c, d\}$ and $u = \{e, f, g, h\}$.

```
i38 : R1 = QQ[e,f,g,h,a,b,c,d, MonomialOrder=>ProductOrder{4,4}];

i39 : L = substitute(J,R1)

o39 = ideal (- b*c + a*d, - f*g + e*h, - e*b + f*a)

o39 : Ideal of R1

i40 : transpose gens gb L

o40 = {-2} | bc-ad   |
      {-2} | eb-fa   |
      {-2} | fg-eh   |
      {-3} | ead-fac |

                4       1
o40 : Matrix R1  <--- R1
```

So $J \cap k[a, b, c, d] = (bc - ad)$, since this is the only element whose lead term bc is in the subring $k[a, b, c, d]$. This whole process can be accomplished more easily using the *Macaulay 2* "eliminate" package.

```
i41 : load "eliminate.m2"
```

This next command ensures that e, f, g, h refer to elements of the ring R.

```
i42 : use R;

i43 : eliminate(J,{e,f,g,h})

o43 = ideal(b*c - a*d)

o43 : Ideal of R
```

5.3.3 Tool: Birational projections

Suppose that J contains an element f which is linear in a variable, say, x_1. Write $f = gx_1 + h$, where g, h don't involve x_1. If g is a non-zero divisor on J, then the projection map $k[t]/J_1 \longrightarrow k[t, x_1]/J$ is *birational* (where $t = \{x_2, \ldots x_n\}$ and $J_1 = \ker(k[t] \longrightarrow k[x_1, t]/J)$. Geometrically, this means that for almost all points p of $V(J_1) \subset k^{n-1}$, there is a unique point $(p_1, p) \in V(J)$ which maps to it. If $g(p) \neq 0$, then this value is $p_1 = -\frac{h(p)}{g(p)}$.

Birational maps are well-behaved with respect to primary decompositions:

Proposition 5.3.4. *Let $J \subset k[x_1, \ldots, x_n]$ be an ideal, containing a polynomial $f = gx_1 + h$, with g, h not involving x_1, and g a non-zero divisor modulo J. Let $J_1 = J \cap k[x_2, \ldots, x_n]$ be the elimination ideal. Then*

(a) $J = (\langle J_1, gx_1 + h \rangle : g^\infty)$,

(b) J is prime if and only if J_1 is prime.

(c) J is primary if and only if J_1 is primary.

(d) Any irredundant primary decomposition of J_1 lifts to an irredundant primary decomposition of J.

This tool may often be used to prove that an ideal is prime (if it is!), and can sometimes simplify the work to look for zero divisors. However, caution is required: the resulting ideal J_1, although it is an ideal in one fewer variable, can sometimes be much more complicated than J.

Exercise 5.3.5. Prove this proposition. (There are at least two related methods to do this: Use pseudo-division by $gx_1 + h$; or use localization by powers of g).

Example 5.3.6. Continuing Example 5.3.1, all variables occur linearly, and so we may choose any one we wish, e.g. a. The corresponding coefficient is d.

```
i44 : use R;
```

In this example, d is not a zero divisor:

```
i45 : J : d == J

o45 = true
```

As above, we use the *Macaulay 2* "eliminate" package for eliminating variables.

```
i46 : I1 = eliminate(J,a)

o46 = ideal (f*g - e*h, b*d*e - b*c*f)

o46 : Ideal of R
```

The variable f occurs linearly, with coefficient g. It so happens that g is also a non-zero-divisor:

```
i47 : I1 : g == I1

o47 = true
```

So I_1 is birational to

```
i48 : I2 = eliminate(I1,f)

o48 = ideal(b*d*e*g - b*c*e*h)

o48 : Ideal of R
```

This single element has three factors:

```
i49 : time factor I2_0
      -- used 0.01 seconds

o49 = (b)(- d*g + c*h)(e)(-1)

o49 : Product
```

The original ideal J is birational to I_2. Another way to factor this is to find the primary decomposition of I_2!

```
i50 : time primaryDecomposition I2
      -- used 0.43 seconds

o50 = {ideal e, ideal b, ideal(d*g - c*h)}

o50 : List
```

Therefore, the original ideal has three components, all prime.

Exercise 5.3.7. Use this factorization, and ideal quotients, to produce the three primary components.

5.3.4 Tool: The flattener of a projection

One method to find a splitting polynomial is to compute the *flattener* of a projection. We develop this method now. This method has many other applications, some of which which we will see later.

Let $J \subset k[x_1, \ldots, x_n]$ be an ideal. A subset of variables $t = \{x_{i_1}, \ldots, x_{i_d}\}$ is called a *maximal independent set* of J if $J \cap k[t] = (0)$ and t has maximal cardinality over all such subsets with this property.

Proposition 5.3.8. *Let $in(J)$ be the initial monomial ideal of $J \subset k[x]$ with respect to some arbitrary term order. Then every maximal independent set of $in(J)$ is also a maximal independent set of J.*

The cardinality d of a maximal independent set of J is called the *dimension* of J.

Geometrically, if $J \cap k[t] = (0)$, the map $V(J) \longrightarrow k^d$ is *dominant*, i.e. the closure of the image is all of k^d. In this case, every component of J which also maps dominantly to k^d must have the same dimension d as J. A component of J which maps into a subvariety of k^d (algebraically: a primary ideal Q of J for which $Q \cap k[t] \neq (0)$) can either have dimension d, or have smaller dimension.

Suppose that $t \subset x$ is a maximal independent set for $in(J)$ and therefore for J, let $u = x \setminus t$, and let $>$ be the product order $u \gg t$ defined above. Let $\{g_1, \ldots, g_r\}$ be a reduced Gröbner basis for the ideal J, where

$$g_i = \alpha_i(t)u^{A_i} + \text{lower terms in the } u \text{ variables.}$$

Since $J \cap k[t] = (0)$, each of the monomials $u^{A_i} \neq 1$. Define $in_u(J) = (u^{A_1}, \ldots, u^{A_r}) \subset k[u]$.

Let $h \in k[t]$ be any non-zero element such that for each minimal generator u^A of the monomial ideal $in_u(J)$, there is an element g of J, such that $g = h(t)u^A + \text{lower terms in } u$. For example, we could take $h = lcm\{\alpha_1, \ldots, \alpha_r\} \in k[t]$. Any such element h is called a *flattener* for J with respect to t.

The reason that h is called a flattener comes from commutative algebra. One can prove that the inclusion of localized rings $k[t]_h \subset k[u, t]_h/J$ is a flat extension. Caution though: our element h enjoys more properties than an arbitrary element that satisfies this flatness.

The key properties of a flattener, for our purposes, is the following observations.

Proposition 5.3.9. *If $h \in k[t]$ is a flattener for J with respect to t, and if P is an associated prime ideal of J, then $h \in P$ if and only if $P \cap k[t] \neq (0)$.*

Since a component P of J which satisfies $P \cap k[t] = (0)$ must have dimension at least the cardinality of t, this implies:

Corollary 5.3.10. *If $h \in k[t]$ is a flattener for J with respect to t, then $(J : h^\infty)$ is equidimensional of dimension d, and in particular has no embedded components.*

So, either h is a splitting polynomial, or J is equidimensional. In the first case, we may split J. We will discuss the second situation later.

Example 5.3.11. Let's use the flattener method to compute the primary decomposition of the ideal of Example 5.3.1. Even though this is a simple example, it highlights several possible efficiency problems.

First, we find a maximal independent set of J. (The *Macaulay 2* routine `independentSets` returns the maximal independent sets of the initial monomial ideal of J. Each monomial represents one independent set. For example, the first set found is $t = \{a, b, d, f, h\}$).

```
i51 : independentSets J

o51 = {a*b*d*f*h, a*c*d*f*h, a*c*e*f*h, c*d*e*f*h, a*b*d*g*h, a*c*d*g* ···

o51 : List
```

We find 8 maximal independent sets.

```
i52 : R1 = QQ[c,e,g,  a,b,d,f,h,MonomialOrder=>ProductOrder{3,5}];

i53 : L = substitute(J,R1)

o53 = ideal (- c*b + a*d, e*h - g*f, - e*b + a*f)

o53 : Ideal of R1

i54 : gens gb L

o54 = | eh-gf eb-af cb-ad gbf-afh caf-ead |

                    1       5
o54 : Matrix R1   <--- R1
```

By examining the lead terms and coefficients, we see that $in_u(J) = (c, e, g)$, and that the lead coefficients of c are af and b, the lead coefficients of e are b and h, and the lead coefficient of g is bf. Therefore abf is a flattener. Let $F = abf$. A better choice for a flattener would be bf. We choose abf instead to show some of the complexities which arise when you choose a flattener which is not the simplest. As an exercise, you should do the same computation here with the flattener bf.

```
i55 : use R

o55 = R

o55 : PolynomialRing

i56 : J1 = saturate(J,a*b*f)

o56 = ideal (f*g - e*h, d*g - c*h, b*g - a*h, d*e - c*f, b*e - a*f, b* ···

o56 : Ideal of R

i57 : J1 == J : (a*b*f)

o57 = true
```

So $J = J_1 \cap J_2$, where

```
i58 : J2 = trim(J + ideal(a*b*f))

o58 = ideal (f*g - e*h, b*e - a*f, b*c - a*d, a*b*f)

o58 : Ideal of R

i59 : J == intersect(J1,J2)

o59 = true
```

As it turns out, J_1 is a prime ideal. How can we see this? Since the initial ideal $in_u(J_1) = (c, e, g)$, this means that the projection map is birational, and therefore the ideal J_1 is prime and even more, is rational.

```
i60 : Q1 = J1;

o60 : Ideal of R
```

Now let's decompose J_2.

```
i61 : independentSets J2

o61 = {c*d*e*f*h, a*b*d*g*h, a*c*d*g*h, c*d*e*g*h}

o61 : List
```

We'll use the first one.

```
i62 : R1 = QQ[a,b,g,  c,d,e,f,h,MonomialOrder=>ProductOrder{3,5}];

i63 : L = substitute(J2,R1)

o63 = ideal (g*f - e*h, - a*f + b*e, - a*d + b*c, a*b*f)

o63 : Ideal of R1

i64 : gens gb L

o64 = | gf-eh af-be ad-bc bde-bcf bge-aeh b2e b2cf abeh a2eh2 |

                  1       9
o64 : Matrix R1  <--- R1
```

In this case $in_u(J_2) = (a, b, g)$ (So the saturation will again be rational and prime, as before). One choice for a flattener is $f(de - cf)$.

```
i65 : use R

o65 = R

o65 : PolynomialRing

i66 : Q2 = saturate(J2,f*(d*e-c*f))

o66 = ideal (b, a, f*g - e*h)

o66 : Ideal of R

i67 : J3 = trim(J2 + ideal(f*(d*e-c*f)))

                                                      2
o67 = ideal (f*g - e*h, b*e - a*f, b*c - a*d, d*e*f - c*f , a*b*f)

o67 : Ideal of R

i68 : J == intersect(Q1,Q2,J3)

o68 = true
```

One more time. Let's decompose J_3.

```
i69 : independentSets J3

o69 = {a*b*d*g*h, a*c*d*g*h}

o69 : List

i70 : R1 = QQ[c,e,f,  a,b,d,g,h, MonomialOrder=>ProductOrder{3,5}]

o70 = R1

o70 : PolynomialRing

i71 : L = substitute(J3,R1)

                                                            2
o71 = ideal (- e*h + f*g, e*b - f*a, c*b - a*d, - c*f  + e*f*d, f*a*b)

o71 : Ideal of R1

i72 : transpose gens gb L

o72 = {-2} | eh-fg   |
       {-2} | eb-fa   |
       {-2} | cb-ad   |
       {-3} | fbg-fah |
       {-3} | cfa-ead |
       {-3} | fab     |
       {-3} | cf2-efd |
       {-4} | fa2h    |
       {-4} | f2a2    |
       {-4} | fa2d    |
       {-5} | ea2d2   |

                11          1
o72 : Matrix R1    <--- R1
```

This time, $in_u(J_3) = (c, e, f)$, and so once again the saturation will be a prime rational ideal. A flattener that works this time is ab. Notice that there are other choices for flatteners, but the others are more complicated and would add extra work.

```
i73 : use R

o73 = R

o73 : PolynomialRing

i74 : Q3 = saturate(J3,a*b)

o74 = ideal (f, e, b*c - a*d)

o74 : Ideal of R

i75 : Q3 == J3 : (a*b)

o75 = false

i76 : Q3 == J3 : (a*b)^2

o76 = true
```

This time,

```
i77 : J4 = trim(J3 + ideal(a^2*b^2))
```

$$o77 = \text{ideal } (f*g - e*h, \ b*e - a*f, \ b*c - a*d, \ d*e*f \overset{2}{-} c*f, \ a*b*f, \ a\,b\,)$$

```
o77 : Ideal of R
```

But notice that

```
i78 : J == intersect(Q1,Q2,Q3)
```

```
o78 = true
```

Therefore, we may avoid the primary decomposition of J_4, since it will only consist of redundant terms. You should check, but the primary decomposition of J_4 has seven primary ideals, and J_4 is not a radical ideal.

Exercise 5.3.12. Apply this technique to other Bayesian network examples, such as the example from the contraction lemma. Consider the cases when $d_i > 2$ for a nice challenge.

5.3.5 Primary decomposition of equidimensional ideals

Here is the situation: Suppose that $J \subset k[x_1, \dots, x_n] = k[u, t]$, where t is a maximal independent set, as above, and that $h \in k[t]$ is a flattener, and $J : h^\infty = J$. How can we tell if J is prime, or primary? And, if not, how do we find a primary decomposition of J?

In the previous example, we used the following fact, which we leave (as usual!) for you to prove as an exercise.

Proposition 5.3.13. *Suppose that t is a maximal independent set for J, $u = x \setminus t$, and h is a flattener for J with respect to t. If $in_u(J)$ is generated by the set of indeterminates t, then $(J : h^\infty)$ is a prime ideal, and is also rational.*

With certain kinds of ideals, such as Markov ideals, this happens quite frequently, as we saw in the previous example. If not, what do we do then? Once again, the flattener comes to the rescue. Algebraically, the flattener h allows us to compute the "generic fiber":

Proposition 5.3.14. *If $h \in k[t]$ is a flattener as defined above, then*

$$(J : h^\infty) = J \, k(t)[u] \cap k[u, t].$$

But notice! Since t is a maximal independent set, $J \, k(t)[u]$ is a zero dimensional ideal of $k(t)[u]$. This means that if we can find the primary decomposition of zero dimensional ideals, then we can compute a primary decomposition of the equidimensional ideal J.

Proposition 5.3.15. *If $Jk(t)[u] \cap k[u, t] = J$, and if $\tilde{Q}_1 \cap \cdots \cap \tilde{Q}_s$ is an irredundant primary decomposition of $Jk(t)[u]$, and if $Q_i = \tilde{Q}_i \cap k[u, t]$, then $J = Q_1 \cap \cdots \cap Q_s$ is an irredundant primary decomposition of J.*

This is great! It allows us to use the results from David Cox's lectures (Chapter 2) on computing the primary decomposition of zero dimensional ideals. These techniques tell us in particular that, if $Jk(t)[u]$ is prime or primary, then J will have the same property.

Exercise 5.3.16. Refer back to Chapter 2, and write an algorithm for computing the primary decomposition of an equidimensional ideal. Apply your algorithm to the equidimensional ideal $J = (ad^2 + bde + ce^2, ad^2 + bdf + cf^2, ae^2 + bef + cf^2) \subset \mathbb{Q}[a, \ldots, f]$.

5.4 Lecture 4: Putting it all together

In this lecture we apply all of our techniques and present a relatively complete algorithm. Keep in mind though that if you have a difficult ideal whose decomposition you desire, canned algorithms often will not finish. Using the techniques we have discussed may make it possible to find its decomposition, by applying them in novel ways, or using some extra information you have regarding your ideal.

5.4.1 Useful subroutines

Some of the techniques that we have considered so far can be summarized by the following routines.

`saturation(J,f)` , returns the pair $(\ell, J : f^\infty)$, where $(J : f^\infty) = (J : f^\ell)$.

`independentSet(J)` , returns a maximal independent set for the ideal J.

We did not discuss an algorithm for this, but it is a good exercise to find one. See also [DGP99].

`flattener(J,t)` , given a maximal independent set t of J, returns the pair $(h, in_u(J))$, where h is a flattener of J with respect to t, $u = x \setminus t$, and $in_u(J)$ are all as in the last lecture.

`equidimensionalPD(J, t, h)`, where t is a maximal independent set of J, h is a flattener, and $J : h = J$. This routine returns a list of pairs $\{P, Q\}$ such that Q is P-primary, and the intersection of all of the Q is J. Note that the P are all of the same dimension, and there are no embedded primes.

As discussed later, in order to handle redundancy, we instead use the following variant.

`equidimensionalPD(J, t, h, L)` , where t is a maximal independent set of J, h is a flattener, and $J : h = J$, and L is an ideal. This routine returns a pair (PQ, L'), where PQ is a list of pairs $\{P, Q\}$ as above except only the pairs (P, Q) with $L \not\subset P$ are returned, and L' is the intersection of L with all of the Q's.

5.4.2 Fighting redundancy

Suppose that we have an algorithm for finding a splitting polynomial for an ideal J, if one exists. Call these routines `thereIsASplittingPolynomial` and `splittingPolynomial`. Recall that if no splitting polynomial exists, then J is primary.

Here is the most naive method based on splitting principles:

```
PDsplit = (J) -> (
    -- input: an ideal J
    -- output: a list of primary ideals whose intersection
    --      is a primary decomposition of J
    if thereIsASplittingPolynomial(J)
    then (
        f := splittingPolynomial(J);
        (d,J1) := saturation(J,f);
        J2 := J + ideal(f^d);
        return join(PDsplit(J1), PDsplit(J2)) -- join the 2 lists
        )
    else
        -- J is already primary
        -- So, return a list with one element.
        return {J}
    );
```

This algorithm is written in *Macaulay 2* form for convenience, but in order to run, routines `thereIsASplittingPolynomial`, `splittingPolynomial` must be provided.

There are several problems with this algorithm. One difficulty is that the algorithm does not return an irredundant primary decomposition. It is far worse: it computes the primary decomposition for a potentially large number of useless redundant ideals.

We may think of this computation as a binary tree, where J is the root, J_1 the left node, and J_2 the right node. By choosing splitting polynomials for each leaf, we continue to build a larger and larger binary tree, until no splitting polynomials can be found, and then each leaf is a primary ideal.

At any time during the construction of this tree, the intersection of all of the leaf ideals is equal to J. The problem of redundancy is: many leaves or whole subtrees are not needed. How can we detect this?

There are two simple methods that help.

Method #1

If we don't mind computing extra ideal quotients, then we can remove many redundant components from this tree.

Lemma 5.4.1. *If* $(J : f^\infty) = J$ *and* $(J : g^\infty) = (J : g^\ell)$, *then*

$$J = (J : g^\infty) \cap ((J + (g^\ell)) : f^\infty).$$

The proof is almost identical to the proof of Lemma 5.1.14.

Method #2

The second trick is to process this tree from left to right, and keep the intersection L of all primary components found so far. Ignore a node J_r if $L \subset J_r$.

Exercise 5.4.2. Convince yourself that both of these methods are valid.

5.4.3 The overall algorithm

We now describe one way to put these techniques together into an algorithm. This is a version of the Gianni-Trager-Zacharias (GTZ) algorithm, which often works quite well.

Before we present the algorithm, you should spend some time working on the following exercise. If you can solve it quickly, then find other, better, solutions!

Exercise 5.4.3. Write an algorithm which computes a primary decomposition of an ideal $J \subset k[x] = k[x_1, \ldots, x_n]$. You may use any of the techniques presented so far, and any of the subroutines `saturation`, `independentSet`, etc.

Try to implement your solution, and try it on several examples, including some Markov ideals, as in the second lecture. Some questions you should consider are: (1) How to process and compute with as few redundant components as possible? (2) The precise element which one splits the ideal by has a dramatic effect on the complexity of the computation. Is there any way to control this?

The GTZ algorithm, as we have it here, splits the ideal J by using a flattener. It is relatively easy to compute the primary decomposition of the left hand side (ideal J1 below), since this ideal is equidimensional. We have indicated that one way to do this is to use multiplication maps, as in Chapter 2. Another way is to use a change of coordinates to bring the ideal into a nice position. See [GTZ88] and [DGP99] for more details. The right hand side (ideal J2 in the algorithm) causes more problems, since by adding f^d to the ideal, the number of components can become quite large. This is the reason for using the redundancy control method #2 above. The ideal L which is an argument to GTZ and `equidimensionalPD` implements this method.

The final algorithm is here:

```
PD = (J) -> (
    -- input: J is an ideal
    -- output: a list of pairs {P,Q} such that Q is P-primary
    --     and the Q form an irredundant primary decomposition
    --     of J.
    L := ideal(1_(ring(J)));
    (PQ,L) = GTZ(J,L);
    -- at this point, L should equal J.
    PQ
    );
```

```
GTZ = (J, L) -> (
    -- input: J is an ideal.
    --        L is an ideal, which is the intersection of all
    --            primary components found so far.
    -- output: a pair (PQ, L'), where
    --        PQ is a list of {P,Q}'s, where P is prime, Q is
    --            primary to P
    --        L' is the intersection of L with all of the Q's in PQ.
    -- The set of Q's in PQ form that part of the primary
    -- decomposition of J with primary ideals not containing L.
    if isSubset(L,J) then return ({}, L);
    t := independentSet(J);
    (f,inJ) := flattener(J,t);
    (d,J1) := saturation(J,f);
    if degree inJ == 1
    then (
        -- J1 is prime
        PQ1 = {J1,J1};
        L = intersect(L, J1);
        )
    else
        -- This also replaces the L with the intersection
        (PQ1,L) = equidimensionalPD(J1,u,L);
    if d == 0 then return (PQ1,L);
    J2 := J + ideal(f^d);
    (PQ2,L) = GTZ(J2,L);
    (join(PQ1,PQ2), L)
    );
```

5.4.4 A harder example: the primary decomposition of a Markov ideal

Exercise 5.4.4. Consider the graph G with 5 vertices, and directed edges $5 \rightarrow 4, 5 \rightarrow 3, 4 \rightarrow 2, 3 \rightarrow 2, 2 \rightarrow 1$ (see Figure 5.2). Find a primary decomposition of the global Markov ideal of this graph, in the case when the five random variables are all binary (i.e. $d_1 = \ldots = d_5 = 2$).

Before looking at the answer below, try this on your own. You should use the `marginMap` trick or something similar to make the resulting ideal as simple as possible.

We now present the answer to this exercise using *Macaulay 2*. We do not use canned algorithms, although we do use routines `independentSet`, `saturation0`, and `flattener`. `saturation0` is the same as in the description of `saturation`, except only the ideal is returned. As of this writing, no system can do these primary decompositions in a small amount of time using canned algorithms. However, by the time you read this, specific computations might be much faster. So you should read this solution with a skeptical eye: we have chosen a technique that seems to work faster at the moment. Your solution might be different, and possibly more elegant, or more efficient in use of time and computer memory.

First, let's generate the ideal, and see what we are up against. You can find this next file at the same web location as `markov.m2`.

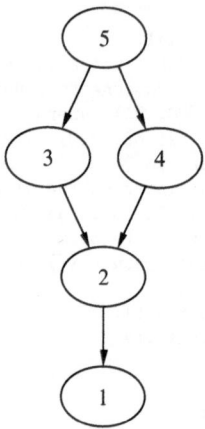

Fig. 5.2. The Bayesian network in Exercise 5.4.4

```
i82 : load "cimpa-tools.m2";

i83 : G = makeGraph {{},{1},{2},{2},{3,4}};

i84 : GM = globalMarkovStmts G;

i85 : GM/print;
{Set {1}, Set {4, 5, 3}, Set {2}}
{Set {1}, Set {4, 5}, Set {2, 3}}
{Set {5, 3}, Set {1}, Set {4, 2}}
{Set {5}, Set {1}, Set {4, 2, 3}}
{Set {1}, Set {4, 3}, Set {5, 2}}
{Set {1}, Set {4}, Set {5, 2, 3}}
{Set {3}, Set {1}, Set {4, 5, 2}}
{Set {5}, Set {2}, Set {4, 1, 3}}
{Set {5}, Set {1, 2}, Set {4, 3}}
{Set {3}, Set {4}, Set {5}}
```

These global Markov statements are all consequences of $X_1 \perp\!\!\!\perp \{X_3, X_4, X_5\} \mid X_2$, $X_5 \perp\!\!\!\perp \{X_1, X_2\} \mid \{X_3, X_4\}$, and $X_3 \perp\!\!\!\perp X_4 \mid X_5$.

```
i86 : R = markovRing(2,2,2,2,2);
```

As mentioned previously, it is often better to change coordinates using the following margin map.

```
i87 : F = marginMap(1,R);

o87 : RingMap R <--- R

i88 : J = trim F markovIdeal(R,GM);

o88 : Ideal of R

i89 : betti J

o89 = generators: total: 1 74
                       0: 1  .
                       1: . 74
```

The ideal J is minimally generated by 74 homogeneous quadrics, 72 of which are binomials, and the last 2 have 28 monomials each. These last two generators increase the difficulty of the computation, especially for canned algorithms. This ideal has codimension 20 and degree 2240:

```
i90 : (codim J, degree J)

o90 = (20, 2240)

o90 : Sequence
```

Our plan is to determine a primary decomposition of the ideal J. We could start our computation in a number of ways. We choose to use the GTZ method, at least for this first step.

```
i91 : u = independentSet J;

i92 : (inJ,h) = flattener(J,u);

i93 : degree inJ

o93 = 1
```

Since the degree is one, the projection map is birational, and so the ideal J0, which we will compute now, is a rational prime.

```
i94 : J0 = saturation0(J,h);

o94 : Ideal of R

i95 : betti J0

o95 = generators: total: 1 175
                      0: 1   .
                      1: .  74
                      2: .   .
                      3: . 101

i96 : (codim J0,degree J0)

o96 = (20, 1496)

o96 : Sequence
```

The prime component J0 of J coincides with the canonical component coming from the factorization theorem. This ideal is generated by J, together with 101 quartics.

We would like to perform the ideal quotient (J : J0), but J0 has 101 quartics, and so the computation takes some time. Often it is the case that such an ideal quotient is equal to (J : f), for some f. This happens with the first element we try:

```
i97 : Jrest = J : J0_74;

o97 : Ideal of R

i98 : J == intersect(J0,Jrest)

o98 = true
```

Since these are equal, and since we have computed `Jrest` as an ideal quotient, an irredundant primary decomposition of `Jrest` will give an irredundant primary decomposition of J. As an exercise, you should try to compute the primary decomposition of (J, h). It has many components which are not needed.

The ideal `Jrest` is generated by J and the following 8 quadrics.

```
i99 : L = ideal(p_(1,1,1,1,2)*p_(1,1,2,2,2)+p_(1,1,2,2,2)*p_(1,2,1,1,2)
            +p_(1,1,1,1,2)*p_(1,2,2,2,2)+p_(1,2,1,1,2)*p_(1,2,2,2,2),
            p_(1,1,1,1,1)*p_(1,1,2,2,2)+p_(1,1,2,2,2)*p_(1,2,1,1,1)
            +p_(1,1,1,1,1)*p_(1,2,2,2,2)+p_(1,2,1,1,1)*p_(1,2,2,2,2),
            p_(1,1,1,1,2)*p_(1,1,2,2,1)+p_(1,1,2,2,1)*p_(1,2,1,1,2)
            +p_(1,1,1,1,2)*p_(1,2,2,2,1)+p_(1,2,1,1,2)*p_(1,2,2,2,1),
            p_(1,1,1,1,1)*p_(1,1,2,2,1)+p_(1,1,2,2,1)*p_(1,2,1,1,1)
            +p_(1,1,1,1,1)*p_(1,2,2,2,1)+p_(1,2,1,1,1)*p_(1,2,2,2,1),
            p_(1,1,1,1,2)*p_(1,1,2,2,2)+p_(1,1,2,2,2)*p_(1,2,1,1,2)
            +p_(1,1,1,1,2)*p_(1,2,2,2,2)+p_(1,2,1,1,2)*p_(1,2,2,2,2),
            p_(1,1,1,2,1)*p_(1,1,2,1,2)+p_(1,1,2,1,2)*p_(1,2,1,2,1)
            +p_(1,1,1,2,1)*p_(1,2,2,1,2)+p_(1,2,1,2,1)*p_(1,2,2,1,2),
            p_(1,1,1,2,2)*p_(1,1,2,1,1)+p_(1,1,2,1,1)*p_(1,2,1,2,2)
            +p_(1,1,1,2,2)*p_(1,2,2,1,1)+p_(1,2,1,2,2)*p_(1,2,2,1,1),
            p_(1,1,1,1,1)*p_(1,1,2,2,1)+p_(1,1,2,2,1)*p_(1,2,1,1,1)
            +p_(1,1,1,1,1)*p_(1,2,2,2,1)+p_(1,2,1,1,1)*p_(1,2,2,2,1));

o99 : Ideal of R

i100 : Jrest == J + L

o100 = true
```

The next step is to decompose `Jrest`. We proceed in the same manner.

```
i101 : u = independentSet Jrest;

i102 : (inJrest,h) = flattener(Jrest,u);

i103 : degree inJrest

o103 = 1
```

Once again, the component J1 which we now compute is a rational prime ideal, since the projection is birational.

```
i104 : factors h

o104 = {p            + p            , p            + p            , p            , ...
          1,1,2,2,2    1,2,2,2,2    1,1,2,1,2    1,2,2,1,2    1,2,2,2,2   ...

o104 : List

i105 : J1 = saturation0(Jrest,h);

o105 : Ideal of R
```

The ideal J1 has the following generators, in addition to those of J or `Jrest`.

```
i106 : M = ideal(
                  p_(1,1,1,2,2)+p_(1,2,1,2,2),
                  p_(1,1,1,2,1)+p_(1,2,1,2,1),
                  p_(1,1,1,1,2)+p_(1,2,1,1,2),
                  p_(1,1,1,1,1)+p_(1,2,1,1,1),
                  p_(2,1,1,2,2)*p_(2,2,1,1,2)-p_(2,1,1,1,2)*p_(2,2,1,2,2),
```

```
                p_(2,1,1,2,1)*p_(2,2,1,1,2)-p_(2,1,1,1,2)*p_(2,2,1,2,1),
                p_(2,1,1,2,2)*p_(2,2,1,1,1)-p_(2,1,1,1,1)*p_(2,2,1,2,2),
                p_(2,1,1,2,1)*p_(2,2,1,1,1)-p_(2,1,1,1,1)*p_(2,2,1,2,1),
                p_(1,2,1,2,2)*p_(2,1,1,1,2)-p_(1,2,1,1,2)*p_(2,1,1,2,2),
                p_(1,2,1,2,1)*p_(2,1,1,1,2)-p_(1,2,1,1,2)*p_(2,1,1,2,1),
                p_(1,2,1,2,2)*p_(2,1,1,1,1)-p_(1,2,1,1,1)*p_(2,1,1,2,2),
                p_(1,2,1,2,1)*p_(2,1,1,1,1)-p_(1,2,1,1,1)*p_(2,1,1,2,1));
```

o106 : Ideal of R

i107 : J1 == J + M

o107 = true

i108 : J1 == Jrest + M

o108 = true

i109 : betti J1

```
o109 = generators: total: 1 74
                      0: 1  4
                      1: . 70
```

i110 : (codim J1, degree J1)

o110 = (20, 170)

o110 : Sequence

We could continue in the same manner, taking ideal quotients, hoping that the ideals still intersect to give Jrest. However, we can use symmetry to find three more prime ideals. One symmetry that is evident from the graph is to interchange random variables 3 and 4:

```
i111 : s = map(R,R,toList apply((1,1,1,1,1)..(2,2,2,2,2), x -> (
              p_(x#0,x#1,x#3,x#2,x#4))))

o111 = map(R,R,{p            , p            , p           , p           , p      ···
                 1,1,1,1,1    1,1,1,1,2    1,1,2,1,1    1,1,2,1,2    1,1,1 ····

o111 : RingMap R <--- R
```

i112 : s L == L

o112 = true

A second symmetry is to interchange the values 1 and 2 of the ith random variable. The permutation t changes these values for the 3rd random variable (this is the only one that produces new ideals).

```
i113 : t = map(R,R,toList apply((1,1,1,1,1)..(2,2,2,2,2), x -> (
              p_(x#0,x#1,3-x#2,x#3,x#4))))

o113 = map(R,R,{p            , p            , p           , p           , p      ···
                 1,1,2,1,1    1,1,2,1,2    1,1,2,2,1    1,1,2,2,2    1,1,1 ····

o113 : RingMap R <--- R
```

i114 : t J == J

o114 = true

```
i115 : t (J+L) == J+L
```

```
o115 = true
```

We now produce three new prime ideals from J1.

```
i116 : J2 = trim(J + s M)
```

$$o116 = \text{ideal } (p_{1,1,2,1,2} + p_{1,2,2,1,2}, p_{1,1,2,1,1} + p_{1,2,2,1,1}, p_{1,1,1} \cdots$$

```
o116 : Ideal of R
```

```
i117 : J3 = trim(J + t M)
```

$$o117 = \text{ideal } (p_{1,1,2,2,2} + p_{1,2,2,2,2}, p_{1,1,2,2,1} + p_{1,2,2,2,1}, p_{1,1,2} \cdots$$

```
o117 : Ideal of R
```

```
i118 : J4 = trim(J + s (t M))
```

$$o118 = \text{ideal } (p_{1,1,2,2,2} + p_{1,2,2,2,2}, p_{1,1,2,2,1} + p_{1,2,2,2,1}, p_{1,1,1} \cdots$$

```
o118 : Ideal of R
```

We could attempt to produce other ideals, but we just obtain ones we have already seen.

```
i119 : J4 == J + t (s (t M))
```

```
o119 = true
```

The following ideal quotients remove these components, leaving Jrest5.

```
i120 : Jrest2 = Jrest : M;
```

```
o120 : Ideal of R
```

```
i121 : Jrest3 = Jrest2 : (s M);
```

```
o121 : Ideal of R
```

```
i122 : Jrest4 = Jrest3 : (t M);
```

```
o122 : Ideal of R
```

```
i123 : Jrest5 = Jrest4 : (s (t M));
```

```
o123 : Ideal of R
```

```
i124 : Jrest == intersect(J1,J2,J3,J4,Jrest5)
```

```
o124 = true
```

The intersection is still correct. If there were embedded components, this approach would still find the associated primes, but would not produce the primary decomposition.

So, all we need to do is decompose Jrest5. It is easy to identify the ideal after viewing it.

```
i125 : (codim Jrest5,degree Jrest5)
```

```
o125 = (20, 64)
```

```
o125 : Sequence
```

```
i126 : P = ideal(R_0 .. R_15)

o126 = ideal (p            , P           , P           , P           , P          ...
              1,1,1,1,1    1,1,1,1,2    1,1,1,2,1    1,1,1,2,2    1,1,2,1 ...

o126 : Ideal of R

i127 : Jrest5 == Jrest + P^3

o127 = true

i128 : transpose gens trim(J+L+P)

o128 = {-1} | p_(1,2,2,2,2)                                                     |
        {-1} | p_(1,2,2,2,1)                                                     |
        {-1} | p_(1,2,2,1,2)                                                     |
        {-1} | p_(1,2,2,1,1)                                                     |
        {-1} | p_(1,2,1,2,2)                                                     |
        {-1} | p_(1,2,1,2,1)                                                     |
        {-1} | p_(1,2,1,1,2)                                                     |
        {-1} | p_(1,2,1,1,1)                                                     |
        {-1} | p_(1,1,2,2,2)                                                     |
        {-1} | p_(1,1,2,2,1)                                                     |
        {-1} | p_(1,1,2,1,2)                                                     |
        {-1} | p_(1,1,2,1,1)                                                     |
        {-1} | p_(1,1,1,2,2)                                                     |
        {-1} | p_(1,1,1,2,1)                                                     |
        {-1} | p_(1,1,1,1,2)                                                     |
        {-1} | p_(1,1,1,1,1)                                                     |
        {-2} | p_(2,1,2,2,2)p_(2,2,2,2,1)-p_(2,1,2,2,1)p_(2,2,2,2,2) |
        {-2} | p_(2,1,2,1,2)p_(2,2,2,1,1)-p_(2,1,2,1,1)p_(2,2,2,1,2) |
        {-2} | p_(2,1,1,2,2)p_(2,2,1,2,1)-p_(2,1,1,2,1)p_(2,2,1,2,2) |
        {-2} | p_(2,1,1,1,2)p_(2,2,1,1,1)-p_(2,1,1,1,1)p_(2,2,1,1,2) |

                    20        1
o128 : Matrix R    <--- R
```

In order to show that `Jrest5` is a primary component, it suffices to show that it is equidimensional, since its radical is a prime ideal.

```
i129 : u = independentSet Jrest5;

i130 : (inJrest5,h) = flattener(Jrest5,u);

i131 : Jrest5 == saturation0(Jrest5,h)

o131 = true
```

Therefore, `Jrest5` is a primary ideal.

So finally we have the following primary decomposition of the original ideal J.

```
i132 : J == intersect(J0, J+M, J+(s M),J+(t M),J+((s(t M))),J+L+P^3)

o132 = true
```

The ideal is equidimensional, something that was not a priori obvious. It is also not a radical ideal.

Exercise 5.4.5. This is a more difficult exercise! Let G be the graph with directed edges $5 \to 3, 4 \to 3, 5 \to 1, 4 \to 1, 3 \to 2, 2 \to 1$. Let J be the global Markov ideal of G. Show that J has 23 minimal prime ideals, and 17 embedded prime ideals.

5.4.5 Other algorithms, and what to read next

Good papers to read next include [GTZ88], [SY96], and [DGP99].

The algorithm of Shimoyama and Yokoyama [SY96] (the SY algorithm) uses the following observation. Suppose that $P_1, \ldots, P_r$ are the minimal primes of J, and that $s_1, \ldots, s_r$ are *separators*, i.e. $s_i \in \cap_{j \neq i} P_j$, but $s_i \notin P_i$. If $(J : s_i^{d_i}) = (J : s_i^\infty)$, then

$$ J = (J : s_1^\infty) \cap \ldots \cap (J : s_r^\infty) \cap (J, s_1^{d_1}, \ldots, s_r^{d_r}). $$

The SY algorithm uses this equality to split an ideal. Each $J_i := (J : s_i^\infty)$ has radical P_i, but possibly has embedded primes too (These are so-called *pseudo-primary* ideals). The algorithm proceeds recursively by using flatteners to split these pseudo-primary ideals. This algorithm requires that the minimal primes be computed first. One method is to use characteristic or triangular sets . See [DGP99] for a description. Another method is to keep splitting $\sqrt{J}$, either using ideal quotients or a factorizing Gröbner basis algorithm.

All of these algorithms use similar methods, with some novelties. With more commutative algebra background, the paper by Eisenbud, Huneke and Vasconcelos [EHV92] has very interesting techniques for computing radicals, identifying associated primes, and computing primary decompositions (and more!).

5.4.6 Some open problems

We close with a few open problems. The first challenge is to find better primary decomposition algorithms.

As for Markov ideals, there are many open problems, see [GSS] for the ones presented here. You will find other open problems there as well.

Problem 5.4.6. Find an efficient method for computing the associated primes of an ideal, without first computing the entire primary decomposition.

There is a very nice solution to this problem, in [EHV92], but unfortunately, there are many practical situations where their algorithm uses too much time or memory to be competitive.

Problem 5.4.7. What is the condition on Bayesian networks G, $(d_1, \ldots, d_n)$ for the ideal $I_{\text{global}(G)}$ to be prime? radical? without embedded components? Can one characterize the primary decomposition in terms of G and the d_i?

Problem 5.4.8. Prove that every associated prime of $I_{\text{local}(G)}$ or $I_{\text{global}(G)}$ is rational.

Problem 5.4.9. Find the best way of putting the techniques for primary decomposition together to handle certain classes of ideals, e.g.

(1) local and global Markov ideals.

(2) binomial ideals

Problem 5.4.10. Prove that the degree 2 part of the ideal $\ker(\Phi)$ (from lecture 2) is exactly the same as the degree 2 part of $I_{\mathrm{global}(G)}$.

This is true for binary random variables, with $n = 5$, and any random variables, if $n \leq 4$, see [GSS].

6

Algorithms and their complexities

Juan Sabia

Departamento de Matemática - Facultad de Ciencias Exactas y Naturales - Universidad de Buenos Aires and CONICET, Argentina, jsabia@dm.uba.ar

Summary. This chapter is intended as a brief survey of the different notions and results that arise when we try to compute the algebraic complexity of algorithms solving polynomial equation systems. Although it is essentially self-contained, many of the definitions, problems and results we deal with also appear in many other chapters of this book. We start by considering algorithms which use the dense representation of multivariate polynomials. Some results about the algebraic complexities of the effective Nullstellensatz, of quantifier elimination processes over algebraically closed fields and of the decomposition of algebraic varieties when considering this model are stated. Then, it is shown that these complexities are essentially optimal in the dense representation model. This is the reason why a change in the encoding of polynomials is needed to get better upper bounds for the complexities of new algorithms solving the already mentioned tasks. The straight-line program representation for multivariate polynomials is defined and briefly discussed. Some complexity results for algorithms in the straight-line program representation model are mentioned (an effective Nullstellensatz and quantifier elimination procedures, for instance). A description of the Newton-Hensel method to approximate roots of a system of parametric polynomial equations is made. Finally, we mention some new trends to avoid large complexities when trying to solve polynomial equation systems.

6.0 Introduction and basic notation

The fundamental problem we are going to deal with, as in most other chapters of this book, is to solve (over the field of complex numbers $\mathbb{C}$) a system of multivariate polynomial equations with coefficients in the field of rational numbers $\mathbb{Q}$ algorithmically, but our particular point of view is related to the question of whether we can predict how long our algorithms will take. Of course, we should define what it means to solve such a system. A first possible answer would be to decide whether there are any solutions to the given system, and, in case there are solutions, to describe them in a 'useful' or at least in an 'easy' way.

Many attempts to do this are based on trying to transform our problem into a linear algebra one. The reason for this is that we know how to solve many linear algebra problems effectively.

The focus of our attention will be the *algorithmic* solutions to these problems; so, we are going to define what an algorithm is for us (perhaps a rather inflexible definition but necessary to meet the requirements of our work). Roughly speaking, the less time an algorithm takes to perform a task, the better. This will lead to the definition of *algebraic complexity*, a kind of measure for the time an algorithm takes to perform what we want it to.

One of the problems we have when we deal with multivariate polynomials is that the known effective ways to factorize them take a lot of time, so we will try not to use this tool within our algorithms.

In the different sections of this chapter, we are going to state the problems that will be taken into account and describe (or just mention, if the description is beyond the scope of this survey) some ways of solving them.

Before we begin considering the problems, we need to fix some notation and give some definitions:

A *system of polynomial equations* is a system

$$\begin{cases} f_1(x_1, \ldots, x_n) = 0 \\ \quad \ldots \\ f_s(x_1, \ldots, x_n) = 0 \end{cases}$$

where $f_1, \ldots, f_s$ are polynomials in $\mathbb{C}[X_1, \ldots, X_n]$ and the solutions considered will be vectors $(x_1, \ldots, x_n) \in \mathbb{C}^n$. Whenever we want to speak about a group of variables or a vector, we often use just a capital or lower case letter with no index; for example in this case, we could have written $\mathbb{C}[X]$ or $x \in \mathbb{C}^n$. The set $V \subset \mathbb{C}^n$ of all the solutions of such a system will be called an *algebraic variety* (or simply a *variety* if the context is clear). Its *dimension* is the minimum number of *generic* hyperplanes such that their common intersection with V is empty. For example, a point has dimension zero (a generic hyperplane does not cut it); a line has dimension one (a generic hyperplane cuts it, but two generic hyperplanes do not), etc. For a more precise definition of dimension see, for example, [Sha77] or [CLO97].

From the algorithmic point of view, we deal exclusively with polynomials with coefficients in $\mathbb{Q}$ but we still consider all the solutions to our systems in $\mathbb{C}^n$.

Sometimes it will be useful to take into account fields other than $\mathbb{Q}$ and $\mathbb{C}$. If k is a field, $\overline{k}$ will denote an algebraic closure of k.

6.1 Statement of the problems

In this section we are going to state some of the questions we usually want to answer when dealing with systems of polynomial equations. Some of these problems are also mentioned or studied in other chapters of this book, but we present them here for the sake of this chapter being self-contained.

6.1.1 Effective Hilbert's Nullstellensatz

Let $X = \{X_1, \ldots, X_n\}$ be indeterminates over $\mathbb{Q}$. Given s polynomials $f_1, \ldots, f_s \in \mathbb{Q}[X]$, if we want to solve the system of polynomial equations

$$\begin{cases} f_1(x_1, \ldots, x_n) = 0 \\ \cdots \\ f_s(x_1, \ldots, x_n) = 0 \end{cases}$$

the very first question we would like to answer is whether there exists any point $(x_1, \ldots, x_n) \in \mathbb{C}^n$ satisfying this system (that is to say, if the equations $f_1 = 0, \ldots, f_s = 0$ share a common solution in $\mathbb{C}^n$).

When all the polynomials $f_1, \ldots, f_s$ have degrees equal to 1, the system we are dealing with is a linear system and there is a simple computation of ranks of matrices involving the coefficients of the polynomials which answers our question:

Suppose our linear system is given by $A.x^t = B$ (with $A \in \mathbb{Q}^{s \times n}$ and $B \in \mathbb{Q}^{s \times 1}$). Then

$$\exists x \in \mathbb{C}^n \ / \ A.x^t = B \iff \operatorname{rank}(A) = \operatorname{rank}(A|B)$$

(where $(A|B)$ denotes the matrix we obtain by adding the column B to the matrix A).

The first step towards a generalization of this result when we deal with polynomials of any degree (generalization in the sense that it relates the existence of solutions to some computations involving the coefficients of the polynomials considered) is the following well-known theorem:

Theorem 6.1.1. *(Hilbert's Nullstellensatz) Let $f_1, \ldots, f_s \in \mathbb{Q}[X_1, \ldots, X_n]$. Then the following statements are equivalent:*
i) $\{x \in \mathbb{C}^n \ / \ f_1(x) = \cdots = f_s(x) = 0\} = \emptyset$.
ii) There exist polynomials $g_1, \ldots, g_s \in \mathbb{Q}[X_1, \ldots, X_n]$ such that $1 = \sum\limits_{1 \leq i \leq s} g_i.f_i$.

(See Chapter 4 for other versions of this theorem.)

A proof of this theorem can be found in almost any basic textbook on algebraic geometry (see for example [Har83], [Kun85] or [CLO97]). This result was already known by Kronecker and it essentially shows how a geometric problem (Is the variety defined as the common zeroes of a fixed set of polynomials empty?) is equivalent to an algebraic one (Is 1 an element of the ideal $(f_1, \ldots, f_s)$?).

We will call an algorithm an *effective Hilbert's Nullstellensatz*, if given as input the polynomials $f_1, \ldots, f_s$, the algorithm computes polynomials $g_1, \ldots, g_s$ (in case they exist) such that $\sum\limits_{1 \leq i \leq s} g_i.f_i = 1$.

Later on, we will mention some effective Hilbert's Nullstellensätze.

6.1.2 Effective equidimensional decomposition

Supposing we already know that a particular system of polynomial equations has solutions, we may need to answer some questions about the geometry of the algebraic variety they define in $\mathbb{C}^n$: Does it consist only of finitely many points? Is there a whole curve of solutions? Are there isolated solutions?, etc.

All these questions can be answered by means of geometric decompositions of the algebraic variety defined by the original system of polynomials. These decompositions we are going to define are intimately bound up with the primary decomposition of ideals considered in Chapter 2 and Chapter 5 but they do not coincide because our approach is exclusively geometric while these others are purely algebraic.

Definition 6.1.2. *An algebraic variety $C \subset \mathbb{C}^n$ is called* **irreducible** *if it satisfies*

$$C = C_1 \cup C_2 \text{ where } C_1 \text{ and } C_2 \text{ are algebraic varieties} \Rightarrow C = C_1 \text{ or } C = C_2.$$

The following is a classical result from algebraic geometry. It states that the affine space $\mathbb{C}^n$ is a Noetherian topological space when considering the Zariski topology (that is, the topology in which the algebraic varieties are the closed sets) and its proof can be found, for example, in [Sha77] or [CLO97].

Proposition 6.1.3. *(Irreducible decomposition) Let $V \subset \mathbb{C}^n$ be an algebraic variety. Then, there exist unique irreducible varieties $C_1, \ldots, C_r$ such that $C_i \not\subset C_j$ if $i \neq j$ and*

$$V = \bigcup_{1 \leq i \leq r} C_i.$$

From our point of view and our definitions, the irreducible decomposition is not algorithmically achievable. If this were so, just by considering the case $n = 1$, we would be able to find all the roots of any univariate rational polynomial (note that the irreducible decomposition of $\{x \in \mathbb{C} \ / \ \prod_{1 \leq i \leq d}(x - \alpha_i) = 0\}$ is exactly $\bigcup_{1 \leq i \leq d}\{\alpha_i\}$). This is the reason why we are going to consider a less refined decomposition of a variety.

Let $V = \bigcup_{1 \leq i \leq r} C_i$ be the irreducible decomposition of the variety V and, for every $0 \leq j \leq n$, consider the union of all the irreducible components of V of dimension j

$$V_j := \bigcup_{\substack{\{i \ / \ 1 \leq i \leq r \\ \text{and dim } C_i = j\}}} C_i.$$

It is obvious that $V = \bigcup_{0 \leq j \leq n} V_j$ where, for every $0 \leq j \leq n$, either $V_j = \emptyset$ or $\dim V_j = j$. This unique decomposition is called the *irredundant equidimensional decomposition* (or *equidimensional decomposition* for short) of V.

Note that the information given by this decomposition still allows us to answer all the questions we asked above. For example, a non-empty variety

V consists only of finitely many points if and only if $V_0 \neq \emptyset$ and $V_j = \emptyset$ for every $1 \leq j \leq n$.

The equidimensional decomposition has the following property, nice from the algorithmic point of view:

Proposition 6.1.4. *Let* $f_1, \ldots, f_s \in \mathbb{Q}[X_1, \ldots, X_n]$ *be polynomials and let* $V \subset \mathbb{C}^n$ *be the algebraic variety of their common zeroes. If* V_j *is one of the components appearing in the irredundant equidimensional decomposition of* V, *then there exist polynomials in* $\mathbb{Q}[X_1, \ldots, X_n]$ *defining* V_j.

The core of this result is that there are **rational** polynomials defining V_j, so we have a chance to compute the irredundant equidimensional decomposition algorithmically using only rational coefficients.

We will call an algorithm an *effective equidimensional decomposition algorithm* if given an algebraic variety $V \subset \mathbb{C}^n$ defined by rational polynomials, the algorithm describes the varieties involved in its equidimensional decomposition (i.e. its *equidimensional components*) as separate varieties.

A final comment has to be made about the irreducible decomposition of a variety defined by **rational** polynomials: we could take into account only varieties defined by rational polynomials as closed sets to define the *rational Zariski topology* in $\mathbb{C}^n$. If this is the case, the irreducible components of a variety will be still definable by rational polynomials. For example, in the case of the variety defined by a squarefree polynomial, its rational decomposition will essentially coincide with the factorization of the considered polynomial, but as we have stated before, we do not want to deal with polynomial factorization, and this is why we are not going to consider this problem. (For an algorithm yielding this irreducible decomposition numerically, see Chapter 8.)

6.1.3 Effective quantifier elimination

Many interesting geometric and algebraic problems can be formulated as first order statements over algebraically closed fields and a well-known result from logic states that any first order formula in the language of algebraically closed fields is equivalent to another formula without quantifiers (see [CK90] for details). This is the reason why, in the last decades, special efforts have been made to find efficient algorithms to eliminate quantifiers.

For the sake of simplicity, we will state precisely what elimination of quantifiers means only in a very particular case:

Theorem 6.1.5. *Let* $X_1, \ldots, X_n, Y_1, \ldots, Y_m$ *be indeterminates over* $\mathbb{Q}$ *and let* $f_1, \ldots, f_s, g_1, \ldots, g_t \in \mathbb{Q}[X_1, \ldots, X_n, Y_1, \ldots, Y_m]$ *be polynomials. Let*

$$V := \{x \in \mathbb{C}^n \ / \ \exists y \in \mathbb{C}^m \ : \ f_1(x, y) = 0 \ \wedge \ \ldots \ \wedge \ f_s(x, y) = 0 \ \wedge$$

$$\wedge \ g_1(x, y) \neq 0 \ \wedge \ \ldots \ \wedge \ g_t(x, y) \neq 0\}.$$

Then, there exists a **quantifier free** *formula φ involving only polynomials in $\mathbb{Q}[X_1, \ldots, X_n]$, equalities, inequalities and the symbols $\wedge$ and $\vee$ such that*

$$V = \{x \in \mathbb{C}^n \mathbin{/} \varphi(x)\}.$$

Let us give some simple examples to make this statement clearer.

Example 6.1.6. Suppose we want to describe the set of all the polynomials of degree bounded by d in one variable that have at least a root in $\mathbb{C}$. This set is

$$V := \{(x_0, x_1, \ldots, x_d) \in \mathbb{C}^{d+1} \mathbin{/} \exists\, y \in \mathbb{C} : x_d y^d + x_{d-1} y^{d-1} + \cdots + x_0 = 0\}.$$

Evidently, the Fundamental Theorem of Algebra states that a quantifier-free way of defining V is

$$V = \{(x_0, x_1, \ldots, x_d) \in \mathbb{C}^{d+1} \mathbin{/} x_0 = 0 \vee x_1 \neq 0 \vee x_2 \neq 0 \vee \cdots \vee x_d \neq 0\}.$$

Example 6.1.7. A very well-known example of a quantifier elimination procedure from linear algebra is the use of the determinant. The set

$$V := \{(x_{ij}) \in \mathbb{C}^{n \times n} \mathbin{/} \exists(y_1, \ldots, y_n, y_1', \ldots, y_n') \in \mathbb{C}^{2n} :$$

$$(y_1, \ldots, y_n) \neq (y_1', \ldots, y_n') \wedge \begin{cases} x_{11}y_1 + \cdots + x_{1n}y_n = x_{11}y_1' + \cdots + x_{1n}y_n' \\ \ldots \\ x_{n1}y_1 + \cdots + x_{nn}y_n = x_{n1}y_1' + \cdots + x_{nn}y_n' \end{cases} \}$$

is exactly the subset of $\mathbb{C}^{n \times n}$ defined by the determinant:

$$V = \{(x_{ij}) \in \mathbb{C}^{n \times n} \mathbin{/} \det(x_{ij}) = 0\}.$$

Example 6.1.8. The classical resultant with respect to a single variable Y between two polynomials $f_1, f_2 \in \mathbb{Q}[X_1, \ldots, X_n][Y]$ monic in Y and of degree r and s respectively is another example of eliminating quantifiers (for the definition and basic properties of the classic resultant between two polynomials, some of which will be used later, see, for example, [CLO97], [Mig92], [vdW49] or [Wal62]):

$$\{x \in \mathbb{C}^n \mathbin{/} \exists y \in \mathbb{C} : f_1(x, y) = 0 \wedge f_2(x, y) = 0\} =$$
$$= \{x \in \mathbb{C}^n \mathbin{/} \mathrm{Res}_Y(f_1(x, Y), f_2(x, Y)) = 0\}.$$

For a more general definition of resultants as eliminating polynomials see Chapter 2 and Chapter 1.

As before, we will say that we have an *efficient quantifier elimination procedure* if we have an algorithm that, from a formula of the type

$$\exists y \in \mathbb{C}^m : f_1(x_1, \ldots, x_n, y) = 0 \wedge \ldots \wedge f_s(x_1, \ldots, x_n, y) = 0 \wedge$$
$$\wedge\, g_1(x_1, \ldots, x_n, y) \neq 0 \wedge \ldots \wedge g_t(x_1, \ldots, x_n, y) \neq 0,$$

produces a quantifier-free formula φ defining the same subset of $\mathbb{C}^n$.

6.2 Algorithms and complexity

When we speak about efficiency related to polynomial equation solving, we mean the existence of algorithms performing different tasks. But what do we call an algorithm?

The idea of algorithm we deal with is the following: given some data in a certain way (numbers, formulae, etc), an algorithm will be a sequential list of fixed operations or comparisons that ends in some logical or mathematical 'object' we would like to compute. For example, suppose you want an algorithm to solve the equation $ax = b$ with coefficients in $\mathbb{Q}$ and that you can deal with rational numbers algorithmically (that is to say, comparisons and operations between rational numbers can be performed somehow). A possible algorithm to do this would be the one shown in Figure 6.1.

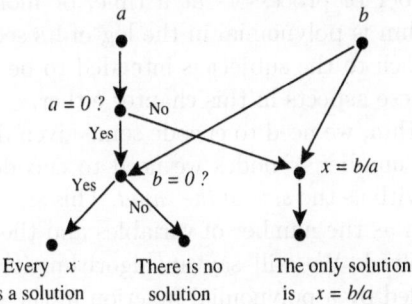

Fig. 6.1. A possible algorithm to solve the equation $ax = b$

Speaking a little more formally, our algorithms are *directed acyclic graphs*. Each node of a graph represents an element of $\mathbb{Q}$, an operation or a comparison between two elements of $\mathbb{Q}$. Each 'incoming' arrow denotes that the previously computed element or condition is needed to perform the following operation. Of course, as any graph, our algorithms have only finitely many nodes. A further comment has to be said about the graphs being 'acyclic'. As we want to predict how long our algorithms will take to compute some object, we will handle a very fixed or restricted family of algorithms: no 'WHILE' instruction is admitted in our algorithms. We can replace each 'WHILE' instruction by a 'FOR', provided we know beforehand how many times we have to repeat the procedure involved. So, an instruction of the kind 'WHILE $x > \epsilon$ DO...' is not acceptable in our algorithms, unless it can be translated into one of the type 'FOR $i = 1$ TO n DO...' and therefore 'disentangled' into a known number of sequential operations to avoid cycles in our graph.

The idea of complexity of an algorithm is related to the time it would take the algorithm to perform the desired task. The more 'complicated' our graph is, the longer the time it will take. So, a first measure of complexity to be taken into account may be the number of nodes in the graph. This will be the notion of *complexity* we are going to use throughout these notes, also known as sequential complexity.

Needless to say, this measure of complexity is not very accurate. For example, it is much simpler for a machine to perform the sum $1 + 1$ than to add two huge numbers but our measure of complexity does not take this into account. Moreover, it is generally quicker to compute a sum than a product. These considerations give place to a number of different kinds of complexities (non-scalar complexity, bit complexity, etc) which we will not take into account. But, of course, if an algorithm has a very high complexity in our terms, then it will be useless to try to run it on any computer.

There are other possible variables to be taken into account when considering the feasibility of an algorithm: for instance, the space in memory needed to perform it or whether it is well-parallelizable (that is to say, roughly speaking, whether it can be run fast enough provided we can use simultaneously a considerable number of processors at a time, or more precisely, that the depth of the algorithm is polynomial in the log of its sequential complexity). However, our approach to the subject is intended to be basic and we are not going to consider these aspects in this chapter either.

To run an algorithm, we need to encode some given data: for the moment, we will refer to the number of nodes we need to encode the input data our algorithm can deal with as the *size of the input*. This size generally depends on some quantities such as the number of variables and the number and degrees of polynomials involved. We will say an algorithm is *polynomial* when its complexity is bounded by a polynomial function in the size of the input.

We will also use the usual $\mathcal{O}$ notation to express orders of complexities: given two functions $f : \mathbb{N} \to \mathbb{N}$ and $g : \mathbb{N} \to \mathbb{N}$, we say that $f = \mathcal{O}(g)$ if and only if there exists $k \in \mathbb{N}$ such that $f(x) \leq k g(x)$ for all $x \in \mathbb{N}$.

6.3 Dense encoding and algorithms

As we are trying to solve algorithmic problems involving polynomials, we need to encode them somehow. The first (and most naive) way of encoding a polynomial is to copy the usual way a polynomial is given: as a sum of monomials. To do this in a way a computer can understand it, we need to know a bound for the degree of the polynomial and the number of variables involved in advance. Then, we should order somehow all the monomials of degree less than or equal to the known bound for the degree in the number of variables involved. Once this is done, we can encode the polynomial as the vector of its coefficients in the preset order.

For example, let $f(X, Y) = X^2 - 2XY + Y^2 + 3$ be a polynomial we want to encode. As we know $\deg(f) = 2$, we only have to store the coefficients of the monomials up to this degree. We previously fix an order for all the monomials up to degree 2 in two variables, for example $(1, X, Y, X^2, XY, Y^2)$ and, using this order, the polynomial f will be encoded as $(3, 0, 0, 1, -2, 1)$.

This way of encoding polynomials is called the *dense encoding*.

Let f be a polynomial of degree bounded by d $(d \geq 2)$ in n variables and let us consider how many coefficients it has, that is to say how many numbers will be needed to encode it (i.e. its size when considered as an input), provided we are given a previous monomial ordering. According to our definition, we have to compute how many monomials in n variables of degree bounded by d there are, and the exact number is $\binom{d+n}{d}$. If we consider that we are working with a fixed number of variables n but that the degrees can change, taking $d \geq 2$, we have that

$$\binom{d+n}{d} = \prod_{1 \leq i \leq n} \frac{d+i}{i} \leq 2d^n.$$

Furthermore, asymptotically in d we have that these two quantities are of the same order because

$$\frac{d^n}{\prod_{1 \leq i \leq n} \frac{d+i}{i}} \leq n!$$

and this is why we say that a polynomial of degree $d \geq 2$ in n variables has $\mathcal{O}(d^n)$ coefficients.

6.3.1 Hilbert's Nullstellensatz and dense encoding

As we have seen in Section 6.1.1, an effective Hilbert's Nullstellensatz is any algorithm that, given as input the polynomials $f_1, \ldots, f_s \in \mathbb{Q}[X_1, \ldots, X_n]$, decides whether there exist polynomials $g_1, \ldots, g_s \in \mathbb{Q}[X_1, \ldots, X_n]$ such that

$$\sum_{1 \leq i \leq s} g_i . f_i = 1 \tag{6.1}$$

and computes a particular solution $(g_1, \ldots, g_s)$ to this identity.

The first step may be to find a bound for the possible degrees of some polynomial solutions $g_1, \ldots, g_s$ to Equation (6.1) as a function of s, n and a bound d for the degrees of the polynomials $f_1, \ldots, f_s$. If we are able to do so, our problem can be easily transformed into a linear algebra problem: we could write new variables for the coefficients of the polynomials $g_1, \ldots, g_s$ up to the degree we found as a bound and Equation (6.1) would turn into a linear system by identifying the coefficients on the left with those on the right. That is why some authors consider the following problem an effective Hilbert's Nullstellensatz:

Show explicitly a function $\varphi : \mathbb{N}^3 \to \mathbb{N}$ *satisfying the following property:*
Let $f_1, \ldots, f_s \in \mathbb{Q}[X_1, \ldots, X_n]$ *such that* $\deg(f_i) \leq d$ $(1 \leq i \leq s)$. *If* $1 \in (f_1, \ldots, f_s)$, *there exist polynomials* $g_1, \ldots, g_s \in \mathbb{Q}[X_1, \ldots, X_n]$ *with* $\deg(g_i) \leq \varphi(n, s, d)$ $(1 \leq i \leq s)$ *such that* $\sum_{1 \leq i \leq s} g_i . f_i = 1$.

In the case the polynomials we obtain by homogenizing $f_1, \ldots, f_s$ have no common zeros at infinity, the Fundamental Theorem of Elimination Theory (see [Laz77] and Chapter 1, for example) shows that $\varphi(n, s, d) \leq n(d-1)+1$, but in the general case this bound does not work.

Just as an example, we are going to show a very elementary result of this kind, where we obtain bounds similar to the ones obtained by G. Hermann [Her26], whose proof was corrected in [MW83].

Theorem 6.3.1. *Let* $f_1, \ldots, f_s \in \mathbb{Q}[X_1, \ldots, X_n]$ *such that* $\deg(f_i) \leq d$ $(1 \leq i \leq s)$. *If* $1 \in (f_1, \ldots, f_s)$, *there exist polynomials* $g_1, \ldots, g_s \in \mathbb{Q}[X_1, \ldots, X_n]$ *with* $\deg(g_i) \leq (3d)^{2^{n-1}}$ $(1 \leq i \leq s)$ *such that* $\sum_{1 \leq i \leq s} g_i \cdot f_i = 1$.

Proof. We shall prove this theorem using induction on n.

For $n = 1$, let $f_1, \ldots, f_s \in \mathbb{Q}[X]$ and suppose $\deg f_1 = d \geq \deg f_i$ $(2 \leq i \leq s)$. If $1 = \sum_{1 \leq i \leq s} h_i \cdot f_i$, applying the division algorithm by f_1 in $\mathbb{Q}[X]$, we have

$$h_i = f_1 \cdot q_i + r_i \qquad (2 \leq i \leq s).$$

Then we obtain, rearranging the sum, that

$$1 = f_1 \cdot (h_1 + \sum_{2 \leq i \leq s} q_i \cdot f_i) + \sum_{2 \leq i \leq s} f_i \cdot r_i.$$

As $\deg r_i \leq d-1$ $(2 \leq i \leq s)$, we have that $\deg(f_1 \cdot (h_1 + \sum_{2 \leq i \leq s} q_i \cdot f_i)) \leq 2d-1$. Therefore, calling $g_1 = h_1 + \sum_{2 \leq i \leq s} q_i \cdot f_i$ and $g_i = r_i$ $(2 \leq i \leq s)$ we get that $1 = \sum_{1 \leq i \leq s} g_i \cdot f_i$ and $\deg g_i \leq d-1$ $(1 \leq i \leq s)$.

Suppose now the result is true for n. Let $f_1, \ldots, f_s \in \mathbb{Q}[X_1, \ldots, X_{n+1}]$ be such that $\deg(f_i) \leq \deg(f_1) = d$ $(2 \leq i \leq s)$.

We want to deal with polynomials which are monic with respect to a variable. To do so, consider the following change of variables (where $\lambda_2, \ldots, \lambda_n$ are new parameters): $X_1 = Y_1, X_2 = Y_2 + \lambda_2 Y_1, \ldots, X_n = Y_n + \lambda_n Y_1$. The polynomials we obtain when applying this change of variables have maximum degree in Y_1 and their leading coefficients in this variable are the homogeneous parts of maximum degree of the original polynomials evaluated in $(1, \lambda_2, \ldots, \lambda_n)$. Choosing a suitable $n-1$-tuple such that these homogeneous parts do not vanish, we get the desired linear change of variables.

So, without loss of generality, we can suppose every polynomial f_i is monic in X_1. Introduce new variables $U_1, \ldots, U_s, V_1, \ldots, V_s$ and consider the polynomials

$$F := \sum_{1 \leq i \leq s} U_i f_i \quad \text{and} \quad G := \sum_{1 \leq i \leq s} V_i f_i \quad \text{in} \quad \mathbb{Q}[U, V][X_1, \ldots, X_{n+1}].$$

The resultant of these polynomials with respect to the variable X_1,

$$\mathrm{Res}_{X_1}(F, G) \in \mathbb{Q}[U, V][X_2, \ldots, X_{n+1}]$$

is bi-homogeneous in the groups of variables (U, V) of bi-degree (d, d). We are going to prove that, if we write

$$\text{Res}_{X_1}(F, G) = \sum_{\alpha, \beta} h_{\alpha, \beta}(X_2, \ldots, X_{n+1}) U^\alpha V^\beta,$$

$f_1, \ldots, f_s$ have a common root in $\mathbb{C}^{n+1}$ if and only if $(h_{\alpha, \beta})_{|\alpha|=d, |\beta|=d}$ have a common root in $\mathbb{C}^n$.

If $(x_1, \ldots, x_{n+1}) \in \mathbb{C}^{n+1}$ is a common root of $f_1, \ldots, f_s$, then

$$\text{Res}_{X_1}(F, G)(x_2, \ldots, x_{n+1})(U, V) = 0$$

and therefore, $(h_{\alpha, \beta})_{|\alpha|=d, |\beta|=d}$ have a common root in $\mathbb{C}^n$.

On the other hand, if $(x_2, \ldots, x_{n+1})$ is a common root of $(h_{\alpha, \beta})_{|\alpha|=d, |\beta|=d}$, consider the polynomials F and G in $\mathbb{Q}(U_1, \ldots, U_s, V_1, \ldots, V_s)[X_1, X_2, \ldots, X_{n+1}]$. Then, $F(X_1, x_2, \ldots, x_{n+1})$ and $G(X_1, x_2, \ldots, x_{n+1})$ share a common root in $\overline{\mathbb{Q}(U, V)}$. But, as the roots of $F(X_1, x_2, \ldots, x_{n+1})$ lie in $\overline{\mathbb{Q}(U)}$ and the roots of $G(X_1, x_2, \ldots, x_{n+1})$ lie in $\overline{\mathbb{Q}(V)}$, the common root must be in $\mathbb{C}$. That is, there exists $x_1 \in \mathbb{C}$ such that $F(x_1, \ldots, x_{n+1}) = 0$ and $G(x_1, \ldots, x_{n+1}) = 0$. As the variables U, V are algebraically independent, we conclude that $(x_1, \ldots, x_{n+1})$ is a common root of the polynomials $f_1, \ldots, f_s$.

Then we have reduced the number of variables by one. Note that, because of Hilbert's Nullstellensatz, we have shown that

$$1 \in (f_1, \ldots, f_s) \iff 1 \in (h_{\alpha, \beta})_{|\alpha|=d, |\beta|=d}.$$

$\text{Res}_{X_1}(F, G)$ can be written as a linear combination of F and G. Taking into account the degrees of the polynomials involved, we can state that there exist polynomials R and S in $\mathbb{Q}[U, V][X]$ of degree bounded by $2d^2$ in the variables X such that $\text{Res}_{X_1}(F, G) = RF + SG$. Rewriting this identity into powers of U and V, we have that

$$h_{\alpha, \beta} = \sum_{1 \le i \le s} p_i^{(\alpha, \beta)} f_i$$

where the polynomials $p_i^{(\alpha, \beta)}$ have degrees bounded by $2d^2$. Using the inductive hypothesis for the polynomials $(h_{\alpha, \beta})_{|\alpha|=d, |\beta|=d}$ whose degrees are bounded by $2d^2$, the theorem follows. $\square$

Evidently, this kind of bound is not good for algorithmic purposes. There are much better bounds for the degrees of the polynomials appearing in the Nullstellensatz but the proofs are beyond the scope of this survey. Brownawell, in [Bro87], obtained the first single exponential bound $\varphi(d, n, s) = 3 \min\{n, s\} n d^{\min\{n, s\}}$ in the characteristic zero case. Then, in [Kol88] and [FG90] the most precise bounds known up to now for any characteristic were found: $\varphi(d, n, s) = (\max\{3, d\})^n$. In [SS95] a better bound for the particular case when $d = 2$, namely $\varphi(2, n, s) = n2^{n+2}$, was shown.

More precise bounds involving other parameters than d, n and s were obtained in [Som97], [KSS97] and [GHM$^+$98] (see Section 6.6.2).

Let us make a final comment on the complexity of an algorithm that, using the dense encoding of polynomials, decides whether the variety they define is empty or not and, if it is empty, gives as output a linear combination of the input polynomials equal to 1.

If the input polynomials $f_1, \ldots, f_s$ have degrees bounded by d and the bound for the degrees of the polynomials involved in the linear combination given by the Nullstellensatz is $\varphi(d, n, s)$, then we only need to solve a system of $\mathcal{O}\left((\varphi(d, n, s) + d)^n\right)$ linear equations in $\mathcal{O}(s\varphi(d, n, s)^n)$ variables (or to prove that this system has no solution). The complexity of doing this, using the techniques in [Ber84] and [Mul87], is of order $\mathcal{O}(s^4.(\varphi(d, n, s) + d)^{4n})$.

Therefore, using the best Effective Nullstellensäzte known up to now, that essentially state $\varphi(d, n, s) = d^n$, the complexity of any algorithm using dense encoding will be at least of order $\mathcal{O}(sd^{n^2})$ (see Proposition 6.3.4 below).

6.3.2 Quantifier elimination and dense encoding

Suppose now we are given $s + t$ polynomials in $\mathbb{Q}[X_1, \ldots, X_n][Y_1, \ldots, Y_m]$ of degrees bounded by d and we want to give algorithmically a quantifier-free formula equivalent to

$$\exists y \in \mathbb{C}^m : f_1(x, y) = 0 \wedge \cdots \wedge f_s(x, y) = 0 \wedge g_1(x, y) \neq 0 \wedge \cdots \wedge g_t(x, y) \neq 0. \tag{6.2}$$

Rabinowicz's trick allows us to consider only equalities by means of a new indeterminate Z and therefore, the previous formula is equivalent to

$$\exists y \in \mathbb{C}^m \, \exists z \in \mathbb{C} : f_1(x, y) = 0 \wedge \cdots \wedge f_s(x, y) = 0 \wedge \left(1 - z. \prod_{1 \leq i \leq t} g_i(x, y)\right) = 0.$$

For a fixed $x \in \mathbb{C}^n$, using Hilbert's Nullstellensatz, this last formula is equivalent to

$$\nexists p_1, \ldots, p_s, p_{s+1} \in \mathbb{C}[Y_1, \ldots, Y_m, Z] \,/\, 1 = \sum_{1 \leq i \leq s} p_i f_i + p_{s+1}\left(1 - Z. \prod_{1 \leq i \leq t} g_i\right).$$

Any effective Hilbert's Nullstellensatz providing upper bounds for the degrees of the polynomials p_i involved allows us to translate this last formula into a quantifier-free formula in the coefficients of the polynomials f_i and g_j by means of linear algebra. Suppose the linear system involved is $A.X^t = B$ where $A \in \mathbb{C}^{\ell \times k}$ and $B \in \mathbb{C}^\ell$. The non-existence of solutions is equivalent to the condition $\text{rank}(A) \neq \text{rank}(A|B)$. Using that the rank of a matrix can be computed by means of the determinants of its minors, this last condition can be translated into a (very long) formula involving $\wedge$, $\vee$, equalities and inequalities to zero. This formula works for every $x \in \mathbb{C}^n$ and therefore, this formula is equivalent to (6.2).

It is evident that the better the effective Nullstellensatz we are using, the smaller the complexity of this kind of algorithm will be, provided we compute the rank of the matrices involved in a smart way (for example, using the algorithm in [Mul87]).

Given a first order prenex formula φ ('prenex' meaning that there are several blocks of existential and universal quantifiers placed at the beginning of the formula) with coefficients over an algebraically closed field, let $|\varphi|$ be its length, i.e. the number of symbols needed to encode φ, let n be the number of indeterminates involved, let D be one plus the sum of the degrees of the polynomials that appear in φ and let r be the number of blocks of quantifiers. Heintz and Wüthrich (see [Hei83] and [HW75]) exhibited elimination algorithms for algebraically closed fields of given characteristic with complexity bounded by $|\varphi|D^{n^{O(n)}}$. In fact, in the 1940s, Tarski already knew the existence of elimination algorithms but he did not describe them explicitly (see [Tar51]). Later, using the fundamental techniques described in [CG83] and [Hei83], Chistov and Grigor'ev considered the problem for prenex formulae and obtained in [CG84] and [Gri87] more precise complexity bounds of order $|\varphi|D^{n^{O(r)}}$. However, these bounds depend on arithmetic properties of the base field involved because polynomial factorization algorithms are used as subalgorithms. None of the algorithms mentioned before are efficiently well-parallelizable. Finally, in [FGM90], a well-parallelizable elimination algorithm within the same sequential complexity bounds obtained in [CG84] and [Gri87] is constructed combining the methods in [Hei83] with some effective versions of Hilbert's Nullstellensatz (see Section 6.3.1). Moreover, the complexity of this algorithm does not depend on particular properties of the base field k. Later, the same result was obtained in [Ier89]. In the context of quantifier elimination, it is also worth mentioning the work of Renegar (see [Ren92]) on elimination over real closed fields since the bounds obtained there are very sharp and imply the bounds for elimination over complex numbers.

6.3.3 Equidimensional decomposition and dense encoding

Different algorithms describing decompositions of an algebraic variety V have been given. Chistov and Grigor'ev (see [CG83]) exhibit an algorithm for the computation of the irreducible decomposition provided an algorithm that factorizes multivariate polynomials with coefficients in the base field is given. Giusti and Heintz (see [GH91]) present an algorithm for the equidimensional decomposition of algebraic varieties which is well-parallelizable. Although we do not include the proof of this last result here, we can state their main theorem and the complexity obtained:

Theorem 6.3.2. *Let $f_1, \ldots, f_s$ be polynomials in $\mathbb{Q}[X_1, \ldots, X_n]$ of degree bounded by d and let V be the variety they define. There exists an algorithm of complexity $s^5 d^{O(n^2)}$ which computes, for every $0 \le i \le n$, $d^{O(n^2)}$ polynomi-*

als of degree bounded by d^n defining the equidimensional component of V of dimension i.

A more recent algorithm to decompose an algebraic variety using Bézoutian matrices can be found in [EM99a]. However, the decomposition obtained there may not be minimal (embedded components may appear) and the algorithm is probabilistic (see Section 6.6.1).

6.3.4 A lower bound

In this section, we are going to show that the better bounds already obtained (and mentioned before) for the efficient Hilbert's Nullstellensatz are of the best possible order.

To do so, we are going to state a very well-known example by Masser and Philippon (see [Bro87]) that gives a very high lower bound for the degrees of the polynomials appearing in the Nullstellensatz:

Example 6.3.3. Take the following polynomials in $\mathbb{Q}[X_1, \ldots, X_n]$:

$$f_1 = X_1^d, f_2 = X_1 - X_2^d, \ldots, f_{n-1} = X_{n-2} - X_{n-1}^d, f_n = 1 - X_{n-1}X_n^{d-1}.$$

If $g_1, \ldots, g_n \in \mathbb{Q}[X_1, \ldots, X_n]$ are polynomials such that $1 = \sum_{1 \leq i \leq n} g_i f_i$, consider a new variable T and evaluate the polynomials in the following vector of elements in $\mathbb{Q}(T)$:

$$(T^{(d-1)d^{n-2}}, \ldots, T^{d-1}, 1/T).$$

Note that, under such evaluation, all the polynomials f_i vanish for $2 \leq i \leq n$ and so we have that

$$1 = g_1\big(T^{(d-1)d^{n-2}}, \ldots, T^{d-1}, 1/T\big)T^{(d-1)d^{n-1}}.$$

This identity implies that $\deg_{X_n}(g_1) \geq (d-1)d^{n-1}$ and therefore $\deg g_1 \geq (d-1)d^{n-1}$.

This simple example shows that, with the notation above, a lower bound for the degrees of the polynomials g_i appearing in the expression $1 = \sum_{1 \leq i \leq s} g_i \cdot f_i$ is $d^{\mathcal{O}(n)}$, and therefore we have

Proposition 6.3.4. *Any general algorithm that, from an input of s polynomials $f_1, \ldots, f_s \in \mathbb{Q}[X_1, \ldots, X_n]$ of degrees bounded by d, computes (provided they exist) polynomials $g_1, \ldots, g_s \in \mathbb{Q}[X_1, \ldots, X_n]$ such that $1 = \sum_{1 \leq i \leq s} g_i \cdot f_i$ and encodes them in dense form must have complexity of order at least $\mathcal{O}(d^{n^2})$.*

Moreover, in [FGM90], it is shown that, from the point of view of overall complexity, the complexities they attain for the quantifier elimination algorithm are optimal when using dense encoding. In fact, they prove the following

Theorem 6.3.5. *There exists a sequence of first order formulae (containing quantifiers and two free variables) φ_k ($k \in \mathbb{N}$) over an algebraically closed field with the following properties:*

- $|\varphi_k| = \mathcal{O}(k)$
- *For each quantifier free formula θ equivalent to φ_k involving the polynomials $F_1, \ldots, F_s$, there exists i, $1 \leq i \leq s$, such that $\deg F_i \geq 2^{2^{ck}}$, where $c > 0$ is a suitable constant.*

Note that this theorem states lower bounds for the degrees of the polynomials appearing in the output formula, and the greater the degrees, the greater the number of nodes needed to encode them.

6.4 Straight-line Program encoding for polynomials

6.4.1 Basic definitions and examples

The comments in Section 6.3.4 show us that it is impossible to obtain more efficient *general* algorithms when dealing with dense encoding of polynomials. There are at least two ways of avoiding this problem: the first one is to change the form the polynomials are encoded (that is to say, to try to find a shorter way for encoding polynomials) while the second one is to design non-general algorithms which can only solve special problems but within a lower complexity. We will now discuss the first of these: changing the way we encode polynomials.

One attempt that has been made to change the representation of polynomials is the so-called 'sparse' encoding, which consists in specifying which monomials of a given polynomial have non-zero coefficients and which are these coefficients. Suppose a polynomial P has only a few monomials with respect to its degree. The sparse encoding will consist of a number of vectors which specify the (non-zero) coefficient of every monomial appearing in P. For example, if $P = 2X^{15}Y^4 + 2X^7Y^3 - 3X^2 + 1$, it can be encoded by a vector of four three-tuples, one for each of the monomials appearing in P. In each three-tuple, the first coefficient would stand for the degree of the monomial in X, the second one for the degree of the monomial in Y and the third one would be the coefficient of the monomial, that is, P would be encoded in the following way

$$P := ((15, 4, 2); (7, 3, 2); (2, 0, -3); (0, 0, 1))$$

instead of using a vector of $\binom{21}{2} = 210$ coordinates.

This way of encoding polynomials has proved to be efficient when dealing with particular families of polynomials (see, for example, Chapter 7 and Chapter 3) and there is a lot of theory and many algorithms that use the sparse encoding. For a complete background of this theory (including sparse resultants, Newton polytopes, toric varieties and Bernstein theorem, among other interesting and very useful notions) we suggest the reader refer to [CLO98], [GKZ94] and [Ful93].

However, it is not clear whether it is worth it to use this sparse encoding in a *general* algorithm: the output polynomials may have too many monomials. Moreover, the sparse encoding does not behave well under linear changes of coordinates in the sense that a 'short' polynomial in the sparse form can change into a very 'long' one by means of a linear change of variables: note that

$$(X + Y)^{100} = \sum_{0 \leq i \leq 100} \binom{100}{i} X^i Y^{100-i}$$

(that is, a single monomial may turn into a polynomial with many monomials under a linear change of variables).

An alternative way to encode polynomials (the one we are going to study here) is based on the following idea:

Let P be the polynomial $P := (X + Y)^{100} - 1$. Why can we define this polynomial so easily (that is to say, using a small number of symbols) but it takes so much space to encode it for a machine (in both the sparse and the dense encoding)?

The answer perhaps is that we are used to thinking of a polynomial as a 'formal expression' rather than a function that can be evaluated. But, as far as fields of characteristic zero are concerned, polynomial functions and polynomials can be considered as the same objects. Therefore, if we define a polynomial function by defining its exact value at every point (that is to say, by means of describing how to evaluate it), we will be defining a polynomial. In the previous example, the polynomial P would be the only polynomial in $\mathbb{Q}[X, Y]$ such that, to evaluate it at a pair (x, y), you have to compute the sum of x and y to the 100-th power and subtract 1 from the result. This way of encoding a polynomial will be called a *straight-line program*. Let us put these ideas more precisely:

Definition 6.4.1. *Let $X_1, \ldots, X_n$ be indeterminates over $\mathbb{Q}$ and let $R \in \mathbb{N}$. An element $\beta := (Q_1, \ldots, Q_R) \in \mathbb{Q}(X_1, \ldots, X_n)^R$ is a* **straight-line program** *(slp for short) if each Q_ρ satisfies one of the following two conditions:*

- $Q_\rho \in \mathbb{Q} \cup \{X_1, \ldots, X_n\}$ *or*
- $\exists \rho_1, \rho_2 < \rho$ *and* $* \in \{+, -, \cdot, \div\}$ *such that* $Q_\rho = Q_{\rho_1} * Q_{\rho_2}$.

We say β is a **division-free slp** *if $Q_\rho = Q_{\rho_1} \div Q_{\rho_2} \Rightarrow Q_{\rho_2} \in \mathbb{Q} - \{0\}$.*

From now on, we are only going to deal with **division-free** slp's. Note that, in this case, each element Q_ρ is a polynomial in $\mathbb{Q}[X_1, \ldots, X_n]$. If $F \in \{Q_\rho \mid 1 \leq \rho \leq R\}$, we say that β *computes* or *calculates* F.

There are several measures of complexity that can be taken into account when considering slp's. For example:

- The *total length* of β (denoted by $L(\beta)$) is the quantity of operations performed during the slp β (more precisely, it is the number of coordinates Q_ρ defined as the result of an operation between two previous coordinates).

- The *additive length* of β $(L_\pm(\beta))$ is the quantity of sums and subtractions performed during the slp.
- The *non-scalar length* of β $(L_\mathbb{Q}(\beta))$ is the number of products between two non-rational elements performed during the slp.

Given any polynomial $F \in \mathbb{Q}[X_1, \ldots, X_n]$ we will define its *total length* (also called *total complexity*) as

$$L(F) := \min\{L(\beta) \ / \ \beta \text{ is an slp computing } F\}.$$

We can respectively define $L_\pm(F)$ and $L_\mathbb{Q}(F)$.

Exercise 6.4.2. Prove that, for any $F \in \mathbb{Q}[X_1, \ldots, X_n]$, $L(F) = \mathcal{O}((L_\mathbb{Q}(F))^2)$.

From now on, unless it expressly stated, we will only consider the total length of an slp of a polynomial and, for the sake of shortness, we will simply call it *its length*.

As an example, we are going to show an slp that calculates the polynomial $F(X) = 1 + X + X^2 + X^3 + \cdots + X^{2^j - 1}$ efficiently. Of course, we can compute every power of X and then add them up, but it would yield an slp of length $2^{j+1} - 3$. A better slp computing F, based on the binary expansion of any positive integer up to $2^j - 1$ is the following one :

$$\beta := \left(1, X, X^2, X^4, \ldots, X^{2^{j-1}}, 1 + X, 1 + X^2, 1 + X^4, \ldots, 1 + X^{2^{j-1}}, \right.$$

$$\left. (1 + X)(1 + X^2), (1 + X)(1 + X^2)(1 + X^4), \ldots, \prod_{0 \leq i \leq j-1} (1 + X^{2^i}) \right)$$

and $L(\beta) = 3j - 2$.

Another well-known example of slp encoding a polynomial is Horner's rule for univariate polynomials:

$$a_0 + a_1 X + a_2 X^2 + \cdots + a_d X^d = (a_0 + X(a_1 + X(a_2 + X(\ldots(a_{d-1} + a_d X)\ldots)))).$$

The length of this slp is $2d$ and it involves d products and d sums. It can be proved that the number of sums and the number of products involved in *any* slp computing this polynomial are bounded by d when the elements $X, a_0, \ldots, a_d$ are algebraically independent (see, for example, [BCS97]).

Exercise 6.4.3. Let $P \in \mathbb{Q}[X, Y]$ be a polynomial whose sparse encoding is $P = ((m_1, n_1, c_1), \ldots, (m_s, n_s, c_s))$. Find a bound for $L(P)$.

Exercise 6.4.4. Find an infinite family of polynomials in $\mathbb{Q}[X, Y]$ such that the number of nodes needed to encode each of them into the sparse form is (much) greater than its length.

Exercise 6.4.5. Given a generic polynomial of degree d in $\mathbb{Q}[X, Y]$, find an upper bound for its length.

Exercise 6.4.6. Try to generalize the three previous exercises to the case of n-variate polynomials.

A last comment has to be made about the complexity of algorithms when dealing with straight-line programs. An slp can be obviously considered as a directed acyclic graph without branchings and, therefore, it has nodes. The complexity of an algorithm using the slp encoding will be the total number of nodes, that is to say, the ones arising as operations or comparisons *plus* the internal nodes of the slp's involved.

6.4.2 Some apparent disadvantages

When we are dealing with slp's to encode polynomials, we face a fundamental problem: the same polynomial may be encoded by means of many different slp's. So, it is not straightforward to verify a polynomial identity.

Suppose you are given an slp of length L that evaluates a polynomial F in n variables of degree bounded by d. If you want to know whether $F \equiv 0$, a naive attempt would be to interpolate F, but it would take so many points to do so that the complexity of doing this would again be too large (within the same order as the number of nodes needed in the dense representation of F).

Another way to solve the problem is to find a smaller particular set of points such that two polynomials of bounded length and degree coincide if and only if they coincide when evaluated in all these points. Luckily, there is a result due to Heintz and Schnorr stating the existence of this set:

Theorem 6.4.7. *(see [HS82]) Let $\widetilde{W}(d, n, L) \subset \mathbb{Q}[X_1, \ldots, X_n]$ be the set of polynomials of degree bounded by d that can be calculated by means of an slp of length L. Let $\Gamma \subset \mathbb{Q}$ be a set of $2L(1 + d)^2$ elements. Then, there exists a set of points $\{\alpha_1, \ldots, \alpha_m\} \subset \Gamma^n$ with $m = 6(L + n)(L + n + 1)$ satisfying*

$$F \in \widetilde{W}(d, n, L) \text{ such that } F(\alpha_i) = 0 \; \forall \; 1 \le i \le m \Rightarrow F \equiv 0.$$

The set $\{\alpha_1, \ldots, \alpha_m\}$ is called a *correct test sequence* or a *set of questors*. Unfortunately, we do not know how to construct such a set within a reasonable cost. A way to avoid this problem is to consider probabilistic algorithms (which we will briefly discuss later in Section 6.6.1).

Another question we can ask is how many polynomials can be evaluated easily (that is to say, can be calculated by means of short slp's). The answer again, as we are going to see now, is not very encouraging (see [Sch78] and [HS80]):

For fixed n, d and L, let us consider the set of all the polynomials $F \in \mathbb{Q}[X_1, \ldots, X_n]$ with $\deg(F) \le d$ and non-scalar length $L_{\mathbb{Q}}(F) \le L$.

Observe that each of these polynomials can be computed by a 'non-scalar' slp (that is to say, the only coordinates we are taking into account in this slp are the products between non-scalar elements)

$$\beta := (\beta_{-n+1}, \dots, \beta_0, \beta_1, \dots, \beta_L)$$

where $\beta_{-n+i} = X_i$ $(1 \le i \le n)$ and, defining $\beta_{-n} := 1$,

$$\beta_k = \left(\sum_{-n \le j \le k-1} a_j^{(k)} \cdot \beta_j \right) \cdot \left(\sum_{-n \le j \le k-1} b_j^{(k)} \beta_j \right).$$

Considering new variables $A_j^{(k)}$ and $B_j^{(k)}$ $(1 \le k \le L; \; -n \le j \le k - 1)$, there exist polynomials $Q_\alpha^{(k)} \in \mathbb{Q}[A_j^{(k)}, B_j^{(k)}]$ such that the coefficients of any polynomial F_k that can be computed in the k-th step of β are the specializations of these polynomials in some rational vectors a and b, that is to say

$$F_k = \sum_\alpha Q_\alpha^{(k)}(a, b) X^\alpha.$$

So we have

Proposition 6.4.8. *For every $L, n \in \mathbb{N}$ there exist $Q_\alpha \in \mathbb{Z}[T_1, \dots, T_m]$ polynomials with $m = (L+n)(L+n+1)$, $\alpha \in (\mathbb{N}_0)^n$, $|\alpha| \le 2^L$, $\deg Q_\alpha \le 2|\alpha|L$ such that for every $F \in \mathbb{Q}[X_1, \dots, X_n]$ satisfying $L_\mathbb{Q}(F) \le L$,*

$$F = \sum_\alpha Q_\alpha(t) X^\alpha \text{ for some } t \in \mathbb{Q}^m.$$

Now, we can consider the morphism obtained by evaluating the family of polynomials $(Q_\alpha : |\alpha| \le d)$:

$$(Q_\alpha : |\alpha| \le d) : \mathbb{C}^{(L+n)(L+n+1)} \to \mathbb{C}^{\binom{n+d}{n}}.$$

Therefore, if $F := \sum_\alpha c_\alpha X^\alpha \in \mathbb{Q}[X_1, \dots, X_n]$ is any polynomial that has $\deg(F) \le d$ and non-scalar length $L_\mathbb{Q}(F) \le L$, considering it as the vector $(c_\alpha) \in \mathbb{Q}^{\binom{n+d}{n}}$, it turns out that $F \in Im(Q_\alpha : |\alpha| \le d)$.

As a consequence, we have that, for fixed $d, n, L \in \mathbb{N}$, the set

$$W(n, d, L) := \overline{Im(Q_\alpha : \alpha \in (\mathbb{N}_0)^n, |\alpha| \le d)} \subset \mathbb{C}^{\binom{n+d}{n}}$$

is a closed set that contains all the vectors of coefficients of polynomials $F \in \mathbb{Q}[X_1, \dots, X_n]$ such that $\deg F \le d$ and $L_\mathbb{Q} \le L$.

A very important remark is that, as $W(n, d, L)$ is defined by means of a polynomial function in $(L+n)(L+n+1)$ variables, its dimension is bounded by $\dim W(n, d, L) \le (L+n)(L+n+1)$. This can be interpreted in the following way: the polynomials of degree bounded by d with non-scalar complexity bounded by L considered in $\mathbb{C}^{\binom{n+d}{n}}$ (a space of dimension $\binom{n+d}{n}$), lie in a variety of dimension $(L+n)(L+n+1)$. So, as long as L satisfies $(L+n)(L+n+1) < \binom{n+d}{n}$, there are very few polynomials easy to evaluate since the complement of the variety they lie in is a non-empty open set in the Zariski topology. Therefore, most polynomials are difficult to evaluate.

Taking these last observations into account, one may wonder if it would be useful to deal with slp's when trying to solve polynomial equations. The answer is affirmative as we will see in the following sections.

6.4.3 A fundamental result

In [GH93], Giusti and Heintz obtain a fundamental result using for the first time straight-line programs to solve algorithmically a problem related to solving a system of polynomial equations. In that paper, they give a **polynomial algorithm** that can decide whether a given algebraic variety V is empty or not from the polynomials defining V encoded in dense form. In fact, they go a little further: given polynomials, encoded in dense form and defining a variety V, they can find the dimension of V algorithmically in polynomial time.

In a first step, they design an algorithm that, given polynomials defining a variety V, computes a variety Z, either zero-dimensional or empty, satisfying the following conditions (V_0 will denote, as usual, the zero-dimensional equidimensional component of V):

- $V_0 \subset Z \subset V$ (that is to say, all the isolated points of V are in Z and all the points of Z are points in the variety.)
- The way Z is presented makes it 'easy' to decide whether it is empty or not (a more precise description of this way of presenting Z will be given in Section 6.4.4).

Note that, if we already know that the variety V is either empty or has dimension 0, we can decide if it is empty by means of this result ($V = \emptyset \iff Z = \emptyset$).

The general idea of the algorithm computing the dimension of V is the following: suppose the variety V is defined by $f_1, \ldots, f_s \in \mathbb{Q}[X_1, \ldots, X_n]$. Generally, if it is not empty, when we cut it with a hyperplane H_1, we will obtain a variety $V \cap H_1$ of dimension $\dim V - 1$. Continuing this process with 'generic' hyperplanes, we have that, after $\dim V + 1$ steps, by reducing the dimension by one in each step, we get the empty set. Then, as $\dim V \leq n$, when we cut it with $n + 1$ 'generic' hyperplanes we obtain the empty set:

$$V \cap H_1 \cap \cdots \cap H_{n+1} = \emptyset.$$

So, we have that $V \cap H_1 \cap \cdots \cap H_n$ is either the empty set or a variety consisting only of isolated points and we are under the required hypotheses to decide whether it is empty or not. If it is not empty, then $\dim V = n$. If it is empty, we consider the variety $V \cap H_1 \cap \cdots \cap H_{n-1}$ and repeat the process. After at most $n + 1$ steps we will know the dimension of V (because it is equal to the minimum number of 'generic' hyperplanes we have to cut V with to obtain the empty set minus one).

The sets of $n+1$ hyperplanes that do not satisfy the desired conditions can be considered as elements of a proper closed set in a proper affine space $\mathbb{C}^N$, that is to say the whole construction we have made works for almost every set of $n + 1$ hyperplanes. This is what we meant by 'generic' hyperplanes in the last paragraph.

The proof of the result by Giusti and Heintz is beyond the scope of this survey, but we are going to take into account some of the ideas used there.

6.4.4 An old way of describing varieties: the Shape lemma

In [GH93], Giusti and Heintz use a particular way of defining zero-dimensional varieties which was already used by Kronecker (see [Kro82]). This way of presenting the variety is called *a shape lemma presentation* or a *geometric resolution* of the variety. (This same description is presented under different names in other chapters of this book: single variable representation in Chapter 2, univariate representation in Chapter 3 and shape lemma in Chapter 4.) The idea of this presentation is quite simple:

Suppose we are given a zero-dimensional variety $Z \subset \mathbb{C}^n$ defined by polynomials in $\mathbb{Q}[X_1, \ldots, X_n]$ and consisting of D points

$$x^{(1)} = (x_1^{(1)}, \ldots, x_n^{(1)}), \ldots, x^{(D)} = (x_1^{(D)}, \ldots, x_n^{(D)}).$$

Suppose also that their first coordinates are all different from one another. Therefore, we can obtain a polynomial $Q \in \mathbb{Q}[T]$ of degree D whose zeroes are exactly these first coordinates; namely

$$Q = \prod_{1 \leq i \leq D} (T - x_1^{(i)}).$$

Moreover, using interpolation, fixing an index j, $(2 \leq j \leq n)$, there exists a unique polynomial $P_j \in \mathbb{Q}[T]$ of degree bounded by $D-1$ such that $P_j(x_1^{(i)}) = x_j^{(i)}$ for every $1 \leq i \leq D$. Then,

$$Z = \{x \in \mathbb{C}^n \ / \ Q(x_1) = 0 \wedge x_2 - P_2(x_1) = 0 \wedge \cdots \wedge x_n - P_n(x_1) = 0\}.$$

This parametric description of Z (note that all coordinates are parametrized as functions of x_1) has the additional property of telling us how many points are in Z (this quantity coincides with the degree of Q).

The only inconvenience of this description is that we need the first coordinates of the points to be different from one another and this is not always the case. The way to solve this is to consider an affine linear form $\ell(X) = u_0 + u_1 X_1 + \cdots + u_n X_n$ in $\mathbb{Q}[X_1, \ldots, X_n]$ such that $\ell(x^{(i)})$ are all different from one another (in this case we say either that ℓ is a *primitive element* of Z or that ℓ *separates the points* in Z).

Now, we are able to define what we call a *geometric resolution* of a zero dimensional variety:

Definition 6.4.9. *Let* $Z = \{x^{(1)}, \ldots, x^{(m)}\} \subset \mathbb{C}^n$ *be a zero-dimensional variety defined by polynomials in* $\mathbb{Q}[X_1, \ldots, X_n]$. *A* **geometric resolution** *of* Z *consists of an affine linear form* $\ell(X) = u_0 + u_1 X_1 + \cdots + u_n X_n$ *in* $\mathbb{Q}[X_1, \ldots, X_n]$, *and polynomials* $Q, P_1, \ldots, P_n \in \mathbb{Q}[T]$ *(where* T *is a new variable) such that:*

- $\ell(x^{(i)}) \neq \ell(x^{(k)})$ *if* $i \neq k$.
- $Q(T) = \prod_{1 \leq i \leq D}(T - \ell(x^{(i)}))$
- *For* $1 \leq j \leq n$, $\deg P_j \leq D - 1$ *and*

$$Z = \{(P_1(\xi), \ldots, P_n(\xi)) \ / \ \xi \in \mathbb{C} \text{ such that } Q(\xi) = 0\}.$$

As this description of Z is uniquely determined up to ℓ we call it **the** geometric resolution of Z associated to ℓ.

For the sake of simplicity, we also define the notion of geometric resolution for the empty set, and in this case, the polynomial Q is 1.

Although this definition is quite easy to understand, the problem underlying it is to find (given the zero-dimensional variety $Z \subset \mathbb{C}^n$ defined by polynomials $f_1, \ldots, f_s$) a proper linear form and the polynomials $Q, P_1, \ldots, P_n$ (note that our definition is based on the coordinates of the points in Z!).

In [GH93] Giusti and Heintz do not find the exact geometric resolution of the isolated points of a variety V but they are able to find a linear form ℓ which separates the isolated points of V, a polynomial which vanishes over the specialization of ℓ in the isolated points of V and, by means of them, they find a geometric resolution of a zero-dimensional variety Z, satisfying $V_0 \subset Z \subset V$. Given the polynomials defining V, in a first step they introduce a new variable to make a deformation in order to reduce the problem to the case of a zero-dimensional projective variety. Then, using some ideas and results of [Laz77] about the regularity of the Hilbert function of a suitable graded ring and some linear algebra algorithms ([Ber84] and [Mul87]), they obtain the characteristic polynomials of several linear maps which allow them to get the desired geometric resolution.

Note that, if the variety V is zero-dimensional, the algorithm of [GH93] computes a geometric resolution of V. For an improved and more detailed version of this construction of a geometric resolution of a zero-dimensional variety from polynomials defining it, see [KP96].

6.4.5 Newer algorithms, lower bounds

The paper we have already mentioned ([GH93]) is a milestone in the development of algorithms solving polynomial equations symbolically. The main theorem proved there is:

Theorem 6.4.10. *There exists an algorithm that, given polynomials* $f_1, \ldots, f_s$ *in* $\mathbb{Q}[X_1, \ldots, X_n]$ *of degrees bounded by* d *in the dense encoding defining an algebraic variety* $V \subset \mathbb{C}^n$, *computes* $\dim V$ *within complexity* $s^{\mathcal{O}(1)} d^{\mathcal{O}(n)}$.

Note that this result allows us to answer the first question concerning a polynomial equation system (whether its set of solutions is empty or not) just by computing its dimension.

Some of the problems stated have since been solved within polynomial time (that is to say, by means of polynomial algorithms), by using different tools.

In [GHS93], given polynomials $f_1, \ldots, f_s \in \mathbb{Q}[X_1, \ldots, X_n]$ such that the variety they define is empty, a family of polynomials $g_1, \ldots, g_s \in \mathbb{Q}[X_1, \ldots, X_n]$ such that $1 = \sum_{1 \leq i \leq s} g_i . f_i$ holds is constructed. The polynomials $g_1, \ldots, g_s$ have degree bounded by $d^{\mathcal{O}(n^2)}$ and are obtained in an slp encoding. The complexity of the whole algorithm is $s^{\mathcal{O}(1)} d^{\mathcal{O}(n)}$ (compare with the end of Subsection 6.3.1). In [FGS95] the same problem is re-considered by using duality theory and a complete different algorithm is designed so that the new polynomials $g_1, \ldots, g_s$ obtained have degree bounded by $d^{\mathcal{O}(n)}$.

A quantifier elimination algorithm using slp's was obtained in [PS98]. The main result there is more general than the one we have stated above, but adapted to our case it would essentially mean that the elimination stated before can be done in polynomial time in the size of the input.

We can also mention polynomial algorithms for the equidimensional decomposition of varieties(see [JS02] and [Lec00]). However, these algorithms are probabilistic (see Section 6.6.1 for a brief account on probabilistic algorithms).

6.5 The Newton-Hensel method

The use of slp's as a way of encoding polynomials made it possible to adapt algorithmically a very well-known concept, the Newton-Hensel method, which can be seen as a particular version of the implicit function theorem. (Compare with the Hensel operator defined in Chapter 9.)

Let $T_1, \ldots, T_m, X_1, \ldots, X_n$ be indeterminates over a field $\mathbb{Q}$. Given $t \in \mathbb{C}^m$, $T - t$ will represent the vector $(T_1 - t_1, \ldots, T_m - t_m)$.

Let $f_1, \ldots, f_n \in \mathbb{Q}[T, X]$ be polynomials. We will denote by f the vector of polynomials $(f_1, \ldots, f_n)$, by Df the Jacobian matrix of f with respect to the indeterminates X and by Jf its determinant.

Lemma 6.5.1. *Let $f_1, \ldots, f_n \in \mathbb{Q}[T, X]$ and let $(t, \xi) \in \mathbb{C}^m \times \mathbb{C}^n$ such that*

$$f_1(t, \xi) = 0, \ldots, f_n(t, \xi) = 0 \text{ and } Jf(t, \xi) \neq 0.$$

Then, there exists a unique n-tuple of formal power series $\mathcal{R} = (\mathcal{R}_1, \ldots, \mathcal{R}_n) \in \mathbb{C}[[T - t]]^n$ such that:

- $f_1(T, \mathcal{R}) = 0, \ldots, f_n(T, \mathcal{R}) = 0$
- $\mathcal{R}(t) := (\mathcal{R}_1(t), \ldots, \mathcal{R}_n(t)) = \xi.$

Proof. (This is only a sketch; for a very detailed proof of this fact, see for example [HKP+00].)

Given $f(X) = (f_1(T, X), \ldots, f_n(T, X))$, we define the Newton-Hensel operator associated to it as

$$N_f(X)^t := X^t - Df(X)^{-1}.f(X)^t.$$

Note that $Jf(X)$ is not the zero polynomial (from our hypothesis, $Jf(t, \xi) \neq 0$) and, therefore, our definition makes sense.

We define the following sequence of rational functions:

$$\begin{cases} \mathcal{R}^{(0)} := \xi \\ \mathcal{R}^{(k)} := N_f(\mathcal{R}^{(k-1)}) = N_f^k(\xi) \ \text{ for } \ k \in \mathbb{N} \end{cases}$$

The first thing to take into account is whether we can define this sequence (that is to say, if we do not try to divide by zero) but this fact can be inductively proved using that $\mathcal{R}^{(k)}(t) = \xi$.

The following conditions are fulfilled (this can be proved recursively):

- $f_i(T, \mathcal{R}^{(k)}) \in (T - t)^{2^k} \subset \mathbb{C}[[T - t]]$ for every $1 \leq i \leq n$
- $\mathcal{R}_j^{(k+1)} - \mathcal{R}_j^{(k)} \in (T - t)^{2^k} \subset \mathbb{C}[[T - t]]$ for every $1 \leq j \leq n$

where $(T - t)$ indicates the ideal in $\mathbb{C}[[T - t]]$ generated by $T_1 - t_1, \ldots, T_m - t_m$.

Therefore, the sequences $(\mathcal{R}_j^{(k)})_{k \in \mathbb{N}}$ are convergent $(1 \leq j \leq n)$ and the n-tuples of their limits $\mathcal{R} := (\mathcal{R}_1, \ldots, \mathcal{R}_n)$ is the vector we are looking for. $\qquad\square$

Just to show how this works, we are going to discuss an example briefly.

Example 6.5.2. Given n polynomials of degrees $d_1, \ldots, d_n$ in n variables defining a zero-dimensional variety V and for a generic linear form ℓ, we show how to compute, in many cases at least, the polynomial $Q(T)$ of Definition 6.4.9 that leads to a geometric resolution of V:

We consider generic polynomials $f_1, \ldots, f_n$ of degrees $d_1, \ldots, d_n$ in the variables $X_1, \ldots, X_n$:

$$f_1(T, X) = \sum_{|\alpha| \leq d_1} T_\alpha^{(1)} X^\alpha$$

$$\ldots$$

$$f_n(T, X) = \sum_{|\alpha| \leq d_n} T_\alpha^{(n)} X^\alpha$$

(note that each coefficient of f_i is a new variable $T_\alpha^{(i)}$).

Consider the variety $W := \{(x_1, \ldots, x_n) \in \mathbb{C}^n \ / \ x_1^{d_1} - 1 = 0, \ldots, x_n^{d_n} - 1 = 0\}$. Of course, we know all the points in this set: they are n-tuples of roots of unity. Let t be the vector of coefficients of the polynomials defining W. Therefore we are under the conditions needed to apply Lemma 6.5.1 because we have that, for every $\xi \in W$, (t, ξ) is a particular instance of (T, X) that satisfy the needed hypotheses (it is easy to see that in this instance, $Jf(t, \xi) \neq 0$). Then, by applying the Newton-Hensel algorithm we can approximate vectors of power series in $T - t$ which will be roots of the original system and we can do it as precisely as we want.

We will have then $\prod_{1 \leq i \leq n} d_i$ different (approximations of) vectors of power series that should be all the roots of the original system in $\overline{\mathbb{C}(T)}$ (it can be seen that the system we are dealing with has dimension zero when we think of T as a set of parameters and Bézout's theorem states that the number of solutions is bounded by $\prod_{1 \leq i \leq n} d_i$).

Suppose that, from every $\xi \in W$, we obtain the associated solution $\mathcal{R}_\xi \in \mathbb{C}[[T - t]]^n$ of the original system. Then

$$\prod_{\xi \in W} (Y - \ell(\mathcal{R}_\xi))$$

is a polynomial in $\mathbb{Q}[[T - t]][Y]$ that vanishes at every point $\mathcal{R}_\xi$. In fact, this polynomial is the polynomial of minimal degree defining the image of our original variety under the morphism

$$\overline{\mathbb{Q}(T)}^n \to \overline{\mathbb{Q}(T)}$$
$$w \mapsto \ell(w)$$

As our original variety is definable with polynomials in $\mathbb{Q}(T)[X]$, this polynomial we obtain must be in $\mathbb{Q}(T)[Y]$ and therefore, by multiplying it by a fixed polynomial $h \in \mathbb{Q}[T]$ we obtain a polynomial $M \in \mathbb{Q}[T][Y]$ satisfying the following:

$$M(T, \ell(X_1, \ldots, X_n)) \in (f_1, \ldots, f_n)$$

(here we are using that the ideal the polynomials $f_1, \ldots, f_n$ define is radical).

Therefore, given n polynomials in n variables defining a zero-dimensional variety V, provided the vector of their coefficients t_0 do not lie in a hypersurface, we can obtain by evaluating $M(T, Y)$ in t_0 a non-zero polynomial $M(t_0, Y) \in \mathbb{Q}[Y]$ which specialized in the linear form ℓ vanishes over the zeroes of V. This is a fundamental step we mentioned before (see Sections 6.4.3 and 6.4.4).

Of course a lot of work has to be done to succeed in finding this polynomial. For example one should know somehow up to what precision the Newton-Hensel algorithm is needed, how to compute the polynomial h, and so on, but this is just an example of how things work.

There are two main features to be taken into account when considering the Newton-Hensel algorithm. The first one is that an approximation of the power series vector up to a given precision can be obtained in very few steps (note that to obtain the series we are looking for up to degree θ we only have to apply $\log_2 \theta$ steps of our iteration). The second one is that the Newton-Hensel method deals essentially with slp's. In fact, an algorithmic statement of the Newton-Hensel method is the following lemma (see [GHH+97] for a proof):

Lemma 6.5.3. *Under the same hypotheses and notation of Lemma 6.5.1, suppose the polynomials $f_1, \ldots, f_n$ have degree bounded by d and are given by an slp of length L. Let $\kappa \in \mathbb{N}$, then there exists an slp of length $O(\kappa d^2 n^7 L)$ which evaluates polynomials $g_1^{(\kappa)}, \ldots, g_n^{(\kappa)}, h^{(\kappa)} \in \mathbb{Q}[T][X]$ with $h^{(\kappa)}(t, \xi) \neq 0$ which represent the numerators and the denominator of the rational functions obtained in the κ-th iteration of the Newton-Hensel operator.*

The Newton-Hensel method has been successfully used to obtain more efficient algorithms to solve polynomial equation systems. This tool has been introduced in this framework for the first time in [GHM+98], where an algorithm solving zero-dimensional systems was designed and an effective Nullstellensatz was stated. However, these procedures required computing with algebraic numbers. In [GHH+97], the first completely rational algorithm using the Newton-Hensel method was obtained and the complexity bounds were improved in [GLS01] and in [HMW01]. The Newton-Hensel method has been extensively applied to other problems: for example, to solve parametric systems (see [HKP+00] and [Sch03b]) and to obtain equidimensional decompositions of varieties (see [Lec00] and [JKSS04]). Some of these algorithms work under certain particular hypotheses while the others work for any given input probabilistically (see Section 6.6.1).

Moreover, in [Lec02] an extension of the Newton-Hensel operator adapted to the non-reduced case was presented. Then, this extension was applied to obtain an algorithm that computes the equidimensional decomposition of a variety (see [Lec03]).

All these algorithms share an important feature: they all use the Newton-Hensel operator, and therefore they can deal with input polynomials codified by means of slp's.

6.6 Other trends

In this last section, we would like to discuss briefly some ideas involved in algorithmic procedures which have been mentioned earlier.

6.6.1 Probabilistic algorithms

Sometimes our algorithms may depend on the choice of an object satisfying certain conditions (a linear form separating points, a point where a polynomial does not vanish, etc). These choices may be very expensive from the algorithmic point of view. Think of a polynomial f in n indeterminates of degree d. If we want to get a n-tuple v such that $f(v) \neq 0$ we have to check through many points. Sometimes, they may even involve a procedure we do not know how to accomplish (for example, we know we have to look for a point that is not a root of certain polynomial of bounded degree, but we cannot compute exactly the involved polynomial). To avoid this, one can choose a random point v to go on. Of course, this may lead to an error. Then, a probabilistic algorithm would be an algorithm that 'generally' performs the task we want accurately, but with a bounded probability of error.

Most algorithms involving slp's can be considered as probabilistic algorithms if we do not know an adequate correct test sequence for the kind of slp's involved. In this case, if we want to decide whether an slp represents the 0 polynomial or not, we just choose a random point and evaluate the slp in it. If the result is not zero, we are sure that the polynomial is not the zero polynomial but if it is zero, we can suppose that the polynomial is the zero one.

A clear example of this is the following (already mentioned in Section 6.4.3): We have a non-empty variety V and we want to compute its dimension. We cut it with a random hyperplane and consider what happens. Suppose that this intersection is empty. We would assume that the original variety is of dimension 0. It is generally the case, but if we are unlucky and the original variety was lying in a hyperplane parallel to the one we chose, our deduction would be false.

In most of the probabilistic algorithms we consider, the generic condition a random point should satisfy is that it is not a zero of a given polynomial $f \in \mathbb{Q}[X_1, \ldots, X_n]$ of bounded degree. The random point we choose has integer coordinates taken from a finite subset of $\mathbb{N}$ big enough. The estimation of the probability of success is done by means of the following well-known result (see [Sch80] and [Zip79]):

Lemma 6.6.1. *Let $R \subset \mathbb{N}$ be a finite subset. Let $f \in \mathbb{Q}[X_1, \ldots, X_n] - \{0\}$ be a polynomial. Then, for random choices of elements $a_1, \ldots, a_n \in R$, we have that*

$$\mathrm{Prob}(f(a_1, \ldots, a_n) = 0) \leq \frac{\deg f}{\#R}.$$

For example, some of the equidimensional decomposition algorithms already mentioned ([Lec00], [JS02], [JKSS04]) are probabilistic.

6.6.2 Non-general algorithms

In Section 6.4, we have mentioned that a possible way to avoid the high complexities involved in dense encoding was to design specific algorithms that would not work for every polynomial system but only for some of them. This is already being done, in the sense that some of the algorithms being produced in computer algebra may be general but work better (have lower complexity) in special cases. This leads us to consider other invariants (not only the degree, quantity and number of variables of the polynomials involved) to compute the complexity of the algorithms. Roughly speaking, the new invariants involved have to do somehow with the geometry of the varieties involved (that is the *semantic* features of the problem) and not with the way the variety is presented (the *syntactic* ones). For a further discussion on this topic see, for example [GHM$^+$98]. Many of the previously mentioned results deal with this new kind of invariants (see, for example, [GHH$^+$97], [KSS97], [HKP$^+$00], [Lec00] and [JKSS04]).

Acknowledgment

I would like to thank Alicia Dickenstein and Ioannis Emiris for inviting me to take part in the writing of this book. I would also like to thank Gabriela Jeronimo for her help in the writing of this chapter, and Mike Stillman for having read it thoroughly and for his comments and advice.

7

Toric resultants and applications to geometric modelling

Ioannis Z. Emiris

Department of Informatics & Telecommunications, National Kapodistrian University of Athens, Panepistimiopolis 15784 Greece, emiris@di.uoa.gr

Summary. Toric (or sparse) elimination theory uses combinatorial and discrete geometry to exploit the structure of a given system of algebraic equations. The basic objects are the Newton polytope of a polynomial, the Minkowski sum of a set of convex polytopes, and a mixed polyhedral subdivision of such a Minkowski sum. Different matrices expressing the toric resultant shall be discussed, and effective methods for their construction will be described based on discrete geometric operations, namely the subdivision-based methods and the incremental algorithm. The former allows us to produce Macaulay-type formulae of the toric resultant by determining a matrix minor that divides the determinant in order to yield the precise resultant. Toric resultant matrices exhibit a quasi-Toeplitz structure, which may reduce complexity by almost one order of magnitude in terms of matrix dimension.

We discuss perturbation methods to avoid the vanishing of the matrix determinant, or of the toric resultant itself, when the coefficients, which are initially viewed as generic, take specialized values. This is applied to the problem of implicitizing parametric (hyper)surfaces in the presence of base points. Another important application from geometric modelling concerns the prediction of the support of the implicit equation, based on toric elimination techniques.

Toric resultant matrices reduce the numeric approximation of all common roots of a polynomial system to a problem in numerical linear algebra. In addition to a survey of recent results, this chapter points to open questions regarding the theory and practice of toric elimination methods.

7.0 Introduction

Toric (or sparse) elimination theory uses combinatorial and discrete geometry to model the structure of a given system of algebraic equations. In particular, we consider algebraic equations with a specific monomial structure. It is thus possible to describe certain algebraic properties of the given system by combinatorial means. This chapter provides a comprehensive state-of-the-art introduction to the theory of toric elimination and toric resultants, paying special attention to the algorithmic and computational issues involved. Different matrices expressing the toric resultants shall be discussed, and effective methods for their construction will be defined based on discrete geometric operations, as well as linear algebra. Toric resultant matrices exhibit a structure close to that of Toeplitz matrices, which may reduce complexity by almost one order of magnitude. These matrices reduce the numeric approximation of all common roots to a problem in numerical linear algebra, as described in Section 7.5 and, in more depth, in Chapters 2 and 3. A relevant feature of resultant matrices in general, is their continuity with respect to small perturbations in the input coefficients.

Our goal is to exploit the fact that systems encountered in engineering applications are, more often than not, characterized by some structure. This claim shall be substantiated by examples in geometric modelling and computer-aided design as well as robotics; further applications exist in vision, and structural molecular biology (cf. [Emi97, EM99b]). A specific motivation comes from systems that must be repeatedly solved for different coefficients, in which case the resultant matrix can be computed exactly once. This occurs, for instance, in parallel robot calibration, see e.g. [DE01c], where 10,000 instances may have to be solved.

This chapter is organized as follows. The next section describes briefly the main steps in the theory of toric elimination, which aspires to generalize the results and algorithms of its mature counterpart, classical elimination. Section 7.2 presents the construction of toric resultant matrices of Sylvester-type. The following section offers a method for implicitizing parametric (hyper)surfaces, including the case of singular inputs, by means of perturbed toric resultants. Section 7.4 applies the tools of toric elimination for predicting the support of the implicit equation. The last section reduces solution of arbitrary algebraic systems to numerical linear algebra, thus yielding methods which avoid any issues of convergence.

This chapter will be of particular interest to graduate students and researchers in theoretical computer science or applied mathematics wishing to combine discrete and algebraic geometry. Some basic knowledge of discrete geometry for polyhedral objects in arbitrary dimension is assumed.

Previous work and open questions are mentioned in the corresponding sections. All algorithms discussed have been implemented either in Maple and/or in C, and are publicly available through the author's webpage. Most are also available in the Maple library MULTIRES or the C++ library SYNAPS, both accessible on the Internet[1].

7.1 Toric elimination theory

Toric elimination generalizes several results of classical elimination theory on multivariate polynomial systems of arbitrary degree by considering their structure. This leads to stronger algebraic and combinatorial results in general

---

[1] http://www-sop.inria.fr/galaad/logiciels/

[CLO98, GKZ94, Stu94a, Stu02]. Assume that the number of variables is n; roots in $(\overline{K}^*)^n$ are called *toric*, where $\overline{K}$ is the algebraic closure of the coefficient field. We use x^e to denote the monomial (or power product) $x_1^{e_1} \cdots x_n^{e_n}$, where $e = (e_1, \ldots, e_n) \in \mathbb{Z}^n$; note that we allow integer exponents. Let the input *Laurent polynomials* be

$$f_1, \ldots, f_n \in K[x_1^{\pm 1}, \ldots, x_n^{\pm 1}]. \tag{7.1}$$

Let the *support* $A_i = \{a_{i1}, \ldots, a_{im_i}\} \subset \mathbb{Z}^n$ denote the set of exponent vectors corresponding to monomials in f_i with nonzero coefficients:

$$f_i = \sum_{j=1}^{m_i} c_{ij} x^{a_{ij}}, \quad \text{for } c_{ij} \neq 0.$$

The *Newton polytope* $Q_i \subset \mathbb{R}^n$ of f_i is the convex hull of support A_i, in other words, the smallest convex polytope that includes all points in A_i. This is a bounded subset of $\mathbb{R}^n$, of dimension up to n. Newton polytopes provide a bridge from algebra to geometry since they permit certain algebraic problems to be cast in geometric terms. For background information and algorithms on polytope theory, the reader may refer to [Ewa96, Sch93]. For arbitrary sets A and $B \subset \mathbb{R}^n$, their *Minkowski sum* is

$$A + B = \{a + b \mid a \in A, b \in B\},$$

where $a + b$ represents the vector sum of points in $\mathbb{R}^n$. For convex polytopes A and B, $A + B$ is a convex polytope.

Definition 7.1.1. *Given convex polytopes* $A_1, \ldots, A_n, A_k' \subset \mathbb{R}^n$, *the* mixed volume $MV(A_1, \ldots, A_n)$ *is the unique real-valued non-negative function, invariant under permutations, such that,*

$$MV(A_1, \ldots, \mu A_k + \rho A_k', \ldots, A_n)$$

is equal to

$$\mu MV(A_1, \ldots, A_k, \ldots, A_n) + \rho MV(A_1, \ldots, A_k', \ldots, A_n),$$

for $\mu, \rho \in \mathbb{R}_{\geq 0}$. *Moreover, we set*

$$MV(A_1, \ldots, A_n) := n! \operatorname{Vol}(A_1), \quad \text{when } A_1 = \cdots = A_n,$$

where $\operatorname{Vol}(\cdot)$ *denotes euclidean volume in* $\mathbb{R}^n$.

If the polytopes have integer vertices, their mixed volume takes integer values. Two equivalent definitions are the following.

Proposition 7.1.2. *For* $\lambda_1, \ldots, \lambda_n \in \mathbb{R}_{\geq 0}$ *and for convex polytopes* $Q_1, \ldots, Q_n$ *lying in* $\mathbb{R}^n$, *the* mixed volume $MV(Q_1, \ldots, Q_n)$ *is precisely the coefficient of* $\lambda_1 \lambda_2 \cdots \lambda_n$ *in*

$$\text{Vol}(\lambda_1 Q_1 + \cdots + \lambda_n Q_n),$$

when the latter is expanded as a polynomial in $\lambda_1, \ldots, \lambda_n$. *Equivalently,*

$$MV(Q_1, \ldots, Q_n) = \sum_{I \subset \{1, \ldots, n\}} (-1)^{n-|I|} \, \text{Vol}\left(\sum_{i \in I} Q_i\right).$$

In the last equality, I ranges over all subsets of $\{1, \ldots, n\}$, so for $n = 2$ this gives $MV(Q_1, Q_2) = \text{Vol}(Q_1 + Q_2) - \text{Vol}(Q_1) - \text{Vol}(Q_2)$.

Exercise 7.1.3. Prove both formulae for the mixed volume from Proposition 7.1.2, in the case $n = 2$, using Definition 7.1.1. You may start by proving that $\text{Vol}(\lambda_1 Q_1 + \lambda_2 Q_2)$ lies in $\mathbb{Z}[\lambda_1, \lambda_2]$ and prove the first part of Proposition 7.1.2. Then prove the second part of the proposition for $n = 2$.

One may verify that mixed volume scales in the same way as the number of common roots of a well-constrained polynomial system with generic coefficients. In particular, when some Newton polytope is expressed as a Minkowski sum, this means that the corresponding polynomial equals the product of two polynomials $f_i f_i'$. So, the mixed volume can be written as a sum of mixed volumes, which corresponds to the fact that the generic number of common roots is given by a sum of root counts, each count corresponding to a system of polynomials including either f_i or f_i'.

Such properties were used by Kushnirenko in proving a restricted version of the following theorem, for the unmixed case [Kus75]. Then, Bernstein (also spelled Bernshteïn) stated, in [Ber75], the now-famous generalization, also known as the Bernstein-Kushnirenko-Khovanskii (BKK) bound. We are now ready to state a slight generalization of this theorem.

Theorem 7.1.4. *Given system (7.1), the cardinality of common isolated zeros in* $(\overline{K}^*)^n$, *counting multiplicities, is bounded by* $MV(Q_1, \ldots, Q_n)$, *regardless of the dimension of the variety. Equality holds when a certain subset of the coefficients corresponding to the vertices of the* Q_i's *are generic.*

Newton polytopes provide a "sparse" counterpart of total degree. The same holds for mixed volume vis-à-vis Bézout's bound, which is equal to the product of all total degrees. The two bounds coincide for completely dense polynomials, because each Newton polytope is an n-dimensional unit simplex scaled by $\deg f_i$. By definition, the mixed volume of the dense system is

$$MV((\deg f_1)S, \ldots, (\deg f_n)S) = \prod_{i=1}^{n} \deg f_i \; MV(S, \ldots, S) = \prod_{i=1}^{n} \deg f_i,$$

where S is the unit simplex in $\mathbb{R}^n$ with vertex set $\{(0,\ldots,0),(1,0,\ldots,0),\ldots,$ $(0,\ldots,0,1)\}$.

There is an intermediate bound between the classical Bézout bound and mixed volume. It is called the m-homogeneous or, simply, m-Bézout bound, and holds for multihomogeneous polynomials. Suppose that the n variables are partitioned into $r \geq 1$ sets of n_j variables each, for $j = 1,\ldots,r$. Then, $n_1 + \cdots + n_r = n$. We may assume that there is a homogenizing variable for each variable subset j such that polynomial f_i becomes homogeneous with respect to each subset, and has degree d_{ij} for $i = 1,\ldots,n$ and $j = 1,\ldots,r$. Then, the m-Bézout number is given by

$$\text{the coefficient of } \prod_{j=1}^{r} x_j^{n_j} \text{ in polynomial } \prod_{i=1}^{n} \left(\sum_{j=1}^{r} d_{ij} x_j \right).$$

This number lies always between the classical Bézout bound and the mixed volume. For a general discussion see [MSW95].

Exercise 7.1.5 (combinatorial). If all d_{ij} are equal to d_j then recover the classical Bézout's bound. Furthermore, show that the mixed volume of a system of multihomogeneous polynomials is given by the m-Bézout bound. For this, write every Newton polytope as $Q_i = \sum_j d_{ij} S_j$, where S_j is the unit simplex in n_j dimensions.

Mixed volume is usually significantly smaller than Bézout's bound for systems encountered in engineering applications. One example is the simple and generalized eigenproblems on $k \times k$ matrices. By adding an equation to ensure unit length of vectors, the Bézout bound in both cases is 2^{k+1}, whereas the number of right eigenvector and eigenvalue pairs is $2k$. This is precisely the mixed volume. We might, alternatively, employ the m-Bézout bound to the $k \times k$ system and obtain the exact count, namely k.

It is possible to generalize the notion of mixed volume to that of *stable mixed volume*, thus extending the bound to affine roots [HS97b].

The mixed volume computation is tantamount to enumerating all *mixed cells* in a mixed (tight coherent) subdivision of $Q_1 + \cdots + Q_n$. The term "decomposition" is also used in the literature, instead of "subdivision". We express the operation of Minkowski addition on n polytopes as a many-to-one function from $(\mathbb{R}^n)^n$ onto $\mathbb{R}^n$:

$$(Q_1,\ldots,Q_n) \to \sum_{i=1}^{n} Q_i \; : \; (p_1,\ldots,p_n) \mapsto \sum_{i=1}^{n} p_i.$$

To define an inverse function, i.e., a unique tuple for every point in the sum, *lifting* is a standard geometric method. Select n generic linear lifting forms $l_i : \mathbb{R}^n \to \mathbb{R}$, $i = 1,\ldots,n$. Then define the lifted polytopes

$$\widehat{Q}_i = \{(p_i, l_i(p_i)) : p_i \in Q_i\} \subset \mathbb{R}^{n+1}, \qquad i = 1,\ldots,n.$$

Now consider the Minkowski sum $\widehat{Q}_1 + \cdots + \widehat{Q}_n$, which is a convex polytope in $\mathbb{R}^{n+1}$. The lower hull of this Minkowski sum is an n-dimensional (convex) polyhedral complex, i.e. a family of convex faces of varying dimensions that includes all subfaces, such that the intersection of any two faces is itself a face of both intersecting faces. The lower hull is defined with respect to the unit vector along the x_{n+1}-axis: It is equal to the union of all n-dimensional faces, or facets, whose inner normal vector has positive last component.

Each facet of $\sum_{i=1}^n \widehat{Q}_i$ can be written itself as a Minkowski sum $\sum_{i=1}^n \widehat{F}_i$ where every $\widehat{F}_i$ is a face of $\widehat{Q}_i$, $i = 1, \ldots, n$. The genericity of the l_i ensures two things: First, that the lower hull projects bijectively onto the Minkowski sum $\sum_{i=1}^n Q_i$ of the original polytopes. Second, it guarantees tightness, which is the formal term for expressing the fact that every lower hull facet is a unique sum of faces $\widehat{F}_i$ so that $\sum_{i=1}^n \dim \widehat{F}_i$ equals the dimension of the facet, namely n. Note that for an arbitrary lifting we would have $\sum_{i=1}^n \dim \widehat{F}_i \geq n$, but tightness means that equality holds.

The subdivision of the lower hull into faces of dimensions from 0 to n induces a subdivision of the Minkowski sum $\sum_{i=1}^n Q_i$ into cells of respective dimensions. Such a subdivision is called regular and is defined by projecting each lower-hull face onto one cell. In particular facets, whose dimension is n, are projected onto n-dimensional (hence, maximal) cells. Furthermore, each (maximal) cell σ is expressed as the Minkowski sum of faces from the Q_i: Each Minkowski sum

$$\sigma = F_1 + \cdots + F_n$$

is unique, where each F_i is a face of Q_i, so that $\sum_i \dim F_i = \dim \sigma$. Each F_i corresponds to $\widehat{F}_i$ that appears in the unique sum defining the corresponding lower-hull facet that projects onto σ. This sum is said to be optimal since it minimizes the aggregate lifting function over the given cell.

The regularity of the subdivision implies its coherence, i.e., a continuous change of the optimal expressions of every cell σ as a sum of faces. This cell complex is, therefore, a tight coherent *mixed subdivision*. We define the *mixed cells* to be precisely those where all summand faces are one-dimensional.

Proposition 7.1.6. *The mixed volume equals the sum of the volumes of all mixed cells in the mixed subdivision.*

Example 7.1.7. Consider the system

$$f_1 = c_{10} + c_{11}x_1x_2 + c_{12}x_1^2x_2 + c_{13}x_1, \quad f_2 = c_{20} + c_{21}x_2 + c_{22}x_1x_2 + c_{23}x_1.$$

These polynomials have Newton polytopes and Minkowski sum as shown in Figure 7.1. The shown subdivision is achieved with $l_1 = -x_1 - 2x_2, l_2 = 4x_1 + x_2$.

It is clear that the mixed volume equals 3, which is the exact number of common roots for two generic polynomials with these supports. However, the system's Bézout number equals 4.

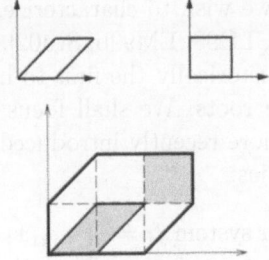

Fig. 7.1. The Newton polytopes and mixed subdivision in Example 7.1.7.

In the sequel, we shall see more examples of mixed subdivision. Some of the simplest instances appear in Examples 7.2.2 and 7.4.5.

Exercise 7.1.8. Compute the mixed volume of

$$A_1 = \{(0,0), (1,0), (2,0)\}, \ A_2 = \{(0,0), (0,1), (0,2)\}.$$

Can you find a linear lifting that yields a single mixed cell, so that the mixed volume equals the volume of a single cell?

In terms of complexity classes, the computation of mixed volume is #P-complete. This computation identifies the integer points comprising a monomial basis of the quotient ring of the ideal defined by the input polynomials. Mixed, or stable mixed, cells also correspond to start systems (of binomial equations, hence with an immediate solution) for a *toric homotopy* to the original system's roots. Such issues go beyond the scope of this chapter; see Chapter 8 or [GLW99, Li97, VG95].

7.1.1 The toric resultant

For a more general introduction to resultants, one may consult Sections 1.3 and 1.6 of Chapter 1, Section 2.3 of Chapter 2, or Chapter 3. The *resultant* of a polynomial system of $n + 1$ polynomials with indeterminate coefficients in n variables is a polynomial in these indeterminates, whose vanishing provides a necessary and sufficient condition for the existence of common roots of the system. Simple examples and a formal definition follow.

The resultant can be expressed by Poisson's formula, namely $C \prod_\alpha f_0(\alpha)$, where f_0 is one of the polynomials, evaluated at all common roots α of the other n equations, and C is a function of the coefficients of these n polynomials. It is then easy to see that the resultant is homogeneous in the coefficients of each polynomial.

The history of resultants (and elimination theory) includes such luminaries as Euler, Bézout, Cayley, and Macaulay. Different resultants exist depending on the space of the roots we wish to characterize, namely projective, affine, toric or residual [BEM01, CLO98, EM99b, Stu02]. Projective resultants (also known as classical) were historically the first to be studied and characterize the existence of projective roots. We shall focus on toric resultants below. Residual resultants were more recently introduced in order to study roots in the difference of two varieties.

Example 7.1.9. The bilinear system $f_i = c_{i0}+c_{i1}x_1+c_{i2}x_2+c_{i3}x_1x_2$, $i = 0, 1, 2$ is used in modelling a bilinear surface in $\mathbb{R}^3$ as the set of values $(f_0, f_1, f_2) \in \mathbb{R}^3$; see Figure 7.2.

Fig. 7.2. A bilinear surface patch.

The bivariate system of the f_i's has toric resultant equal to

$$\text{Res} = \det \begin{bmatrix} c_{00} & c_{01} & c_{02} & c_{03} & 0 & 0 \\ c_{10} & c_{11} & c_{12} & c_{13} & 0 & 0 \\ c_{20} & c_{21} & c_{22} & c_{23} & 0 & 0 \\ 0 & c_{00} & 0 & c_{02} & c_{01} & c_{03} \\ 0 & c_{10} & 0 & c_{12} & c_{11} & c_{13} \\ 0 & c_{20} & 0 & c_{22} & c_{21} & c_{23} \end{bmatrix},$$

assuming the matrix $[c_{ij}]_{i,j\geq 0}$ is regular. Notice that the first three matrix rows correspond to the input polynomials, whereas the last three rows correspond to the same polynomials multiplied by x_1. This determinant has degree 2 per polynomial, which is precisely the mixed volume of two input polynomials; remark that this is the generic number of roots. Hence the determinant equals the toric resultant.

In the following sections, we shall discuss ways to construct this matrix and, ultimately, the resultant. Two alternative ways are presented in Chapter 1.

If our only tool were the projective (classical) resultant, one would consider 3 bivariate polynomials, each of total degree 2. The resultant has degree 4 per polynomial, hence 12 in total in the c_{ij}'s. For the bilinear system, certain coefficients must be specialized to zero. One can show that the projective (classical) resultant vanishes identically in this case.

The simplest case, where the classical projective and toric resultants coincide, is that of a linear system of $n+1$ equations in n variables. The determinant of the coefficient matrix is the system's resultant and, under the assumption on the non-vanishing of certain minors, it becomes zero exactly when there is a common root. Due to the linearity of the equations, this root is then unique.

Exercise 7.1.10. Using linear algebra, prove that the resultant of a linear system vanishes precisely when there exists a unique common root, provided that certain minors are nonzero. Moreover, apply Cramer's rule in order to compute each coordinate of this root as a ratio of determinants.

The question of whether two polynomials $f_1(x), f_2(x) \in K[x]$ have a common root leads to a condition that has to be satisfied by the coefficients of both polynomials; again classical and toric resultants coincide. The system's *Sylvester matrix* is of dimension $\deg f_1 + \deg f_2$ and its determinant is the system's resultant, provided the leading coefficients are nonzero. This matrix rows contain the coefficient vectors of polynomials $x^k f_j$, for $k = 0, \ldots, \deg f_i - 1$ and $\{i, j\} = \{1, 2\}$.

Bézout developed a method for computing the resultant as a determinant of a matrix of dimension equal to $\max\{\deg f_1, \deg f_2\}$. Its construction goes beyond the scope of this chapter; the reader may refer to Chapters 1 and 3.

For an illustration, consider $f_1 = a_{d_1} x^{d_1} + \cdots + a_0, f_2 = b_{d_2} x^{d_2} + \cdots + b_0$, with all coefficients nonzero. Their resultant is the determinant of the Sylvester matrix, namely

$$\begin{bmatrix} a_{d_1} & a_{d_1-1} & \cdots & a_0 & 0 & \cdots & & 0 \\ 0 & a_{d_1} & a_{d_1-1} & \cdots & a_0 & 0 & \cdots & 0 \\ \vdots & & \ddots & & & & \ddots & \\ 0 & & & a_{d_1} & a_{d_1-1} & \cdots & & a_0 \\ b_{d_2} & b_{d_2-1} & \cdots & & b_0 & 0 & \cdots & 0 \\ 0 & b_{d_2} & b_{d_2-1} & \cdots & & b_0 & 0 & 0 \\ \vdots & & \ddots & & & & \ddots & \\ 0 & & & b_{d_2} & b_{d_2-1} & \cdots & & b_0 \end{bmatrix}.$$

The interested reader may refer to Section 1.3 of Chapter 1 for a more detailed discussion on resultants of univariate polynomials.

Exercise 7.1.11. Using the greatest common divisor of f_1, f_2 prove that the resultant of these two polynomials vanishes precisely when they have a common root. Can you compute the coordinates of this root from the kernel vectors of the Sylvester matrix?

Toric resultants express the existence of toric roots. Formally,

$$f_0, \ldots, f_n \in K[x_1^{\pm 1}, \ldots, x_n^{\pm 1}], \tag{7.2}$$

f_i corresponding to generic point $c_i = (c_{i1}, \ldots, c_{im_i})$ in the space of polynomials with support A_i. This space is identified with projective space $\mathbb{P}_K^{m_i - 1}$. Then system (7.2) can be thought of as point $c = (c_0, \ldots, c_n)$. Let Z denote the Zariski closure, in the product of projective spaces, of the set of all c such that the system has a solution in $(\overline{K}^*)^n$. Note that Z is an irreducible variety.

A technical assumption is that, without loss of generality, the affine lattice generated by $\sum_{i=1}^{n+1} A_i$ is n-dimensional. This lattice is identified with $\mathbb{Z}^n$ possibly after a change of variables, which can be implemented by computing the appropriate Smith's Normal form [Stu94a].

Definition 7.1.12. *The toric (or sparse) resultant* Res $=$ Res$(A_0, \ldots, A_n)$ *of system (7.2) is a polynomial in* $\mathbb{Z}[c]$. *If* $codim(Z) = 1$ *then* Res *is the defining irreducible polynomial of the hypersurface* Z. *If* $codim(Z) > 1$ *then* Res $= 1$.

An additional assumption we make is that the family $A_0, \ldots, A_n$ is *essential*. This means that, for every proper index subset $I \subset \{0, \ldots, n\}$ with cardinality $|I|$, the following holds for the dimension of certain Minkowski sums:

$$\dim \left(\sum_{i \in I} A_i \right) \geq |I|.$$

Essential support families are also discussed in Section 1.6 of Chapter 1.

Then, the toric resultant Res$(A_0, \ldots, A_n)$ is homogeneous in the coefficients of f_i with $\deg_{f_i}$ Res$(A_i) = MV_{-i}$. The vanishing of Res$(A_0, \ldots, A_n)$ is a necessary and sufficient condition for the existence of roots in the projective toric variety X, corresponding to the Minkowski sum of the $n+1$ Newton polytopes. A projective toric variety is the closure of the image of the following map of the torus:

$$(\mathbb{C}^*)^n \to \mathbb{P}^m : t \mapsto \left(t^{b_0} : \cdots : t^{b_m} \right),$$

where the $b_i \in \mathbb{Z}^n$ are the vertices of the Minkowski sum. If all Newton polytopes are identical, then these are simply the vertices of the Newton polytope. For instance, when this polytope is the unit simplex, the toric resultant coincides with $\mathbb{P}^n$. In the case of bilinear systems (see Example 7.1.9), $X = \mathbb{P}^1 \times \mathbb{P}^1$. Toric varieties are also discussed in Chapter 3 as well as in [Cox95, GKZ94, KSZ92].

Some fundamental properties of the toric resultant are as follows.

- The toric resultant subsumes the classical resultant in the sense that they coincide if the polynomials are dense.
- Just as in the classical case, when all coefficients are generic, the resultant is irreducible.

- While the classical resultant is invariant under linear transformations of the variables, the toric resultant is invariant under transformations that preserve the polynomial support.
- In the case of non-generic coefficients, certain divisibility properties hold. In particular, when a system of polynomials lies in the ideal generated by another system, then the latter resultant is divisible by the former resultant.

7.2 Matrix formulae

Different means of expressing each resultant are possible, distinguished into Sylvester, Bézout and hybrid-type formulae [BEM01, CLO98, DE03, EM99b, Stu02]. Ideally, we wish to express it as a matrix determinant, a quotient of two determinants, or a divisor of a determinant where the quotient is a nontrivial extraneous factor. This section discusses matrix formulae for the toric resultant known as *toric resultant matrices*.

We restrict ourselves to Sylvester-type matrices; such matrices for the toric resultant are also known as *Newton matrices* because they depend on the input Newton polytopes. Sylvester-type matrices generalize the coefficient matrix of a linear system and Macaulay's matrix. The latter extends Sylvester's construction to arbitrary systems of homogeneous polynomials, and its determinant is a nontrivial multiple of the projective resultant. Other types of resultant matrices are discussed in Chapter 3.

The transpose of a Sylvester-type matrix corresponds to the following linear transformation:

$$(g_0, \ldots, g_n) \quad \mapsto \quad \sum_{i=0}^{n} g_i f_i, \tag{7.3}$$

where the support of each polynomial g_i is related to the matrix. If we expressed the g_i's in the monomial basis, then $(g_0, \ldots, g_n)$ would be a vector that multiplies from the left the transposed matrix (or from the right, the resultant matrix itself). The support of each g_i is the set of monomials multiplying f_i in order to define the rows that correspond to f_i. These rows contain shifted copies of the f_i coefficients. The shift is performed in such a way so as to obtain $g_i f_i$ as the product of g_i-block of the vector, multiplied by the block of rows corresponding to f_i. The reader should consult the examples of resultant matrices given above as well as in the sequel.

Overall, each row expresses the product of a monomial with an input polynomial; its entries are coefficients of that product, each corresponding to the monomial indexing the corresponding column. The degree of $\det M$ in the coefficients of f_i equals the number of rows with coefficients of f_i. This must be greater than or equal to $\deg_{f_i} \text{Res}$. It is possible to pick any one polynomial so that there is an optimal number of rows containing its coefficients; this number is obviously $\deg_{f_i}$ Res. This is true both in the case of Macaulay's matrix and in the case of the Newton matrix constructions below.

7.2.1 Subdivision-based construction

There are two main approaches to construct a well-defined, square, generically nonsingular matrix M, such that Res $|\det M$. The second algorithm is incremental and shall be presented later. The first approach (cf. [CE93, CE00, CP93, Stu94a]), relies on a mixed (tight coherent) subdivision of the Minkowski sum

$$Q = Q_0 + \cdots + Q_n,$$

which generalizes the discussion of Section 7.1. It uses $n+1$ generic linear lifting forms $l_i : \mathbb{R}^n \to \mathbb{R}$ to define the lifted polytopes. Maximal cells in the subdivision are written uniquely as $\sigma = F_0 + \cdots + F_n$, where $F_i \subset Q_i$ and $\sum_i \dim F_i = n$. Therefore, at least one face is a vertex. The *mixed cells* are precisely those where all other summand faces are one-dimensional. If this is a vertex from Q_i, then the cell is said to be i-mixed.

It can been shown [Emi96] that the i-mixed cells are the same as the mixed cells in the mixed subdivision the n Newton polytopes $Q_0, \ldots, Q_{i-1}, Q_{i+1}, \ldots, Q_n$, provided that we use the same lifting functions in both cases. A direct consequence is that the mixed volume of $f_0, \ldots, f_{i-1}, f_{i+1}, \ldots, f_n$ is given by the sum of volumes of all i-mixed cells, thus extending Proposition 7.1.6.

The matrix construction algorithm uses a subset of $(Q + \delta) \cap \mathbb{Z}^n$ to index the rows and columns of resultant matrix M, where $\delta \in \mathbb{R}^n$ is an arbitrarily small and *sufficiently generic* vector. This vector must perturb all integer points indexing some row (or column) of the matrix in the strict interior of a maximal cell. It can be chosen randomly and the validity of our choice can be confirmed by the matrix construction algorithm. The probability of error for a vector with uniformly distributed entries is bounded in [CE00].

Now consider an integer point p, such that $p+\delta$ lies in an arbitrary maximal cell σ. The algorithm associates to p the pair (i,j) if and only if $a_{ij} \in Q_i$ is a vertex in the optimal sum of σ and i is the maximum index of any vertex summand. The row of M corresponding to p shall contain the coefficients of polynomial

$$x^{p-a_{ij}} f_i.$$

The entries corresponding to column monomials that do not explicitly appear in the row polynomial are set to zero. If σ is i-mixed, then a_{ij} is the unique vertex summand. For non-mixed cells, the Minkowski sum has more than one vertices, and the above rule defines a matrix with the minimum number of rows with f_0, because in these cases it shall avoid the 0 index.

Therefore, the number of f_0 rows equals the number of integer points in 0-mixed cells, which equals

$$\mathrm{MV}(f_1, \ldots, f_n) = \deg_{f_0} \mathrm{Res}(A_i).$$

As for the number of f_i rows, for $i > 0$, this is larger or equal to the number of integer points in i-mixed cells. The above argument tells us that this is at least as large as $\deg_{f_i}$ Res. Now recall that the degree of the matrix determinant in the coefficients of f_i equals the number of its rows containing shifted copies of the coefficient vector of f_i. The algorithm may use an analogous rule to avoid index i if we wish the matrix to have the minimum number of rows containing f_i, for $i > 0$.

It can be proven that every principal minor of matrix M, including its determinant, is nonzero when the polynomials have generic coefficients [CE00]. The proof of this theorem uses an adequate specialization of the input coefficients, in terms of a new parameter t. In particular, the coefficient in f_i that multiplies the monomial x^{a_j} is specialized to $t^{l_i(a_j)}$, where l_i is the lifting applied to Q_i. Then, each row of the specialized matrix, indexed by some point p, is multiplied by the power $t^{h-l_k(a_s)}$. Here, h denotes the vertical distance of $p \in \mathbb{R}^n$ to the lower hull of $\sum_{i \geq 0} \widehat{Q}_i$ and we have assumed that p has been associated to the pair (k, s). The last step in the proof establishes that the product of all diagonal entries in the new matrix equals the trailing term of its determinant with respect to t.

Moreover, it is not so hard to show that the determinant of M vanishes whenever Res $= 0$. We thus arrive at the following theorem.

Theorem 7.2.1 ([CE93, CE00]). *We are given an overconstrained system with fixed supports. With the above notation, matrix M is well-defined and square. Its determinant is generically nonzero and divisible by the toric resultant Res.*

Example 7.2.2. Let us apply the subdivision-based algorithm to construct Sylvester's matrix. Take

$$f_0 = c_{00} + c_{01}x, \quad f_1 = c_{10} + c_{11}x + c_{12}x^2.$$

There are two possible subdivisions obtained with linear liftings; one is shown in Figure 7.3, along with the δ perturbation.

For illustration, we note that the algorithm associates to point 2 the pair $(1, 2)$, i.e. the matrix row indexed by x^2 shall contain the coefficients of $x^{2-2}f_1 = f_1$. A similar argument builds the other rows of the matrix. The reader may check that this is indeed the well-known Sylvester matrix.

Example 7.2.3. For $n = 2$, let us apply the subdivision-based algorithm in the case of linear polynomials. Take

$$f_i = c_{i0} + c_{i1}x_1 + c_{i2}x_2, \ i = 0, 1, 2.$$

One possible linear lifting induces the subdivision in Figure 7.4. The same figure shows the perturbation of choice, so that we recover the matrix of the system's coefficients, as expected. In fact, any vector $\delta \in \mathbb{R}_{>0}$ would do.

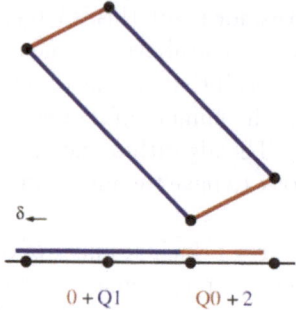

Fig. 7.3. The Minkowski sum of the lifted Newton segments and the induced subdivision in Example 7.2.2.

Then, there are three integer points in the perturbed Minkowski sum, namely $(1,2), (1,1)$, and $(2,1)$. They are associated, respectively, to pairs $[2, (0,1)]$, $[1, (0,0)]$ and $[0, (1,0)]$. For instance, the row indexed by $x_1 x_2^2$ shall contain polynomial $x^{(1,2)-(0,1)} f_2 = x^{(1,1)} f_2$.

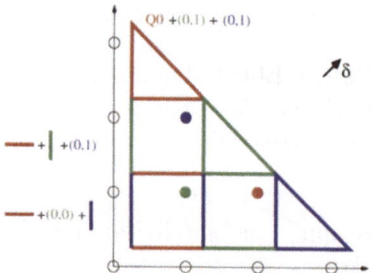

Fig. 7.4. The mixed subdivision and the perturbation with respect to the original Minkowski sum.

The resultant matrix is therefore

$$M = \begin{bmatrix} c_{01} & c_{02} & c_{03} \\ c_{11} & c_{12} & c_{13} \\ c_{21} & c_{22} & c_{23} \end{bmatrix},$$

with rows corresponding to the polynomials $x_1 x_2 f_i$ and columns indexed by $x_1^2 x_2$, $x_1 x_2^2$, $x_1 x_2$.

There is a greedy variant from [CP93] of the subdivision-based algorithm. It starts with a single row, corresponding to some integer point, and proceeds iteratively by adding new rows (and columns) as need be. For a given set of rows, the column set comprises all columns required to express the row polynomials. For a given set of columns, the rows are updated to correspond to the same set. The algorithm continues by adding rows and the corresponding columns until a square matrix has been obtained.

Example 7.2.4. Consider a system of 3 polynomials in 2 unknowns:

$$f_0 = c_{01} + c_{02}xy + c_{03}x^2y + c_{04}x,$$
$$f_1 = c_{11}y + c_{12}x^2y^2 + c_{13}x^2y + c_{14}x,$$
$$f_2 = c_{21} + c_{22}y + c_{23}xy + c_{24}x.$$

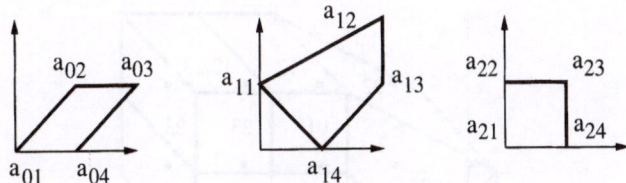

Fig. 7.5. The supports and Newton polytopes in Example 7.2.4.

The Newton polytopes are shown in Figure 7.5. The mixed volumes are $MV(Q_0, Q_1) = 4$, $MV(Q_1, Q_2) = 4$, $MV(Q_2, Q_0) = 3$, so the toric resultant's total degree is 11. Compare this with the Bézout numbers of these subsystems: $8, 6, 12$; hence the projective resultant's total degree is 26.

Assume that the lifting functions are $l_0(x, y) = Lx + L^2y$, $l_1(x, y) = -L^2x - y$, $l_2(x, y) = x - Ly$, where $L \gg 1$. The lifted Newton polytopes and the lower hull of their Minkowski sum is shown below. These functions are sufficiently generic since they define a mixed subdivision where every cell is uniquely defined as the Minkowski sum of faces $F_i \subset Q_i$.

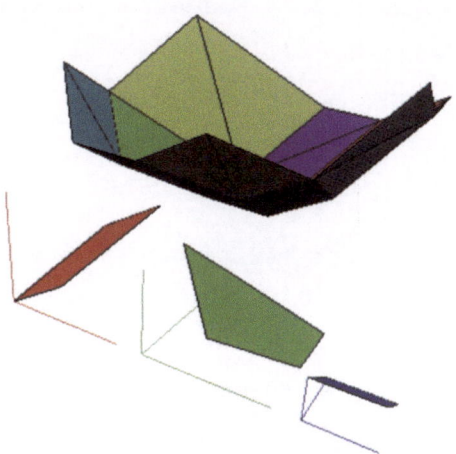

The lower hull of the Minkowski sum of the lifted Q_i's is then projected to the plane, yielding generically a mixed subdivision of Q. Figure 7.6 shows $Q+\delta$ and the integer points it contains; notice that every point belongs to a unique maximal cell. Every maximal cell σ is labeled by the indices of the Q_i vertex or vertices appearing in the unique Minkowski sum $\sigma = F_0 + \cdots + F_n$, with ij denoting vertex $a_{ij} \in Q_i$. For instance, point $(1,0)$ belongs to a maximal cell $\sigma = a_{01} + F + F'$, where F, F' are the edges $(a_{14}, a_{13}) \subset Q_1$ and (a_{21}, a_{24}) respectively. The corresponding row in the matrix will be filled in with the coefficient vector of $x^{(1,0)} f_0$.

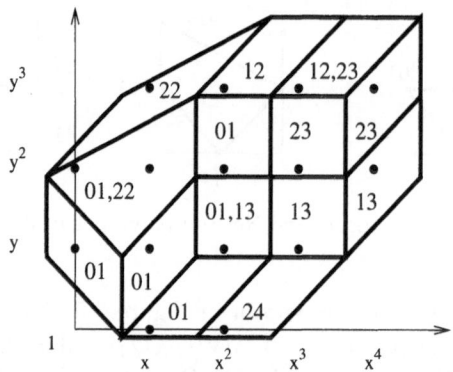

Fig. 7.6. A mixed subdivision of Q perturbed by $(-3/8, -1/8)$, in Example 7.2.4.

The Newton matrix M appears below with rows and columns indexed by the integer points in the perturbed Minkowski sum. M contains, by construction, the minimum number of f_0 rows, namely 4. The total number of rows is $4 + 4 + 7 = 15$, i.e., the determinant degree is higher than optimal by 1 and 3, respectively, in the coefficients of f_1 and f_2.

	1,0	2,0	0,1	1,1	2,1	3,1	0,2	1,2	2,2	3,2	4,2	1,3	2,3	3,3	4,3
1,0	c_{01}	c_{04}	0	0	c_{02}	c_{03}	0	0	0	0	0	0	0	0	0
2,0	c_{21}	c_{24}	0	c_{22}	c_{23}	0	0	0	0	0	0	0	0	0	0
0,1	0	0	c_{01}	c_{04}	0	0	0	c_{02}	c_{03}	0	0	0	0	0	0
1,1	0	0	0	c_{01}	c_{04}	0	0	0	c_{02}	c_{03}	0	0	0	0	0
2,1	c_{14}	0	c_{11}	0	c_{13}	0	0	0	c_{12}	0	0	0	0	0	0
3,1	0	c_{14}	0	c_{11}	0	c_{13}	0	0	0	c_{12}	0	0	0	0	0
0,2	0	0	c_{21}	c_{24}	0	0	c_{22}	c_{23}	0	0	0	0	0	0	0
1,2	0	0	0	c_{21}	c_{24}	0	0	c_{22}	c_{23}	0	0	0	0	0	0
2,2	0	0	0	0	0	0	0	0	c_{01}	c_{04}	0	0	0	c_{02}	c_{03}
3,2	0	0	0	0	c_{21}	c_{24}	0	0	c_{22}	c_{23}	0	0	0	0	0
4,2	0	0	0	0	0	c_{14}	0	0	c_{11}	0	c_{13}	0	0	0	c_{12}
1,3	0	0	0	0	0	0	0	c_{21}	c_{24}	0	0	c_{22}	c_{23}	0	0
2,3	0	0	0	c_{14}	0	0	c_{11}	0	c_{13}	0	0	0	c_{12}	0	0
3,3	0	0	0	0	0	0	0	0	c_{21}	c_{24}	0	0	c_{22}	c_{23}	0
4,3	0	0	0	0	0	0	0	0	0	c_{21}	c_{24}	0	0	c_{22}	c_{23}

The greedy version produces a matrix with dimension 14 which can be obtained by deleting the row and the column corresponding to point $(1, 3)$.

The subdivision-based approach can be coupled with the existence of a minor in the Newton matrix that divides the determinant so as to yield the exact toric resultant [D'A02]. D'Andrea has proposed a recursive lifting procedure that gives a much lower value to a chosen vertex of Q_0. The cells whose optimal sum does not contain this vertex are then further subdivided by assigning this special role to a vertex of Q_1, and so on. This generalizes Macaulay's famous quotient formula that yields the exact projective resultant [Mac02].

The existence of a non-recursive algorithm, relying on a single lifting, is still open in the general case. It is, nonetheless, possible for $n = 2$ and for families of sufficiently different Newton polytopes. A glimpse of what this lifting may look like is offered by the hybrid matrix constructed in [DE01b].

Example 7.2.5 (Continued from Example 7.1.9). The bilinear system $f_i = c_{i0} + c_{i1}x_1 + c_{i2}x_2 + c_{i3}x_1x_2$, $i = 0, 1, 2$, despite its apparent simplicity, does not admit an optimal toric resultant matrix, when we apply the subdivision-based algorithm. In contrast, the greedy variant may yield an optimal matrix and the incremental algorithm of the next section produces the optimal 6×6 matrix in Example 7.1.9. It is possible to construct the following 9×9 numerator matrix, using the subdivision-based algorithm:

$$
M = \begin{bmatrix}
c_{00} & c_{01} & c_{02} & c_{03} & 0 & 0 & 0 & 0 & 0 \\
c_{10} & c_{11} & c_{12} & c_{13} & 0 & 0 & 0 & 0 & 0 \\
c_{20} & c_{21} & c_{22} & c_{23} & 0 & 0 & 0 & 0 & 0 \\
0 & 0 & 0 & c_{00} & c_{01} & c_{02} & 0 & 0 & c_{03} \\
0 & c_{10} & 0 & c_{12} & c_{13} & 0 & c_{11} & 0 & 0 \\
0 & 0 & c_{20} & c_{21} & 0 & c_{23} & 0 & c_{22} & 0 \\
0 & c_{20} & 0 & c_{22} & c_{23} & 0 & c_{21} & 0 & 0 \\
0 & 0 & c_{10} & c_{11} & 0 & c_{13} & 0 & c_{12} & 0 \\
0 & 0 & 0 & c_{10} & c_{11} & c_{12} & 0 & 0 & c_{13}
\end{bmatrix}
\begin{matrix}
f_0 \\
f_1 \\
f_2 \\
x_1 x_2 f_0 \\
x_1 f_1 \\
x_2 f_2 \\
x_1 f_2 \\
x_2 f_1 \\
x_1 x_2 f_1
\end{matrix}
$$

The choice was $\delta = (\frac{2}{3}, \frac{1}{2})$ and the lifting is such that one vertex of the first polytope has an infinitesimal lifting value compared to the other values. It is now possible to define a denominator matrix M', of dimension 3, which is a submatrix of M. It is defined by the rows indexed by polynomials $f_1, f_2, x_1 x_2 f_1$ and the respective columns; these correspond precisely to the integer points in non-mixed cells. The ratio of the determinants yields precisely the toric resultant.

7.2.2 Incremental construction

The second algorithm [EC95], is *incremental* and yields usually smaller matrices and, in any case, no larger than those of the subdivision algorithm. The flexibility of the construction makes it suitable for overconstrained systems. On the downside, there exists a randomized step so certain properties of the subdivision-based construction cannot be guaranteed *a priori*.

The selection of integer points, which correspond to monomials multiplying the row polynomials, uses a vector $v \in (\mathbb{Q}^*)^n$. The goal is to choose an adequate subset of integer points in

$$
Q_{-i} := \sum_{j=0, j \neq i}^{n} Q_j, \ i = 0, \dots, n.
$$

This is achieved by first sorting all points $p \in Q_{-i} \cap \mathbb{Z}^n$ according to their distance, along v, from the boundary. This distance is defined as follows, for point p:

$$
v\text{-distance}(p) := \max\{s \in \mathbb{R}_{\geq 0} : p + sv \in Q_{-i}\}.
$$

The construction is incremental, in the sense that successively larger point sets are considered by decreasing the lower bound on the v-distance of the set's points. For given point sets, a candidate matrix is defined. If the number of rows is at least as large as the number of columns and it has full rank for generic coefficients, then the algorithm terminates and returns a nonsingular maximal square submatrix. The determinant of this submatrix is a nontrivial multiple of the toric resultant; otherwise, new rows (and columns) are added to the candidate.

In those cases where a minimum matrix of Sylvester type provably exists [SZ94, WZ94], the incremental algorithm produces this matrix. For general multi-homogeneous systems, the best vector is obtained in [DE03]. These are precisely the systems for which v can be deterministically specified; otherwise, a random v can be used. Different choices can be tried out so that the smallest matrix may be chosen.

Example 7.2.6 (Continued from Example 7.2.4). Figure 7.7 shows Q_{-0} in bold and randomly chosen vector $v = (20, 11)$. The different point subsets in Q_{-0} with respect to v-distance are shown by the thin-line polygons. In fact, the thin lines represent contours of fixed v-distance. The final point set from Q_{-0} is the following, shown with the respective v-distances:

$$\{(0, 1; 3/20), (1, 0; 1/10), (1, 1; 1/10), (1, 2; 1/11)\}.$$

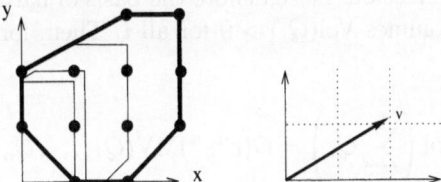

Fig. 7.7. Q_{-0} subsets with different v-distance bounds and vector v.

This v leads to a 13×12 nonsingular matrix M shown below. Deleting the last row defines the 12×12 resultant submatrix.

	1,2	2,2	0,1	1,1	2,1	3,1	1,0	2,0	3,2	2,3	3,3	0,2
0,1	c_{02}	c_{03}	c_{01}	c_{04}	0	0	0	0	0	0	0	0
1,0	0	0	0	0	c_{02}	c_{03}	c_{01}	c_{04}	0	0	0	0
1,1	0	c_{02}	0	c_{01}	c_{04}	0	0	0	c_{03}	0	0	0
1,2	c_{01}	c_{04}	0	0	0	0	0	0	0	c_{02}	c_{03}	0
0,0	0	c_{12}	c_{11}	0	c_{13}	0	c_{14}	0	0	0	0	0
1,0	0	0	0	c_{11}	0	c_{13}	0	c_{14}	c_{12}	0	0	0
1,1	c_{11}	0	0	0	c_{14}	0	0	0	c_{13}	0	c_{12}	0
0,1	0	c_{13}	0	c_{14}	0	0	0	0	0	c_{12}	0	c_{11}
0,1	c_{23}	0	c_{21}	c_{24}	0	0	0	0	0	0	0	c_{22}
1,1	c_{22}	c_{23}	0	c_{21}	c_{24}	0	0	0	0	0	0	0
1,0	0	0	0	c_{22}	c_{23}	0	c_{21}	c_{24}	0	0	0	0
2,1	0	c_{22}	0	0	c_{21}	c_{24}	0	0	c_{23}	0	0	0
2,2	0	c_{21}	0	0	0	0	0	0	c_{24}	c_{22}	c_{23}	0

Other techniques to reduce matrix size (and mixed volumes) include the introduction of new variables to express subexpressions which are common to several input polynomials. For an illustration, see [Emi97].

Clearly, mixed volume captures the inherent complexity of algebraic problems in the context of sparse elimination and thus provides lower bounds on the complexity of algorithms. On the other hand, several toric elimination algorithms rely on Minkowski sums of Newton polytopes. Therefore, a crucial question in deriving output-sensitive upper bounds is the relation between mixed volume and the volume of these Minkowski sums. In manipulating mixed volumes, some fundamental results can be found in [Sch93]. In particular, the Aleksandrov-Fenchel inequality leads to the following bound [Emi96, Lut86]:

$$\mathrm{MV}^n(Q_1, \ldots, Q_n) \geq (n!)^n \mathrm{Vol}(Q_1) \cdots \mathrm{Vol}(Q_n).$$

For a system of Newton polytopes Q_i, define its *scaling factor* s to be the minimum real value so that $Q_i + t_i \subset s\, Q_\mu$ for all Q_i, where Q_μ is the polytope of minimum euclidean volume and the $t_i \in \mathbb{R}^n$ are arbitrary translation vectors. Clearly, $s \geq 1$ and s is finite if and only if all polytopes have an affine span of the same dimension. Let e denote the basis of natural logarithms, and suppose that the volumes $\mathrm{Vol}(Q_i) > 0$ for all i. Then, for a well-constrained system, we have

$$\mathrm{Vol}\left(\sum_{i=1}^{n} Q_i\right) = O(e^n s^n)\mathrm{MV}(Q_1, \ldots, Q_n),$$

whereas for an overconstrained system the same techniques yield

$$\mathrm{Vol}\left(\sum_{i=0}^{n} Q_i\right) = O\left(\frac{e^n s^n}{n}\right) \sum_{i=0}^{n} \mathrm{MV}_{-i},$$

where $\mathrm{MV}_{-i} = \mathrm{MV}(Q_0, \ldots, Q_{i-1}, Q_{i+1}, \ldots, Q_n)$ [Emi96].

As a consequence, the asymptotic bit complexity of both subdivision-based and incremental algorithms is singly exponential in n, proportional to the total degree of the toric resultant, and polynomial in the number of Q_i vertices, provided all $\mathrm{MV}_{-i} > 0$.

Newton matrices, including the candidates constructed by the incremental algorithm, are characterized by a structure that generalizes the Toeplitz structure and has been called *quasi-Toeplitz* [EP02] (cf. [CKL89]). By exploiting this structure, determinant evaluation has quasi-quadratic arithmetic complexity and quasi-linear space complexity in the matrix dimension (here "quasi" means that polylogarithmic factors are ignored). The efficient implementation of this structure is open today and is important for the competitiveness of the entire approach.

7.3 Implicitization with base points

The problem of switching from a rational parametric representation to an implicit, or algebraic, representation of a curve, surface, or hypersurface lies at the heart of several algorithms in computer-aided design and geometric modelling. Given are rational parametric expressions

$$x_i = p_i(t)/q(t) \in K(t) = K(t_1, \ldots, t_n), \quad i = 0, \ldots, n,$$

over some field K of characteristic zero. The implicitization problem consists in computing the smallest algebraic hypersurface in terms of $x = (x_0, \ldots, x_n)$ containing the closure of the image of the parametric map $t \mapsto x$. The most common case is for curve and surface implicitization, namely when $n = 1$ and $n = 2$ respectively. Resultants offer an efficient approach for this problem, but face certain questions due to degeneracy conditions, discussed below. Several other algorithms exist for this problem, including methods based on Gröbner bases, moving surfaces, and residues. Their enumeration goes beyond the scope of this chapter; cf. also, Chapter 3.

Implicitization is equivalent to eliminating all parameters t from the polynomial system

$$f_i(t) = p_i(t) - x_i q(t), \ i = 0, \ldots, n,$$

regarded as polynomials in t. The resultant is well-defined for this system, and shall be a polynomial in x, equal to the implicit expression, provided that it does not vanish and the parametrization is generically one-to-one. Otherwise, the resultant is a power of the implicit equation. More subtle is the case where the resultant is identically zero. This happens precisely when there exist values of t, known as *base points*, for which the f_i vanish for all x_i; in other words, the $p_i(t)$ and $q(t)$ evaluate to zero. Base points forming a component of codimension 1 can be easily removed by canceling common factors in the numerator and denominator of the rational expressions for the x_j's. But higher codimension presents a harder problem.

Besides cases where the (toric) resultant vanishes, another problem with non-generic coefficients is that the resultant matrix may be identically singular. We understand that avoiding degeneracies is an important problem, whose relevance extends beyond the question of implicitization with base points. In [DE01a], a toric (sparse) projection operator is defined by perturbing the subdivision-based matrix such that, after specialization, this operator is not identically zero but vanishes on roots in the proper components of the variety, including all isolated roots.

This is a standard idea in handling degeneracies in the case of resultants. In the classical context, Canny [Can90] perturbed each f_i by adding a new factor $\epsilon x_i^{d_i}$, where $i = 1, \ldots, n$, and f_0 by adding ϵ, where ϵ is a positive infinitesimal indeterminate. Rojas proposed a perturbation scheme for toric resultants in [Roj99a] which yields a perturbed resultant of low degree in ϵ but is, nonetheless, rather expensive to compute. Our scheme generalizes [Can90] and requires virtually no extra computation besides the matrix construction.

Suppose we have a family $p := (p_0(x) \ldots, p_n(x))$ of Laurent polynomials such that $\text{supp}(p_i) \subset A_i$, and $\text{Res}(p_0, \ldots, p_n) \neq 0$. The Toric Generalized Characteristic Polynomial (p-GCP) is

$$C_p(\epsilon) := \mathrm{Res}\,(f_0 - \epsilon p_0, \ldots, f_n - \epsilon p_n)\,.$$

Let $C_{p,k}(y_1, \ldots, y_m)$ be the coefficient of $C_p(\epsilon)$ of lowest degree in ϵ, namely k. The coefficient $C_{p,k}$ is a suitable *projection operator*. In fact, the polynomials p_i may have random coefficients and support including precisely those monomials of f_i which appear on the diagonal of the toric resultant matrix. The perturbation has been implemented in Maple; see also Section 7.5.

Example 7.3.1 (Continued from Example 7.2.4). In the special case

$$f_0 = 1 + x_1\,x_2 + x_1^2\,x_2 + x_1, \;\; f_2 = 1 + x_2 + x_1\,x_2 + x_1,$$

the toric resultant vanishes for all c_{1j} since the variety $V(f_0, f_1)$ has positive dimension: it is formed by the union of the isolated point $(1, -1)$ and the line $\{-1\} \times \mathbb{C}$. For a specific lifting and matrix construction, the trailing coefficient in the perturbed determinant is that of ϵ^2 and equals

$$-(c_{12}c_{13})(c_{14} - c_{11} + c_{12} - c_{13})(c_{14} + c_{11} - c_{12} + c_{13}).$$

So we can recover in the last two factors the value of f_1 at the isolated zero $(1, -1)$ and the point $(-1, -1)$ in the positive-dimensional component.

The next example illustrates the perturbation method in applying toric resultants for system solving.

Example 7.3.2. This is the example of [Roj99a]. To the system

$$f_1 := 1 + 2x - 2x^2y - 5xy + x^2 + 3x^3y, \;\; f_2 := 2 + 6x - 6x^2y - 11xy + 4x^2 + 5x^3y,$$

we add $f_0 := u_1 x + u_2 y + u_0$, which does not have to be perturbed. We use the function `spresultant` from Maple library MULTIRES to construct a 16×16 matrix M in parameters u_0, u_1, u_2, ϵ. The number of rows per polynomial are, respectively, $4, 6, 6$, whereas the mixed volumes of the 2×2 subsystems are all equal to 4. Here is the Maple code for these operations, where e stands for ϵ:

```
M  := spresultant ([f0,f1,f2], [x,y]):
DM := det(M):              # in u0,u1,u2,e
degree (DM,e);             # outputs 12
ldg := ldegree(DM,e);      # outputs 1
phi := primpart(coeff(DM,e,ldg)):
factor(phi);
```

For certain ω and δ, we have used $p_1 := -3x^2 + x^3y$, $p_2 := 2 + 5x^2$. The perturbed determinant has maximum and minimum degree in ϵ, respectively, 12 and 1. The trailing coefficient gives two factors corresponding to isolated solutions $(1/7, 7/4)$ and $(1, 1)$: $(49\,u_2 + 4\,u_1 + 28\,u_0)\,(u_2 + u_1 + u_0)$. Another two factors give points on the line $\{-1\} \times \mathbb{C}$ of solutions, but the specific points are very sensitive to the choice of ω and δ. One such choice yields: $(-u_0 + u_1)\,(27\,u_2 + 40\,u_1 - 40\,u_0)$.

Example 7.3.3. In the robot motion planning implementation of Canny's *roadmap algorithm* in [HP00], numerous "degenerate" systems are encountered. Let us examine a 3×3 system, where we hide x_0 to obtain dense polynomials of degrees $3, 2, 1$:

$$
\begin{aligned}
f_0 ={}& 54x_1{}^3 - 21.6x_1{}^2 x_2 - 69.12x_1x_2{}^2 + 41.472x_2{}^3 + (50.625 + 75.45x_0)\, x_1{}^2 \\
&+ (-92.25 + 32.88x_0)\, x_1 x_2 + (-74.592x_0 + 41.4)\, x_2{}^2 + \\
&+ (131.25 + 19.04x_0{}^2 - 168x_0)x_1 + (-405 + 25.728x_0{}^2 + 126.4x_0)\, x_2 + \\
&+ (-108.8\, x_0{}^2 + 3.75\, x_0 + 234.375), \\
f_1 ={}& -37.725\, x_1{}^2 - 16.44\, x_1 x_2 + 37.296\, x_2{}^2 + (-38.08x_0 + 84)\, x_1 + \\
&+ (-63.2 - 51.456x_0)\, x_2 + (2.304x_0{}^2 + 217.6x_0 - 301.875), \\
f_2 ={}& 15\, x_1 - 12\, x_2 + 16\, x_0.
\end{aligned}
$$

The Maple function `spresultant` applies an optimal perturbation to an identically singular 14×14 matrix in x_0. Now $\det M(\epsilon)$ is of degree 14 and the trailing coefficient of degree 2, which provides a bound on the number of affine roots. We obtain

$$
\phi(x_0) = \left(x_0 - \frac{1434}{625} \right) \left(x_0 - \frac{12815703325}{21336} \right),
$$

the first solution corresponding to the unique isolated solution but the second one is superfluous, hence the variety has dimension zero and degree 1.

Our perturbation method applies directly, since the projection operator will contain, as an irreducible factor, the implicit equation. The extraneous factor has to be removed by factorization. Distinguishing the implicit equation from the latter is straightforward by using the parametric expressions to generate points on the implicit surface.

Example 7.3.4. Let us consider the de-homogenized version of a system defined in [Bus01b]:
$$
p_0 = t_1^2, \; p_1 = t_1^3, \; p_2 = t_2^2, \; q = t_1^3 + t_2^3.
$$

It has one base point, namely $(0,0)$, of multiplicity 4. The toric resultant here does not vanish, so it yields the implicit equation

$$
x_2^3 x_1^2 - x_0^3 x_1^2 + 2x_0^3 x_1 - x_0^3.
$$

But under the change of variable $t_2 \to t_2 - 1$ the new system has zero toric resultant. The determinant of the perturbed 27×27 resultant matrix has a trailing coefficient which is precisely the implicit equation. The degree of the trailing term is 4, which equals in this case, the number of base points in the toric variety counted with multiplicity.

Example 7.3.5. The problem of computing the sparse, or toric, discriminant of a polynomial specified by its support can be formulated as an implicitization problem [DS02, GKZ94]. Let us fix the polynomial support in $\mathbb{Z}^m$, and suppose that the support's cardinality equals $m + 1 + s$, $s \geq 0$. The case $s = 2$ was studied in [DS02] and reduces to curve implicitization, though the approach used in that article was not based on implicitization.

Here $s = 3$, so we have a surface implicitization problem with base points. Base points forming a component of codimension 1 can be easily removed by canceling common factors in the numerator and denominator of the rational expressions for the $x_0, \ldots, x_{s-1}$.

The parametric expressions for the x_i's and the ensuing implicitization problem shall be defined in terms of the entries of some matrix B, specified from the support of the input polynomial. Its row dimension is s and its column dimension equals the cardinality of the polynomial support. We do not go into the technical details of deriving B from the support.

Let us consider a specific example with $m = 3$ and $s = 3$, hence the support cardinality equals 7. The problem reduces to implicitizing the parametric surface given by

$$x_i = \prod_{j=1}^{7} (b_{0j} + t_1 b_{1j} + t_2 b_{2j})^{b_{ij}}, \quad i = 0, 1, 2,$$

where the matrix $B = (b_{ij})$, for $i = 0, \ldots, 2$, $j = 1, \ldots, 7$, is as follows:

$$B = \begin{bmatrix} 1 & 0 & -1 & 0 & 2 & -1 & -1 \\ 0 & 1 & -1 & 2 & 0 & -1 & -1 \\ 1 & 1 & -2 & 1 & 0 & -1 & 0 \end{bmatrix}.$$

There are base points forming components of codimension 2, including a single affine base point $(1, -1)$. Our algorithm constructs a 33×33 matrix, whose perturbed determinant has a trailing term of degree 3 in ϵ. The corresponding coefficient has total degree 14 in x_0, x_1, x_2. When factorized, it yields the precise implicit equation, which is of degree 9 in x_0, x_1, x_2.

7.4 Implicit support

In this section, we exploit information on the support of the toric resultant in order to predict the support of the implicit equation of a parametric (hyper)surface.

Our approach is to consider the extreme monomials i.e., the vertices of the Newton polytope of the toric resultant Res. The output support scales with the sparseness of the parametric polynomials and is much tighter than the one predicted by degree arguments. In many cases, we obtain the exact support of the implicit equation, as seen by applying our Maple program. Moreover, it is possible to specify certain coefficients in this equation. Our motivation comes mainly from two implicitization algorithms which apply interpolation, namely the direct method of [CGKW01] and the one based on perturbations (cf. Section 7.3 or [MC92]).

The initial form $In_\omega(F)$ of a multivariate polynomial F in k variables, with respect to some functional $\omega : \mathbb{Z}^k \to \mathbb{R}$, is the sum of all terms in F which maximize the inner product of ω with the corresponding exponent vector. Let us define

$$k := |A_0| + \cdots + |A_n|,$$

then ω defines a *lifting* function on the input system, by lifting every support point $a \in A_i$ to $(a, \omega(a)) \in \mathbb{Z}^n \times \mathbb{R}$. This generalizes the linear lifting of Section 7.2. The lower hull facets of the lifted Minkowski sum correspond to maximal cells of an induced coherent mixed subdivision of Q. If ω is sufficiently generic, then this subdivision is tight; in the sequel, we assume our mixed subdivision is both coherent and tight and denote it by Δ_ω. If $F_i \in A_i$ is a vertex summand of an i-mixed cell, then the corresponding coefficient in f_i is denoted by c_{iF_i}. We recall our assumption that the A_i span $\mathbb{Z}^n$.

Theorem 7.4.1. *The initial form of the toric resultant* Res *with respect to a generic ω equals the monomial*

$$In_\omega(\text{Res}) = \prod_{i=0}^{n} \prod_F c_{iF_i}^{\text{Vol}(F)}, \tag{7.4}$$

where $\text{Vol}(\cdot)$ *denotes ordinary Euclidean volume and the second product is over all mixed cells of type i in the mixed subdivision Δ_ω.*

For a detailed proof of this theorem, see [Stu94a]. This proof can be obtained from the toric resultant matrix construction, by means of the subdivision-based algorithm. Let us use the same specialization of the coefficients in terms of a new parameter t, as in the discussion that leads to Theorem 7.2.1. Then, the resultant becomes univariate in t and the proof is completed by relating, on the one hand, the degree of $In_\omega(\text{Res})$ in t and, on the other, the sum of all exponents in expression (7.4). The latter, for fixed i, equals $\text{MV}_{-i} = \deg_{f_i} \text{Res}$.

For a generic vector ω, the initial form $In_\omega(\text{Res})$ corresponds to a vertex of the Newton polytope of the resultant Res. It is precisely the vertex with inner normal ω. So, by varying the lifting ω, we can compute all vertices of this Newton polytope, hence a superset of the resultant's support.

A bijective correspondence exists between the extreme monomials and the configurations of the mixed cells of the A_i. So, it suffices to compute all distinct mixed-cell configurations, as discussed in [MC00, MV99].

Another (simpler) means of reducing the number of relevant mixed subdivisions is by bounding the number of cells. This bound is usually straightforward to compute in small dimensions (e.g. when $n = 2, 3$) and reduces drastically the set of mixed subdivisions. For instance, when studying the implicitization of a biquadratic surface, the total number of mixed subdivisions is 19728, whereas those with 8 cells is 62.

In certain special cases, we can be more specific about the Newton polytope of the toric resultant. First, its dimension equals $k - 2n - 1$ [GKZ94, Stu94a]. Certain corollaries follow: For essential support families (defined in [Stu94a]), a 1-dimensional Newton polytope of Res is possible if and only if all polynomials are binomials. The only resultant polytope of dimension 2 is the triangle; in this case the support cardinalities must be 2 and 3. For dimension 3, the possible polytopes are the tetrahedron, the square-based pyramid, and polytope $N_{2,2}$ given in [Stu94a]; the support cardinalities are respectively $2, 2$ and 3.

One corollary of Theorem 7.4.1 (and of its proof) is that the coefficients of all extreme monomials are in $\{-1, 1\}$ [GKZ91, CE00, Stu94a]. Sturmfels [Stu94a] also specifies, for all extreme monomials, a way to compute their precise coefficients. But this requires computing several coherent mixed subdivisions, and goes beyond the scope of the present chapter.

The so-called *Cayley trick* introduces a new point set $C := \{(z, a_{0j}, 1)$: $a_{0j} \in A_0\} \cup \{(e_i, a_{ij}, 1)$: $i = 1, \ldots, n, \ a_{ij} \in A_i\} \subset \mathbb{Z}^{2n+1}$, where $z = (0, \ldots, 0) \in \mathbb{N}^n$ is the zero vector and $e_i = (0, \ldots, 0, 1, 0, \ldots, 0) \in \mathbb{N}^n$ has a unit at the i-th position and $n - 1$ zeroes.

Theorem 7.4.2. *The problem of computing all mixed subdivisions of supports $A_0, \ldots, A_n$, which lie in $\mathbb{Z}^n$, is equivalent to computing all regular triangulations of the set C defined above. This set contains $k_0 + \cdots + k_n$ points, where $k_i = |A_i|$.*

Example 7.4.3 (Continued from Example 7.2.2). The Cayley trick in the univariate case goes as follows. Consider $f_0 = c_{00} + c_{01}x$, $f_1 = c_{10} + c_{12}x^2$, then the points in the set C appear in the columns of matrix

$$\begin{bmatrix} 0 & 0 & 1 & 1 \\ 0 & 1 & 0 & 2 \end{bmatrix}.$$

There are two possible triangulations of these points, namely

$$\left(\begin{bmatrix} 0 \\ 0 \end{bmatrix}, \begin{bmatrix} 0 \\ 1 \end{bmatrix}, \begin{bmatrix} 1 \\ 2 \end{bmatrix} \right), \ \left(\begin{bmatrix} 0 \\ 0 \end{bmatrix}, \begin{bmatrix} 1 \\ 0 \end{bmatrix}, \begin{bmatrix} 1 \\ 2 \end{bmatrix} \right),$$

which is the one shown in Figure 7.3, and

$$\left(\begin{bmatrix} 0 \\ 0 \end{bmatrix}, \begin{bmatrix} 0 \\ 1 \end{bmatrix}, \begin{bmatrix} 1 \\ 0 \end{bmatrix} \right), \ \left(\begin{bmatrix} 0 \\ 1 \end{bmatrix}, \begin{bmatrix} 1 \\ 0 \end{bmatrix}, \begin{bmatrix} 1 \\ 2 \end{bmatrix} \right).$$

Efficient algorithms (and implementations) exist for computing all regular triangulations of a point set [Ram01]. Regular are those triangulations that can be obtained by projection of a lifted triangulation.

We produce a superset of the monomials in the support of the implicit equation of the input. Consider, as in Section 7.3 the polynomials $f_i(t) = p_i(t) - x_i q(t)$, where we ignore the specific values of the coefficients.

This is an interesting feature of the algorithm, namely that it considers the monomials in the parametric equations but not their actual coefficients. This shows that the algorithm is suitable for use as a preprocessing off-line step in CAGD computations, where one needs to compute thousands of examples with the same support structure in real time. This handles the implicitization of (multiparametric) families of (hyper)surfaces, indexed by one or more parameters.

Of course, the generic resultant coefficients are eventually specialized to functions of the x_i. Then, any bounds on the implicit degree in the x_i may be applied, in order to reduce the final support set. One step yields as by-product all partial mixed volumes MV_{-i} for $i = 0, \ldots, n$, and hence the implicit degree separately in the x_i variables.

We examine our method on some small examples, and summarize the results in Table 7.1 below.

Example 7.4.4. We consider the Folium of Descartes, shown also in Figure 7.8. $x = 3t^2/(t^3 + 1)$, $y = 3t/(t^3 + 1)$.

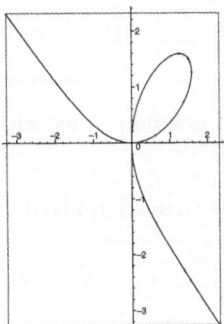

Fig. 7.8. The Folium of Descartes

The output monomials are $\{y^3, x^3, x^3 y^3, x y, y^2 x^2\}$. After applying the degree bound $d = 3$ we obtain the support $\{y^3, x^3, x y\}$, which is optimal, since the implicit equation is $x^3 + y^3 - 3 x y = 0$.

Example 7.4.5. An example in 3 dimensions comes from [Buc88b]; the surface is drawn in Figure 7.9. The parametric expressions are: $x = s\,t$, $y = s\,t^2$, $z = s^2$.

In order to apply toric elimination theory, we consider polynomials

$$f_0 = c_{00} - c_{01}st, \ f_1 = c_{10} - c_{11}st^2, \ f_2 = c_{20} - c_{21}s^2.$$

There are the following two possible mixed subdivisions, each containing exactly three maximal cells, all of which are mixed, see Figure 7.10.

The computed support is optimal and the implicit equation is $x^4 - y^2 z = 0$.

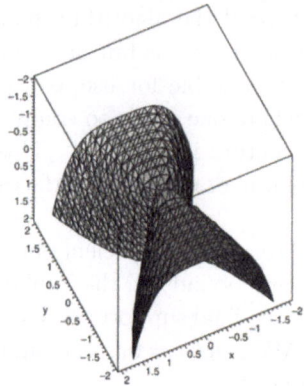

Fig. 7.9. The surface in Example 7.4.5.

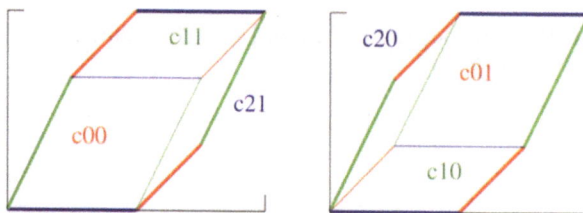

Fig. 7.10. Mixed cells in the subdivisions, with vertex summands shown.

Example 7.4.6. Let us consider a system attributed to Fröberg and discussed in Chapter 1.

$$x = t^{48} - t^{56} - t^{60} - t^{62} - t^{63}, \ y = t^{32}.$$

The Minkowski sum is the segment $Q_0 + Q_1 = [0, 95]$. One type of triangulations, obtained from a non-linear lifting, divides it to the following 3 cells (which are all segments):

$$(Q_0' + 0), \ (a + Q_1), \ (Q_0'' + 32), \ \text{where } Q_0' = [0, a], \ Q_0'' = [a, 63],$$

and $a \in A_0 = \{0, 48, 56, 60, 62, 63\}$. Every such triangulation yields a support point y^a. The triangulation $(0 + Q_1), (Q_0 + 32)$, which is induced from a linear lifting, yields support point x^{32}. Note that only certain of these monomials are extreme when we consider the resultant in terms of all input coefficients, in order for the respective coefficients to lie in $\{-1, 1\}$.

Therefore, we find, as the toric resultant support, the triangle with vertices $(32, 0), (0, 48)$ and $(0, 63)$. Equivalently, it is delimited by the y-axis and the lines $y = -(3/2)x + 48$ and $y = -(63/32)x + 63$, as shown in Figure 7.11.

Counting the points with integer coordinates inside (and on the sides) of the triangle, we see that there are 257 such points, which is seen to be optimal by actually computing the resultant.

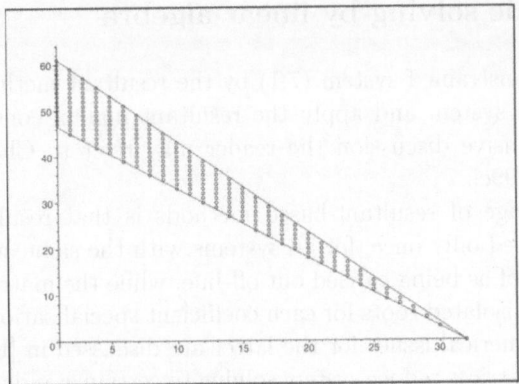

Fig. 7.11. Toric resultant support.

Table 7.1. Predicting the implicit support.

Problem	Input Degree	Degree of Implicit Eq.	General # monomials	# monomials from [EK03]
Unit Circle	2	2	6	3
Descartes Folium, Ex. 7.4.4	3	3	10	3
Fröberg-Dickenstein, Ex. 7.4.6	63	63	1057	257
Buchberger, Example 7.4.5	1,2	4	35	2
Busé, Example 7.3.4	3	5	56	4
Bilinear, Example 7.1.9	1,1	2	10	9

Example 7.4.7. The well-known bicubic surface represents a challenge for our current implementation: $x = 3\,t\,(t-1)^2 + (s-1)^3 + 3\,s$, $y = 3\,s\,(s-1)^2 + t^3 + 3\,t$, $z = -3s(s^2 - 5s + 5)t^3 - 3(s^3 + 6s^2 - 9s + 1)t^2 + t(6s^3 + 9s^2 - 18s + 3) - 3s(s-1)$. We computed 737129 regular triangulations (by TOPCOM) [Ram01]. For illustration purposes, we show one of them:

{2,3,4,7,13},{3,4,5,7,13},{3,5,6,7,13},{3,6,9,13,14},
{6,9,12,13,14},{3,6,9,14,15},{6,9,12,14,15},{6,12,13,14,16},
{6,12,14,15,16},{6,12,15,16,17},{3,6,9,15,18},{6,9,12,15,18},
{6,12,15,17,18},{3,9,15,18,19},{3,6,9,18,19},{6,9,12,18,19},
{6,12,16,17,20},{6,12,17,18,20},{3,6,9,19,23},{6,9,12,19,23},
{6,12,19,22,23},{6,12,22,23,24},{6,12,23,24,25},{3,6,9,23,26},
{6,9,12,23,26},{6,12,23,25,26},{0,2,4,7,13},{3,6,7,9,13},
{6,12,18,19,22},{6,12,18,20,24},{6,7,9,12,13},{6,12,18,22,24}.

The size of the file is 383 MBytes. This underlines the fact that we should not compute all regular triangulations but only the mixed-cell configurations.

7.5 Algebraic solving by linear algebra

To solve well-constrained system (7.1) by the resultant method we define an overconstrained system and apply the resultant matrix construction. For a more comprehensive discussion the reader may refer to Chapters 2 and 3, or [CLO98, EM99c].

One advantage of resultant-based methods is that resultant matrix M need be computed only once, for all systems with the same supports. So this step is thought of as being carried out off-line, while the matrix operations to approximate all isolated roots for each coefficient specialization constitute the online part. Numerical issues for the latter are discussed in [Emi97, EM99c].

Resultant matrices reduce system solving to certain standard operations in computer algebra. In particular, univariate or multivariate determinants can be computed by evaluation and interpolation techniques. However, the determinant development in the monomial basis may be avoided because there are algorithms for univariate polynomial solving as well as multivariate polynomial factorization which require only the values of these polynomials at specific points; cf. e.g. [Pan97]. All of these evaluations would exploit the quasi-Toeplitz structure of Sylvester-type matrices [CKL89, EP02].

We present two ways of defining an overconstrained system. The first method adds to the given system *an extra polynomial*, namely

$$f_0 = u_0 + u_1 x_1 + \cdots + u_n x_n \ \in (K[u_0, \ldots, u_n])[x_1^{\pm 1}, \ldots, x_n^{\pm 1}],$$

thus yielding a well-studied object, the u-resultant. Coefficients $u_1, \ldots, u_n$ may be randomly specialized or left as indeterminates; in the latter case, solving reduces to factorizing the u-polynomial. It is known that the u-resultant factorizes into linear factors $u_0 + u_1 \alpha_1 + \cdots + u_n \alpha_n$ where $(\alpha_1, \ldots, \alpha_n)$ is an isolated root of the original system. This is an instance of Poisson's formula. Now, u_0 is usually an indeterminate that we shall denote by x_0 below for uniformity of notation. Matrix M will describe the multiplication map for f_0 in the coordinate ring of the ideal defined by the system in (7.1).

An alternative way to obtain an overconstrained system is by *hiding* one of the original variables in the coefficient field and consider the system as follows (we modify the previous notation to unify the subsequent discussion):

$$f_0, \ldots, f_n \in (K[x_0]) [x_1^{\pm 1}, \ldots, x_n^{\pm 1}].$$

M is a matrix polynomial in x_0, and may not be linear.

An important issue concerns the degeneracy of the input coefficients. This may result in the trivial vanishing of the toric resultant or of $\det M$ when there is an infinite number of common roots (in the torus or at toric infinity) or simply due to the matrix constructed. An infinitesimal perturbation has been proposed [DE01a] which respects the structure of Newton polytopes and is computed at no extra asymptotic cost, cf. Section 7.3.

The perturbed determinant is a polynomial in the perturbation variable, whose leading coefficient is nonzero whereas the least significant coefficient is det M. Irrespective of which coefficients vanish, there is always a trailing nonzero coefficient which vanishes when x_0 takes its values at the system's isolated roots, even in the presence of positive-dimensional components. This univariate polynomial is known as a *projection operator* because it projects the roots to the x_0-coordinate. Univariate polynomial solving thus yields these coordinates. Again, the u-resultant allows us to recover all coordinates via multivariate factoring.

A basic property of resultant matrices is that right vector multiplication expresses evaluation of the row polynomials. Specifically, multiplying by a column vector containing the values of column monomials q at some $\alpha \in (\overline{K}^*)^n$ produces the values of the row polynomials

$$\alpha^p f_{i_p}(\alpha).$$

Computationally it is preferable to have to deal with as small a matrix as possible. To this end we partition M into four blocks M_{ij} so that the upper left submatrix M_{11} is square, independent of x_0, and of maximal dimension so that it remains well-conditioned.

If the matrix is obtained from the subdivision-based algorithm, then we know that M_{11} corresponds to the integer points in the 0-mixed cells. More precisely, the columns of M_{11} are indexed by those points, whereas its rows contain the multiples of f_0 with the corresponding monomials. It can be proven that these monomials form a basis of the quotient ring defined by the ideal of $f_1, \ldots, f_n$, namely $K[x_1^{\pm 1}, \ldots, x_n^{\pm 1}]/\langle f_1, \ldots, f_n \rangle$. For a proof, see [Emi96, PS96].

Once M_{11} is specified, let $A(x_0) = M_{22}(x_0) - M_{21}(x_0)M_{11}^{-1}M_{12}(x_0)$. To avoid computing M_{11}^{-1}, we may use its LU (or QR) decomposition to solve $M_{11}X = M_{12}$ and compute $A = M_{22} - M_{21}X$.

Let $\mathcal{E}$ be the monomial set indexing the rows and columns of M and let $B \subset \mathcal{E}$ index A. If $(\alpha_0, \alpha) \in \overline{K}^{n+1}$ is a common root with $\alpha \in \overline{K}^n$, then $\det A(\alpha_0) = 0$ and, for any vector $v' = [\cdots \alpha^q \cdots]$, where q ranges over B, $A(\alpha_0)v' = 0$. Moreover,

$$\begin{bmatrix} M_{11} & M_{12}(\alpha_0) \\ 0 & A(\alpha_0) \end{bmatrix} \begin{bmatrix} v \\ v' \end{bmatrix} = \begin{bmatrix} 0 \\ 0 \end{bmatrix} \Rightarrow M_{11}v + M_{12}(\alpha_0)v' = 0,$$

determines v once v' has been computed. Vector $[v, v']$ contains the values of every monomial in $\mathcal{E}$ at α.

It can be shown that $\mathcal{E}$ affinely spans $\mathbb{Z}^n$ and an affinely independent subset can be computed in polynomial time [Emi96]. Given v, v' and these points, we can compute the coordinates of α. If all independent points are in B then v' suffices for solving. To find the vector entries that will allow us to recover the root coordinates, it is typically sufficient to search in B for pairs of entries corresponding to q_1, q_2 such that $q_1 - q_2 = (0, \ldots, 0, 1, 0, \ldots, 0)$. This lets us compute the i-th coordinate, if the unit appears at the i-th position. In general, the problem of choosing the best vector entries for computing the root coordinates is open, and different choices may lead to different accuracy.

To reduce the problem to an eigendecomposition, let r be the dimension of $A(x_0)$, and $d \geq 1$ the highest degree of x_0 in any entry. We wish to find all values of x_0 at which

$$A(x_0) = x_0^d A_d + x_0^{d-1} A_{d-1} + \cdots + x_0 A_1 + A_0$$

becomes singular. These are the eigenvalues of the *matrix polynomial*. Furthermore, for every eigenvalue λ, there is a basis of the kernel of $A(\lambda)$ defined by the right eigenvectors of the matrix polynomial associated to λ. If A_d is nonsingular then the eigenvalues and right eigenvectors of $A(x_0)$ are the eigenvalues and right eigenvectors of monic matrix polynomial $A_d^{-1} A(x_0)$. This is always the case when adding an extra linear polynomial, since $d = 1$ and $A_1 = I$ is the $r \times r$ identity matrix; then

$$A(x_0) = -A_1(-A_1^{-1} A_0 - x_0 I).$$

Generally, the *companion matrix* of a monic matrix polynomial is a square matrix C of dimension rd. The eigenvalues of C are precisely the eigenvalues λ of $A_d^{-1} A(x_0)$, whereas its right eigenvector $w = [v_1, \ldots, v_d]$ contains a right eigenvector v_1 of $A_d^{-1} A(x_0)$ and $v_i = \lambda^{i-1} v_1$, for $i = 2, \ldots, d$.

We now address the question of a singular A_d. The following *rank balancing* transformation in general improves the conditioning of A_d. If matrix polynomial $A(x_0)$ is not identically singular for all x_0, then there exists a transformation $x_0 \mapsto (t_1 y + t_2)/(t_3 y + t_4)$ for some $t_i \in \mathbb{Z}$, that produces a new matrix polynomial of the same degree and with nonsingular leading coefficient. If A_d is ill-conditioned for all linear rank balancing transformations, then we build the matrix pencil and apply a *generalized eigendecomposition* to solve $C_1 x + C_0$. This returns pairs (α, β) such that matrix $C_1 \alpha + C_0 \beta$ is singular with an associated right eigenvector.

Acknowledgment

I acknowledge financial support by INRIA Sophia-Antipolis (France) through the bilateral collaboration "Calamata" between the Galaad Group and the Department of Informatics & Telecommunications of the National Kapodistrian University of Athens.

Introduction to numerical algebraic geometry

Andrew J. Sommese[1]*, Jan Verschelde[2]**, and Charles W. Wampler[3]

[1] Department of Mathematics, University of Notre Dame, Notre Dame, IN
46556-4618, USA, sommese@nd.edu, http://www.nd.edu/~sommese
[2] Department of Mathematics, Statistics, and Computer Science, University of
Illinois at Chicago, 851 South Morgan (M/C 249), Chicago, IL 60607-7045, USA,
jan@math.uic.edu, jan.verschelde@na-net.ornl.gov,
http://www.math.uic.edu/~jan
[3] General Motors Research and Development, Mail Code 480-106-359, 30500
Mound Road, Warren, MI 48090-9055, USA, Charles.W.Wampler@gm.com

Summary. In a 1996 paper, Andrew Sommese and Charles Wampler began developing a new area, "Numerical Algebraic Geometry", which would bear the same relation to "Algebraic Geometry" that "Numerical Linear Algebra" bears to "Linear Algebra".

To approximate all isolated solutions of polynomial systems, numerical path following techniques have been proven reliable and efficient during the past two decades. In the nineties, homotopy methods were developed to exploit special structures of the polynomial system, in particular its sparsity. For sparse systems, the roots are counted by the mixed volume of the Newton polytopes and computed by means of polyhedral homotopies.

In Numerical Algebraic Geometry we apply and integrate homotopy continuation methods to describe solution components of polynomial systems. In particular, our algorithms extend beyond just finding isolated solutions to also find all positive dimensional solution sets of polynomial systems and to decompose these into irreducible components. These methods can be considered as symbolic-numeric, or perhaps rather as numeric-symbolic, since numerical methods are applied to find integer results, such as the dimension and degree of solution components, and via interpolation, to produce symbolic results in the form of equations describing the irreducible components.

Applications from mechanical engineering motivated the development of Numerical Algebraic Geometry. The performance of our software on several test problems illustrates the effectiveness of the new methods.

* This material is based upon work supported by the National Science Foundation
under Grant No. 0105653; and the Duncan Chair of the University of Notre Dame.
** This material is based upon work supported by the National Science Foundation
under Grant No. 0105739 and Grant No. 0134611.

8.0 Introduction

The goal of this chapter is to provide an overview of the main ideas developed so far in our research program to implement numerical algebraic geometry, initiated in [SW96].

We are concerned with numerically solving polynomial systems. While the homotopy continuation methods of the past were limited to approximating only the isolated roots, we developed tools to describe all positive dimensional irreducible components of the solution set of a polynomial system. In particular, our algorithms produce for every irreducible component a *witness set*, whose cardinality equals the degree of the component, as this set is obtained by intersecting the component with a general linear space of complementary dimension. A point of a witness set corresponds to what is known in algebraic geometry as a generic point. Our main results [SV00, SVW01a, SVW01b, SVW01c, SVW02c, SVW02b, SVW02a, SVW03, SVW, SVW04] can be summarized in four items:

1. In [SV00] we presented a cascade of homotopies (extended in [SVW]) to find candidate witness points for every component of the solution set. Separating the junk from the candidate witness points was done in [SVW01a], where factorization methods based on interpolation implemented a numerical irreducible decomposition. The use of central projections and a homotopy membership test to filter junk were the improvements of [SVW01b].
2. The treatment of high-degree components and components of multiplicity greater than one can present numerical challenges. The use of monodromy [SVW01c] followed by the validation by the linear trace [SVW02c] enabled us to deal with high degree components of multiplicity one, using only machine floating point numbers. In [SVW02b], we presented an approach to tracking paths on sets of multiplicity greater than one, which in theory makes the algorithm for irreducible decomposition completely general, although in practice this portion of the framework needs further refinement. However, for the case of the factorization of a single multivariate polynomial, we can use differentiation to reduce the treatment of higher multiplicity components to nonsingular path tracking, as we described in [SVW04]. This addresses an open problem in symbolic-numeric computing: the factorization of multivariate polynomials with approximate coefficients [Kal00].
3. Our new homotopy algorithms have been implemented and tested using the path trackers in the software package PHCpack [Ver99a]. In [SVW03] we outlined the new tools in PHCpack and described a simple interface to Maple. Our software found the degrees of all irreducible components of the cyclic 8 and 9 roots problems, which previously could only be done via Gröbner bases (and only by the very best implementation [Fau99]).
4. Polynomial systems with positive dimensional components occur naturally when designing mechanical devices which permit motion. We inves-

tigated a special case of a moving platform, discovering through a numerical irreducible decomposition [SVW02c] a component not reported by experts [HK00]. This and other applications of our tools to systems coming from mechanical design are described in [SVW02a].

In this chapter we will introduce these results, first explaining homotopy methods for isolated solutions. We can only mention some recent and exciting new developments in fields related to numerical algebraic geometry: numerical Schubert calculus ([HSS98], [HV00], [LWW02], [SS01], [VW02]) and numerical jet geometry [RSV02].

8.1 Homotopy continuation methods – an overview

Homotopy continuation methods operate in two stages. Firstly, homotopy methods exploit the structure of the system $f(\mathbf{x}) = 0$ to find a root count and to construct a start system $g(\mathbf{x}) = 0$ that has exactly as many regular solutions as the root count. This start system is embedded in the *homotopy*

$$h(\mathbf{x}, t) = \gamma(1 - t)g(\mathbf{x}) + tf(\mathbf{x}) = 0, \quad t \in [0, 1], \tag{8.1}$$

with $\gamma \in \mathbb{C}$ a random number. Secondly, as t moves from 0 to 1, numerical continuation methods trace the paths that originate at the solutions of the start system towards the solutions of the target system. The good properties we expect from a homotopy are (borrowed from [Li97, Li03]):

1. (*triviality*) The solutions for $t = 0$ are trivial to find.
2. (*smoothness*) No singularities along the solution paths occur (because of γ).
3. (*accessibility*) An isolated solution of multiplicity m is reached by exactly m paths.

Continuation or path-following methods are standard numerical techniques ([AG90a, AG93, AG97], [Mor87], [Wat86, Wat89]) to trace the solution paths defined by the homotopy using *predictor-corrector* methods. The smoothness property of complex polynomial homotopies implies that paths never turn back, so that during correction the parameter t stays fixed, which simplifies the set up of path trackers. The adaptive step size control determines the step length while enforcing quadratic convergence in Newton's method to avoid path crossing (see also [KX94] for the application of interval methods to control the step size). At the end of the path, end games ([HV98], [MSW91, MSW92a, MSW92b], [SWS96]) deal with diverging paths and paths leading to singular roots.

Following [HSS98], we say that a homotopy is *optimal* if every path leads to one solution. The classification in Table 8.1 (from [Ver99b]) contains key words for three classes of polynomial systems for which optimal homotopies are available in PHCpack [Ver99a]. These homotopies have no diverging paths for generic instances of polynomial systems in their class.

system	model	theory	space	
dense	highest degrees	Bézout	$\mathbb{P}^n$	projective
sparse	Newton polytopes	Bernshteĭn	$(\mathbb{C}^*)^n$	toric
determinantal	localization posets	Schubert	G_{mr}	Grassmannian

Table 8.1. Key words of the three classes of polynomial systems.

The earliest applications of homotopies for solving polynomial systems ([CMPY79], [Dre77], [GZ79], [GL80], [Li83], [LS87] [Mor83], [Wri85], [Zul88]) belong to the dense class, where the number of paths equals the product of the degrees in the system. Multi-homogeneous homotopies were introduced in [MS87b, MS87a] and applied in [WMS90, WMS92], see also [Wam92]. Similar are the random product homotopies [LSY87a, LSY87b], see also [Li87] and [LW91]. Methods to construct linear-product start systems were introduced in [VH93], and extended in [VC93, VC94], [LWW96], and [WSW00]. A general approach to exploit product structures was developed in [MSW95].

Almost all systems have fewer terms than allowed by their degrees. Implementing constructive proofs of Bernshteĭn's theorems [Ber75], polyhedral homotopies were introduced in [HS95] and [VVC94] to solve sparse systems more efficiently. These methods provided ways to start cheater's homotopies ([LSY89], [LW92]) and special instances of coefficient-parameter polynomial continuation ([MS89, MS90]). The root count requires the calculation of the mixed volume[4], for which a lift-and-prune approach was presented in [EC95]. Exploitation of symmetry was studied in [VG95] and the dynamic lifting of [VGC96] led to incremental polyhedral continuation. See [Ver00] for a Toric Newton. Extensions to count all affine roots (also those with zero components) were proposed in [EV99], [GLW99], [IIS97b], [LW96], [Roj94, Roj99b], and [RW96]. Very efficient calculations of mixed volumes are described in [DKK03], [GL00, GL03], [KK03b], [LL01], and [TKF02].

Determinantal systems (with equations like $\det(A|X) = 0$) arise in problems of enumerative geometry. The homotopies in numerical Schubert calculus first appeared explicitly in [HSS98], originating from questions in real enumerative geometry [Sot97a, Sot97b]. While real enumerative geometry [Sot03] is interesting on its own, these homotopies solve the pole placement problem ([Byr89], [RRW96, RRW98], [Ros94], [RW99]) in control theory. Recent improvements and applications can be found in [HV00], [LWW02], [SS01], and [VW02].

We end this section noting that homotopies have a wider application range than "just" solving polynomial systems, see for instance [Wat02] for a survey, [WBM87], and [WSM+97] for a description of HOMPACK. The

---

[4] The mixed volume was nicknamed in [CR91] as the BKK bound to honor Bernshteĭn [Ber75], Kushnirenko [Kus76], and Khovanskiĭ [Kho78b].

speedup of continuation methods on multi-processor machines has been addressed in [ACW89, CARW93, HW89].

8.2 Homotopies to approximate all isolated solutions

We first prove the regularity and boundedness of the solution paths defined by homotopies, before surveying path following techniques. We obtain more efficient homotopies by exploiting product structures and using Newton polytopes to model the sparsity of the system.

8.2.1 Regularity and boundedness of solution paths

To illustrate how homotopy methods work, let us consider a simple example of solving two quadrics:

$$f(x,y) = \begin{pmatrix} x^2 + 4y^2 - 4 \\ 2y^2 - x \end{pmatrix}.$$

To solve $f(x,y) = 0$, we match it with a start system of two easily solved quadrics:

$$g(x,y) = \begin{pmatrix} x^2 - 1 \\ y^2 - 1 \end{pmatrix},$$

with which we form the following homotopy:

$$h(x,y,t) = \begin{pmatrix} x^2 - 1 \\ y^2 - 1 \end{pmatrix}(1-t) + \begin{pmatrix} x^2 + 4y^2 - 4 \\ 2y^2 - x \end{pmatrix} t. \tag{8.2}$$

At $t = 1$, $h(x,y,t = 1) = 0$ is $f(x,y) = 0$, the system we wish to solve while at $t = 0$, $h(x,y,t = 0) = 0$ is the start system $g(x,y) = 0$ we can easily solve. As we usually move t from 0 to 1 when we solve the system, we may view the movement of t from 1 to 0 as a degeneration of the system, i.e., we deform the general hypersurfaces into degenerate products of hyperplanes.

But does this work? We will see in a moment that it does not, but that there is a simple maneuver that fixes the trouble once and for all. For numerical solving, we would need the solution paths to be free of singularities. A singularity occurs where the Jacobian matrix J_h of the homotopy $h(x,y,t) = 0$ has a zero determinant. The singularities along the solution paths are solutions of the system

$$\begin{cases} h(x,y,t) = \mathbf{0} \\ \det(J_h(x,y,t)) = 0 \end{cases} \quad \text{where} \quad J_h = \begin{bmatrix} 2x & 8yt \\ -t & 2y + 2yt \end{bmatrix}. \tag{8.3}$$

If this "discriminant system" has any roots with $t \in [0, 1)$, there is at least one homotopy solution path with singularities. To explore this situation, let's solve this system by elimination. This is not a step that we normally perform in the course of solving $f(x) = 0$, but we do it here to reveal the flaw in the naive homotopy of (8.2) and to illustrate how we fix the flaw. To solve this discriminant system, we will eliminate from the system the variables x and y to obtain one polynomial in the continuation parameter t. The roots of this polynomial define the singularities along the solution paths.

While there are many ways to perform this elimination, we let Maple compute a lexicographical Gröbner basis of the discriminant system. Below are the Maple commands, to save space we suppressed most of the output.

```
> f := [x^2 + 4*y^2 - 4,2*y^2- x];        # target system
> g := [x^2 - 1, y^2 - 1];                # start system
> h := t*f + (1-t)*g;                     # the homotopy
> eh := expand(h);                        # expanded homotopy
> jh := matrix(2,2,                       # Jacobian matrix
          [[diff(eh[1],x),diff(eh[1],y)],
           [diff(eh[2],x),diff(eh[2],y)]]);
> sys := [eh[1],eh[2],       # discriminant system solved by
          linalg[det](jh)]; # pure lex Groebner basis in gb
> gb := grobner[gbasis](sys,[x,y,t],plex);
> gb[nops(gb)];                 # discriminant polynomial
                  3      5       4      2      7      6
    -1 + t + 10 t  + 29 t  + 13 t  - 5 t  + 12 t  + 21 t
```

As the degree of this "discriminant polynomial" is seven, we have seven roots:

```
> fsolve(gb[nops(gb)],t,complex);          # numerical solving
  -.8818537646 - .9177002576 I, -.8818537646 + .9177002576 I,
  -.2011599690 - .8877289373 I, -.2011599690 + .8877289373 I,
 .006853764567 - .3927967328 I, .006853764567 + .3927967328 I,
    .4023199381
```

We are troubled by the root around 0.4, because, as t moves from 0 to 1, we will encounter a singularity. So our homotopy in (8.2) does not work!

We can fix this problem by the choice of a random constant $\gamma = e^{\theta\sqrt{-1}}$, for some random angle θ. Now, consider the homotopy

$$h(x,y,t) = \gamma \begin{pmatrix} x^2 - 1 \\ y^2 - 1 \end{pmatrix} (1-t) + \begin{pmatrix} x^2 + 4y^2 - 4 \\ 2y^2 - x \end{pmatrix} t. \qquad (8.4)$$

The random choice of γ will cause all roots of the discriminant polynomial to lie outside the interval $[0,1)$. That $t = 0$ is excluded is obvious (because the start system has only regular roots), but at $t = 1$ we may find singular solutions of the given system f.

Exercise 8.2.1. Modify the homotopy in the sequence of Maple commands above taking h := t*f + (1+I)*(1-t)*g; and verify that none of the roots of the discriminant polynomial is real. The choice of γ as $1 + \sqrt{-1}$ does not give the Gröbner package of Maple a hard time. If Maple is unavailable, then another computer algebra system should do just as well.

The above example illustrates the general idea behind the regularity of solution paths defined by a homotopy. The main theorem of elimination theory says that the projection of an algebraic set in complex projective space is again an algebraic set. Consider the discriminant system as a polynomial system in $\mathbf{x}, t$, and γ. If we eliminate $\mathbf{x}$, we obtain a polynomial in t and γ. This polynomial does not vanish entirely as the start system (at $t = 0$) has no singular roots. Thus it has only finitely many roots for general γ. Furthermore, a random complex choice of γ will insure that all those roots miss the interval $[0, 1)$. A schematic (as in [Mor87]) illustrating what cannot and what can happen is in Figure 8.1.

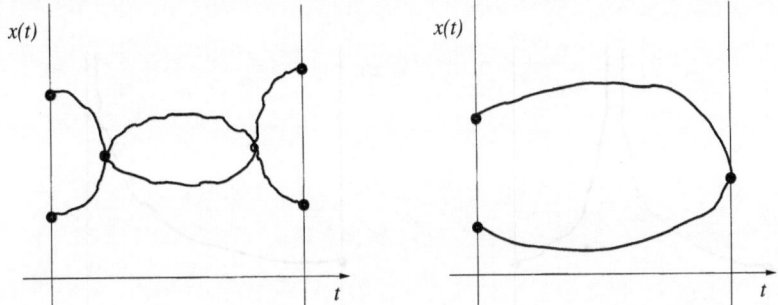

Fig. 8.1. By a random choice of a complex constant γ, singularities will not occur for all $t \in [0, 1)$ as on the left, but they may occur at the end, for $t = 1$.

The same random constant γ ensures that all paths stay bounded for all $t \in [0, 1)$. By this we mean that no path diverges to infinity for some $t \in [0, 1)$. Equivalently, for all $t \in [0, 1)$, the system $h(\mathbf{x}, t) = \mathbf{0}$ has no solutions at infinity (see Figure 8.2). To see this, invoke a homogeneous coordinatetransformation introducing one extra coordinate, and consider the system in projective space. That is, consider the homogenized system $H(X, Y, Z, t) = 0$ obtained by clearing Z from denominators in the expression $h(X/Z, Y/Z, t) = 0$. Now, instead of the discriminant system of (8.3) our concern is the system

$$\begin{cases} H(X, Y, Z, t) = 0 \\ Z = 0 \end{cases}$$

Since h is homogeneous in X, Y, Z, the solutions live in projective space, which we can restate to say that all solutions to $H(X, Y, 0, t) = 0$ must either satisfy $H(X/Y, 1, 0, t) = 0$ or $H(1, Y/X, 0, t) = 0$ (or both, if neither X or Y is zero). Either of these is a system of two polynomials in two variables and γ and so one can again apply elimination and see that, except for special choices of γ, there will be no solutions at infinity for $t \in [0, 1)$.

Note that if the polynomials in the start system $g(x, y) = 0$ have lower degrees than their counterparts in $f(x, y) = 0$, then $H(X, Y, Z, t) = 0$ could have solutions at infinity for $t = 0$. By matching the degrees of the polynomials in g and f, we avoid this, which is key in proving the third property of a good homotopy: accessibility.

Exercise 8.2.2. Consider the homotopy

$$h(x, y, t) = \left(\left\{ \begin{array}{l} x^2 - 1 = 0 \\ y^2 - 1 = 0 \end{array} \right. \right) (1 - t) + \left(\left\{ \begin{array}{l} y^2 - 1 = 0 \\ x^2 - 3 = 0 \end{array} \right. \right) t.$$

For which values of t do we have diverging paths? Show that with a random complex constant γ in $h(x, y, t) = \mathbf{0}$ (as in (8.4)) there are no divergent paths.

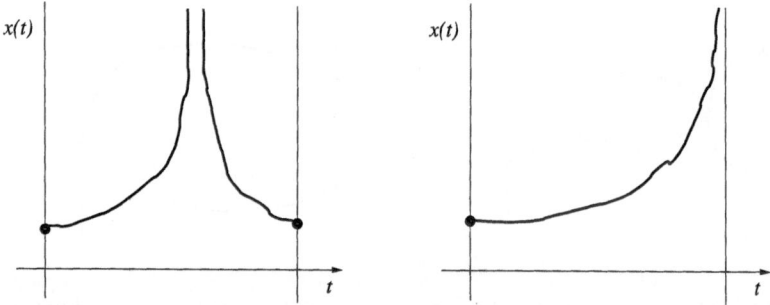

Fig. 8.2. By a random choice of a complex constant γ, divergence will not occur for all $t \in [0, 1)$ as on the left, but may occur at the end, for $t = 1$.

To understand why the homotopy has the accessibility property (defined in Section 8.1), consider that whenever the number of equations is equal to the number of variables $\mathbf{x}$, continuity implies that an isolated root at $t = 1$ must be approached by at least one isolated root as $t \to 1$. Since there are no singularities or solutions at infinity for t in $[0, 1)$, we can carry this argument backwards all the way to $t = 0$, where we know we are starting with all the solutions of the homotopy.

The arguments described above can be found in [BCSS98], see also [LS87].

8.2.2 Path following techniques

Consider any homotopy $h_k(x(t), y(t), t) = 0$, $k = 1, 2$. Since we are interested to see how x and y change as t changes, we apply the operator $\frac{\partial}{\partial t}$ on the homotopy. Via the chain rule, we obtain

$$\frac{\partial h_k}{\partial x}\frac{\partial x}{\partial t} + \frac{\partial h_k}{\partial y}\frac{\partial y}{\partial t} + \frac{\partial h_k}{\partial t} = 0, \quad k = 1, 2.$$

Denote $\Delta x := \frac{\partial x}{\partial t}$ and $\Delta y := \frac{\partial y}{\partial t}$. For fixed t (after incrementing $t := t + \Delta t$), for $k = 1, 2$, we solve the linear system

$$\begin{bmatrix} \frac{\partial h_1}{\partial x} & \frac{\partial h_1}{\partial y} \\ \frac{\partial h_2}{\partial x} & \frac{\partial h_2}{\partial y} \end{bmatrix} \begin{bmatrix} \Delta x \\ \Delta y \end{bmatrix} = - \begin{bmatrix} \frac{\partial h_1}{\partial t} \\ \frac{\partial h_2}{\partial t} \end{bmatrix}$$

and obtain $(\Delta x, \Delta y)$, the tangent to the path. For some step size $\lambda > 0$, the updates $x := x + \lambda \Delta x$ and $y := y + \lambda \Delta y$ give the Euler predictor.

To avoid solving a linear system at each predictor step, we may use a secant predictor. A secant predictor is less accurate and will require more corrector steps, but the total amount of work for the prediction can be less. Cubic interpolation, using the tangent vectors at two points along the path, leads to the Hermite predictor. See Figure 8.3 for a comparison.

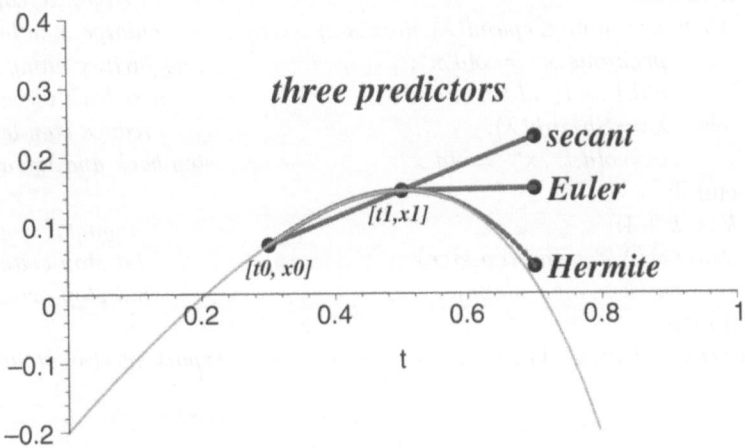

Fig. 8.3. Three predictors: secant, Euler, and Hermite.

The *predictor* delivers at each step of the method a new value of the continuation parameter and predicts an approximate solution of the corresponding new system in the homotopy. Then, the predicted approximate solution is corrected by applying the *corrector*, e.g., by Newton's method. With a good homotopy, the solution paths never turn back as t increases. Therefore, the continuation parameter can remain fixed while correcting the predicted solution. This leads to so-called *increment-and-fix* path following methods. In practice, determining the step length during the prediction stage is done by a hit-or-miss method, which can be implemented by means of an *adaptive step size control*, as done in the algorithm below.

Algorithm 8.2.3 Following one solution path by an increment-and-fix predictor-corrector method with an adaptive step size control strategy.

Input: $h(\mathbf{x}, t),\ \mathbf{x}^* \in \mathbb{C}^n\colon h(\mathbf{x}^*, 0) = \mathbf{0},$ *homotopy and root*
           $\epsilon > 0,\ max\_it,\ max\_steps,$ *defines stop criteria*
           $min\_step\_size, max\_step\_size.$ *for step size control*
Output: $\mathbf{x}^*$, success if $\|h(\mathbf{x}^*, 1)\| \le \epsilon.$ *approximate root at end*

$t := 0;\ \ k := 0;$	*initialization*
$\lambda := max\_step\_size;$	*step length*
$old\_t := t;\ \ old\_\mathbf{x}^* := \mathbf{x}^*$	*back up for t and $\mathbf{x}^*$*
$previous\_\mathbf{x}^* := \mathbf{x}^*;$	*previous solution*
stop := false;	*combines stop criteria*
while $t < 1$ and not stop loop	
$t := \min(1, t + \lambda);$	*secant predictor for t*
$\mathbf{x}^* := \mathbf{x}^* + \lambda(\mathbf{x}^* - previous\_\mathbf{x}^*);$	*secant predictor for $\mathbf{x}^*$*
Newton$(h(\mathbf{x}, t), \mathbf{x}^*, \epsilon, max\_it,$success$);$	*correct with Newton*
if success	*step size control*
then $\lambda := \min(Expand(\lambda), max\_step\_size);$	*enlarge step length*
$previous\_\mathbf{x}^* := old\_\mathbf{x}^*;$	*go further along path*
$old\_t := t;\ \ old\_\mathbf{x}^* := \mathbf{x}^*;$	*new back up values*
else $\lambda := Shrink(\lambda);$	*reduce step length*
$t := old\_t;\ \ \mathbf{x}^* := old\_\mathbf{x}^*;$	*step back and try again*
end if;	
$k := k + 1;$	*augment counter*
stop := $(\lambda < min\_step\_size)$	*1st stop criterion*
or $(k > max\_steps);$	*2nd stop criterion*
end loop;	
success := $(\|h(\mathbf{x}^*, 1)\| \le \epsilon).$	*report success or failure*

The path following algorithm contains three key ingredients in its loop: the predictor, the corrector and the step size control. The step size λ is controlled by the functions *Shrink* and *Expand* which respectively reduce and enlarge λ, depending on the outcome of the corrector.

The algorithm is still abstract because we did not specify particular values for the constants, such as tolerances on the solutions, minimal and maximal step size, maximum number of iterations of Newton's method, etc.

8.2.3 Homotopies exploiting product structures

A typical homotopy looks as follows:

$$h(\mathbf{x}, t) = \gamma g(\mathbf{x})(1 - t) + f(\mathbf{x})t = 0, \quad \gamma \in \mathbb{C},$$

where a random γ ensures the regularity and boundedness of the paths.

In general, for a system $f = (f_1, f_2, \ldots, f_n)$, with $d_i = \deg(f_i)$, we set up a start system $g(\mathbf{x}) = \mathbf{0}$ as follows:

$$g(\mathbf{x}) = \begin{cases} \alpha_1 x_1^{d_1} - \beta_1 = 0 \\ \alpha_2 x_2^{d_2} - \beta_2 = 0 \\ \vdots \\ \alpha_n x_n^{d_n} - \beta_n = 0 \end{cases}$$

where the coefficients α_i and β_i, for $i = 1, 2, \ldots, n$, are chosen at random in $\mathbb{C}$. Therefore $g(\mathbf{x}) = \mathbf{0}$ has exactly as many regular solutions as the total degree $D = \prod_{i=1}^{n} d_i$. So this homotopy defines D solution paths. The theorem of Bézout (which can be proven constructively via a homotopy) indeed predicts D as the number of solutions *in complex projective space*.

Exercise 8.2.4. Consider the following polynomial system:

$$\begin{cases} x^{108} + 1.1 y^{54} - 1.1 y = 0 \\ y^{108} + 1.1 x^{54} - 1.1 x = 0 \end{cases}.$$

This system was constructed by Bertrand Haas [Haa02] who provided with this system a counterexample to the conjecture of Kushnirenko on the number of real roots of sparse systems. Use **phc** (available via [Ver99a]) to determine[5] how many solutions of this system are complex. How many are real?

In almost all applications, the systems have far fewer solutions than the total degree (most solutions lie at infinity and are of no interest). Consider the eigenvalue problem $A\mathbf{x} = \lambda \mathbf{x}$, $A \in \mathbb{C}^{n \times n}$. To make the system square, we can add one general hyperplane to obtain a unique $\mathbf{x}$ for every λ. If we apply Bézout's theorem in a straightforward manner, we consider $A\mathbf{x} = \lambda \mathbf{x}$ as a system of n quadrics and obtain a homotopy with $D = 2^n$ to trace, whereas we know there can be at most n solutions! This is a highly wasteful computation, as $2^n - n$ of our solution paths are certain to diverge to infinity.

Let us examine the smallest nontrivial case: $n = 2$. We consider a general 2-by-2 matrix A and scale the components of the eigenvector with a random hyperplane $c_0 + c_1 x_1 + c_2 x_2 = 0$. So we look at the system

$$f(x_1, x_2, \lambda) = \begin{cases} a_{11} x_1 + a_{12} x_2 - \lambda x_1 = 0 \\ a_{21} x_1 + a_{22} x_2 - \lambda x_2 = 0 \\ c_0 + c_1 x_1 + c_2 x_2 = 0 \end{cases}.$$

To compute the solutions at infinity, we go to homogeneous coordinates, replacing x_1 by x_1/x_0, x_2 by x_2/x_0, and λ by λ/x_0. Clearing denominators:

---

[5] This may take some time (especially on slower machines)...

$$f(x_0, x_1, x_2, \lambda) = \begin{cases} a_{11}x_0x_1 + a_{12}x_0x_2 - \lambda x_1 = 0 \\ a_{21}x_0x_1 + a_{22}x_0x_2 - \lambda x_2 = 0 \\ c_0x_0 + c_1x_1 + c_2x_2 = 0 \end{cases}.$$

Solutions at infinity are solutions of the homogeneous system with $x_0 = 0$ and not all components equal to zero. If $\lambda = 0$, then $(x_0, x_1, x_2, \lambda) = (0, 1, -c_1/c_2, 0)$ represents one point at infinity. If $\lambda \neq 0$, then the other solution at infinity is represented by $(x_0, x_1, x_2, \lambda) = (0, 0, 0, 1)$. So we found where two of the four paths are diverging to.

Now we embed our problem in *multi-projective* space: $\mathbb{P} \times \mathbb{P}^2$, separating λ from $\mathbf{x}$. To go to 2-homogeneous coordinates, we replace x_2 by x_2/x_0, x_1 by x_1/x_0 (as before), and λ by λ_1/λ_0 (this is new), clearing denominators:

$$f(x_0, x_1, x_2, \lambda_0, \lambda_1) = \begin{cases} a_{11}\lambda_0 x_1 + a_{12}\lambda_0 x_2 - \lambda_1 x_1 = 0 \\ a_{21}\lambda_0 x_1 + a_{22}\lambda_0 x_2 - \lambda_1 x_2 = 0 \\ c_0x_0 + c_1x_1 + c_2x_2 = 0 \end{cases}. \qquad (8.5)$$

Looking for roots at infinity of (8.5) we see that $\lambda_0 = 0$ implies $x_1 = 0$, $x_2 = 0$, and thus $x_0 = 0$, so we have no proper solution at infinity with $\lambda_0 = 0$. For the solutions at infinity of (8.5) with $x_0 = 0$, considering (8.5) back in affine coordinates for λ (as λ_0 cannot be zero), we are looking at a homogeneous system of three equations in three unknowns: x_1, x_2, and λ. For general matrices, the trivial zero solution is the only solution. Thus in $\mathbb{P} \times \mathbb{P}^2$, the general eigenvalue problem has no solutions at infinity.

To arrive at a version of Bézout's theorem for polynomial systems over multi-projective spaces, we need to define our root count. Continuing our running example, we record the degrees in λ and $\{x_1, x_2\}$ of every equation in a table. Corresponding to this degree table is a linear-product start system, written in (8.6) in table format.

	$\{\lambda\}$	$\{x_1, x_2\}$
(1)	1	1
(2)	1	1
(3)	0	1

degree table

$\Longleftrightarrow$

	$\{\lambda\}$	$\{x_1, x_2\}$
(1)	$\alpha_{10} + \alpha_{11}\lambda$	$\beta_{10} + \beta_{11}x_1 + \beta_{12}x_2$
(2)	$\alpha_{20} + \alpha_{21}\lambda$	$\beta_{20} + \beta_{21}x_1 + \beta_{22}x_2$
(3)	1	$\beta_{30} + \beta_{31}x_1 + \beta_{32}x_2$

linear-product start system   (8.6)

The coefficients α_{ij} and β_{ij} in (8.6) are randomly chosen complex numbers. Except for a special choice of these numbers, the linear-product start system will always have two regular solutions. We derive a formal root count following the moves we make to solve the linear-product start system:

$$\begin{array}{ccccccccccccccc} B = & 1 & \times & 1 & \times & 1 & + & 1 & \times & 1 & \times & 1 & + & 0 & \times & 1 & \times & 1. \\ & (1)_\lambda & & (2)_\mathbf{x} & & (3)_\mathbf{x} & & (2)_\lambda & & (1)_\mathbf{x} & & (3)_\mathbf{x} & & (3)_\lambda & & (1)_\mathbf{x} & & (2)_\mathbf{x} \end{array} \qquad (8.7)$$

The labels in (8.7) show the navigation through the table at the right of (8.6).

Exercise 8.2.5. The matrix polynomial

$$p(\lambda) = A_d\lambda^d + A_{d-1}\lambda^{d-1} + \cdots + A_1\lambda + A_0, \quad A_i \in \mathbb{C}^{n \times n},$$

defines the generalized eigenvalue problem $p(\lambda)\mathbf{x} = \mathbf{0}$. How many generalized eigenvalue-eigenvector pairs can we expect for randomly chosen matrices A_i?

To show that B is an upper bound for the number of isolated solutions of a polynomial system, we show the regularity and boundedness of the solution paths in a typical homotopy, using a linear-product start system.

For many applications (like the eigenvalue problem) it is obvious how best to separate the variables into a partition. But for black-box solvers and systems with no apparent product structure, we need to find that partition which leads to the smallest Bézout number. One strategy is to enumerate all partitions and retain the partition with the smallest Bézout number. While the number of partitions grows faster than 2^n, finding the smallest Bézout number for $n = 8$ by enumeration takes less than a second of CPU time.

Instead of using one partition of the variables to model the product structure of the system, we may use different partitions for different equations, and extend this even further to construct in this way general linear-product start systems. The solving of the start system now involves more work, but we may expect the homotopy to be more efficient. Schematically, a hierarchy of homotopies (and root counting methods) is given in Figure 8.4.

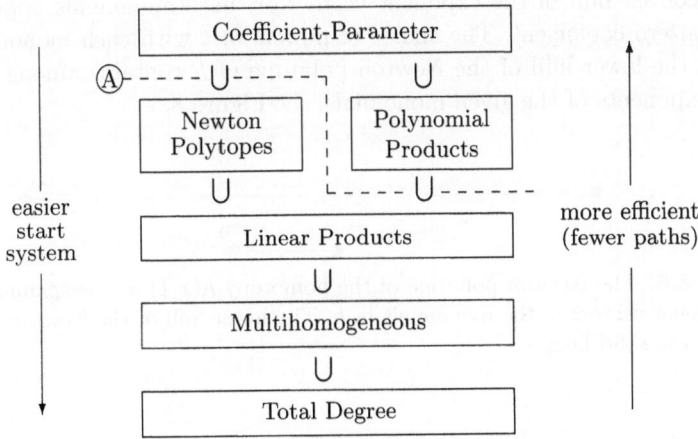

Fig. 8.4. A hierarchy of homotopies. All homotopies below the dashed line A can be done automatically. Above the line, apply special ad-hoc methods or bootstrapping. Homotopies at the bottom of the hierarchy are often used to find solutions for generic instances of parameters in a coefficient-parameter homotopy.

We will not address the "polynomial products" of Figure 8.4 here; for this, see [MSW95]. We introduce the Newton polytopes in the following two sections.

For the relation between Newton polytopes and resultants, see Chapter 7.

8.2.4 Polyhedral homotopies to glue real solutions

The purpose of this section is to introduce Newton polytopes and polyhedral homotopies, but without mixed volumes. So we restrict ourselves to polynomials in one variable. Instead of "just" solving a polynomial in one variable, we consider a different problem:

> Input: k distinct monomials in one variable x:
> $x^{a_1}, x^{a_2}, \ldots, x^{a_k}$, with $a_i \neq a_j$ for $i \neq j$.
> Output: coefficients $c_{a_1}, c_{a_2}, \ldots, c_{a_k}$ such that
> $f(x) = c_{a_1} x^{a_1} + c_{a_2} x^{a_2} + \cdots + c_{a_k} x^{a_k}$
> has $k - 1$ positive real roots.

For example, take $1, x^5, x^7, x^{11}$ as monomials on input. Then the problem is to find $c_0, c_5, c_7,$ and c_{11} such that $f(x) = c_0 1 + c_5 x^5 + c_7 x^7 + c_{11} x^{11}$ has three positive real solutions. We will show that we can reduce this four dimensional problem in that of one dimension, considering the homotopy

$$h(x, t) = t - x^5 + x^7 - x^{11} t = 0, \quad \text{for } t \geq 0.$$

The alternation of signs in the coefficients is a deliberate choice to maximize the number of positive real roots. The Newton polytope of a polynomial is the convex hull of the exponent vectors of those monomials appearing with a nonzero coefficient. The choice of powers of t with each monomial is such that the lower hull of the Newton polytope of h contains among its vertices all exponents of the given monomials, see Figure 8.5.

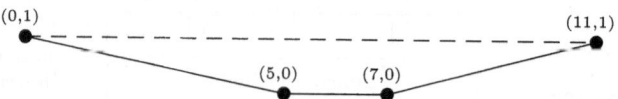

Fig. 8.5. The Newton polytope of the homotopy $h(x, t) = 0$ is spanned by by the exponent vectors of the monomials in h. The lower hull of the Newton polytopes is drawn in solid lines.

At $t = 0$, the homotopy $h(x, 0) = -x^5 + x^7 = x^5(-1 + x^2) = 0$ has one positive real root: $x = 1$. The idea is to choose $t = \Delta t > 0$ such that Newton's method applied to $h(x, \Delta t) = 0$ converges quadratically to a positive real root starting at $x = 1$. (Notice that by the fortunate choice of the powers of t in the example, Δt can be chosen arbitrarily large as $h(1, t) \equiv 0$, for any value of t.)

Observe that the monomials in $h(x, 0)$ correspond to the lowest middle edge on the lower hull of the Newton polytope of h in Figure 8.5. For every edge of the lower hull of the Newton polytope we will use one homotopy to find one positive real root. Each time, the start system in the homotopy has its two monomials as vertices of an edge of the lower hull. To find the homotopies with the other two edges, we need to consider the vectors orthogonal to the edges (we call those vectors *inner normals*), see Figure 8.6.

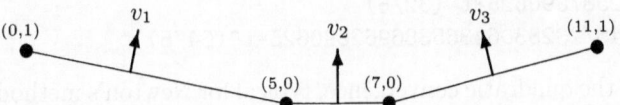

Fig. 8.6. Inner normals $v_1 = (\frac{1}{5}, 1)$, $v_2 = (0, 1)$, $v_3 = (-\frac{1}{4}, 1)$ on the edges of the lower hull of the Newton polytope of the homotopy $h(x, t) = 0$.

The inner normal v_1 attains the minimal inner product with those vertices on the first edge of the lower hull. Consider the four values of the inner product of v_1 with the four vertices of the lower hull:

$$\left\langle \left(\frac{1}{5}, 1 \right), \{(0,1), (5,0), (7,0), (11,1)\} \right\rangle = \left\{ 1, 1, \frac{7}{5}, \frac{16}{5} \right\}.$$

Indeed, the minimal values occur with the first two vertices which span the first edge. This geometric construction motivates the following change of co-ordinates: let $x = yt^{1/5}$, we obtain

$$h(y, t) = t - y^5 t + y^7 t^{7/5} - y^{11} t^{16/5} \tag{8.8}$$

$$= t \left(1 - y^5 + y^7 t^{2/5} - y^{11} t^{11/5} \right). \tag{8.9}$$

We see that $\frac{1}{t} h(y, 0) = 1 - y^5 = 0$ has one positive real root: $y = 1$. Now we can choose $t = \Delta t > 0$ such that Newton's method converges quadratically to a positive real root starting at $y = 1$. Let $y^*: h(y^*, \Delta t) = 0$, then we find the corresponding root in the original coordinates as $x^* = y^*(\Delta t)^{1/5}$.

We can even explicitly construct the fractional power series using Newton's method in a computer algebra system like Maple. The following sequence of Maple commands achieve this:

```
> h := t-x^5 + x^7 - x^(11)*t:
> hy := subs(x = y*t^(1/5),h):
> hyt := simplify(hy/t):
> newton := x -> x - subs(y=x,hyt/diff(hyt,y)):
> x[0] := 1:
> for k from 1 to 6 do
>    x[k] := newton(x[k-1]):
>    s[k] := series(x[k],t=0,15):
>    lprint(op(1,s[k]-s[k-1]));
> end do:
```

The output of the loop (done in Maple 9) shows the errors between two consecutive series expansions:

```
1
-301/15625*t^2
-84/3125*t^2
-2112/1953125*t^(18/5)
-32768/152587890625*t^(32/5)
-2147483648/23283064365386962890625*t^(64/5)
```

We observe the quadratic convergence, typical for Newton's method. While the particular values for the errors shows above may differ on other platforms with different versions of Maple, the computed fractional power series expansion is "exact", here we see the series up to third order:

```
> series(x[6],t=0,3);
```

```
        2/5     4/5         6/5           8/5            2       11/5
       t       t       34 t         266 t        11284 t       t
   1 + ---- + ---- + ------- + -------- + -------- - -----
        5       5       125           625          15625         5


              12/5         13/5        14/5
        100947 t       14 t        12 t                3
     + ------------ - -------- + -------- + O(t )
           78125           25            5
```

To find the third positive real root, we proceed in a similar fashion, using the third inner normal $v_3 = (-1/4, 1)$ in the coordinate change $x = yt^{-1/4}$. As it turns out, we can take Δt quite large. For $\Delta t = 0.1$, $h(x, 0.1) = 0$ has the following three positive (approximate) real roots: 0.73, 1.0, and 1.56. As Δt grows larger, the real roots collide into multiple roots before escaping to the complex plane.

Exercise 8.2.6. Compute the fractional power series for the third positive real root, using Newton's method like shown above. Make sure enough terms in the series expansions are used so that the quadratic convergence is obvious.

In numerical implementations of polyhedral homotopies, we only use the first term of the fractional power series (also known as Puiseux series). The connection between these fractional power series and Newton polygons is classical for polynomials in two variables, see for example [Lef53] or [Wal62]. The generalization to systems of equations can be found in [McD02].

Using Newton polytopes to construct real curves and hypersurfaces with a prescribed topology is done by Viro's method [IS03, IV96]. This homotopy to glue real roots can be generalized to the case of complete intersections by the use of mixed subdivisions, see [Stu94b, Stu94c]. We will define these mixed subdivisions in the next section. We apply these co-called polyhedral homotopies to solve generic polynomial systems with given fixed Newton polytopes.

8.2.5 The Cayley trick and Minkowski's theorem

Mixed volumes were defined by Minkowski who showed that the volume of a linear combination of polytopes is a homogeneous polynomial in the factors of of the combination. The coefficients of this polynomial are mixed volumes. We will visualize this theorem on a simple example by the Cayley trick.

The Cayley trick [GKZ94, Proposition 1.7, page 274] is a method to rewrite a certain resultant as a discriminant of one single polynomial with additional variables. The polyhedral version of this trick as in [Stu94a, Lem. 5.2] is due to Bernd Sturmfels. See [HRS00] for another application of this trick.

Consider the following system:

$$f = (f_1, f_2) \qquad\qquad \mathcal{A} = (A_1, A_2)$$
$$= \begin{cases} x_1^3 x_2 + x_1 x_2^2 + 1 = 0 & A_1 = \{(3,1),(1,2),(0,0)\} \\ x_1^4 + x_1 x_2 + 1 = 0 & A_2 = \{(4,0),(1,1),(0,0)\} \end{cases}$$

The sparse structure of f is modeled by the tuple $\mathcal{A} = (A_1, A_2)$, where A_1 and A_2 are the *supports* of f_1 and f_2 respectively. The Newton polytopes are the convex hulls of the supports. The Cayley polytope of r polytopes is the convex hull of the polytopes placed at the vertices of an $(r-1)$-dimensional unit simplex. Figure 8.7 illustrates this construction for our example.

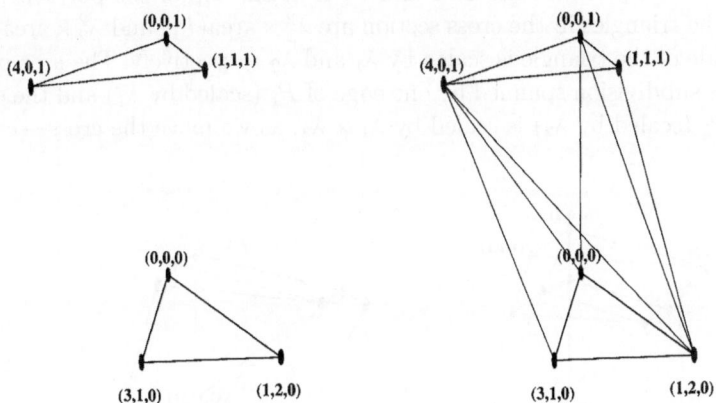

Fig. 8.7. The Cayley polytope of two polygons. The first polygon is placed at the vertex $(0,0,0)$, the second polygon is placed at $(0,0,1)$.

For our example, the Cayley polytope is so simple that a triangulation is obvious (see Figure 8.8). As every simplex has four vertices, either the simplex has three vertices from the same polygon (and the fourth one of the other polygon), or the simplex has two vertices of each polygon. A simplex of the first type is called unmixed, a simplex of the second type is mixed. Imagine taking slices parallel to the base of the Cayley polytope. These slices produce scaled copies of the original polygons in the unmixed simplices. In the mixed simplex we find one scaled edge from the first and another scaled edge from the second polygon, see Figure 8.8.

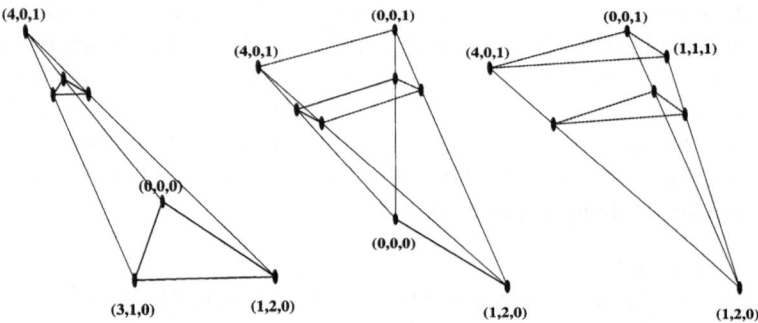

Fig. 8.8. A triangulation of the Cayley polytope. The middle simplex is mixed, the other two simplices are unmixed.

On Figure 8.9 we see in the cross section of the Cayley polytope a mixed subdivision of the convex combination $\lambda_1 P_1 + \lambda_2 P_2$, $\lambda_1 + \lambda_2 = 1$, $\lambda_1 \geq 0$ and $\lambda_2 \geq 0$, where P_1 defines the base and P_2 is at the top of the polytope. The areas of the triangles in the cross section are $\lambda_1^2 \times \text{area}(P_1)$ and $\lambda_2^2 \times \text{area}(P_2)$, as each side of the triangle is scaled by λ_1 and λ_2 respectively. The area of the cell in the subdivision spanned by one edge of P_1 (scaled by λ_1) and the other edge of P_2 (scaled by λ_2) is scaled by $\lambda_1 \times \lambda_2$, as we move the cross section.

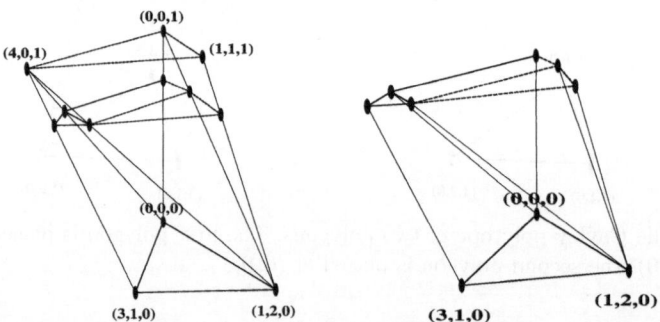

Fig. 8.9. A mixed subdivision induced by a triangulation of the Cayley polytope.

In Figure 8.10 we show the Minkowski sum of the two polygons P_1 and P_2, with their mixed subdivision corresponding to the triangulation of the Cayley polytope. For this example, Minkowski's theorem becomes

$$\text{area}(\lambda_1 P_1 + \lambda_2 P_2) = V(P_1, P_1)\lambda_1^2 + V(P_1, P_2)\lambda_1\lambda_2 + V(P_2, P_2)\lambda_2^2 \qquad (8.10)$$
$$= 3\lambda_1^2 + 8\lambda_1\lambda_2 + 2\lambda_2^2.$$

The coefficients in the polynomial (8.10) are mixed volumes (or areas in our example): $V(P_1, P_1)$ and $V(P_2, P_2)$ are the respective areas of P_1 and P_2, while $V(P_1, P_2)$ is the mixed area.

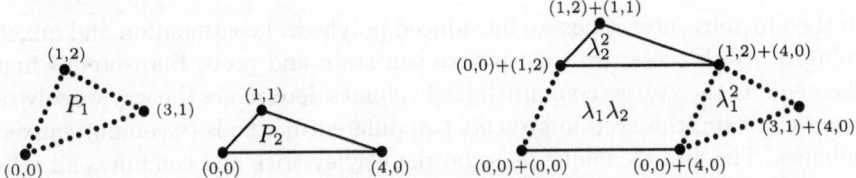

Fig. 8.10. A subdivision of the sum of two polygons P_1 and P_2. The sum is the convex hull of all sums of the vertices of the polygons. The cells in the subdivision are labeled by the multipliers for the area of $\lambda_1 P_1 + \lambda_2 P_2$.

The subdivisions we need are induced by a lifting. Such subdivisions are called regular, they define polyhedral homotopies. For the example, the lifted supports are $\widehat{\mathcal{A}} = (\widehat{A}_1, \widehat{A}_2)$, with

$$\widehat{A}_1 = \{(3,1,1), (1,2,0), (0,0,0)\} \text{ and } \widehat{A}_2 = \{(4,0,0), (1,1,1), (0,0,0)\}.$$

Figure 8.11 shows the mixed subdivision of Figure 8.10 as induced by the lower hull of the sum of the lifted polytopes.

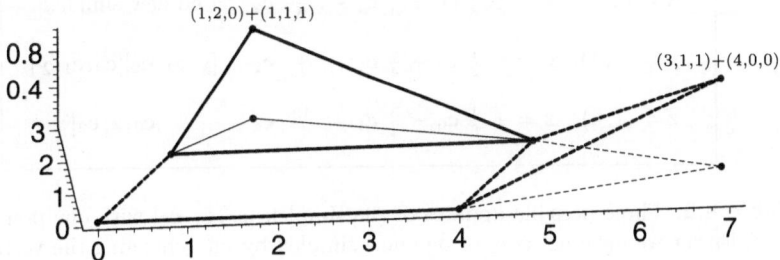

Fig. 8.11. A mixed subdivision is regular if it is induced by a lifting.

As there is only one mixed cell in the mixed subdivision of the Newton polytopes of our example, there is only one homotopy to consider, for example:

$$h(\mathbf{x}, t) = \begin{cases} x_1^3 x_2 t + x_1 x_2^2 + 1 = 0 \\ x_1^4 + x_1 x_2 t + 1 = 0 \end{cases} \qquad (8.11)$$

The powers of the t in $h(\mathbf{x}, t) = \mathbf{0}$ are the lifting values of the supports which induced the mixed subdivision shown in Figure 8.11.

Exercise 8.2.7. Verify that the start system $h(\mathbf{x}, t = 0) = \mathbf{0}$ in the polyhedral homotopy (8.11) has indeed eight $(= V(P_1, P_2))$ regular solutions. Show that any system with exactly two monomials in every equation has always as many regular roots as its mixed volume, for any nonzero choice of the coefficients.

8.2.6 Computing mixed volumes and polyhedral continuation

In the previous subsections we introduced polyhedral continuation and mixed volumes. With these two concepts we can state and prove Bernshteĭn's first theorem. As the way we compute mixed volumes determines the way we solve a generic system, this section presents two different methods to compute mixed volumes. The first technique relies on the Cayley trick and computes all cells in a mixed subdivision. The second method uses linear programming and leads to an efficient enumeration of all mixed cells in a mixed subdivision.

With the Cayley trick we can obtain a regular mixed subdivision as a regular triangulation of the Cayley polytope. We next introduce a method to compute a regular triangulation of any polytope. Our method will construct the triangulation incrementally, adding the points one after the other. The key operation is to decompose one point with respect to one simplex. Consider for example the simplex $[\mathbf{c}_0, \mathbf{c}_1, \mathbf{c}_2]$ spanned by $\mathbf{c}_0 = (0,0)$, $\mathbf{c}_1 = (3,2)$, and $\mathbf{c}_2 = (2,4)$. If we take one extra point, three possible updates can occur, illustrated by Table 8.2.

point	barycentric decomposition	pivoting
$\mathbf{x} = (2,3)$:	$\mathbf{x} = +\frac{1}{8}\mathbf{c}_0 + \frac{1}{4}\mathbf{c}_1 + \frac{5}{8}\mathbf{c}_2$	no new simplex
$\mathbf{y} = (5,1)$:	$\mathbf{y} = -\frac{1}{3}\mathbf{c}_0 + \frac{9}{4}\mathbf{c}_1 - \frac{7}{8}\mathbf{c}_2$	$[\mathbf{y}, \mathbf{c}_1, \mathbf{c}_2][\mathbf{c}_0, \mathbf{c}_1, \mathbf{y}]$
$\mathbf{z} = (1,5)$:	$\mathbf{z} = +\frac{1}{8}\mathbf{c}_0 - \frac{3}{4}\mathbf{c}_1 + \frac{13}{8}\mathbf{c}_2$	$[\mathbf{c}_0, \mathbf{z}, \mathbf{c}_2]$

Table 8.2. Three possible updates of the simplex $[\mathbf{c}_0, \mathbf{c}_1, \mathbf{c}_2]$ with one point, $\mathbf{x}$, $\mathbf{y}$, or $\mathbf{z}$. Either we have no, two, or one new simplex by interchanging the vertex with negative coefficient with the point.

Solving a linear system we can write any point as a linear combination of the vertices of a simplex, requiring the coefficients in that linear combination to sum up to one. We call this linear combination a barycentric decomposition of a point with respect to a simplex. The negative signs of the coefficients in this barycentric decomposition tell which vertices of the simplex to interchange with the new point to create new simplices in the triangulation of the convex hull of the original simplex and the point. As we can see from Fig-

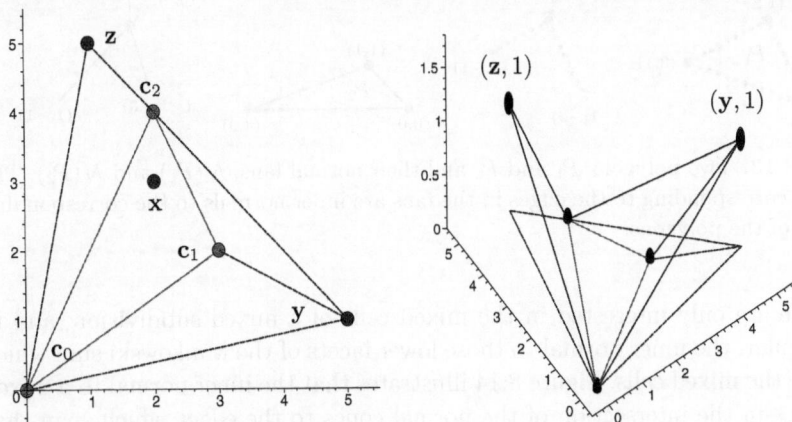

Fig. 8.12. Pivoting to obtain a regular triangulation of a polygon. The construction on the right shows how the triangulation can be obtained as the lower hull of $\mathbf{y}$ and $\mathbf{z}$ lifted at height one, with $[\mathbf{c}_0, \mathbf{c}_1, \mathbf{c}_2]$ sitting at level zero.

ure 8.12, any triangulation obtained by placing points (see [Lee91] for more on triangulations) in this way is regular.

The algorithm to compute regular triangulations incrementally leads to an incremental polyhedral solver, which solves polynomial systems adding one monomial after the other, see [VGC96]. If the structure of a polynomial system is such that most polynomials share the same support (or more generally span the same Newton polytope), and thus there are only few distinct Newton polytopes to consider, then the Cayley trick is not too wasteful.

The complexity of computing volumes and mixed volumes is discussed respectively in [DF88] and [DGH98].

Theorem 8.2.8. (Bernshteĭn's theorem A) *The number of roots of a generic system equals the mixed volume of its Newton polytopes.*

In his proof of this theorem, Bernshteĭn [Ber75] used a homotopy (implemented in [VVC94]), based on a recursive formula for computing mixed volumes. This proof idea was generalized by Huber and Sturmfels in [HS95]. Note that the theorem concerns "generic systems", which are systems with randomly chosen coefficients. These generic systems serve as start system in a coefficient-parameter homotopy to solve any specific polynomial system with the same Newton polytopes.

For the coordinate changes in the polyhedral homotopies, we need to know the inner normals to the mixed cells. Therefore, we use a dual representation of polytopes, see Figure 8.13. The normal fan of a polytope is the collection of the normal cones to all faces of the polytope. The normal cone to a face contains all inner normals which define the face.

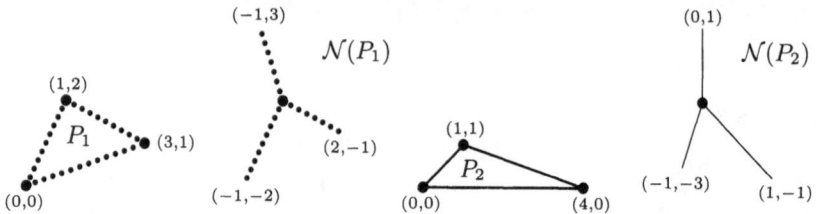

Fig. 8.13. Two polygons P_1 and P_2 and their normal fans, $\mathcal{N}(P_1)$ and $\mathcal{N}(P_2)$. The labels corresponding to the edges in the fans are inner normals to the corresponding edges of the polygons.

We are only interested in the mixed cells of a mixed subdivision, and in particular, the inner normal to those lower facets of the Minkowski sum which define the mixed cells. Figure 8.14 illustrates that the inner normal to a mixed cell lies in the intersection of the normal cones to the edges which span that mixed cell.

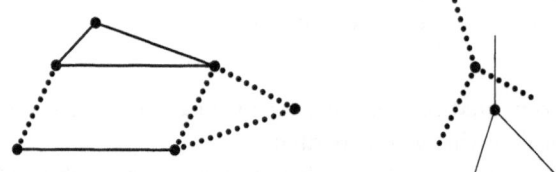

Fig. 8.14. The dual representation of a mixed subdivision.

The search for all inner normals to the mixed cells in a mixed subdivision naturally leads to a system of linear equalities and inequalities. For a tuple of n supports $(A_1, A_2, \ldots, A_n)$, consider an edge of the kth polytope, spanned by $\{\mathbf{a}, \mathbf{b}\} \subseteq A_k$. Then the inner normal $\mathbf{v}$ to this edge satisfies

$$\begin{cases} \langle \mathbf{a}, \mathbf{v} \rangle = \langle \mathbf{b}, \mathbf{v} \rangle \\ \langle \mathbf{a}, \mathbf{v} \rangle \leq \langle \mathbf{c}, \mathbf{v} \rangle, \quad \text{for all } \mathbf{c} \in A_k. \end{cases} \tag{8.12}$$

Enumerating all edges of a polytope is thus equivalent to enumerating all feasible solutions to the system (8.12). Letting k range from 1 to n in (8.12) applied to the lifted point sets $\widehat{A}_k$ provides the dual linear-programming model to enumerate all inner normals to the mixed cells in a regular mixed subdivision.

A lift-and-prune strategy to enumerate all mixed cells in a regular mixed subdivision was proposed in [EC95] and dualized in [VGC96]. Recently, insight in the linear programming methods has led to very efficient calculations of mixed volumes, as developed in [DKK03], [GL00, GL03], [KK03b], [LL01], and [TKF02].

8.2.7 Bernshteĭn's second theorem

When tracing solution paths diverging to infinity, one may wonder when to stop. After all, infinity is pretty far off, and even if good knowledge of the application domain gives us good bounds on the size of the solutions, we do not want to miss valid solutions with large components. If a path seems to diverge, we must know whether we have true divergence or convergence to a root with large components. Bernshteĭn's second theorem [Ber75] will provide us with a certificate of divergence.

For a system $f(\mathbf{x}) = 0$, supported by $\mathcal{A} = (A_1, A_2, \ldots, A_n)$, we can write its equations $f = (f_1, f_2, \ldots, f_n)$ as

$$f_i(\mathbf{x}) = \sum_{\mathbf{a} \in A_i} c_{i\mathbf{a}} \mathbf{x}^{\mathbf{a}}, \quad i = 1, 2, \ldots, n.$$

The Newton polytopes of f are denoted by $\mathcal{P} = (P_1, P_2, \ldots, P_n)$, with $P_i := \mathrm{conv}(A_i)$, $i = 1, 2, \ldots, n$. Then for any $\omega \neq \mathbf{0}$, we define the tuple of faces $\partial_\omega \mathcal{P} = (\partial_\omega P_1, \partial_\omega P_2, \ldots, \partial_\omega P_n)$, as $\partial_\omega P_i := \mathrm{conv}(\partial_\omega A_i)$, with

$$\partial_\omega A_i := \{ \, \mathbf{a} \in A_i \mid \langle \mathbf{a}, \omega \rangle = \min_{\mathbf{a}' \in A_i} \langle \mathbf{a}', \omega \rangle \, \}. \tag{8.13}$$

The set $\partial_\omega A_i$ is the support of the face of the ith polynomial f_i:

$$\partial_\omega f_i(\mathbf{x}) = \sum_{\mathbf{a} \in \partial_\omega A_i} c_{i\mathbf{a}} \mathbf{x}^{\mathbf{a}}.$$

We write $\partial_\omega f = (\partial_\omega f_1, \partial_\omega f_2, \ldots, \partial_\omega f_n)$ as the face of the system f determined by $\omega \neq \mathbf{0}$. The mixed volume of $\mathcal{P}$ is denoted by $V(\mathcal{P})$ and $\mathbb{C}^* = \mathbb{C} \setminus \{0\}$.

Theorem 8.2.9. (Bernshteĭn's theorem B) *If $\forall \omega \neq \mathbf{0}$, $\partial_\omega f(\mathbf{x}) = \mathbf{0}$ has no solutions in $(\mathbb{C}^*)^n$, then $V(\mathcal{P})$ is exact and all solutions are isolated. Otherwise, for $V(\mathcal{P}) \neq 0$: $V(\mathcal{P}) > \#isolated\ solutions$.*

Interestingly, the Newton polytopes may often be in general position, i.e.: $V(\mathcal{P})$ is exact for every nonzero choice of the coefficients. Consider for example the following system:

$$f(\mathbf{x}) = \begin{cases} c_{111} x_1 x_2 + c_{110} x_1 + c_{101} x_2 + c_{100} = 0 \\ c_{222} x_1^2 x_2^2 + c_{210} x_1 + c_{201} x_2 = 0 \end{cases}$$

We show the tuple of Newton polytopes in Figure 8.15.

Exercise 8.2.10. Verify that the mixed volume $V(P_1, P_2)$ of the polygons P_1 and P_2 is indeed equal to four.

While the observation in Figure 8.15 would let us believe that the mixed volume always provides a sharp root count, we have to keep in mind that the vertices of the polytopes are not randomly chosen. The vertices occur as

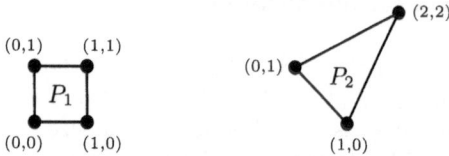

Fig. 8.15. Two Newton polygons in general position: $\forall w \neq \mathbf{0} : \partial_w A_1 + \partial_w A_2 \leq 3 \Rightarrow V(P_1, P_2) = 4$ is always exact, for all nonzero choices of the coefficients of f, because we need at least four monomials for $\partial_w f(\mathbf{x}) = \mathbf{0}$ to have all its roots in $(\mathbb{C}^*)^2$.

the exponents in the polynomials. For instance, general Newton polytopes are almost never simplicial, we usually find k-dimensional faces spanned by far more than $k + 1$ vertices.

Following Bernshteĭn we look at what happens when we consider the solution paths in a homotopy going from a generic to a specific polynomial system. At the limit of the paths, we look at the power series expansion, using the following result.

Theorem 8.2.11. $\forall \mathbf{x}(t)$, $h(\mathbf{x}(t), t) = (1 - t)g(\mathbf{x}(t)) + tf(\mathbf{x}(t)) = \mathbf{0}$,
$$\exists s > 0, \ m \in \mathbb{N} \setminus \{0\}, \ w \in \mathbb{Z}^n : \begin{cases} x_i(s) = b_i s^{w_i}(1 + O(s)), & i = 1, 2, \dots, n \\ t(s) = 1 - s^m & \text{for } t \approx 1, s \approx 0 \end{cases}$$

The number m is called the *winding number* of the solution at the end of the path (not to be confused with the multiplicity). The winding number is the smallest number so that $\mathbf{z}(2\pi m) = \mathbf{z}(0)$, if we consider $\mathbf{z}(\theta)$ a solution path of $h(\mathbf{z}(\theta), t(\theta)) = \mathbf{0}$, winding around 1 with values for the continuation parameter t defined by $t = 1 + (t_0 - 1)e^{i\theta}$, as $t_0 \approx 1$.

At the end of a path, when does $\lim_{t \to 1} x_i(t) \in \mathbb{C}^*$? From Theorem 8.2.11, we can characterize the divergence of the path $\mathbf{x}(t)$ by the leading exponents w in the power series:

$$x_i(t) \begin{cases} \to \infty \\ \in \mathbb{C}^* \\ \to 0 \end{cases} \Leftrightarrow w_i \begin{cases} < 0 \\ = 0 \\ > 0 \end{cases}$$

From this simple observation we see that a solution at infinity and a solution with zero components are regarded (or disregarded) equally.

Next we show the relation between face systems and power series. Assuming $\lim_{t \to 1} x_i(t) \notin \mathbb{C}^*$, and $w_i \neq 0$, we consider a diverging path.

First we substitute the power series $x_i(s) = b_i s^{w_i}(1 + O(s))$, $i = 1, 2, \dots, n$, $t(s) = 1 - s^m$, $s \approx 0$ into the homotopy $h(\mathbf{x}, t) = (1 - t)g(\mathbf{x}) + tf(\mathbf{x}) = \mathbf{0}$. We find
$$h(\mathbf{x}(s), t(s)) = \underbrace{f(\mathbf{x}(s))}_{\text{dominant as } s \to 0} + s^m(g(\mathbf{x}(s)) - f(\mathbf{x}(s))) = \mathbf{0}.$$

Thus (as expected), the choice of the start system $g(\mathbf{x}) = \mathbf{0}$ plays no role in what happens as s approaches zero. Let us now see what the substitution does to the ith polynomial:

$$f_i(\mathbf{x}) = \sum_{\mathbf{a} \in A_i} c_{i\mathbf{a}} \mathbf{x^a} \to f_i(\mathbf{x}(s)) = \underbrace{\sum_{\mathbf{a} \in A_i} c_{i\mathbf{a}} \prod_{i=1}^{n} b_i^{a_i} s^{\langle \mathbf{a}, \omega \rangle} (1 + O(s))}_{\partial_\omega f_i(\mathbf{x}(s)) \text{ dominant}}.$$

Arranging the monomials in $f(\mathbf{x}(s))$ in increasing order of powers of s, we see that the monomials that become dominant as $s \to 0$ have exponents whose inner product is minimal with ω. Recall that we characterize these exponents by the face of the support A_i in the direction of ω, see (8.13). Moreover, as $f_i(\mathbf{x}(s)) = 0$ for $s \to 0$, we see from the result of the substitution that then $\partial_\omega f_i(\mathbf{b}) = 0$, and thus $\partial_\omega f(\mathbf{b}) = \mathbf{0}$ for some $\mathbf{b} \in (\mathbb{C}^*)^n$.

This is the key idea in the proof of Bernshteĭn's second theorem. Like his first theorem, his idea is very constructive: follow the direction of a diverging path and (in addition to a solution at infinity) we find a face system which has solutions in $(\mathbb{C}^*)^n$. This face system forms a certificate for the mixed volume to overshoot the actual number of roots.

That Richardson extrapolation is useful to find ω is not so surprising. A closer inspection of the errors of the error expansion reveals that a similar extrapolation scheme can be applied to approximate the winding number m.

As we get closer to our target system, we have to decrease our step size when dealing with a difficult path. For the purpose of extrapolation, we better decrease the step size geometrically, i.e., for some λ, $0 < \lambda < 1$, consecutive values $t_0, t_1, \ldots t_k$ of the continuation parameter t satisfy $1 - t_k = \lambda(1 - t_k) = \cdots = \lambda^k(1 - t_0)$ and for the corresponding sequence of s-values we have $s_k = \lambda^{1/m} s_{k-1} = \cdots = \lambda^{k/m} s_0$.

Recall the form of the power series for a solution path $\mathbf{x}(s)$ for s approaching zero: $x_i(s) = b_i s^{\omega_i}(1 + O(s))$ with $t(s) = 1 - s^m$. Sampled along $s_0, s_1, \ldots, s_k$, we obtain

$$x_i(s_k) = b_i \lambda^{k\omega_i/m} s_0^{\omega_i} (1 + O(\lambda^{k/m} s_0)). \tag{8.14}$$

Since we are interested in the leading powers ω_i, we take the logarithms of the magnitudes of the points sampled along the path:

$$\log |x_i(s_k)| = \log |b_i| + \frac{k\omega_i}{m} \log(\lambda) + \omega_i \log(s_0) + \log \left| 1 + \sum_{j=0}^{\infty} b'_j (\lambda^{k/m} s_0)^j \right|.$$

A first-order approximation for ω_i is given by v_{kk+1} with the general extrapolation formula in $v_{k..l}$:

$$v_{kk+1} := \log |x_i(s_k + 1)| - \log |x_i(s_k)|, \quad v_{k..l} = v_{k..l-1} + \frac{v_{k+1..l} - v_{k..l-1}}{1 - \lambda}$$

which results in $\omega_i = m\frac{v_{0..r}}{\log(\lambda)} + O(s_0^r)$. While we can make the order r of the extrapolation as high as we like (thereby increasing the accuracy of ω_i). Notice that the formula assumes we know the winding number m.

If we examine the expansion of the errors:

$$e_i^{(k)} = (\log|x_i(s_k)| - \log|x_i(s_{k+1})|) \tag{8.15}$$
$$-(\log|x_i(s_{k+1})| - \log|x_i(s_{k+2})|) \tag{8.16}$$
$$= c_1\lambda^{k/m}s_0(1 + O(\lambda^{k/m})), \tag{8.17}$$

we find similar extrapolation formulas to approximate m:

$$e_i^{(kk+1)} := \log(e_i^{(k+1)}) - \log(e_i^{(k)}), \quad e_i^{(k..l)} = e_i^{(k+1..l)} + \frac{e_i^{(k..l-1)} - e_i^{(k+1..l)}}{1 - \lambda_{k..l}}$$

with $\lambda_{k..l} = \lambda^{(l-k-1)/m_{k..l}}$. So we obtain $m_{k..l} = \frac{\log(\lambda)}{e_i^{(k..l)}} + O(\lambda^{(l-k)k/m})$

The system of Cassou-Noguès is a very nice example. It illustrates how symbolic results can be obtained by purely numerical means.

$$f(b,c,d,e) = \begin{cases} 15b^4cd^2 + 6b^4c^3 + 21b^4c^2d - 144b^2c - 8b^2c^2e \\ -28b^2cde - 648b^2d + 36b^2d^2e + 9b^4d^3 - 120 = 0 \\ 30c^3b^4d - 32de^2c - 720db^2c - 24c^3b^2e - 432c^2b^2 + 576ec \\ -576de + 16cb^2d^2e + 16d^2e^2 + 16e^2c^2 + 9c^4b^4 + 5184 \\ +39d^2b^4c^2 + 18d^3b^4c - 432d^2b^2 + 24d^3b^2e - 16c^2b^2de - 240c = 0 \\ 216db^2c - 162d^2b^2 - 81c^2b^2 + 5184 + 1008ec - 1008de \\ +15c^2b^2de - 15c^3b^2e - 80de^2c + 40d^2e^2 + 40e^2c^2 = 0 \\ 261 + 4db^2c - 3d^2b^2 - 4c^2b^2 + 22ec - 22de = 0 \end{cases}$$

Root counts: $D = 1344$, $B = 312$, $V(\mathcal{P}) = 24$, but there are only 16 finite roots.

$$\partial_{(0,0,0,-1)}f(b,c,d,e) = \begin{cases} -8b^2c^2e - 28b^2cde + 36b^2d^2e = 0 \\ -32de^2c + 16d^2e^2 + 16e^2c^2 = 0 \\ -80de^2c + 40d^2e^2 + 40e^2c^2 = 0 \\ 22ec - 22de = 0 \end{cases}$$

The winding number is $m = 2$. See [HV98] for more about polyhedral end games.

8.3 Homotopies for positive dimensional solution sets

To introduce the numerical representation of positive dimensional solution sets, we start off with a dictionary, linking concepts in algebraic geometry to data and algorithms in numerical analysis. Witness sets form the central data and are obtained by a cascade of homotopies. The companion algorithms to the witness sets are membership tests to decide whether any given point belongs to a certain component of the solution set. We illustrate a numerical irreducible decomposition on a simple example and give an overview of our numerical factorization methods.

8.3.1 A dictionary

Kempf writes in [Kem93] that "Algebraic geometry studies the delicate balance between the geometrically plausible and the algebraically possible". With our numerical tools, we feel closer to the geometrical than to the algebraic side, because we are not calculating with polynomials in the algebraic sense. In [SVW03] we outlined the structure of a dictionary, presented as Table 8.3.

Numerical Algebraic Geometry Dictionary		
Algebraic Geometry	example in 3-space	Numerical Analysis
variety	collection of points, algebraic curves, and algebraic surfaces	polynomial system + union of witness sets, see below for the definition of a witness point
irreducible variety	a single point, or a single curve, or a single surface	polynomial system + witness set + probability-one membership test
generic point on an irreducible variety	random point on an algebraic curve or surface	point in a witness set; a witness point is a solution of the polynomial system on the variety and on a random slice whose codimension is the dimension of the variety
pure dimensional variety	one or more points, or one or more curves, or one or more surfaces	polynomial system + set of witness sets of same dimension + probability-one membership tests
irreducible decomposition of a variety	several pieces of different dimensions	polynomial system + array of sets of witness sets and probability-one membership tests

Table 8.3. Dictionary to translate algebraic geometry into numerical analysis.

8.3.2 Witness sets and a cascade of homotopies

A witness set is the basic concept of numerical algebraic geometry as it allows us to apply numerical methods for isolated solutions to positive dimensional solution components.

Every irreducible component of a solution set is presented by a witness set whose cardinality equals the degree of the irreducible component. To reduce a solution set of dimension k to a set of isolated points, we cut the k degrees of freedom by adding k random hyperplanes $L(\mathbf{x}) = \mathbf{0}$ to the system $f(\mathbf{x}) = \mathbf{0}$ which defines the entire solution set.

One obstacle is that we have to deal with systems whose number of equations in not necessarily the same as the number of unknowns. If there are fewer equations than unknowns, we simply add enough random hyperplanes to make up for the difference, so underdetermined systems are easy to handle.

Let us consider overdetermined systems, say f consists of 5 equations in 3 variables. To turn f into a system of N equations in N variables where N is either 3 or 5, we can respectively apply the following techniques:

randomization: Choosing random complex numbers a_{ij}, we add random combinations of the last two polynomials to the first three polynomials:

$$\begin{cases} f_1(\mathbf{x}) + a_{11} f_4(\mathbf{x}) + a_{12} f_5(\mathbf{x}) = 0 \\ f_2(\mathbf{x}) + a_{21} f_4(\mathbf{x}) + a_{22} f_5(\mathbf{x}) = 0 \\ f_3(\mathbf{x}) + a_{31} f_4(\mathbf{x}) + a_{32} f_5(\mathbf{x}) = 0 \end{cases}$$

slack variables: We introduce two new variables z_1 and z_2 (so-called slack variables) and add random multiples of these variables to every equation:

$$\begin{cases} f_1(\mathbf{x}) + a_{11} z_1 + a_{12} z_2 = 0 \\ f_2(\mathbf{x}) + a_{21} z_1 + a_{22} z_2 = 0 \\ f_3(\mathbf{x}) + a_{31} z_1 + a_{32} z_2 = 0 \\ f_4(\mathbf{x}) + a_{41} z_1 + a_{42} z_2 = 0 \\ f_5(\mathbf{x}) + a_{51} z_1 + a_{52} z_2 = 0 \end{cases}$$

While the randomization technique might seem at first more attractive because we are left with fewer equations, working with slack variables provides a cascade of homotopies to compute candidate witness points on all positive dimensional components.

In particular, considering f_4 and f_5 as hyperplanes L_1 and L_2 to cut the solution set of the first three equations in f, we consider a cascade of three systems. To get witness points on the two dimensional solution sets, we first solve

$$\begin{cases} f_1(\mathbf{x}) + a_{11} z_1 + a_{12} z_2 = 0 \\ f_2(\mathbf{x}) + a_{21} z_1 + a_{22} z_2 = 0 \\ f_3(\mathbf{x}) + a_{31} z_1 + a_{32} z_2 = 0 \\ \qquad\qquad L_1(\mathbf{x}) + z_1 = 0 \\ \qquad\qquad L_2(\mathbf{x}) + z_2 = 0 \end{cases}$$

Solutions with $z_1 = 0$ and $z_2 = 0$ define witness points on the two dimensional solution components. Solutions with $z_1 \neq 0$ and $z_2 \neq 0$ provide start points in the homotopy which removes L_2 from the system, which leads to the next system in the cascade:

$$\begin{cases} f_1(\mathbf{x}) + a_{11}z_1 + a_{12}z_2 = 0 \\ f_2(\mathbf{x}) + a_{21}z_1 + a_{22}z_2 = 0 \\ f_3(\mathbf{x}) + a_{31}z_1 + a_{32}z_2 = 0 \\ \qquad\qquad L_1(\mathbf{x}) + z_1 = 0 \\ \qquad\qquad\qquad\qquad z_2 = 0 \end{cases}$$

The paths defined by this move end at witness points on the one dimensional components, picked out by $z_1 = 0$. Solutions with $z_1 \neq 0$ are used in the homotopy which removes L_1 to lead to the isolated solutions of the system. The last system in the cascade is

$$\begin{cases} f_1(\mathbf{x}) + a_{11}z_1 = 0 \\ f_2(\mathbf{x}) + a_{21}z_1 = 0 \\ f_3(\mathbf{x}) + a_{31}z_1 = 0 \\ \qquad\qquad z_1 = 0 \\ \qquad\qquad z_2 = 0 \end{cases}$$

In the next section we give a specific example of this cascade.

The idea of slicing a solution set by hyperplanes to determine its dimension appeared in [GH93] to prove that the theoretical complexity of this problem is polynomial.

Exercise 8.3.1. Consider the adjacent minors of a general 2×4-matrix:

$$\begin{bmatrix} x_{11} & x_{12} & x_{13} & x_{14} \\ x_{21} & x_{22} & x_{23} & x_{24} \end{bmatrix} \qquad f(\mathbf{x}) = \begin{cases} x_{11}x_{22} - x_{21}x_{12} = 0 \\ x_{12}x_{23} - x_{22}x_{13} = 0 \\ x_{13}x_{24} - x_{23}x_{14} = 0 \end{cases}$$

Verify that $\dim(f^{-1}(\mathbf{0})) = 5$ and $\deg(f^{-1}(\mathbf{0})) = 8$. This is the simplest instance of a general family of problems introduced in [DES98], see [HS00] for special decomposition methods.

8.3.3 A probability-one membership test

A probability-one membership test determines whether a given point $\mathbf{p}$ lies on a pure dimensional solution set. Suppose we have witness points defined by a polynomial system $f(\mathbf{x}) = \mathbf{0}$ and hyperplanes $L(\mathbf{x}) = \mathbf{0}$. A homotopy method implements the probability-one membership test:

1. Define $K(\mathbf{x}) = L(\mathbf{x}) - L(\mathbf{p})$. As $K(\mathbf{p}) = \mathbf{0}$, the hyperplanes K pass through $\mathbf{p}$.
2. Consider the homotopy

$$h(\mathbf{x}, t) = \begin{pmatrix} f(\mathbf{x}) \\ K(\mathbf{x}) \end{pmatrix}(1 - t) + \begin{pmatrix} f(\mathbf{x}) \\ L(\mathbf{x}) \end{pmatrix} t = \mathbf{0}.$$

At $t = 1$ we start tracking paths at the witness set and find their end points at $t = 0$.

3. If **p** belongs to the solution set of $h(\mathbf{x}, 0) = \mathbf{0}$, then it is also a witness point of the pure dimensional solution set.

Notice that this test does not move the point **p**, which may be a highly singular point. This observation is important for the numerical stability of this test. The test is illustrated in Figure 8.16.

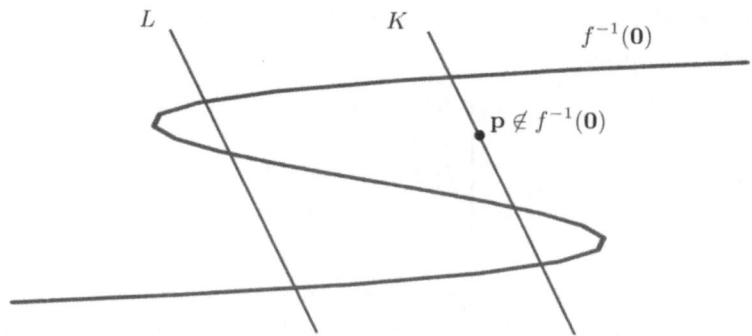

Fig. 8.16. Illustration of a probability-one membership test using a homotopy. The homotopy moves the line L of the witness set for $f^{-1}(0)$ to the line K, which passes to the test point **p**. As none of the witness points on K equals **p**, $\mathbf{p} \notin f^{-1}(0)$.

8.3.4 A numerical irreducible decomposition

Consider the following example:

$$f(\mathbf{x}) = \begin{cases} (x_1 - 1)(x_2 - x_1^2) = 0 \\ (x_1 - 1)(x_3 - x_1^3) = 0 \\ (x_1^2 - 1)(x_2 - x_1^2) = 0 \end{cases}$$

From its factored form we see that $f(\mathbf{x}) = \mathbf{0}$ has two solution components: the two dimensional plane $x_1 = 1$ and the twisted cubic $\{ (x_1, x_2, x_3) \mid x_2 - x_1^2 = 0, \; x_3 - x_1^3 = 0 \}$.

To describe the solution set of this system, we use a cascade of homotopies, the chart in Figure 8.17 illustrates the flow of data for this example.

Because the top dimensional component is of dimension two, we add two random hyperplanes to the system and make it square again by adding two slack variables z_1 and z_2:

$$e(\mathbf{x}, z_1, z_2) = \begin{cases} (x_1 - 1)(x_2 - x_1^2) + a_{11}z_1 + a_{12}z_2 = 0 \\ (x_1 - 1)(x_3 - x_1^3) + a_{21}z_1 + a_{22}z_2 = 0 \\ (x_1^2 - 1)(x_2 - x_1^2) + a_{31}z_1 + a_{32}z_2 = 0 \\ c_{10} + c_{11}x_1 + c_{12}x_2 + c_{13}x_3 + z_1 = 0 \\ c_{20} + c_{21}x_1 + c_{22}x_2 + c_{23}x_3 + z_2 = 0 \end{cases}$$

where all constants a_{ij}, $i = 1, 2, 3$, $j = 1, 2$, and c_{kl}, $k = 1, 2$, $l = 0, 1, 2, 3$ are randomly chosen complex numbers. Observe that when $z_1 = 0$ and $z_2 = 0$ the solutions to $e(\mathbf{x}, z_1, z_2) = \mathbf{0}$ satisfy $f(\mathbf{x}) = \mathbf{0}$. So if we solve $e(\mathbf{x}, z_1, z_2) = \mathbf{0}$ we will find a single witness point on the two dimensional solution component $x_1 = 1$ as a solution with $z_1 = 0$ and $z_2 = 0$. Using polyhedral homotopies, this requires the tracing of six solutions paths.

The embedding was proposed in [SV00] to find generic points on all positive dimensional solution components with a cascade of homotopies. In [SV00] it was proven that solutions with slack variables $z_i \neq 0$ are regular and, moreover, that those solutions can be used as start solutions in a homotopy to find witness points on lower dimensional solution components. At each stage of the algorithm, we call solutions with nonzero slack variables *nonsolutions*.

In the solution of $e(\mathbf{x}, z_1, z_2) = \mathbf{0}$, one path ended with $z_1 = 0 = z_2$, the five other paths ended in regular solutions with $z_1 \neq 0$ and $z_2 \neq 0$. These five "nonsolutions" are start solutions for the next stage, which uses the homotopy

$$
h_2(\mathbf{x}, z_1, z_2, t)
= \left\{
\begin{array}{l}
(x_1 - 1)(x_2 - x_1^2) + a_{11}z_1 + a_{12}z_2 = 0 \\
(x_1 - 1)(x_3 - x_1^3) + a_{21}z_1 + a_{22}z_2 = 0 \\
(x_1^2 - 1)(x_2 - x_1^2) + a_{31}z_1 + a_{32}z_2 = 0 \\
c_{10} + c_{11}x_1 + c_{12}x_2 + c_{13}x_3 + z_1 = 0 \\
z_2(1 - t) + (c_{20} + c_{21}x_1 + c_{22}x_2 + c_{23}x_3 + z_2)t = 0
\end{array}
\right.
$$

where t goes from one to zero, replacing the last hyperplane with $z_2 = 0$. Of the five paths, four of them converge to solutions with $z_1 = 0$. Of those four solutions, one of them is found to lie on the two dimensional solution component $x_1 = 1$, the other three are generic points on the twisted cubic. As there is one solution with $z_1 \neq 0$, we have one candidate left to use as a start point in the final stage, which searches for isolated solutions of $f(\mathbf{x}) = \mathbf{0}$. The homotopy for this stage is

$$
h_1(\mathbf{x}, z_1, t) = \left\{
\begin{array}{l}
(x_1 - 1)(x_2 - x_1^2) + a_{11}z_1 = 0 \\
(x_1 - 1)(x_3 - x_1^3) + a_{21}z_1 = 0 \\
(x_1^2 - 1)(x_2 - x_1^2) + a_{31}z_1 = 0 \\
z_1(1 - t) + (c_{10} + c_{11}x_1 + c_{12}x_2 + c_{13}x_3 + z_1)t = 0
\end{array}
\right.
$$

which as t goes from 1 to 0, replaces the last hyperplane $z_1 = 0$. At $t = 0$, the solution is found to lie on the twisted cubic, so there are no isolated solutions.

The calculations are summarized in Figure 8.17. The breakup into irreducibles will be explained in the next section.

8.3.5 Factorization methods

A recent trend in computer algebra is the adaptation of symbolic methods to deal with approximate input data, which leads to the use of hybrid methods [CKW02]. One such problem is the factorization of multivariate polynomials, listed as a challenge in [Kal00]. Recent papers on this problem are [CGvH+01, CGKW02], [GR01, GR02], [HWSZ00], and [Sas01].

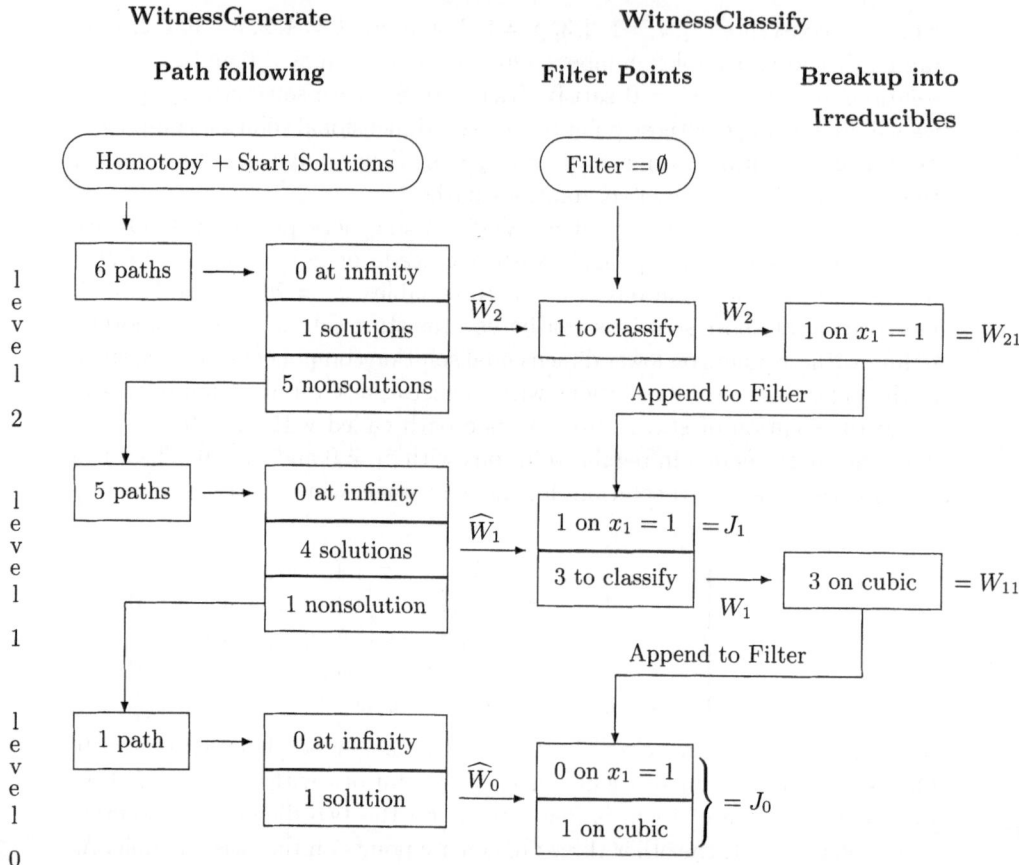

Fig. 8.17. Numerical Irreducible Decomposition of a system whose solutions are the 2-dimensional plane $x_1 = 1$ and the twisted cubic. At level i, for $i = 2, 1, 0$, we filter candidate witness sets $\widehat{W}_i$ into junk sets J_i and witness sets W_i. The sets W_i are partitioned into witness sets W_{ij} for the irreducible components.

Monodromy to partition witness point sets

We can see whether a curve factors or not by looking at its plot in complex space, i.e.: we consider the curve as a Riemann surface. Figure 8.18 was made with Maple (see [CJ98] for instructions).

Looking at Figure 8.18, imagine a line which intersects the surface in three points. Taking one complete turn of the line around the vertical axis $z = 0$ will cause the points to permute. For example, the point which was lowest will have moved up, while another point will have come down. Such a permutation can only happen if the corresponding algebraic curve is irreducible.

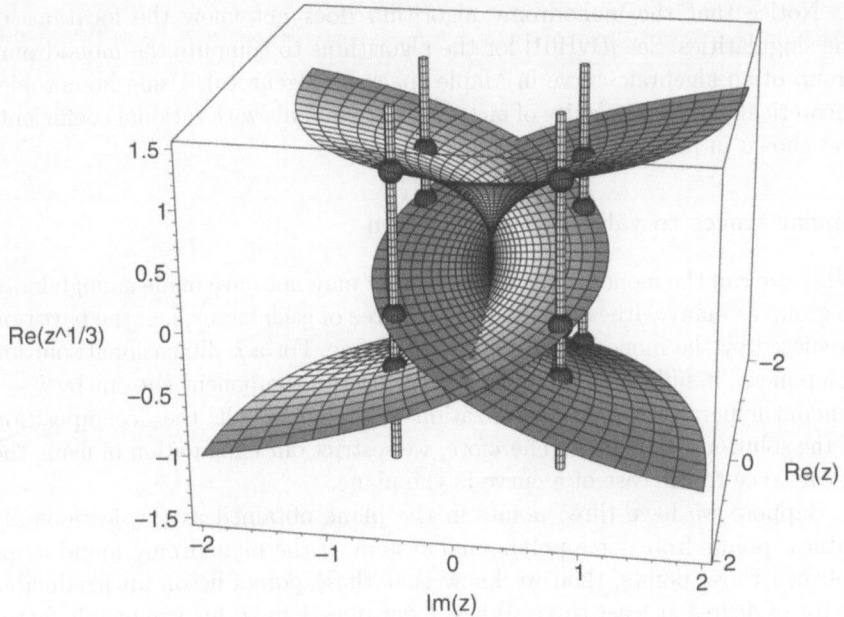

Fig. 8.18. The Riemann surface of $z^3 - w = 0$. The height of the surface is the real part of $w = z^{1/3}$, while the gray scale corresponds to the imaginary part of $w = z^{1/3}$. Observe that a loop around the origin permutes the order of points.

Based on this observation, we can decompose any pure dimensional set into irreducible components. Our monodromy algorithm returns a partition of the witness set for a pure dimensional component: points in the same subset of the partition belong to the same irreducible component. Recall that witness points are defined by a system $f(\mathbf{x}) = \mathbf{0}$ and a set of hyperplanes $L(\mathbf{x}) = \mathbf{0}$. With the homotopy

$$h_{KL}(\mathbf{x}, t) = \lambda \begin{pmatrix} f(\mathbf{x}) \\ K(\mathbf{x}) \end{pmatrix} (1 - t) + \begin{pmatrix} f(\mathbf{x}) \\ L(\mathbf{x}) \end{pmatrix} t = \mathbf{0}, \quad \lambda \in \mathbb{C},$$

we find new witness points on the hyperplanes $K(\mathbf{x}) = \mathbf{0}$, starting at those witness points satisfying $L(\mathbf{x}) = \mathbf{0}$, letting t move from one to zero. Choosing another random constant $\mu \neq \lambda$, we move back from K to L, using the homotopy

$$h_{LK}(\mathbf{x}, t) = \mu \begin{pmatrix} f(\mathbf{x}) \\ L(\mathbf{x}) \end{pmatrix} (1 - t) + \begin{pmatrix} f(\mathbf{x}) \\ K(\mathbf{x}) \end{pmatrix} t = \mathbf{0}, \quad \mu \in \mathbb{C}.$$

The homotopies $h_{KL}(\mathbf{x}, t) = \mathbf{0}$ and $h_{LK}(\mathbf{x}, t) = \mathbf{0}$ implement one loop in the monodromy algorithm, moving witness points from L to K and then back from K to L. At the end of the loop we have the same witness set as the set we started with, except possibly permuted. Permuted points belong to the same irreducible component.

Notice that the monodromy algorithm does not know the locations of the singularities. See [DvH01] for the algorithms to compute the monodromy group of an algebraic curve in Maple (package algcurves). Using homotopies theoretically, the complexity of factoring polynomials with rational coefficients was shown in [BCGW93] to be in NC.

Linear traces to validate the partition

When we run the monodromy algorithm, we may not have made enough loops to group as many witness points as the degree of each factor, i.e.: the partition predicted by the monodromy might be too fine. For a k-dimensional solution component, it suffices to consider a curve on the component cut out by $k - 1$ random hyperplanes. The factorization of the curve tells the decomposition of the solution component. Therefore, we restrict our explanation of using the linear trace to the case of a curve in the plane.

Suppose we have three points in the plane obtained as (projections of) witness points from some polynomial system. If the monodromy found loops between those points, then we know that these points lie on an irreducible factor of degree at least three. Whence our question: is this irreducible factor on which the given three points lie of degree three?

To answer this question we represent the factor by a cubic polynomial f in the form

$$
\begin{aligned}
f(x, y(x)) &= (y - y_1(x))(y - y_2(x))(y - y_3(x)) \\
&= y^3 - t_1(x)y^2 + t_2(x)y - t_3(x)
\end{aligned}
$$

Since $\deg(f) = 3$, $\deg(t_1) = 1$, so t_1 is the *linear trace*: $t_1(x) = c_1 x + c_0$.

We now proceed as follows. Via interpolation we find the coefficients c_0 and c_1. We first sample the cubic at $x = x_0$ and $x = x_1$. The samples are $\{(x_0, y_{00}), (x_0, y_{01}), (x_0, y_{02})\}$ and $\{(x_1, y_{10}), (x_1, y_{11}), (x_1, y_{12})\}$. To find c_0 and c_1 we then solve the linear system

$$
\begin{cases}
y_{00} + y_{01} + y_{02} = c_1 x_0 + c_0 \\
y_{10} + y_{11} + y_{12} = c_1 x_1 + c_0
\end{cases}
$$

With t_1 we can predict the sum of the y's for a fixed choice of x. For example, samples at $x = x_2$ are $\{(x_2, y_{20}), (x_2, y_{21}), (x_2, y_{22})\}$, see Figure 8.19. So our test consists in computing $t_1(x_2)$ in two ways:

$$
c_1 x_2 + c_0 \stackrel{?}{=} y_{20} + y_{21} + y_{22}.
$$

If the equality holds, then the answer to our question is yes.

Efficiency and numerical stability

The validation with the linear trace is fast. Therefore, our implementation does this validation each time a new loop with the monodromy algorithm

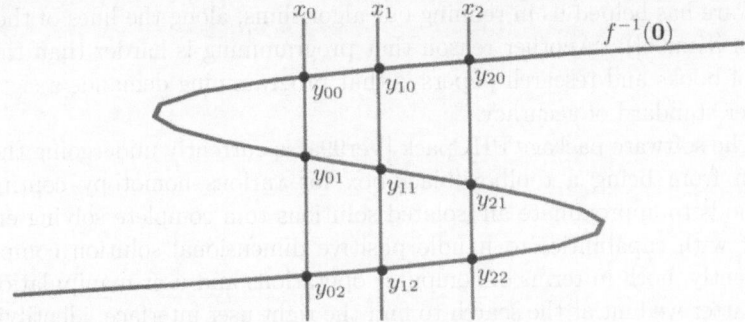

Fig. 8.19. The linear trace test on a planar cubic. To find the trace we interpolate through the samples at $x = x_0$ and $x = x_1$. Samples at $x = x_2$ are used in the test.

is found. Even as we do not know the locations of the singularities, practical experiences on many systems all lead to a rapid finding of permutations. While this approach is suitable for irreducible factors of very large degree (e.g., one thousand), strategies based purely on traces often perform better for smaller degrees.

Related to the efficiency is good numerical stability: if we can compute witness points with standard machine arithmetic, then we can also factor using standard machine arithmetic. This feature is very important when the accuracy of coefficients of the polynomial system is limited.

Exercise 8.3.2. Apply phc -f to factor

```
x**6 - x**5*y + 2*x**5*z - x**4*y**2 - x**4*y*z+x**3*y**3
    - 4*x**3*y**2*z + 3*x**3*y*z**2 - 2*x**3*z**3 + 3*x**2*y**3*z
    - 6*x**2*y**2*z**2 + 5*x**2*y*z**3 - x**2*z**4 + 3*x*y**3*z**2
    - 4*x*y**2*z**3 + 2*x*y*z**4+y**3*z**3 - y**2*z**4;
```

which is a polynomial in a format accepted by phc.

Exercise 8.3.3. Consider again the system of adjacent minors from Exercise 8.3.1. Determine the number of irreducible factors and their degrees.

See Chapter 9 for more on factorization methods.

8.4 Software and applications

8.4.1 Software for polynomial homotopy continuation

We agree with the statement: "It can be argued that the 'mission' of numerical analysis is to provide the scientific community with effective software tools." (taken from the preface to [GVL83]). Aside from our missionary intentions, software has helped us in refining our algorithms, along the lines of the quote (from [Knu96]): "Another reason that programming is harder than the writing of books and research papers is that programming demands a significant higher standard of accuracy."

The software package PHCpack [Ver99a] is currently undergoing the transition from being a toolbox/black-box for various homotopy continuation methods to approximate all isolated solutions to a complete solving environment with capabilities to handle positive dimensional solution components efficiently, both in terms of computer operations and user manipulations. By the latter we hint at the search to find the right user interface, identifying the right data flow and trying to balance the toolbox with the black-box approach.

While PHCpack offered the first reliable implementation of polyhedral homotopies, its efficiency is currently surpassed by the implementations described in [GL00, GL03, LL01] and [DKK03, GKK+04, KK03b, TKF02]. To interact better with other codes, we are currently developing an interface from the Ada routines in PHCpack to routines written in C. Another (but related) interface concerns the interaction with computer algebra software. In [SVW03] we describe a very simple interface to Maple.

8.4.2 Applications

A benchmark suite for systems with positive dimensional solution components is gradually taking shape. Rather than listing summaries of a benchmark, we choose to treat two very typical applications: the cyclic n-roots problem from computer algebra and a special Stewart-Gough platform from mechanical design.

The cyclic n-roots problem. This problem is already interesting not only by its compact formulation and widespread fame in the computer algebra community, but also by known theoretical results concerning the number of isolated roots when n is prime [Haa96].

For $n = 8$, there are 16 one dimensional irreducible components: eight quadrics and eight curves of degree 16. While approximations to all 1,152 isolated cyclic 8-roots were found already in the first release of PHCpack, monodromy was needed to factor the curve of degree 144 into irreducibles. To compute all witness points for the cyclic 9-roots problem, the software of [LL01] was essential. While the factorization of a two dimensional component of degree 18 into six cubics posed no difficulty, the homotopy membership test was required to certify that among the 6,642 isolated ones 162 cyclic 9-roots occurred with multiplicity four. In addition, multi-precision arithmetic was used to confirm this result.

The isolated cyclic n-roots (up to $n = 13$, for which 2,704,156 paths were traced) can be found on the Internet[6] These roots have been computed with PHoM [GKK+04].

---

[6] http://www.is.titech.ac.jp/~kojima/polynomials/cyclic13.

A special Stewart-Gough platform. The Stewart-Gough platform is a parallel robot which attracted lots of interest from computational kinematicians and researchers in computer algebra. That the platform has forty isolated solutions was first established computationally by continuation [Rag93] and elimination methods [Laz92, Mou93], and later proved analytically [Hus96], [RV95], and [Wam96].

A six-legged platform (similar to the general Stewart-Gough platform) which permits motion was presented by Griffis and Duffy in [GD93] and first analyzed in [HK00]. It is called the Griffis-Duffy platform. Instead of forty isolated solutions we now consider a curve. In our formulation of the two cases we studied, twelve lines corresponded to degenerate cases deemed uninteresting from a mechanisms point of view. In the first case we were then left with one irreducible component of degree 28, while in the second case we found five components, four of degree six (one sextic was not reported in the analysis of [HK00]), and one component of degree four, see Figure 8.20.

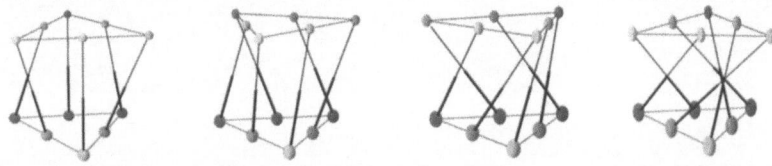

Fig. 8.20. One component of the Griffis-Duffy platform. Starting at the configuration at the left, we see the clockwise rotation of the end platform.

It is interesting to note that the running times for the factorization with the monodromy-traces method do not seem to depend on the particular geometry of the system, i.e.: the execution times are about the same in both cases, when we deal with one irreducible factor of high degree or with several factors of smaller degrees.

Acknowledgments

The authors thank Alicia Dickenstein and Ioannis Emiris for their invitation to present their work at the summer school. We are grateful to Dan Bates for his careful reading and comments. The revision benefited greatly from the stimulating questions from Olga Kashcheyeva, Anton Leykin, Yusong Wang, and Ailing Zhao at the MCS 595 graduate seminar. Some of the exercises were first presented at the RAAG summer school on Computer Tools for Real Algebraic Geometry, June 30-July 5, 2003, organized by Michel Coste, Laureano Gonzalez-Vega, Fabrice Rouillier, Marie-Françoise Roy, and Markus Schweighofer, whom we thank for their invitation.

Four lectures on polynomial absolute factorization

Guillaume Chèze and André Galligo

Laboratoire de mathématiques J.A. Dieudonné,
Université de Nice Sophia-Antipolis, France.
cheze@math.unice.fr, galligo@math.unice.fr

Summary. Polynomial factorization is one of the main chapters of Computer Algebra. Recently, significant progress was made on absolute factorization (i.e., over the complex field) of a multivariate polynomial with rational coefficients, with two families of algorithms proposing two different strategies of computation. One is represented by Gao's algorithm and is explained in Lecture 2. The other is represented by the Galligo-Rupprecht-Chèze algorithm, presented in Lectures 4 and 5. The latter relies on an original use of the monodromy map attached to a generic projection of a plane curve on a line. It also involves zero-sums relations (introduced by Sasaki and his collaborators) with efficient semi-numerical computations to produce a certified exact result.

9.0 Introduction, definitions and examples

9.0.1 Rational and absolute factorization

A system of polynomial equations $I = (g_1, \ldots, g_n)$ corresponds to an algebraic variety $V = V(g_1, \ldots, g_n)$. When the dimension of V is zero a natural question is: What is the cardinality of V, and what are the coordinates of the points of V? When the dimension is not zero this natural question becomes: What is the number of irreducible components of V, and what are the equations of these components?

In the special case of one polynomial $f(X, Y) \in \mathbb{Q}[X, Y]$, the answer to these questions is given by the absolute factorization of $f(X, Y)$. The absolute factorization of f is the factorization $f = f_1 \ldots f_s$, where the f_i are irreducible in $\mathbb{C}[X, Y]$. This provides the decomposition into irreducible components $V(f) = V(f_1) \cup \ldots \cup V(f_s)$. Now, let us tell a short story about absolute polynomial factorization.

Polynomial factorization is one of the main chapters of Computer Algebra. The implementation of basic algorithms, derived from classical and elementary commutative algebra, appeared in the first Computer Algebra systems in the 60's. During the last 30 years, at every international conference in

Computer Algebra, there have been new contributions on factorization algorithms and their complexity. Several papers and books, including [vzG85] and [Kal90], relate the early history of these topics and provide a comprehensive bibliography till the end of the 80's.

The early authors and implementors, like Berlekamp, Musser, Wang, Zassenhaus and others, introduced many ideas that are now classical such as reduction and Hensel liftings, genericity and randomization. The LLL lattice basis reduction algorithm of Lenstra, Lenstra and Lovász (1982) allowed for the first time a polynomial-time algorithm to be established for factoring univariate polynomials over the rational numbers. In the early 80's, this was followed by many complexity results on univariate or multivariate factorization algorithms (over different fields), which are either polynomial-time or probabilistic polynomial-time. One of the first such results for multivariate absolute irreducibility testing, e.g. over the complexes, for a polynomial with rational coefficients, is due to Heintz-Sieveking [HS81]. It popularized in the community of Computer Algebra and Complexity the use of Bertini's theorem (see Lecture 1) and was followed by many authors, including von zur Gathen and Kaltofen.

The early works of Berlekamp (1967 and 1970), or Cantor and Zassenhaus (1981) for univariate polynomials over finite fields could run in quadratic or even subquadratic time. In practice, rational factorization of most polynomials can be computed efficiently using Hensel lifting; see e.g. Musser (1975) and Wang (1978). Lauder and Gao [Gao03] proved that the average running time of a Hensel lifting based algorithm for factoring bivariate polynomials over finite fields is almost linear. There are, however, infinitely many polynomials that need exponential time via Hensel lifting (see [Kal85b]). Although this can be improved, we can say that there are good algorithms and rather satisfactory implementations to perform rational polynomial bivariate (and multivariate) factorization.

Absolute factors of a polynomial with rational coefficients have coefficients which are algebraic numbers. These can be represented either by elements in a precisely described extension $\mathbb{Q}(\alpha)$ of $\mathbb{Q}$ or in $\mathbb{C}$ by imprecise floating point numbers which approximate them. This distinction gives rise to two families of algorithms: one kind which ultimately relies on linear algebra and can be developed on $\overline{\mathbb{Q}}$, e.g. the algorithms by Trager-Traverso, Kaltofen, Duval, Gao, Cormier-Singer-Trager-Ulmer (see Lecture 3), and another kind which uses topological properties of $\mathbb{C}^2$, Newton approximation or so-called homotopy methods and for which floating point approximations are better suited, e.g. the algorithms of Sasaki, Galligo-Rupprecht, Sommese-Verschelde-Wampler (see Lecture 4). Once such an approximate absolute factorization algorithm is available, it is still necessary to compute the exact factors. This has been done by Chèze-Galligo and will be discussed in Lecture 5. One can say today that the best algorithms were all discovered within the past ten years and there is still progress to be made. Another important preliminary topic is absolute irreducibility testing (see Lecture 2).

Now we introduce basic definitions and statements.

9.0.2 Facts and definitions

Definition 9.0.1. *Let A be a domain. We say that A is a unique factorization domain if for all $a \in A - \{0\}$ we can write $a = u.p_1 \ldots p_s$ where u is a unit and $p_1, \ldots, p_s$ are irreducible in A and this decomposition is unique up to reordering and multiplication by units.*

Example 9.0.2. $\mathbb{Z}$ and all fields are unique factorization domains.

Theorem 9.0.3. *If A is a unique factorization domain then $A[X]$ is a unique factorization domain.*

Corollary 9.0.4. *Let k be a field, then $k[X_1, \ldots, X_n]$ is a unique factorization domain.*
    This means: for all $P \in k[X_1, \ldots, X_n], P = P_1 \ldots P_s$ (factorization), with P_i irreducible in $k[X_1, \ldots, X_n]$ and this decomposition is unique up to reordering and multiplication by constant factors.

Remark 9.0.5. Let $k \subset K$ be an inclusion of fields and $P \in k[X_1, \ldots, X_n]$. P can be irreducible in $k[X_1, \ldots, X_n]$ but reducible in $K[X_1, \ldots, X_n]$. For example: $k = \mathbb{Q}$, $K = \mathbb{C}$ and $X^2 + Y^2 = (X + iY)(X - iY)$.

Definition 9.0.6. *Let $K = \overline{k}$ be the algebraic closure of the field k, and $P \in k[X_1, \ldots, X_n]$. The factorization of P in $K[X_1, \ldots, X_n]$ is called the absolute factorization of P.*

Exercise 9.0.7. Let $P(X, Y) \in \mathbb{Q}[X, Y]$, and $P(X, Y) = \prod_{i=1}^{s} P_i(X, Y)$ its factorization in $\mathbb{C}[X, Y]$. Show that this factorization is the absolute factorization (i.e. $P_i(X, Y) \in \overline{\mathbb{Q}}[X, Y]$).
a) Set $P_1(X, Y) = a_m(X)Y^m + \cdots + a_0(X)$. Show that for all $x \in \mathbb{Q}$, $a_i(x)$ belongs to $\overline{\mathbb{Q}}$.
b) Let $p(T) = \sum_{i=0}^{k} p_i T^i \in \mathbb{C}[T]$ such that for all $x \in \mathbb{Q}$, $p(x)$ belongs to $\overline{\mathbb{Q}}$. Prove that $p_i \in \overline{\mathbb{Q}}$. (Hints: Write a Vandermonde system, and use Cramer's rule.)

There exist simple algorithms which compute absolute factorizations but are not efficient for degree ≥ 15. For example, in Maple the command

```
evala(AFactor(.))
```

implements an algorithm which we will explain below.

```
evala(AFactor(X^2-2*Y^2));
```

gives

```
(X-RootOf(Z^2-2)Y)(X+RootOf(Z^2-2)Y).
```

That means $X^2 - 2Y^2 = (X - \sqrt{2}Y)(X + \sqrt{2}Y)$. We remark that the two factors have the same monomials and their coefficients are conjugate over $\mathbb{Q}$. The next lemma generalizes this remark.

Lemma 9.0.8 (Fundamental Lemma). *Let $P \in \mathbb{Q}[X,Y]$ be a monic and irreducible polynomial in $\mathbb{Q}[X,Y]$. $P(X,Y) = Y^n + a_{n-1}(X)Y^{n-1} + \cdots + a_0(X)$ with $\deg(a_i(X)) \le n - i$.*
Let $P = P_1 \ldots P_s$ be a factorization of P into irreducible polynomials P_i in $\mathbb{C}[X,Y]$. Denote by $\mathbb{K} = \mathbb{Q}[\alpha]$ the extension of $\mathbb{Q}$ generated by all the coefficients of P_1. Then each P_i can be written:
$$P_i(X,Y) = Y^m + b_{m-1}(\alpha_i, X)Y^{m-1} + \cdots + b_0(\alpha_i, X),$$
with $b_k \in \mathbb{Q}[Z,X]$, $\deg_X(b_k) \le m - k$, and where $\alpha_1, \ldots, \alpha_s$ are the different conjugates over $\mathbb{Q}$ of $\alpha = \alpha_1$.

Proof. We can suppose that each P_i is monic in Y, because P is monic in Y. We set $P_i(X,Y) = Y^{n_i} + a_{n_i-1}^{(i)}Y^{n_i-1} + \cdots + a_0^{(i)}(X)$ with $a_k^{(i)} \in \overline{\mathbb{Q}}[X]$ and $\deg_X(a_k^{(i)}(X)) \le n_i - k$. Let $\mathbb{K}$ be the field generated by all the coefficients of P_1; by the primitive element theorem we can set $\mathbb{K} = \mathbb{Q}[\alpha]$. α is an algebraic number over $\mathbb{Q}$ and we denote by $\alpha_1 = \alpha, \alpha_2, \ldots, \alpha_k$ its k different conjugates over $\mathbb{Q}$, and by $\sigma_1, \ldots, \sigma_k$ the $\mathbb{Q}$-homomorphism from $\mathbb{Q}[\alpha]$ into $\mathbb{C}$ such that $\sigma_i(\alpha) = \alpha_i$.

Now we prove that $k \le s$. Let $\mathbb{M}$ be the extension of $\mathbb{Q}$ generated by the coefficients of $P_1, \ldots, P_s$; $\mathbb{M}$ is a finite extension of $\mathbb{Q}$, and we have
$$\mathbb{C} \supset \mathbb{M} \supset \mathbb{K} \supset \mathbb{Q}.$$
We can extend to $\mathbb{M}$ all the σ_i. Then we extend σ_i to $\mathbb{M}[X,Y]$, and we denote this map by $\tilde{\sigma}_i$. We have $\tilde{\sigma}_i(P) = \tilde{\sigma}_i(P_1) \ldots \tilde{\sigma}_i(P_s) = P$. Since $\overline{\mathbb{Q}}[X,Y]$ is a unique factorization domain, there exists an index j_0 such that $\tilde{\sigma}_i(P_1) = P_{j_0}$. Furthermore, if $\tilde{\sigma}_i(P_1) = \tilde{\sigma}_j(P_1)$ then $\sigma_i = \sigma_j$. So the map:
$$ev_{P_1} : \{\tilde{\sigma}_1, \ldots, \tilde{\sigma}_k\} \to \{P_1, \ldots, P_s\} : \tilde{\sigma}_i \mapsto \tilde{\sigma}_i(P_1)$$
is injective and $k \le s$.

If $k < s$ we get an absurd result. Indeed, consider $F = \prod_{i=1}^{k} \tilde{\sigma}_i(P_1)$; this polynomial divides P so if we prove that $F \in \mathbb{Q}[X,Y]$, we are done.

Write $P_1(X,Y) = \sum_{a,b} c_{a,b} X^a Y^b$ where $c_{i,j}(T) \in \mathbb{Q}[T]$. Thus
$$F(X,Y) = \prod_{i=1}^{k} (\sum_{a,b} (\alpha_i) X^a Y^b).$$
The coefficient of $X^a Y^b$ is written

$$\sum_{\substack{i_1 + \cdots + i_k = a \\ j_1 + \cdots + j_k = b}} c_{i_1, j_1}(\alpha_1) \ldots c_{i_k, j_k}(\alpha_k).$$

It is a symmetric polynomial in $\alpha_1, \ldots, \alpha_k$, so it is rational; we deduce that $F(X,Y) \in \mathbb{Q}[X,Y]$.

Remark 9.0.9. For each P it suffices to get P_1 to describe the absolute factorization of P.

A first method

Here we describe the absolute factorization algorithm implemented in Maple. It consists of the following 4 steps.

Algorithm 9.0.10 TRAGER-TRAVERSO ALGORITHM.

Input: $f(X_1, \ldots, X_n) \in \mathbb{Z}[X_1, \ldots, X_n]$.

1. *Compute a factorization in $\mathbb{Z}[X_1, \ldots, X_n]$ and reduce to 2 variables (see below).*
2. *For each irreducible factor $P \in \mathbb{Z}[X, Y]$: Fix an integer value α of X such that $\text{disc}_Y P(\alpha) \neq 0$. Factorize $P(\alpha, Y)$ over $\mathbb{Z}[Y]$, choose an irreducible factor q and make an alias: $\beta = RootOf(q)$.*
3. *Compute a factorization of P in $\mathbb{Q}(\beta)[X, Y]$, i.e. apply $factor(P, \beta)$. This does not provide a complete absolute factorization, but splits the polynomials into (at least) two factors in $K[X, Y]$ with $K = \dfrac{\mathbb{Q}[t]}{q(t)}$.*
4. *Lift the factorization.*

Output: An absolute factor of f.

The first step uses Hilbert's or Bertini's theorem and the last step uses Hensel's theorem. In Lecture 1 we will study these theorems. Step 3 is validated by the following theorem (cf. [Kal85a] or [DT89]).

Definition 9.0.11. *Let $\overline{k}$ be the algebraic closure of the field k, and let $(\alpha, \beta) \in \overline{k}^2$. We say that (α, β) is a simple solution of $P(X, Y) \in k[X, Y]$ when $P(\alpha, \beta) = 0$ and either $\frac{\partial P}{\partial X}(\alpha, \beta)$ or $\frac{\partial P}{\partial Y}(\alpha, \beta)$ is nonzero.*

Remark 9.0.12. It is easy to see that if (α, β) is a simple solution of $P(X, Y)$ then (α, β) is a simple solution of just one absolute factor of P.

In step 2 of the algorithm we get a simple point (α, β) of $P(X, Y)$.

Theorem 9.0.13. *Let (α, β) be a simple solution of $P(X, Y)$. Then one absolute factor of $P(X, Y)$ belongs to $k[\alpha, \beta][X, Y]$.*

Proof. Let $P = F_1 F_2 \ldots F_t$ be the factorization of P in $k[\alpha, \beta][X, Y]$, where F_1 is such that $F_1(\alpha, \beta) = 0$. F_1 is the only factor with this property. Suppose that F_1 is reducible in $\overline{k}[X, Y]$. We are going to show that this is absurd.

Write $F_1 = \prod_{j=i_1}^{i_k} P_j$ where P_j are absolute factors of P and suppose that $P_{i_1}(\alpha, \beta) = 0$. With the same kind of arguments as those used in the proof of the fundamental lemma, we can show that there exist a $k(\alpha, \beta)$-homomorphism σ and an index $i_l \neq i_1$ such that $\sigma(P_{i_1}) = P_{i_l}$. As $P_{i_1}(\alpha, \beta) = 0$. we have $\sigma(P_{i_1}(\alpha, \beta)) = 0$; we deduce that (α, β) is not a simple solution and this contradicts the hypothesis.

Example 9.0.14 (see [Rag97]). Let $P(X,Y) = Y^{10} - 2X^2Y^4 + 4X^6Y^2 - 2X^{10}$. It is an irreducible polynomial in $\mathbb{Q}[X,Y]$ since $P(1,Y) = Y^{10} - 2Y^4 + 4Y^2 - 2$ is irreducible in $\mathbb{Q}[Y]$. Let $K = \dfrac{\mathbb{Q}[T]}{P(1,T)}$. The factorization of P over $K[X,Y]$ is the following:

```
>alias(beta=RootOf(x^{10}-2*x^4+4*x^2-2));
>factor(P,beta);
```

$(Y^5 - 2Y^2X\beta + 2Y^2X\beta^3 - Y^2X\beta^5 - Y^2X\beta^7 - Y^2X\beta^9 + 2X^5\beta - 2X^5\beta^3 + X^5\beta^5 + X^5\beta^7 + X^5\beta^9)(Y^5 + 2Y^2X\beta - 2Y^2X\beta^3 + Y^2X\beta^5 + Y^2X\beta^7 + Y^2X\beta^9 - 2X^5\beta + 2X^5\beta^3 - X^5\beta^5 - X^5\beta^7 - X^5\beta^9)$

The time needed by Maple for this computation is 691.531 seconds (on a small PC). Here $P_1(X,Y) = Y^5 + (\beta^9 + \beta^7 + \beta^5 - 2\beta^3 + 2\beta)Y^2X + (-\beta^9 - \beta^7 - \beta^5 + 2\beta^3 - 2\beta)X^5$ is an absolute factor of P and it satisfies $P_1(\alpha,\beta) = 0$.

The method is simple but the drawback is that K is too big. A smaller extension will work better and faster, in most cases. In our example, an absolute factor is $G(X,Y) = Y^5 - \sqrt{2}XY + \sqrt{2}X^5$. Computing `factor(P,sqrt(2))` takes only 0.27 seconds.

Note that this first method relies on a rational factorization algorithm.

9.0.3 Rational factorization

We can quickly summarize the polynomial factorization process over a finite extension of $\mathbb{Q}$ with the following diagram.

$P \in \mathbb{Q}[\alpha][X,Y]$ with $q(\alpha) = 0$ and $q \in \mathbb{Z}[T]$ is irreducible.

$\downarrow y_0$ generic   Hensel lifting $\uparrow$
                              in $(y - y_0)^j$

$f(X) = P(X,y_0) \in \mathbb{Q}(\alpha)[X], \quad f \in \dfrac{1}{D}\mathbb{Z}[\alpha][X]$

$\downarrow$ p a generic   Hensel lifting $\uparrow$
   prime number   in p^j

$\overline{Df} \in \dfrac{\mathbb{Z}}{p\mathbb{Z}}(\overline{\alpha})[X]$ with $\overline{q}(\overline{\alpha}) = 0$ and $\overline{q} \in \dfrac{\mathbb{Z}}{p\mathbb{Z}}[T]$ is irreducible

We see that Hensel's lifting is a very useful tool; we recall Hensel's theorem in Lecture 1.

9.1 Lecture 1: Theorems of Hilbert and Bertini, reduction to the bivariate case, irreducibility tests

9.1.1 Hilbert's irreducibility theorem

We present a simple version of Hilbert's theorem (see [Lan83] and [Zip93]):

Theorem 9.1.1 (Hilbert's irreducibility theorem). *Let $\mathbb{K}$ be a finite algebraic extension of $\mathbb{Q}$, and let $f(T_1, \ldots, T_r, X_1, \ldots, X_s)$ be an irreducible polynomial in $\mathbb{Q}[T_1, \ldots, T_r, X_1, \ldots, X_s]$. Then almost all points $(t_1, \ldots, t_r) \in \mathbb{K}^r$ are such that $f(t_1, \ldots, t_r, X_1, \ldots, X_s)$ is irreducible in $\mathbb{K}[X_1, \ldots, X_s]$.*

Remark 9.1.2. This theorem is false if we replace $\mathbb{K}$ by a finite field $\mathbb{F}$. Consider $f(X, Y) = X^2 - Y$, then for all points t of $\mathbb{F} = \mathbb{Z}/2$, $f(X, t)$ is reducible. Below we will study Bertini's theorem which works also for finite fields. Furthermore, Bertini's theorem gives rise to a probabilistic statement.

9.1.2 Hensel's lemma

The idea of Hensel's lemma is to mimic Newton's method in an algebraic setting. Newton's iteration method gives an exact solution from an approximate one; here the approximation is I-adic, where I is an ideal in a ring A (see [Eis95], [Gou97], [Zip93] for the definition of the complete ring A_I and its properties).

Hereafter all rings are commutative, unique factorization domains, and Noetherian. Indeed in our setting A is one of the following rings:

$\mathbb{Z}, \mathbb{Z}[X_1, \ldots, X_n], \mathbb{Q}[X_1, \ldots, X_n], \mathbb{C}[X_1, \ldots, X_n]$, where $n \geq 1$.

Theorem 9.1.3. *Let I be an ideal of a ring A. Let*
$$\mathbf{F} = (F_1(X_1, \ldots, X_n), \ldots, F_n(X_1, \ldots, X_n))$$
be polynomials over A, and denote their Jacobian with respect to the X_i by Jac. Let $(x_1, \ldots, x_n)$ be a zero of $\mathbf{F}$ modulo I, such that the determinant $Jac(x_1, \ldots, x_n)$ has an inverse in $\dfrac{A}{I}$.

Then there exist unique elements $(\hat{x}_1, \ldots, \hat{x}_n)$ of A_I, $\hat{x}_i = x_i$ mod I for which $\mathbf{F}(\hat{x}_1, \ldots, \hat{x}_n) = 0$.

This implies (see [Zip93]):

Theorem 9.1.4 (Hensel's lemma). *Let $f(X)$ be a monic polynomial over A, and I be an ideal of A. Assume there exist monic polynomials $g_1(X), h_1(X)$ in $\dfrac{A}{I}[X]$ which are relatively prime and such that $f(X) = g_1(X)h_1(X)$ mod I. Then for every $n \in \mathbb{N} - \{0\}$ there exist monic polynomials $g_n(X), h_n(X)$ over $\dfrac{A}{I^n}[X]$ such that $g_n(X) = g_1(X)$ mod I, $h_n(X) = h_1(X)$ mod I and $f(X) = g_n(X)h_n(X)$ mod I^n.*

Furthermore, there exist unique polynomials $\hat{g}(X)$ and $\hat{h}(X)$ over A_I such that $\hat{g}(X) = g_1(X)$ mod I, $\hat{h}(X) = h_1(X)$ mod I, and for all $n \in \mathbb{N}$, $f(X) = \hat{g}(X)\hat{h}(X)$ mod I^n.

Lifting a factorization

Now we show how to get a factorization in $\mathbb{C}[X_1, \ldots, X_n]$ from a factorization in $\mathbb{C}[X_1, X_2]$. Let $f(X_1, \ldots, X_n)$, $f_1(X_1, \ldots, X_n)$ and $f_2(X_1, \ldots, X_n)$ be polynomials of $\mathbb{C}[X_1, \ldots, X_n]$ such that

$$f(X_1, \ldots, X_n) = f_1(X_1, \ldots, X_n) f_2(X_1, \ldots, X_n)$$

is the absolute factorization of f. We know f and we want to find the f_i. We set $d_i = \deg_{X_i}(f)$. Let $J = (X_3 - x_3, \ldots, X_n - x_n)$ be an ideal of $\mathbb{C}[X_1, \ldots, X_n]$ where $x_i \in \mathbb{C}$ for $i = 1, \ldots, n$. We set

$$\bar{f}^{(2)} = f(X_1, X_2, x_3, \ldots, x_n)$$

the image of f in $\dfrac{\mathbb{C}[X_1, \ldots, X_n]}{J}$, similarly

$$\bar{f}^{(3)} = f(X_1, X_2, X_3, x_4, \ldots, x_n)$$

and so on. We will get recursively a factorization for all $\bar{f}^{(k)}$ $f(X_1, \ldots, X_n)$. So we start with an absolute factorization for $\bar{f}^{(2)}$ namely $\bar{f}^{(2)} = g_1 h_1$. Applying Hensel's lemma (with $A = \mathbb{C}[X_2, X_3]$ and $I = (X_3 - x_3)$), we can lift this factorization to $\mathbb{C}[X_2, X_3][X_1]$ and get:

$$\bar{f}^{(3)} = g_{d_3+1} h_{d_3+1} \mod (X_3 - x_3)^{d_3+1}.$$

The degree condition with d_3 and the unicity property in Hensel's lemma imply the following:

$$\text{if} \quad \begin{cases} f_1(X_1, X_2, x_3, \ldots, x_n) = g_1(X_1, X_2) \\ \\ f_2(X_1, X_2, x_3, \ldots x_n) = h_1(X_1, X_2) \end{cases} \quad (1)$$

$$\text{then} \quad \begin{cases} g_{d_3+1}(X_1, X_2, X_3) = f_1(X_1, X_2, X_3, x_4, \ldots, x_n) \\ \\ h_{d_3+1}(X_1, X_2, X_3) = f_2(X_1, X_2, X_3, x_4, \ldots x_n). \end{cases} \quad (2)$$

So we obtain a factorization of $\bar{f}^{(3)}$. Now we restart with $A = \mathbb{C}[X_2, X_3, X_4]$, $I = (X_4 - x_4)$ and

$$\begin{aligned} \bar{f}^{(4)}(X_1, \ldots, X_n) &= f(X_1, X_2, X_3, X_4, x_5, \ldots, x_n) \\ &= g_{d_3+1}(X_1, X_2, X_3) h_{d_3+1}(X_1, X_2, X_3) \mod (X_4 - x_4) \end{aligned}$$

Then, after $n - 2$ liftings, we obtain the factorization of f.

Remark 9.1.5. We supposed that condition (1) is true, that is to say: $f_1(X_1, X_2, x_3, \ldots, x_n)$ and $f_2(X_1, X_2, x_3, \ldots, x_n)$ are the absolute factors of $f(X_1, X_2, x_3, \ldots, x_n)$. In other words we supposed that the $f_i(X_1, X_2, x_3, \ldots, x_n)$ are irreducible, but this is not always the case as we might have the following kind of phenomenon (see [Kal85b]):

$f(X, Y) = X^4 + (12Y^3 - 18Y^2 - 18Y + 12)X^3 + (30Y^3 - 72Y^2 + 42Y - 36)X^2 + (-432Y^3 + 648Y^2 + 648Y - 432)X - 432Y^3 + 2592Y^2 - 2160Y$ is absolutely irreducible but $f(X, 0) = (X - 6)X(X + 6)(X + 12)$ is reducible.

In order to avoid this situation which gives rise to an algorithm with an exponential complexity, we want to obtain a reduction from multivariate polynomials to bivariate polynomials preserving irreducibility. Using Hilbert's theorem we do not know the probability for a polynomial to remain irreducible after a substitution. Bertini's theorem, which we now study, provides this knowledge because it implies a useful probabilistic statement.

9.1.3 Bertini's theorem

In the beginning of the 20th century, Bertini proved two important theorems which bear his name; S. Kleiman gave in [Kle98] a comprehensive report on the history, evolution and impact of these theorems. Although it was phrased in the terminology of his time, Bertini's second theorem says the following: The intersection of an irreducible algebraic set in $\mathbb{C}^n$ of dimension $r \geq 2$, with a "generic" $(n-r+1)$-plane is an irreducible curve. Of course the meaning of the adjective generic has to be specified. Together with considerations of more general varieties this gives rise to several versions of the theorem. A complete treatment is provided by Jouanolou's book [Jou83].

In the community of Computer Algebra and Complexity theory, Bertini's (second) theorem was first used and popularized by Heintz-Sieveking [HS81] in 1981, soon followed by Kaltofen, von zur Gathen and many others. In [HS81], the following "General hyperplane section lemma" is stated, and a short algebraic proof is given in an appendix.

Lemma 9.1.6. *Let k be an algebraically closed field and $X_1, \ldots, X_n$ be indeterminates over k. Let $\mathcal{P}$ be a prime ideal of $k[X_1, \ldots, X_n]$ which defines an affine subvariety of k^n of dimension $r \geq 2$. Let $A_{i,j}$, A_i, $i = 1$ to $r - 1$, $j = 1$ to n, be transcendental quantities over k, and let K be an algebraically closed field containing k, $A_{i,j}$ and A_i, $i = 1$ to $r - 1$, $j = 1$ to n. Then, the ideal $\mathcal{P} + (X_1 - \sum_{j>1} A_{1,j}X_j - A_1, \ldots, X_{r-1} - \sum_{j>r-1} A_{r-1,j}X_j - A_{r-1})$ is a prime ideal in $K[X_1, \ldots, X_n]$.*

The proof relies on College algebra (i.e. ring and field extensions). The work [HS81] applies this lemma to the case $k = \mathbb{C}$, $\mathcal{P} = (f)$, $f \in \mathbb{Q}[X_1, \ldots, X_n]$, $r = n - 1$. By successive substitutions of X_i by $\sum_{j>i} A_{i,j}X_j + A_i$, $i = 1$ to $n - 2$, in $f(X_1, \ldots, X_n)$, one obtains a bivariate polynomial $f_0(X_{n-1}, X_n)$, whose coefficients are polynomials in $A_{i,j}$ and A_i.

Corollary 9.1.7. *f is absolutely irreducible $\iff$ f_0 is absolutely irreducible.*

Then it is proven in [HS81] that this claim still holds if we replace the indeterminates $A_{i,j}$ and A_i by random values. They also provide bounds which allow to control this randomness.

In 20 years, these bounds and the presentation of the result have been improved by several authors including Kaltofen, von zur Gathen, and Gao. The sharpest and best result (as far as we know) is due to Gao and is a consequence of his method that we will present in the next section.

Theorem 9.1.8 (Gao's 2000 improved version of Bertini's theorem).
Let $\mathbb{K}$ be a field and S a finite subset of $\mathbb{K}$. Let $f \in \mathbb{K}[X_1, \ldots, X_n]$ of total degree d and $f_0(x, y) = f(a_1x + b_1y + c_1, \ldots, a_nx + b_ny + c_n)$. Suppose $\mathbb{K}$ is either of characteristic zero (e.g. $\mathbb{K} = \mathbb{Q}$) or of characteristic larger than $2d^2$. Then, for random choices of a_i, b_i, c_i in S, with probability at least $1 - \frac{2d^3}{|S|}$, all the absolute irreducible factors of f remain absolutely irreducible factors of f_0 in $\mathbb{K}[x, y]$.

We give a sketch of the proof in exercise 9.2.14 below.

This theorem and an absolute bivariate factorization algorithm, give a randomized algorithm for factoring absolutely multivariate polynomials in $\mathbb{Q}[X_1, \ldots, X_n]$. The strategy is as follows. If d is the total degree of the input polynomial $f(X_1, \ldots, X_n)$, one chooses random values $a_1, b_1, c_1, \ldots, a_n, b_n, c_n$ from a set S in $\mathbb{Q}$ with $|S| \geq 4d^3$ and factors the bivariate polynomial $f_0(a_1x + b_1y + c_1, \ldots, a_nx + b_ny + c_n)$ over $\overline{\mathbb{Q}}$. With probability at least $1/2$, the factors of f_0 correspond to the factors of f evaluated at $X_i = a_ix + b_iy + c_i$, $i = 1$ to n.

A finer general hyperplane section lemma than the one stated in [HS81], is given by part 4 of Theorem 6.3, page 67, in the book of Jouanolou [Jou83].

Theorem 9.1.9 (Jouanolou's 1983 version of Bertini's theorem). *Let k be an infinite field, X a scheme of finite type and $f : X \to k^n$ a $k-$morphism. Suppose that $\dim \overline{f(X)} \geq 2$ and X is geometrically irreducible. Then, for almost all affine hyperplanes H in k^n, $f^{-1}(H)$ is geometrically irreducible.*

In our setting, $k = \mathbb{Q}$, X is an irreducible variety embedded in an affine space k^m and defined by a prime ideal $\mathcal{P}$ of $k[X_1, \ldots, X_m]$ of dimension $r \geq 2$. For m such that $m \geq r \geq n \geq 2$, we project X surjectively onto the affine space $k^n \times 0$ included in k^m, by f which is here the canonical projection. X being geometrically irreducible means $\mathcal{P}$ absolutely prime. H is defined by linear equations L which involve only the first n coordinates, and $f^{-1}(H)$ geometrically irreducible means $\mathcal{P} + L$ is absolutely prime.

Then we see that this statement improves Lemma 9.1.6 because if we choose $n = 2$ and get a surjective projection, then the generic linear equations of H depend only on the first 2 coordinates.

9.1.4 Irreducibility testing

In the factorization process an important preliminary task is to test if a polynomial is already irreducible. This is the case for rational factorization as well as for absolute factorization.

For example the polynomial $P(x) = 5x^4 + 17x^3 - 15x^2 + 12x + 19 \in \mathbb{Z}[X]$ is irreducible because modulo 3 we get $\overline{P}(X) = 2x^4 + 2x^3 + 1 = -(x^4 + x^3 - 1) \in \frac{\mathbb{Z}}{3\mathbb{Z}}[X]$ which is irreducible. Indeed, if we had $P = P_1.P_2$ in $\mathbb{Z}[X]$ with $\deg(P_1) > 0$ and $\deg(P_2) > 0$, then $\mathrm{lc}(P_1).\mathrm{lc}(P_2) = 5 \equiv -1 \mod 3$, $\overline{P}_1$ and $\overline{P}_2$ would satisfy $\overline{P} = \overline{P}_1.\overline{P}_2$ and $\deg \overline{(P}_1) > 0$, $\deg(\overline{P}_2) > 0$. This reduction argument is very general and allows to reduce absolute irreducibility testing to the bivariate case using Bertini's theorem.

One can use a similar argument to derive the following simple absolute irreducibility test which was successfully studied by J.F. Ragot in his Ph.D thesis [Rag97]. It is efficient, easy to implement and works in most cases. Let $P \in \mathbb{Z}[X, Y]$ be irreducible. The idea is to find a simple point of the curve $P(x, y) = 0$ over $\frac{\mathbb{Z}}{p\mathbb{Z}}$ for some prime p (see Definition 9.0.11); it proceeds by an extensive sieve.

We denote by $(*)$ the following condition:

$$(*) \quad \begin{cases} f \text{ is irreducible in } \dfrac{\mathbb{Z}}{p\mathbb{Z}}[X, Y], \\[2mm] \text{there exists a simple point } (a, b) \in \left(\dfrac{\mathbb{Z}}{p\mathbb{Z}}\right)^2 \text{ of } f \mod p, \\[2mm] \text{the degree of } f \mod p \text{ is equal to the degree of } f. \end{cases}$$

Algorithm 9.1.10 RAGOT'S ALGORITHM

Input: $f(X, Y) \in \mathbb{Z}[X, Y]$
For p from 2 to (say) 101 do
      if $f \mod p$ satisfies $(*)$ then return ("f is absolutely irreducible") end if.
end for.
return ("I don't know")
Output: "f is absolutely irreducible" or "I don't know".

Often, the mathematical idea behind an irreducibility test can be extended to get a factorization algorithm. This was the case for the irreducibility test of Ruppert [Rup99], whose idea was later reused by Gao [Gao03] as we will see in Lecture 2. This was also the case for the absolute irreducibility test of Galligo and Watt [GW97], which was developed into a factorization algorithm by Galligo and Rupprecht and was later improved by Chèze as we will see in Lectures 3 and 4.

Before concentrating in the next subsection on generalizations of Eisenstein's criterion, let us mention another active direction of investigation. It deals with multivariate polynomials with complex coefficients, which are known only with a given precision. See e.g. the works of [Nag02] and [KM03].

Eisenstein's classical theorem (see e.g. [Eis95]) states that:

Theorem 9.1.11 (Eisenstein's criterion). *Let R be a unique factorization domain and $f = f_0 + f_1 X + \cdots + f_n X^n \in R[X]$. If there is a prime $p \in R$*

such that all the coefficients except f_n of f are divisible by p, but f_0 is not divisible by p^2, then f is irreducible in $R[X]$.

Example 9.1.12. Let $R = \mathbb{C}[Y]$, $p = Y$ and $f(X,Y) = X^5 + 2YX^4 + (Y^2 + 3Y)X^3 + (Y^3 + 6Y)X^2 + (Y^4 + 10Y^2 + Y)X + Y + 1$, then f is absolutely irreducible by Eisenstein's criterion.

Several mathematicians (Dumas [Dum06], Kurscak [Kur23], Ore [Ore23], [Ore24b], [Ore24a], Rella [Rel27]) have generalized this criterion by using Newton polygons.

Construct a polygon in the Euclidean plane as follows. Suppose that the coefficient f_i is divisible by p^{a_i} but not by any higher power, where $a_i \geq 0$ and a_i is undefined if $f_i = 0$. Plot the points $(0, a_0), (1, a_1), \ldots, (n, a_n)$ in the Euclidean plane and form the lower convex hull of these points. This results in a sequence of line segments starting at the y-axis and ending at the x-axis, called the Newton polygon of f (with respect to the prime p). Dumas [Dum06] determines the degrees of all the possible nontrivial factors of f in terms of the widths of the line segments on the Newton polygon of f. Consequently a simple criterion for the irreducibility of f is established.

Theorem 9.1.13 (Eisenstein-Dumas criterion). *Let R be a unique factorization domain and $f = f_0 + f_1X + \cdots + f_nX^n \in R[X]$ with $f_0f_n \neq 0$. Assume that f is primitive, i.e. $\gcd(f_0, \ldots, f_n) = 1$. If the Newton polygon of f, with respect to some prime $p \in R$, consists only of a line segment from $(0, m)$ to $(n, 0)$, and if $\gcd(n, m) = 1$, then f is irreducible in $R[X]$.*

The condition on the Newton polygon means that $a_i \geq (n - i)m/n$ for $0 \leq i \leq n$ where p^{a_i} exactly divides f_i. When $m = a_0 = 1$, this condition is the same as in Eisenstein's criterion.

A somehow related criterion is used for local irreducibility of Weierstrass polynomials for instance in the proof of the Newton-Puiseux theorem. A series in Weierstrass form $f = Y^n + \sum_{i=0}^{n-1} a_i(X)Y^i \in \mathbb{C}[[X]][Y]$, with valuation $(a_i(X)) = v_i$ and $v_0 = m$, is irreducible in $\mathbb{C}[[X]][Y]$ when $\gcd(n, m) = 1$ and $v_i > \frac{(n-i)m}{n}$, for i=1 to $n - 1$. In that case, the upper Newton polygon of f with respect to Y has only one segment $[(0, n); (m, 0)]$. Of course, this also holds if we replace the series by polynomials vanishing at 0.

These two criteria have been generalized by Gao. After Ostrowski (1921 and 1970), he considered not only lower or upper Newton polygon but the complete convex hull of the support $\{(i, j)/a_{i,j} \neq 0\}$ of $P = \sum_{i,j} a_{i,j}X^iY^j$. As application of this theorem, Gao obtained for instance the following nice special criterion for absolute irreducibility for bivariate polynomials.

Theorem 9.1.14. *Let F be any field and $f = aX^n + bY^m + cX^uY^v + \sum c_{i,j}X^iY^j \in F[X,Y]$ with a, b, c nonzero. Suppose that the Newton polytope of f is the triangle with vertices $(n, 0)$, $(0, m)$ and (u, v). If $\gcd(m, n, u, v) = 1$ then f is absolutely irreducible over F.*

Example 9.1.15. $P(X,Y) = X^3 + Y^2 + X^5Y^4 + X^4Y^2 + X^3Y^2 + X^2Y + XY^2$ and $Q(X,Y) = 2X^3 + 7Y^2 + X^5Y^4 + 10X^4Y^2 + X^3Y^2 + X^2Y + 3XY^2$ are absolutely irreducible. Here $n = 3$, $m = 2$, $u = 5$, $v = 4$. The Newton polytope of these two polynomials is shown in Figure 9.1.15:

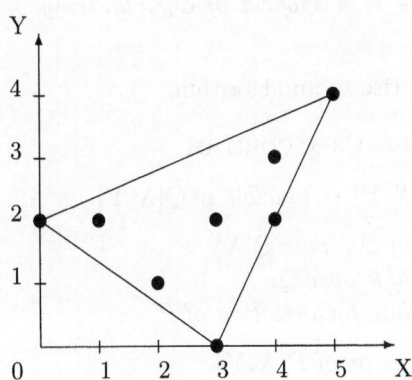

Fig. 9.1. The Newton polytope of an absolute irreducible polynomial.

Gao and his coworkers wrote several papers on these topics (e.g. [Gao01]). They contain an extensive bibliography on the subject.

9.2 Lecture 2: Factorization algorithms via computations in algebraic number fields

The first algorithms by Trager-Traverso, which we explained in the introduction, and the early ones by Kaltofen were used for getting complexity bounds. But they were hardly efficient for degrees greater than 15 because they required the solution of huge linear systems over large algebraic number fields. Then in the late 80's, D. Duval presented in her PhD thesis an algorithm relying on classical algebraic geometry of complex curves and algebraic function fields. This algorithm was able to compute, in a first step, the number of absolute irreducible factors of a polynomial, and a minimal extension which contains the coefficients of one factor.

9.2.1 Duval's algorithm (1987)

Let $P(X,Y) \in \mathbb{Q}[X,Y]$ be irreducible. k is the algebraic closure of $\mathbb{Q}$ in $\mathbb{Q}(x,y) = \dfrac{\mathbb{Q}(X)[Y]}{P(X,Y)} = K$. Let $\mathcal{C}$ be the curve in $\mathbb{C}^2$ defined by $P(X,Y) = 0$,

$K = \mathbb{Q}(\mathcal{C})$ is the field of $\mathbb{Q}$-rational functions on $\mathcal{C}$ and k is called its subfield of constants.

Theorem 9.2.1. *The number of absolute irreducible factors of P is $[k : \mathbb{Q}]$. Moreover, P factorizes as follows: $P = \prod_i \sigma_i(G)$ where G is the minimal polynomial of y over $k(X)$, and σ_i are the $\mathbb{Q}$-isomorphisms of k in $\overline{\mathbb{Q}}$.*

Proposition 9.2.2. *k is a subfield of $\mathcal{O}_K$, the ring of integers of K over $\mathbb{Q}[X]$.*

These results give rise to an algorithm.

Algorithm 9.2.3 DUVAL'S ALGORITHM

Input: $P(X, Y) \in \mathbb{Q}[X, Y]$ *irreducible in $\mathbb{Q}[X, Y]$.*

1. *Compute a basis of $\mathcal{O}_K$ over $\mathbb{Q}[X]$.*
2. *Compute a basis of k over $\mathbb{Q}$.*
3. *Compute an absolute factorization of P.*

Output: An absolute factor of $P(X, Y)$.

The first step relies on the work of Ford-Zassenhaus or Dedekind-Weber. The second step relies on the study of parameterizations of $\mathcal{C}$ by $\mathbb{Q}[[t]]$ and valuations, in order to get special bases for $\mathcal{O}_K$ on which one reads the result. This study is related to normalization. The third step results from a gcd computation in $\mathbb{Q}(\alpha)(X)[Y]$.

This algorithm was implemented by J.F. Ragot in 1994, in Maple. It was more efficient than the former ones. But Gao's algorithm, which we now present, is simpler and more efficient.

9.2.2 Gao's algorithm for absolute factorization

Gao's algorithm is based on a geometric idea inspired by the proof of an irreducibility theorem of W. Ruppert [Rup99] and by the work of H. Niederreiter [Nie93] on factorization. In these notes we specialize Gao's algorithm to the following input and output.
Input: $P(X, Y) \in \mathbb{Q}[X, Y]$, irreducible in $\mathbb{Q}[X, Y]$.
Output:

1. The number d of absolute irreducible factors.
2. A minimal polynomial q of α, $\deg(q) = d$.
3. An absolute irreducible factor of $P = P_1(X, Y) \in \mathbb{Q}(\alpha)[X, Y]$.

Briefly, the algorithm will produce a $\mathbb{Q}$-vector space F of dimension d, whose elements are some rational solutions of a partial differential equation. A basis of this F is computed by solving a rather large system of linear equations. As we will see, a basis of $\overline{F} = F \otimes_{\mathbb{Q}} \overline{\mathbb{Q}}$ consists of s polynomials closely related to the d factors of $P(X, Y)$.

Remark 9.2.4. A version of Gao's algorithm can be also used for finding a rational factorization. Its complexity is polynomial and efficient ("almost" quadratic for dense inputs). It extends to an effective Bertini irreducibility theorem.

In order to ease the exposition of Gao's approach, we point out two key observations in his method.

First observation

Let $P(X,Y) = \prod_{i=1}^{s} P_i(X,Y)$. Taking logarithms, this implies $\log(P(X,Y)) = \sum_{i=1}^{s} \log(P_i(X,Y))$. Set

$$\frac{\partial P}{\partial X} = \sum_{i=1}^{s} \underbrace{(\prod_{j\neq i} P_j) \cdot \frac{\partial P_i}{\partial X}}_{g_i}, \quad \frac{\partial P}{\partial Y} = \sum_{i=1}^{s} \underbrace{(\prod_{j\neq i} P_j) \cdot \frac{\partial P_i}{\partial Y}}_{h_i}.$$

Then we have $\dfrac{\partial}{\partial X}(\log P_i) = \dfrac{1}{P_i} \cdot \dfrac{\partial P_i}{\partial X} = \dfrac{g_i}{P}$, and $\dfrac{\partial}{\partial Y}(\log P_i) = \dfrac{h_i}{P}$.

The following relation expresses the classical Schwartz equality on the second derivatives:

$$(*)\frac{\partial}{\partial Y}(\frac{g_i}{P}) = \frac{\partial}{\partial X}(\frac{h_i}{P}) \text{ for } i = 1 \text{ to } s, \text{ and } \sum_{i=1}^{s} g_i = \frac{\partial P}{\partial X}, \sum_{i=1}^{s} h_i = \frac{\partial P}{\partial Y}.$$

Moreover, we define the bidegree of a polynomial $f \in \mathbb{Q}[X,Y]$ by:
$$\text{bideg}(f) = (\deg_X(f), \deg_Y(f)) = (m,n).$$
If the previous factors P_i are in $\mathbb{C}[X,Y]$, we have with respect to the natural partial ordering:
$$(**) \begin{cases} \text{bideg}(g_i) \leq (m-1,n) \\ \text{bideg}(h_i) \leq (m,n-1). \end{cases}$$

Therefore, to a polynomial factorization of P we naturally attach a set of polynomials (g_i, h_i) which satisfy $(*)$ and $(**)$. In 1986, W. Ruppert derived a condition for absolute irreducibility from similar data.

Let us note for algorithmic purposes that $(*)$ can be rewritten linearly in g_i and h_i as

$$(*') \ P \cdot (\frac{\partial g_i}{\partial Y} - \frac{\partial h_i}{\partial X}) + h_i \frac{P}{\partial X} - g_i \frac{\partial P}{\partial Y} = 0.$$

Definition 9.2.5. *Let F be the $\mathbb{Q}-$vector space of solutions $(v,w) \in \mathbb{Q}[X,Y]^2$ of the PDE*

$$(*) \ \frac{\partial}{\partial Y}(\frac{v}{P}) = \frac{\partial}{\partial X}(\frac{w}{P})$$

such that $\text{bideg}(v) \leq (m-1,n)$, $\text{bideg}(w) \leq (m,n-1)$.
Moreover we set $\overline{F} = F \otimes_\mathbb{Q} \overline{\mathbb{Q}}$.

Remark 9.2.6. A basis of F can be computed by solving a linear system with $m(n+1) + n(m+1)$ unknowns and $4mn$ equations.

Theorem 9.2.7 (Gao's theorem). *Let P be irreducible in $\mathbb{Q}[X,Y]$, and $\prod_{i=1}^{s} P_i$ be an absolute factorization of P. Let $g_i = \prod_{j \neq i} P_j \cdot \dfrac{\partial P_i}{\partial X}$ and $h_i = \prod_{j \neq i} P_j \cdot \dfrac{\partial P_i}{\partial Y}$. Then $\dim_{\mathbb{Q}} F = \dim_{\overline{\mathbb{Q}}} \overline{F}$ is the number of factors s of P.*

Moreover $\{(g_1, h_1), \ldots, (g_s, h_s)\}$ is a basis of $\overline{F}$.

The proof of this theorem uses partial fraction decompositions over $\overline{\mathbb{Q}[X]}(Y)$ of $\dfrac{\partial}{\partial Y}\left(\dfrac{v}{P}\right)$ and of $\dfrac{\partial}{\partial X}\left(\dfrac{w}{P}\right)$. It relies on the following lemma, which allows grouping of terms by conjugacy classes.

Lemma 9.2.8. *Let $k = \mathbb{Q}(X)$, let the derivation $\dfrac{d}{dx}$ extend to $K = \overline{k}$, and assume we have $\dfrac{d}{dx}(K) \subset K$. Let α be algebraic over $\mathbb{Q}(X)$; if $\dfrac{d\alpha}{dX} = 0$ then $\alpha \in \overline{\mathbb{Q}}$.*

Proof. Let $T = T(z, X) = z^l + v_{l-1}(X)z^{l-1} + \cdots + v_0(X)$, $v_i \in \mathbb{Q}(X)$ be the unique minimal polynomial of α.

Since $T(\alpha, X) = 0$, taking the derivative (with respect to X) of this composition of functions gives $\dfrac{\partial T}{\partial z}(\alpha, X)\dfrac{d\alpha}{dX} + \dfrac{\partial T}{\partial X}(\alpha, X) = 0$. As $\dfrac{d\alpha}{dX} = 0$ we get $\dfrac{\partial T}{\partial X}(\alpha, X) = 0$ which can be written $\dfrac{\partial v_{l-1}}{\partial X}(X)\alpha^{l-1} + \cdots + \dfrac{\partial v_0}{\partial X}(X) = 0$.

If this were not identically zero, it would contradict the fact that T is the minimal polynomial of α. So we get $\dfrac{\partial v_i}{\partial X} = 0$, therefore $v_i \in \mathbb{Q}$ and $\alpha \in \overline{\mathbb{Q}}$.

Proof (Gao's theorem). Let n be the degree of P in Y. Consider the factorization $P = \prod_{i=1}^{s} P_i$ where $P_i \in \overline{\mathbb{Q}}[X,Y]$. We can decompose this factorization further over $K[Y]$, with $K = \overline{\mathbb{Q}[X]}$, and get:

$$P = u_n(X) \prod_{j=1}^{n} (Y - \varphi_j(X)), \quad \varphi_j \in K.$$

We set

$$P_i = u_i(X) \prod_{j \in I_i} (Y - \varphi_j(X))$$

where the disjoint union of the I_i gives $\{1, \ldots, n\}$. The φ_j, $j \in I_i$, are conjugate over $\overline{\mathbb{Q}}[X]$. Now $\dfrac{\partial}{\partial X}$ and $\dfrac{\partial}{\partial Y}$ act on K: $\dfrac{\partial}{\partial X}(K) \subset K$, $\dfrac{\partial}{\partial Y}(K) = 0$. We have a unique partial fraction decomposition in $K[Y]$. As $\deg_Y(w) < \deg_Y(P)$ and $\deg_Y(v) \leq \deg_Y(P)$, we have:

$$\frac{w}{P} = \sum_{j=1}^{n} \frac{A_j}{Y - \varphi_j}, \text{ where } A_j = \frac{w(X, \varphi_j(X))}{\frac{\partial P}{\partial Y}(X, \varphi_j(X))} \in K,$$

$$\frac{v}{P} = \sum_{j=1}^{n} \frac{B_j}{Y - \varphi_j} + v_1, \text{ where } B_j \in K, \text{ and } v_1 \in K.$$

By $(*)$ we get:

$$\frac{\partial}{\partial X}\left(\frac{w}{P}\right) = \sum_{j=1}^{n} \frac{(\partial A_j)/(\partial X)}{Y - \varphi_j} - \sum_{j=1}^{n} \frac{A_j}{(Y - \varphi_j)^2} \frac{\partial \varphi_j}{\partial X} =$$

$$= \frac{\partial}{\partial Y}\left(\frac{v}{P}\right) = \sum_{j=1}^{n} \frac{-B_j}{(Y - \varphi_j)^2}.$$

So for all j we have $\dfrac{dA_j}{dX} = 0$, and the previous lemma implies $A_j \in \overline{\mathbb{Q}}$.

Moreover, for each i, $i = 1$ to s, by conjugation, for all $j \in I_i$ the A_j are conjugated over $\overline{\mathbb{Q}}(X)$. But since $A_j \in \overline{\mathbb{Q}}$ they are equal, so we denote by λ_i their common value. Hence $\dfrac{w}{P} = \sum_{i=1}^{s} \sum_{j \in I_i} \dfrac{\lambda_i}{Y - \varphi_j} = \sum_{i=1}^{s} \lambda_i \dfrac{\frac{\partial P_i}{\partial Y}}{P_i}$ and we get $w = \sum_{i=1}^{s} \lambda_i h_i$ as claimed.

Now we get $\dfrac{\partial}{\partial Y}\left(\dfrac{v - \sum_{i=1}^{s} g_i}{P}\right) = 0$ (because (v, w) and (g_i, h_i) satisfy $(*)$); this implies $\dfrac{v - \sum_{i=1}^{s} g_i}{P} = A(X)$. As $\deg_X(v) \leq m - 1$ we get $A = 0$ and then $v = \sum_{i=1}^{s} \lambda_i g_i$.

We deduce two corollaries. The first is an irreducibility criterion and the second gives rise to Gao's algorithm.

Corollary 9.2.9. *P is absolutely irreducible if and only if $\dim_{\mathbb{Q}} F = 1$.*

Corollary 9.2.10. *If $v = \sum_{i=1}^{s} \lambda_i g_i$, where $\lambda_i \neq \lambda_j$ if $i \neq j$, then*

$$P_i = \gcd\left(P, v - \lambda_i \frac{\partial P}{\partial X}\right).$$

Exercise 9.2.11. Prove Corollary 9.2.10. (Hint: $\dfrac{\partial P}{\partial X} = \sum_{i=1}^{s} g_i$.)

Second observation

Now we are in the following situation: given v in E, we want to know its coordinates $\lambda_1, \ldots, \lambda_s$ in the basis $\{g_1, \ldots, g_s\}$. Call E the first projection of F, $E = \{v \in \mathbb{Q}[X, Y]_{\leq(m-1,n)} | \exists w \text{ such that } (v, w) \in F\}$. Note that E can be embedded in $\mathbb{Q}[X, Y]/P$. We set $\overline{E} = E \otimes_{\mathbb{Q}} \overline{\mathbb{Q}}$, $E = \mathbb{Q} < v_1, \ldots, v_d >$, and $\overline{E} = \overline{\mathbb{Q}} < g_1, \ldots, g_d >$.

We have $g_i = \prod_{j \neq i} P_j \dfrac{\partial P_i}{\partial X}$ and $\sum_{i=1}^{s} g_i = \dfrac{\partial P}{\partial X}$, so

$$\begin{cases} \forall i \neq j & g_i g_j \equiv 0 \mod P \\ \forall i & g_i^2 \equiv g_i \dfrac{\partial P}{\partial X} \mod P. \end{cases}$$

Let $v = \sum_{i=1}^{s} \lambda_i g_i$, then $vg_j = \sum_i \lambda_i g_i g_j \equiv \lambda_j g_j^2 \equiv \lambda_j g_j \dfrac{\partial P}{\partial X}$ mod P. In other words, the λ_i are the eigenvalues of the following linear map:

$$\overline{mult_v} : \overline{E} \to \overline{E}',$$

where $\overline{E}' = \overline{\mathbb{Q}} < g_1 \dfrac{\partial P}{\partial X}, \ldots, g_s \dfrac{\partial P}{\partial X} >$. In other words, the λ_i are the roots of the characteristic polynomial $q_v(t) = P_{char}(\overline{mult_v})$.

We can compute $q_v(t)$ for $v \in E$ from our knowledge of the basis $v_1, \ldots, v_d$ of E. Indeed if B is the matrix of the change of basis from $\{v_1, \ldots, v_d\}$ of $\overline{E}$ to $\{g_1, \ldots, g_d\}$, then it does the same job for $\overline{E}'$. So

$$q_v(t) = \det(B) \cdot P_{char}(mult_v) \det(B^{-1}).$$

Finally we get the following algorithm:

Algorithm 9.2.12 GAO'S ALGORITHM

Input: $P(X, Y) \in \mathbb{Q}[X, Y]$, *irreducible in* $\mathbb{Q}[X, Y]$.

1. *Choose a "generic" v in E.*
2. *Compute $q_v(t) = P_{char}(mult_v)$, irreducible in $\mathbb{Q}[t]$. If $q_v(t)$ has a multiple root then go to the last step.*
3. *Call λ_1 a root of $q_v(t)$, then $v - \lambda_1 P_X \in \overline{\mathbb{Q}} < g_2, \ldots, g_d >$, therefore this polynomial is divisible by P_1.*
4. *$P_1 = \gcd(P, v - \lambda_1 P_X)$ in $\dfrac{\mathbb{Q}[t]}{q_v(t)}[X, Y] = \mathbb{Q}(\lambda_1)[X, Y]$.*

Output: An absolute factor $P_1(X, Y) \in \mathbb{Q}[\lambda_1][X, Y]$, and the minimal polynomial $q_v(t)$ of λ_1 over $\mathbb{Q}$.

Implementation, examples, and exercises

Among the implementations of Gao's algorithm there is one by J. May in Maple. It is well commented and available on the web[1]. We downloaded and tested it on the example described hereafter.

We consider a polynomial $R(X, Y) \in \mathbb{Q}[X, Y]$ with bidegree $(12, 12)$. The PDE is written as a linear system: the number of coefficients of (v, w) is $2 \times (12 \times 13) = 312$, and the number of equations is $24 \times 24 = 576$ but many are identically 0. This gives rise to a vector space E of dimension 4 over $\mathbb{Q}$, $E = < v_1, v_2, v_3, v_4 >$, where each v_i is a polynomial of degree 11.

We choose a generic linear combination $v = 8v_1 - v_2 - v_3 + v_4$, then write the 4×4 matrix $mult_v$: we first compute $v_1 \dfrac{\partial P}{\partial X}, v_2 \dfrac{\partial P}{\partial X}, v_3 \dfrac{\partial P}{\partial X}, v_4 \dfrac{\partial P}{\partial X}$ reduced by P, to get 4 polynomials of degree 11.

---

[1] http://www4.ncsu.edu:8030/jpmay/ECCAD01/

Then we compute the reduction of $v_1 \cdot v, v_2 \cdot v, v_3 \cdot v, v_4 \cdot v$ by P to get 4 polynomials of degree 11. After that, we solve 4 systems with 4 unknowns and 276 linear equations. We first obtain the irreducible polynomial

$$q(t) = t^4 - \frac{1}{3}t^3 + \frac{69791}{2313}t^2 - \frac{73148}{2313}t + \frac{253583}{20817}.$$

And finally we get an absolute factor. The total time (on a small PC) was 95 seconds.

J. May and E. Kaltofen have also developed recently a version of this algorithm adapted to imprecise input data, i.e. polynomials with floating point coefficients. See [KM03].

Theorem 9.2.7 and then Gao's algorithm works for a field $\mathbb{F}$ of characteristic zero and for fields of characteristic $p > (2m - 1)n$. Now we present an exercise which follows the idea of Ruppert (see [Rup99], and [Gao03]). For an explicit bound M in the following exercise you can see the proof in [Rup99].

Exercise 9.2.13 (Absolute irreducibility modulo p).
a) Show that $X^4 + 1$ is an irreducible polynomial of $\mathbb{Z}[X]$.
b) Show that $X^4 + 1 \mod p$ is reducible for every prime p.
c) Show that these kinds of phenomena cannot appear when we study an absolutely irreducible polynomial $P \in \mathbb{Z}[X, Y]$. That is to say: show that if $P(X, Y) \in \mathbb{Z}[X, Y]$ is absolutely irreducible, then there exists an integer M such that for every prime number $p > M$, $P(X, Y) \mod p \in \frac{\mathbb{Z}}{p\mathbb{Z}}[X, Y]$ is absolutely irreducible. (Hints: Apply Theorem 9.2.7 to P, study the rank of the linear system (see Definition 9.2.5) related to P and to $P \mod p$, and apply Theorem 9.2.7 to $P \mod p$.)

Finally, we give an exercise on the proof of Bertini's theorem (Theorem 9.1.8).

Exercise 9.2.14 (Bertini's theorem).
a) Show that we can assume that f is square free.
b) Suppose that f has r absolutely irreducible factors. Show that $f_0 = f(a_1X + b_1Y + c_1, \ldots, a_nX + b_nY + c_n)$ has r absolutely irreducible factors over $\mathbb{L} = \mathbb{F}(a_1, b_1, c_1, \ldots, a_n, b_n, c_n)$. (Hint: Use Corollary 9.1.7).
c) Consider the linear systems for f_0 over $\mathbb{L}$ (see Definition 9.2.5) and let M be the associated matrix. Let N be the number of unknowns of the system. Show that rank $M \leq N - r$, and that $N \leq d(d + 1)$. (Hint: You can replace $\mathbb{Q}$ by $\mathbb{L}$ in Theorem 9.2.7.)
d) Show that there is an $(N - r) \times (N - r)$ submatrix M_1 of M whose determinant is nonzero, and that all the $(N - r + 1) \times (N - r + 1)$ submatrices of M have determinant zero.
e) Apply to $\det(M_1)$ the following lemma (see [Zip93, p.192]) and conclude.

Lemma 9.2.15. *Let* $P \in A[X_1, \ldots, X_n]$ *be a polynomial of total degree D over a domain A. Let S be a subset of A of cardinality B. Then*
$$\mathcal{P}(P(x_1, \ldots, x_n) = 0 | x_i \in S) \leq \frac{D}{B}.$$

9.2.3 An algorithm using an ODE

In the same family of algorithms for computing an absolute factorization, we also mention the work of Cormier et al. [CSTU02]. Their algorithm uses for each polynomial P an adapted ODE which is a suitable generalization of a minimal polynomial but with respect to differential Galois theory. It has been implemented, it does not seem more efficient that Gao's algorithm. Nevertheless it has its own theoretical interest.

9.3 Lecture 3: Factorization algorithms via computations in the complex plane

This section and the next one describe the main contributions of the two authors to absolute factorization. It is divided into three parts.

The first one contains results on topology and algebraic geometry of plane curves. The second describes their absolute factorization algorithm. The third part briefly presents other contributions related to the use of floating point and/or monodromy methods.

9.3.1 Topology and algebraic geometry of plane curves

Basic definitions and classical results

Here we recall some classical results, for the proof we refer e.g. to [Rot88].

Definition 9.3.1. *Let X be a topological space, and x_0 a point of X. Let $\Gamma(X, x_0) = \{\gamma \in C^0([0,1], X) | \gamma(0) = \gamma(1) = x_0\}$ be the loops space on X. Homotopy between loops, denoted by the symbol $\sim$, is an equivalence relation on $\Gamma(X, x_0)$. We denote by $[\gamma]$ the homotopy class of γ, and by $\pi_1(X, x_0)$ the set $\dfrac{\Gamma(X, x_0)}{\sim}$.*

It can be shown that $\pi_1(X, x_0)$ equipped with the concatenation is a group, in general non commutative.

Example 9.3.2. Take for X the complement in the real plane of a set of N points; $X = \mathbb{R}^2 - \{p_1, \ldots, p_N\}$. Then $\pi_1(X, x_0)$ is a free group generated by N small loops around the points p_i. See Figure 9.2 where $\mathbb{C} \cong \mathbb{R}^2$ and $\Delta = \{p_1, \ldots, p_N\}$.

Definition 9.3.3. *$\Pi : Y \to X$ is a covering if Π is continuous and if all $x \in X$ have an open neighborhood U_x such that $\Pi^{-1}(U_x)$ is a disjoint union of open sets V_i in Y, with $\Pi_{/V_i} : V_i \to U_x$ is an homeomorphism for every i. We call $\Pi_{/V_i}^{-1} : U_x \to V_i$ a section of Π.*

Proposition 9.3.4. *Let X be a connected topological space, and let $\Pi : Y \to X$ be a covering. If there exists $x_0 \in X$ such that the cardinality of $\Pi^{-1}(x_0)$ satisfies $|\Pi^{-1}(x_0)| = N$ then for all x in X, we have $|\Pi^{-1}(x)| = N$. In this situation, Π (or Y) is called an $N-$fold covering.*

Exercise 9.3.5. Let $P(X,Y) = Y^n + a_{n-1}(X)Y^{n-1} + \cdots + a_0(X)$ be a polynomial in $\mathbb{C}[X,Y]$, $\mathcal{C} = \{(x,y) \in \mathbb{C}^2 | P(x,y) = 0\}$, $pr_1 : \mathbb{C}^2 \to \mathbb{C}$ be the projection on the first coordinate and $\Delta = \{x \in \mathbb{C} | Disc_Y(P(X,Y))(x) = 0\}$. Show that $pr_{1/\mathcal{C}-pr_1^{-1}(\Delta)} : \mathcal{C} - pr_1^{-1}(\Delta) \to \mathbb{C} - \Delta$ is an $n-$fold covering. (Hint: use the implicit function theorem.)

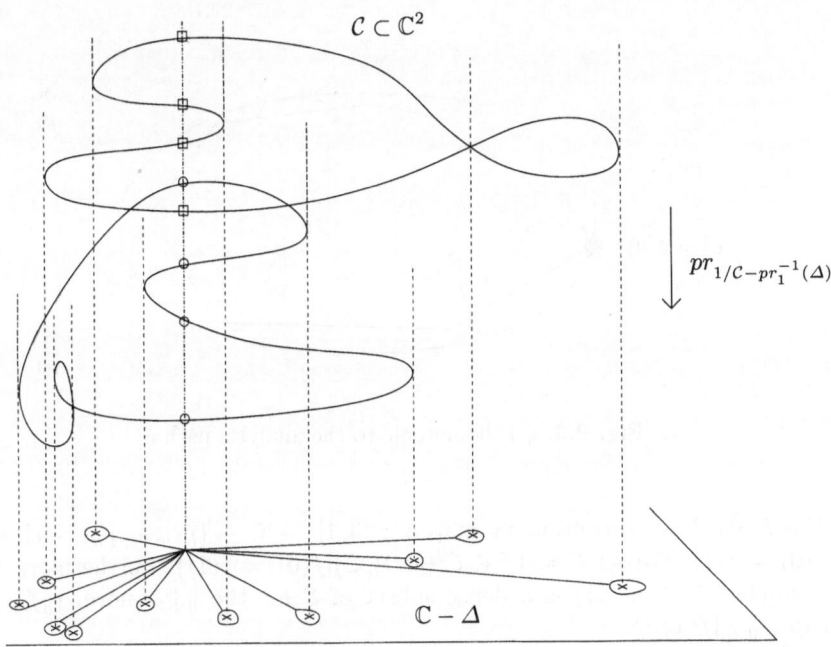

Fig. 9.2. A ramified covering with a smooth generic fiber.

Theorem 9.3.6 (Lifting lemma). *Let X be a connected topological space, $\Pi : Y \to X$ be a covering and $\gamma : [0,1] \to X$ a path such that $\gamma(0) = \gamma(1) = x_0$.*

If y_0 is in the fiber over x_0 (i.e. $\Pi(y_0) = x_0$) then there exists a unique path $\tilde{\gamma}_{y_0} : [0,1] \to Y$ such that $\tilde{\gamma}_{y_0}(0) = y_0$ and $\Pi \circ \tilde{\gamma}_{y_0} = \gamma$.

With this lifting, we can define a group action on the fiber.

Proposition 9.3.7. *Let X be a connected topological space, let $\Pi : Y \to X$ be a N-fold covering, and x_0 a point of X. We denote by $\mathcal{F}$ the fiber over x_0 (i.e. $\mathcal{F} = \Pi^{-1}(x_0)$). We have a group action:*

$$\pi_1(X, x_0) \times \mathcal{F} \to \mathcal{F}$$

$$([\gamma], y_0) \mapsto \tilde{\gamma}_{y_0}(1)$$

and a group homomorphism:

$$\pi_1(X, x_0) \to Aut(\mathcal{F}) = \mathfrak{S}_N$$

where $\mathfrak{S}_N$ is the symmetric group.

Now we give two useful lemmas about analytic paths.

Lemma 9.3.8. *Let γ be a closed path in $\mathbb{C}-\{p_1,\ldots,p_d\}$. Then γ is homotopic to an analytic closed path δ.*

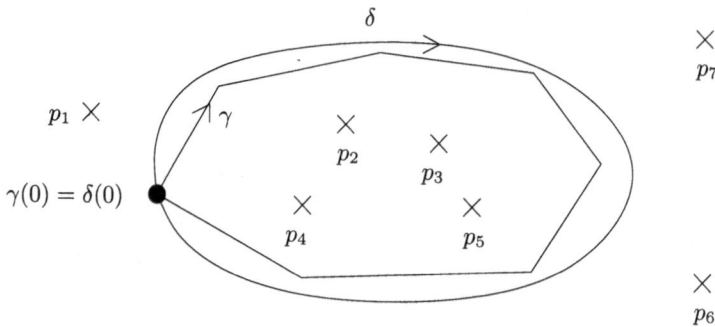

Fig. 9.3. γ is homotopic to the analytic path δ.

Proof. We have a continuous map $\gamma : [0,1] \to \mathbb{C} - \{p_1,\ldots,p_d\}$ such that $\gamma(0) = \gamma(1)$. We set $\mathcal{E} = \{f \in \mathcal{C}^0([0,1], \mathbb{C}) | f(0) = f(1)\}$. Furthermore $S = Span(\{e^{2i\pi n\theta} | n \in \mathbb{Z}\})$ is a dense subset of $\mathcal{E}$ for the $\|.\|_\infty$ norm ($\|f\|_\infty = \sup_{x \in [0,1]} |f(x)|$).

Now the distance between $\gamma([0,1])$ and $\{p_1,\ldots,p_d\}$ is strictly bigger than 0, because these two compact sets are such that $\gamma([0,1]) \cap \{p_1,\ldots,p_d\} = \emptyset$. So we set $d(\gamma([0,1]), \{p_1,\ldots,p_d\})/4 = \epsilon$ and we have $\epsilon > 0$.

Because of the density of S there exists a sequence $(f_n)_n \in S$ with the following property: there exists a number N such that for all $n \geq N$ we have $\|\gamma - f_n\|_\infty < \epsilon$. We set $\delta(t) = f_N(t) - f_N(0) + \gamma(0)$. Then,

$$\|\delta(t) - \gamma(t)\|_\infty \leq \|f_N(t) - \gamma(t)\|_\infty + \|f_N(0) - \gamma(0)\| < d(\gamma([0,1]), \{p_1,\ldots,p_d\}).$$

So δ is homotopic to γ, δ is analytic and $\delta(0) = \delta(1) = \gamma(0) = \gamma(1)$.

Lemma 9.3.9. *The lifting ($\tilde{\gamma}$) of an analytic path γ in Theorem 9.3.6 is analytic.*

Exercise 9.3.10. Prove Lemma 9.3.9. (Hint: Use the implicit function theorem.)

Irreducibility and path connected spaces

Theorem 9.3.11. Let $P(X, Y)$ be a square-free polynomial of $\mathbb{C}[X, Y]$,
$\Delta = \{x | Disc_Y(P(X, Y))(x) = 0\}$, and $\mathcal{C} = \{(x, y) \in \mathbb{C}^2 | P(x, y) = 0\}$.
Then:
P is irreducible in $\mathbb{C}[X, Y]$ $\iff$ $\mathcal{C} - pr_1^{-1}(\Delta)$ is path connected.

We need two lemmas to prove this theorem.

Lemma 9.3.12. Let $\Pi : Y \to X$ be an $n-$fold covering, and Y_1 be a connected component of Y. Then $\Pi_{/Y_1} : Y_1 \to X$ is a $d-$fold covering with $d \le n$.

Exercise 9.3.13. Prove Lemma 9.3.12.

Lemma 9.3.14. Let $P(X, Y) = Y^n + a_1(X)Y^{n-1} + \cdots + a_n(X)$, $x \in \mathbb{C}$, and $y(x)$ be a root of $P(x, Y)$, then $|y(x)| \le \max(1, \sum_{i=1}^{n} |a_i(x)|) \le 1 + \sum_{i=1}^{n} |a_i(x)|$.

Proof. If $|y(x)| \le 1$ then the lemma is true.

If $|y(x)| \ge 1$ then $P(x, y(x)) = 0$ thus $y(x)^n = a_1(x)y(x)^{n-1} + \cdots + a_n(x)$. It follows that $|y(x)|^n \le \sum_{j=1}^{n} |a_i(x)||y(x)|^{n-i} \le \sum_{j=1}^{n} |a_i(x)||y(x)|^{n-1}$. So $|y(x)| \le \sum_{j=1}^{n} |a_i(x)|$. ∎

Proof (Theorem 9.3.11). ⇒) We suppose that P is irreducible. First we remark that $\mathcal{C} - pr_1^{-1}(\Delta)$ is locally path-connected because $\mathbb{C} - \Delta$ is locally path-connected. So it suffices to show that $\mathcal{C} - pr_1^{-1}(\Delta)$ is connected.

Let C_1 be a connected component of $\mathcal{C} - pr_1^{-1}(\Delta)$. Lemma 9.3.12 implies that $pr_{1/C_1} : C_1 \to \mathbb{C} - \Delta$ is a $d-$fold covering with $d \le n$. Thus if we show that $d = n$ then we have $C_1 = \mathcal{C} - pr_1^{-1}(\Delta)$ and we are done.

For every $x_0 \in \mathbb{C} - \Delta$, we have $pr_{1/C_1}^{-1}(x_0) = \{y_1(x_0), \ldots, y_d(x_0)\}$ with $y_i(x_0) \ne y_j(x_0)$ when $i \ne j$. As pr_{1/C_1} is a covering we have a neighborhood V_{x_0} of x_0 such that $y_1, \ldots, y_d$ are defined on V_{x_0}. Furthermore, we have $pr_{1/C_1}^{-1}(x) = \{y_1(x), \ldots, y_d(x)\}$ for every $x \in V_{x_0}$. Hence y_i is analytic on V_{x_0}, by the implicit function theorem applied to $P(x_0, y_i(x_0))$.

We consider the polynomial

$$(Y - y_1(x)) \ldots (Y - y_d(x)) = Y^d + S_1(x)Y^{d-1} + \ldots + S_d(x).$$

The $S_i(x)$ are the elementary symmetric functions in $y_i(x)$. $S_i(x)$ is defined on V_{x_0}, and we now see that $S_i(x)$ is a polynomial. First, we show that $S_i(x)$ is defined on $\mathbb{C} - \Delta$, secondly we show that $S_i(x)$ is defined on $\mathbb{C}$, and bounded by a polynomial.

Let $x_1 \ne x_0$, as before there exit a neighborhood U_{x_1} of x_1 and d analytic functions $\varphi_1, \ldots, \varphi_d$, such that $pr_{1/C_1}^{-1}(x) = \{\varphi_1(x), \ldots, \varphi_d(x)\}$ for every x in V_{x_1}. If $V_{x_1} \cap V_{x_0} \ne \emptyset$ then, as pr_{1/C_1} is a covering and y_i and φ_i are sections of pr_{1/C_1}, we have an element $\sigma \in S_d$ such that $y_i = \varphi_{\sigma(i)}$ on $V_{x_0} \cap V_{x_1}$. Therefore, we have, for example

$$s_1(x) = \sum_{i=1}^{d} y_i(x) = \sum_{i=1}^{d} \varphi_{\sigma(i)}(x) = \sum_{i=d}^{d} \varphi_i(x)$$

on $V_{x_1} \cap V_{x_0}$, thus $S_1(x)$ is defined and analytic in $V_{x_0} \cup V_{x_1}$. So if we repeat this procedure we get d analytic functions $S_i(x)$ defined on $\mathbb{C} - \Delta$.

Now we have to prove that $S_i(x)$ is defined on $\mathbb{C}$. Let us do it for $S_1(x)$. Let x_0 be in Δ, and let $(\epsilon_n)_n$ be a sequence of elements in $\mathbb{C}$ such that $\lim_{n \to \infty} \epsilon_n = 0$.
Then

$$\lim_{n \to \infty} S_1(x_0 + \epsilon_n) = \lim_{n \to \infty} \sum_{i=1}^{d} y_i(x_0 + \epsilon_n)$$

with $y_i(x)$ well defined and continuous in x_0 (as they are the roots of the polynomial $P(x, Y)$). We get

$$\lim_{n \to \infty} S_1(x_0 + \epsilon_n) = \sum_{i=1}^{d} y_i(x_0)$$

(here there exist i_0 and j_0 such that $y_i(x_0) = y_j(x_0)$). Thus there is no singularity on Δ, and we can extend analytically S_1 to $\mathbb{C}$. We proceed similarly for all S_i.

Let x be in $\mathbb{C}$, and $y_i(x)$ be a root of $P(x, Y)$. Lemma 9.3.14 implies $|y_i(x)| \leq 1 + \sum_{j=1}^{n} |a_j(x)|$; this means that $|S_i(x)|$ is bounded by a polynomial, then Liouville's theorem implies that $S_i(x)$ is a polynomial.

Therefore, $P_1(X, Y) = Y^d + S_1(x)Y^{d-1} + \cdots + S_d(X)$ belongs to $\mathbb{C}[X, Y]$. Now we perform the Euclidean division of P by P_1, in $\mathbb{C}(X)[Y]$. As P_1 is monic we get $P(X, Y) = A(X, Y)P_1(X, Y) + R(X, Y)$ with $A(X, Y)$, $R(X, Y) \in \mathbb{C}[X, Y]$ and $R(X, Y) = r_{d-1}(X)Y^{d-1} \ldots + r_0(X)$. For every $x \notin \Delta$, we set $\{y_1(x), \ldots, y_d(x)\} = \{y | P_1(x, y) = 0\}$, where $y_i(x) \neq y_j(x)$ if $i \neq j$. As $(x, y_i(x)) \in C_1 \subset \mathbb{C} - pr_1^{-1}(\Delta)$ we have $P(x, y_i(x)) = 0$ for $i = 1 \ldots d$. Hence $R(x, y_i(x)) = 0$ for $i = 1 \ldots d$, and thus $R(x, Y) = 0$ in $\mathbb{C}[Y]$ for every $x \notin \Delta$. So $r_{d-1}(X) = \ldots = r_0(X) = 0$ in $\mathbb{C}[X]$, and then P_1 divides P. Now as P is irreducible, it follows that $P_1 = P$ and then $d = n$.

$\Leftarrow$) We suppose $P = P_1 \cdot P_2$ where P_i is irreducible in $\mathbb{C}[X, Y]$, and $P_1 \neq P_2$ because P is square free (if we have more than two factors the proof is similar).

We set $\mathcal{V}(P_i) = \{(x, y) \in \mathbb{C}^2 | P_i(x, y) = 0\}$ and $C_i = \mathcal{V}(P_i) \cap (\mathcal{C} - pr_1^{-1}(\Delta))$. C_i is a closed subset of $\mathcal{C} - pr_1^{-1}(\Delta)$.

Furthermore, C_1 and C_2 are distinct, because $pr_1(\mathcal{V}(P_1) \cap \mathcal{V}(P_2)) \subset Disc_Y(P)$. Indeed, we have

$$Disc_Y(P) = Disc_Y(P_1) \cdot Disc_Y(P_2) \cdot Res_Y(P_1, P_2)^2.$$

So we can conclude that $\mathcal{C} - pr_1^{-1}(\Delta) = C_1 \sqcup C_2$, and then that $\mathcal{C} - pr_1^{-1}(\Delta)$ is not connected.

Double point set and transpositions

Lemma 9.3.15 (Change of coordinates). *Let $P(X,Y) \in \mathbb{Q}[X,Y]$ be of total degree n. We consider the change of coordinates $f_\lambda(X,Y) = P(X + \lambda Y, Y)$.*

Let U be the subset of $\mathbb{Q}$ such that for every λ in U we have $\deg_Y(f_\lambda(X,Y)) = n$, and there exists x_0 in $\mathbb{C}$ such that the polynomial $f_\lambda(x_0, Y) \in \mathbb{C}[Y]$ has one root y_0 of multiplicity two, all the other roots have multiplicity one, and $\dfrac{\partial f_\lambda}{\partial X}(x_0, y_0) \neq 0$. Then there exists a finite subset F of $\mathbb{Q}$ such that $U = \mathbb{Q} - F$.

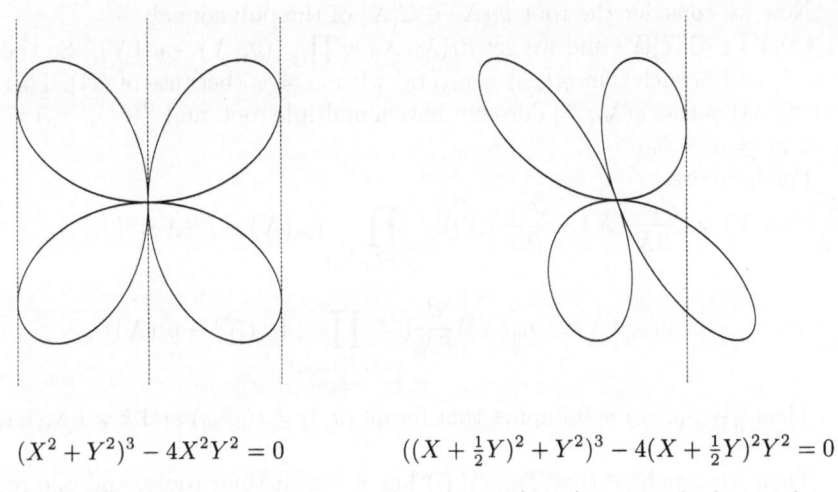

$$(X^2 + Y^2)^3 - 4X^2Y^2 = 0 \qquad\qquad ((X + \tfrac{1}{2}Y)^2 + Y^2)^3 - 4(X + \tfrac{1}{2}Y)^2Y^2 = 0$$

Fig. 9.4. Examples of a bad case $f_0(X,Y) = P(X,Y) = 0$ and of a good case $f_{1/2}(X,Y) = 0$.

Proof. First we show that $V = \{\lambda \in \mathbb{Q} \,|\, \deg_Y(f_\lambda(X,Y)) \neq n\}$ is a finite subset of $\mathbb{Q}$. We set $P(X,Y) = \sum_{k+l \leq n} a_{k,l} X^k Y^l$; then, for $a_n(\lambda) \in \mathbb{Q}[\lambda]$ we have

$$f_\lambda(X,Y) = \sum_{k+l \leq n} a_{k,l} \left(\sum_{i=0}^{k} \binom{k}{i} \lambda^i Y^i X^{k-i} \right) Y^l =$$

$$= a_n(\lambda) Y^n + a_{n-1}(X, \lambda) Y^{n-1} + \cdots + a_0(X, \lambda).$$

Thus $V = \{\lambda \,|\, a_n(\lambda) = 0\}$ hence V is finite.

Now we consider $d_1(\lambda, X) = Disc_Y(f_Y(X,Y)) \in \mathbb{Q}[\lambda, X]$. If (u_0, v_0) is a singular point of $P(X,Y)$ then $(u_0 - \lambda v_0, v_0)$ is a singular point of $f_\lambda(X,Y)$. So $d_1(\lambda, u_0 - \lambda v_0) = 0$ and $X - (x_0 - \lambda v_0)$ divides $d_1(\lambda, X)$. We denote by (u_i, v_i), for $i = 1$ to d the singular points of P, then

$$d_1(\lambda, X) = \prod_{i=1}^{d} (X - (u_i - \lambda v_i))^{e_i} q(\lambda, X)$$

and we set

$$d_2(\lambda) = Disc_X(q(\lambda, X)).$$

Now we claim that if λ is not a root of d_2, and if the change of coordinates preserves the degree in Y of P, then λ belongs to U ; this will prove the lemma. Indeed, we choose (λ_0, x_0) in the following way:

$$\begin{cases} d_2(\lambda_0) \neq 0 & (1) \\ d_1(\lambda_0, x_0) = 0 & (2) \\ x_0 \neq u_i - \lambda_0 v_i & (3) \end{cases}$$ In order to satisfy (2) and (3), we choose a root

x_0 of $q(\lambda_0, X)$ which is not $u_i - \lambda_0 v_i$.

Now we consider the root $y_i(X) \in \overline{\mathbb{C}[X]}$ of the polynomial $f_{\lambda_0}(X, Y) \in \mathbb{C}[X][Y]$ and we get $d_1(\lambda_0, X) = \prod_{i \neq j} (y_i(X) - y_j(X))$. So there exist i_0 and j_0 such that $y_i(x_0) = y_{j_0}(x_0)$ with $i_0 \neq j_0$ (because of (2)). Therefore $d_2(\lambda_0) \neq 0$ so $q(\lambda_0, X)$ does not have a multiple root, and $\frac{\partial d_1}{\partial X}(\lambda_0, x_0) \neq 0$, since $x_0 \neq u_i - \lambda_0 v_i$.

Furthermore,

$$\frac{\partial d_1}{\partial X}(\lambda_0, X) = (\frac{\partial y_{i_0}}{\partial X}(X) - \frac{\partial y_{j_0}}{\partial X}(X)) \prod_{\substack{(i,j) \neq (i_0, j_0) \\ i \neq j}} (y_i(X) - y_j(X))$$

$$+ (y_{i_0}(X) - y_{j_0}(X)) \frac{\partial}{\partial X}(\prod_{\substack{(i,j) \neq (i_0, j_0) \\ i \neq j}} (y_i(X) - y_j(X))).$$

Thus $\frac{\partial d_1}{\partial X}(\lambda_0, x_0) \neq 0$ implies that for all $(k, l) \neq (i_0, j_0)$ and $k \neq l$ we have $y_k(x_0) \neq y_l(x_0)$.

Then we conclude that $f_{\lambda_0}(x_0, Y)$ has $n - 1$ distinct roots, and one root has multiplicity two ($y_{i_0}(x_0) = y_{j_0}(x_0)$). As $\frac{\partial f_{\lambda_0}}{\partial X}(x_0, y_{i_0}) \neq 0$ (because x_0 is not the abscissa of a singular point) the claim is proven.

Theorem 9.3.16. *Let λ_0 be as in Lemma 9.3.15, $\Delta = Disc_Y(f_{\lambda_0}(X, Y))$. Then there exists $X_0 \in \mathbb{C} - \Delta$, γ a path in $\mathbb{C} - \Delta$ such that the monodromy action relative to X_0 of γ on the fiber $f_{\lambda_0}^{-1}(X_0, Y) = \{z_1, \ldots, z_n\}$ is for $i_0 \neq j_0$:*

$$\begin{cases} [\gamma].z_{i_0} = z_{j_0} \\ [\gamma].z_{j_0} = z_{i_0} \\ [\gamma].z_i = z_i \ if \ i \neq i_0 \ and \ i \neq j_0. \end{cases}$$

Proof. a) Let λ_0, x_0 and y_0 be as in Lemma 9.3.15. We have:

$$f_{\lambda_0}(x_0, y_0) = \frac{\partial f_{\lambda_0}}{\partial Y}(x_0, y_0) = 0, \quad \frac{\partial^2 f_{\lambda_0}}{\partial Y^2}(x_0, y_0) \neq 0 \text{ and } \frac{\partial f_{\lambda_0}}{\partial X}(x_0, y_0) \neq 0.$$

We denote by $y_3, \ldots, y_n$ the simple roots of $f_{\lambda_0}(x_0, Y)$. We have

$$\begin{cases} \forall i \geq 3 & f_{\lambda_0}(x_0, y_i) = 0 \\ \forall i \geq 3 & \frac{\partial f_{\lambda_0}}{\partial Y}(x_0, y_i) \neq 0. \end{cases}$$

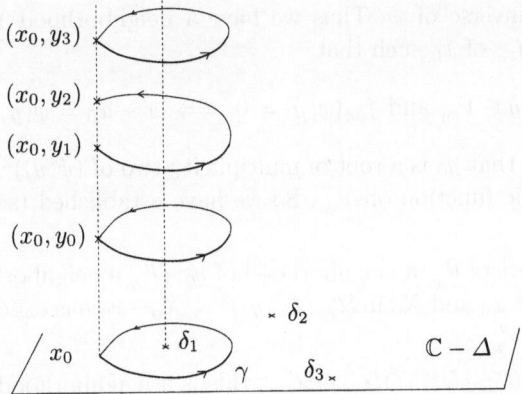

Fig. 9.5. The monodromy action of γ gives the transposition $(y_1 \quad y_2)$.

So we can apply the analytic version of the implicit function theorem to every (x_0, y_i). Thus, there exits a neighborhood V_{y_0} of y_0, a neighborhood $U_{x_0}^0$ of x_0, and an analytic function φ such that

$$x \in U_{x_0}^0, y \in V_{y_0} \text{ and } f_{\lambda_0}(x, y) = 0 \iff x = \varphi(y) \text{ and } y \in V_{y_0}.$$

For $i \geq 3$, there exists a neighborhood $U_{x_0}^i$ of x_0, V_{y_i} a neighborhood of y_i, and p_i an analytic function such that:

$$x \in U_{x_0}^i, y \in V_{y_i} \text{ and } f_{\lambda_0}(x, y) = 0 \iff y = p_i(x) \text{ and } x \in U_{x_0}^i.$$

Now, we consider the parametrization $x = \varphi(y)$. We have $\varphi(y) = x_0 + a(y - y_0) + b(y - y_0)^2 + \dots$ in a neighborhood of y_0, with $a = -\dfrac{\partial f_{\lambda_0}}{\partial Y}(x_0, y_0)(\dfrac{\partial f_{\lambda_0}}{\partial X}(x_0, y_0))^{-1} = 0$, and

$b = -\dfrac{1}{2}\dfrac{\partial^2 f_{\lambda_0}}{\partial Y^2}(x_0, y_0)(\dfrac{\partial f_{\lambda_0}}{\partial X}(x_0, y_0))^{-1} \neq 0$. Thus in a neighborhood of (x_0, y_0) we have

$$(x - x_0) = (y - y_0)^2[b + \sum_{k \geq 1} a_k(y - y_0)^k].$$

b) The equation $z^2 = b$ has two distinct nonzero roots r_1 and r_2. Near $r_1 \neq 0$ the function $c : \mathbb{C} \to \mathbb{C}$ given by $z \mapsto z^2$ has an analytic inverse because $c'(r_1) = 2r_1 \neq 0$. Let r be this inverse, $r : V_b \to W_{r_1}$. Hence, in a neighborhood $V_{y_0}^{(1)}$ of y_0 we define: $\mathcal{R} : V_{y_0}^{(1)} \to W_{r_1} : y \mapsto r(b + \sum_{k \geq 1} a_k(y - y_0)^k)$. Thus in a neighborhood of (x_0, y_0) we have: $x - x_0 = (y - y_0)^2(\mathcal{R}(y))^2$. We denote by ψ the following map: $\psi : V_{y_0}^{(2)} \to V_0$ given by $y \mapsto (y - y_0)\mathcal{R}(y)$ where V_0 is a neighborhood of 0, and $V_{y_0}^{(2)}$ is a neighborhood of y_0 on which ψ is an isomorphism (this is possible because $\psi'(y_0) = \mathcal{R}(y_0) = r(b) \neq 0$). We denote then by ξ the inverse of ψ. Thus we have a neighborhood $\mathcal{V}_{y_0}$ of y_0 and a neighborhood $\mathcal{U}_{x_0}$ of x_0 such that

$$(*) \quad x \in \mathcal{U}_{x_0}, y \in \mathcal{V}_{y_0} \text{ and } f_{\lambda_0}(x,y) = 0 \iff x - x_0 = \psi(y)^2 \text{ and } y \in \mathcal{V}_{y_0}$$

Now we remark that y_0 is a root of multiplicity two of $(\psi(y))^2 = 0$. ψ is a non constant analytic function on $\mathcal{V}_{y_0}$. So we have established (see [Car61] p 97, Prop 4.2):

$(**)$ There exists $\mathcal{V}'_{y_0}$ a neighborhood of y_0, $\mathcal{U}'_{x_0}$ a neighborhood of x_0 such that for all $X_0 \neq x_0$ and X_0 in $\mathcal{U}'_{x_0}$, $(\psi(y))^2 = X_0 - x_0$ has exactly two distinct simple roots in $\mathcal{V}'_{y_0}$.

We set $V = (\cap_{i \geq 3} \mathcal{U}^i_{x_0}) \cap \mathcal{U}_{x_0} \cap \mathcal{U}'_{x_0}$; this is a neighborhood of x_0. Now we choose a real number $\rho > 0$ such that $\mathcal{B}(x_0, \rho) \subset V$ and $\mathcal{B}(0, \sqrt{\rho}) \subset V_0$.

c) Lifting paths

We set: $X_0 = x_0 + \rho$ and $\gamma : [0,1] \to \mathbb{C} - \Delta : t \mapsto x_0 + \rho e^{2i\pi t}$. Thus $f_{\lambda_0}(X_0, Y)$ has n distinct roots: $z_1, \ldots, z_n$. Now we write all these roots with ξ or p_i. If $X_0 \in V$ and $y \in \mathcal{V}'_{y_0}$, then $f_{\lambda_0}(X_0, y)$ has two distinct roots z_1 and z_2 in $\mathcal{V}'_{y_0}$ by $(**)$ and by $(*)$ we can set $\psi(z_1) = \sqrt{\rho}$ and $\psi(z_2) = -\sqrt{\rho}$. Hence $z_1 = \xi(\sqrt{\rho})$ and $z_2 = \xi(-\sqrt{\rho})$. Furthermore, $X_0 \in V$ and $y \in V_{y_i}$, $f_{\lambda}(X_0, y) = 0 \iff y = p_i(X_0)$ for $i = 3 \ldots n$. So we set $z_i = p_i(X_0)$. Now we lift γ above z_1 and z_2. We set

$$\gamma_1(t) = \xi(\sqrt{\rho} e^{i\pi t}) \; ; \; \gamma_2(t) = \xi(-\sqrt{\rho} e^{i\pi t}).$$

These two paths are well defined, continuous and $\gamma_i(0) = z_i$.

For all $t \in [0,1]$, $\gamma(t) - x_0 = \rho e^{2i\pi t} = [\psi(\xi(\sqrt{\rho} e^{i\pi t}))]^2$ because $\psi \circ \xi = \text{id}$, then for all $t \in [0,1]$, $\gamma(t) - x_0 = [\psi(\gamma_1(t))]^2 = [\psi(\gamma_2(t))]^2$.

By $(*)$ we get $f(\gamma(t), \gamma_i(t)) = 0, \forall t \in [0,1]$. Thus γ_1 lifts γ above z_1 and γ_2 lifts γ above z_2. As $\gamma_1(1) = z_2$ and $\gamma_2(1) = z_1$ we get: $[\gamma].z_1 = z_2$, $[\gamma].z_2 = z_1$.

Therefore we set $\gamma_i(t) = p_i(\gamma(t))$ for $i = 3 \ldots n$. We have

$$f(\gamma(t), \gamma_i(t)) = 0, \forall t \in [0,1],$$

by the definition of p_i. Now, γ_i lifts γ above z_i and $\gamma_i(1) = p_i(\gamma(1)) = p_i(\gamma(0)) = z_i$. Hence, for $i = 3, \ldots, n$, $[\gamma].z_i = z_i$.

Transpositions will play an important role for the proof of Harris' lemma and its generalizations (see Theorem 9.3.20). Theorem 9.3.16 will be used with the following lemma.

Definition 9.3.17. *Let $G \times X \to X$ be a group action. If for every two pairs of points x_1, x_2 and y_1, y_2 ($x_1 \neq x_2$ and $y_1 \neq y_2$) there is a group element g such that $g.x_i = y_i$, then the group action is called $2-transitive$.*

Lemma 9.3.18. *Let G be a subgroup of $\mathfrak{S}_n$ such that the action of G on $\{1, \ldots, n\}$ is $2-transitive$ and such that there exists a transposition τ in G, then $G = \mathfrak{S}_n$.*

Proof. We can suppose $\tau = (1 \ \ 2)$. Let x and y be two distinct elements of $\{1, \ldots, n\}$. There exists a permutation $\sigma \in G$ such that $\sigma(1) = x$ and $\sigma(2) = y$ (because the action is 2−transitive). Then $\sigma^{-1}\tau\sigma = (\sigma(1) \ \ \sigma(2)) = (x \ \ y) \in G$. Therefore every transposition belongs to G ; this implies that $G = S_n$.

Monodromy and genericity

We saw in the last subsection that when P is irreducible, the monodromy action on the fiber is transitive ; that is, any two points y_i and y_j of the fiber $\phi^{-1}(x_0)$ can be exchanged following a continuous path on the curve on top of some loop γ. This result also expresses the connectivity of the subspace formed by the curve C minus the ramification points. In fact there is a stronger connectivity result which is a consequence of a lemma due to J. Harris (see [Har80] or [ACGH85]) which was originally used to establish his uniform position theorem on the generic hyperplane section of a projective curve.

We will adapt Harris' lemma to our setting in order to obtain what we call an Affine Harris theorem. This theorem says that if we perform a **generic** change of coordinates before taking the projection, not only the action of the monodromy group is transitive but any permutation of the point of the fiber $\phi^{-1}(x_0)$ can be obtained following a continuous path on the curve on top of some loop γ. This key fact and its application to absolute factorization was first observed by Galligo, stated in [GW97], then in [Gal99] and in [Rup00], [Rup04]. However in these papers it was just indicated that this statement was a consequence of Harris' lemma and classical arguments in algebraic geometry. As Sommese-Verschelde-Wampler needed this statement to improve their algorithm (see below), in [SVW02c] they gave a more complete proof of it and made precise references to two textbooks ([ACGH85] and [GM83]). In the next subsection, we will give a detailed exposition of this result. Let us start by reviewing Harris' lemma and its proof in the case of plane curves.

Lemma 9.3.19 (Harris' lemma). *Let C be an irreducible projective plane curve, possibly singular, and call n its degree. Let U be the Zariski open subset in $\mathbb{P}^2(\mathbb{C})^*$ of lines transverse to C, i.e. cutting C in n simple points. Consider the incidence correspondence graph I and its second projection:*

$$pr_2 : I = \{(p, H) \in C \times U \ \mid \ p \in H\} \to U.$$

Then pr_2 is a n-fold topological covering. We fix $H_0 \in U$ and let Γ_0 denote the set of n intersection points $C \cap H_0$. Then the monodromy map

$$\pi_1(U, H_0) \to S_n$$

is surjective.

Proof. Let G be the image of the monodromy map. We know by application of Theorem 9.3.11 that C minus its singular locus is path connected. As by

Theorem 9.3.16, U contains in its border a line H_1 which is tangent to C at a simple point and transverse to C at all the other $n - 2$ intersection points, we deduce that G contains a transposition. So in order to apply Lemma 9.3.18 we need only to prove that G is 2-transitive. To express this property geometrically let us set

$$I_2 = \{(p_1, p_2, H) \in C \times C \times U \mid p_1 \in H, p_2 \in H, p_1 \neq p_2\}$$

and similarly

$$J_2 = \{(p_1, p_2, H) \in C \times C \times \mathbb{P}^2(\mathbb{C})^* \mid p_1 \in H, p_2 \in H, p_1 \neq p_2\}.$$

With this definition, obviously J_2 is a line bundle over $C \times C - \Lambda$ where Λ is the diagonal of $C \times C$ (indeed two distinct points define a line). Now Λ is a complex subvariety of $C \times C$ of strictly smaller dimension, and $C \times C$ is path connected because C is path connected. Therefore J_2 is also path connected. As I_2 is obtained from J_2 by subtracting a complex subvariety of strictly smaller dimension, it is also path connected. This implies that G is 2-transitive.

The Affine Harris theorem

Theorem 9.3.20 ([GW97]). *Let $P \in \mathbb{Q}[X, Y]$ be a an absolutely irreducible polynomial of total degree n. Let C be the corresponding affine curve in $\mathbb{C}^2$. Then there exists a Zariski open set of affine changes of coordinates such that, the projection on the new first coordinate x:*

$$pr_1 : C \to \mathbb{C}$$

induces on the fiber $pr_1{}^{-1}(0)$ a monodromy map

$$\pi_1(\mathbb{C} - \Delta, 0) \to \mathfrak{S}_n$$

which is surjective.

Proof. The theorem is a corollary of Harris' lemma and a classical theorem of van Kampen, recalled below as Theorem 9.3.21. With the notations of the last subsection we identify $\mathbb{C} - \Delta$ with the set of all lines in $\mathbb{C}^2$ parallel to the Oy-axis and transverse to C. Then we include this set in the intersection of U with the line of $\mathbb{P}^2(\mathbb{C})^*$ formed by all the lines passing through the point at infinity corresponding to the Oy-axis. Moreover we suppose that O is not in Δ and choose $H_0 = Oy$. Then to prove the theorem, it suffices to show that the induced group homomorphism

$$\pi_1(\mathbb{C} - \Delta, 0) \to \pi_1(U, H_0)$$

is surjective.

We view U as the complement of a reduced projective curve in the dual projective plane $\mathbb{P}^2(\mathbb{C})^*$, which is isomorphic to the usual projective plane

$\mathbb{P}^2(\mathbb{C})$. Then our theorem will be a consequence of a classical theorem of E. van Kampen in 1933. A short, rigorous and self-contained exposition of this last result was given in a paper by D. Cheniot in 1973. It contains a precise description (by generators and relations) of the fundamental groups. We summarize it as follows.

Theorem 9.3.21 (van Kampen). *Let $\mathcal{H}$ be a reduced algebraic curve of degree n in $\mathbb{P}^2(\mathbb{C})$ and A be point not on $\mathcal{H}$. Let L_i for $i = 1$ to m and L_∞ be all the line passing through A and not transverse to $\mathcal{H}$. Call λ_i for $i = 1$ to m and λ_∞ their direction in a $\mathbb{P}^1(\mathbb{C})$ complement to A. Let L be a line passing through A and transverse to $\mathcal{H}$, and call λ its direction. Then*

$$\pi_1(\mathbb{P}^1(\mathbb{C}) - \{\lambda_1, \ldots, \lambda_i, \lambda_\infty\}, \lambda) \to \pi_1(\mathbb{P}^2(\mathbb{C}) - \mathcal{H}, A)$$

is surjective.

With our notation, we get

$$\pi_1(\mathbb{P}^1(\mathbb{C}) - \{\lambda_1, \ldots, \lambda_i, \lambda_\infty\}, \lambda) = \pi_1(\mathbb{C} - \Delta, 0)$$

and we are done.

Remark 9.3.22. It would have been more elegant to provide an algebraic proof of our Affine Harris theorem. A natural way to do this, is to adapt the proof recalled in the last section, by using Jouanolou's version of Bertini's theorem applied to the algebraic set I_2. However, this only proves a weaker version of our claim. Indeed, instead of obtaining the monodromy map associated to the lines parallel to a generic direction Oy, we get the monodromy map associated to the lines passing through a generic point of $\mathbb{P}^2(\mathbb{C})$ and we cannot be sure that we can choose such a point on the line at infinity. So we are led to rely on a topological analysis, which in this situation gives more precise information.

Composite Monodromy

To validate the Galligo-Rupprecht algorithm, a result for the composite case is needed.

When P has several factors, then $\mathcal{C}$ has several irreducible components $\mathcal{C}_1, \ldots, \mathcal{C}_s$, and each of them has a monodromy action. So we can relate the monodromy of $\mathcal{C}$ to the monodromies of the $\mathcal{C}_i$. The result obtained in [Gal99] and in [Rup00], says that after a generic change of coordinates, the following group homomorphism is surjective:

$$\pi_1(\mathcal{C} - \Delta) \to \mathfrak{S}_{n_1} \times \mathfrak{S}_{n_2} \times \cdots \times \mathfrak{S}_{n_s}.$$

This result is a straightforward corollary of the proof of Theorem 9.3.20 that we explained above.

9.3.2 The Galligo-Rupprecht's algorithm

The algorithm takes as input a bivariate polynomial with rational coefficients irreducible over $\mathbb{Q}$, performs a generic affine change of coordinates in order to get a new polynomial $P(X, Y)$, and outputs an approximate absolute factorization of P. This will be used in the next section to produce an exact absolute factorization of P.

The algorithm first finds a key combinatorial fact about the target factorization: the partition generated by the factorization on a smooth generic fiber $P(0, Y)^{-1}(0)$.

This is obtained by analyzing the restriction of the factorization modulo X^3 which gives so called 'Zero sum' relations (see below for the history of this concept). Our Affine Harris Theorem proves that these relations indeed provide sufficient and necessary conditions for the absolute factorization of P. The corresponding factorization of $P(0, Y)$ is later lifted to a factorization of P by Hensel liftings. In order to compute efficiently these 'Zero sum' relations and Hensel liftings we rely on good Newton approximation of the roots of $P(0, Y) = 0$.

So, first we perform some reductions on the input polynomial in order to get a monic square-free polynomial which is irreducible (over $\mathbb{Q}$). We consider "generic" affine change of coordinates in 2 variables

$$X = x + ay + b \,;\, Y = y + c.$$

In practice, this means a change of coordinates whose coefficients (a, b, c) are decimal numbers provided by a "random function" that one can find on any computer.

Simplifying, we get a new monic polynomial in $\mathbb{Q}[x, y]$, that we call P :
$y^n + a_{n-1}(x)y^{n-1} + \cdots + a_0(x)$ with $\deg a_i(x) \leq n - i$.

A consequence of the fundamental Lemma 9.0.8 is that the factors all have the same degree.

As there are efficient algorithms for the detection of factors of degree 1, *we suppose that the degree of the factors of P is greater or equal to 2.* This assumption will be used in Lemma 9.3.23. Now we describe the main ideas behind the Zero-sum relations.

Definition of the numbers b_i and their properties

Let P be a square-free polynomial in $\mathbb{Q}[X, Y]$ of total degree n, monic in Y. For $x_0 \in \mathbb{Q}$, we denote by $y_1(x_0), \ldots, y_n(x_0)$ the roots of $P(x_0, Y)$. Then for all but at most $n(n - 1)$ values of x_0, these roots are distinct and the curve defined by P is smooth at the points $(x_0, y_i(x_0))$, for $i = 1, \ldots, n$. If we choose such a value for x_0, then there exist analytic functions $\varphi_i(X)$ in the neighborhood of x_0 (for $i = 1, \ldots, n$) such that

$$\begin{cases} \varphi_i(x_0) = y_i(x_0) \\ P(X, \varphi_i(X)) = 0. \end{cases}$$

There exist complex numbers a_i and b_i (for $i = 1, \ldots, n$) such that

$$\varphi_i(X) = y_i(x_0) + a_i(X - x_0) + b_i(X - x_0)^2 + \cdots.$$

If

$$\alpha(x, y) = \frac{\partial P}{\partial x}(x, y), \quad \beta(x, y) = \frac{\partial P}{\partial y}(x, y),$$

$$\gamma(x, y) = \frac{\partial^2 P}{\partial x^2}(x, y), \quad \delta(x, y) = \frac{\partial^2 P}{\partial y^2}(x, y), \quad \varepsilon(x, y) = \frac{\partial^2 P}{\partial x \partial y}(x, y)$$

then we have

$$a_i = -\frac{\alpha(x_0, y_i(x_0))}{\beta(x_0, y_i(x_0))} \quad \text{and}$$

$$b_i = -\frac{1}{2\beta(x_0, y_i(x_0))} \left(\gamma(x_0, y_i(x_0)) + 2\varepsilon(x_0, y_i(x_0))a_i + \delta(x_0, y_i(x_0))a_i^2 \right).$$

We use these formulas to introduce analytic functions a and b defined on $\mathcal{C} - pr_1^{-1}(\Delta)$ which is a n-fold covering of an open subset of the complex plane (see Subsection 9.3.1). We set

$$a(X, Y) = -\frac{\alpha(X, Y)}{\beta(X, Y)},$$

$$b(X, Y) = -\frac{1}{2\beta(X, Y)} \left(\gamma(X, Y) + 2\varepsilon(X, Y)a + \delta(X, Y)a \right) \in \mathbb{C}(X, Y).$$

Let U be a (small) open neighborhood of x_0 in $\mathbb{C}$ where all the $\varphi_i(X)$ are defined for $i = 1$ to n. Then denote by $V = \cup V_i$ its inverse image by $pr_1^{-1}(U)$ in $\mathcal{C} - pr_1^{-1}(\Delta)$. We also consider the restrictions of $a(X, Y)$ and $b(X, Y)$ to each V_i, and set $b_i(X) = b(X, \varphi_i(X))$.

As $P(X, \varphi_i(X)) = 0$ on U for all i, and P is monic in Y, we can write $P(X, Y) = \prod_{i=1}^{n}(Y - \varphi_i(X))$.

For each $k = 1, \ldots, s$, the factor P_k in the factorization $P(X, Y) = \prod_{k=1}^{s} P_k$ can be written

$$P_k(X, Y) = \prod_{j=i_1}^{i_m} (Y - \varphi_j(X)). \quad (1)$$

The total degree of P_k is m so we can write:

$$P_k(X, Y) = Y^m + (q_1(X))Y^{m-1} + q_2(X)Y^{m-2} + \cdots + q_m(X) \quad (2)$$

where $q_j(X) \in \mathbb{Q}[X]$ and $\deg(q_j(X)) \leq j$. In particular, $\deg(q_1(X)) \leq 1$ so the coefficient of its degree two term is zero. From (1) and (2), we get $\sum_{j=i_1}^{i_m} \varphi_j(X) = q_1(X)$, thus

$$\sum_{j=i_1}^{i_m} b_j(X) = 0.$$

So we have found a necessary condition on each factor P_k of P. Our aim is to prove that such a Zero-sum relation is also a sufficient condition to characterize a factor of P. The next lemma is an intermediate result in that direction.

Lemma 9.3.23. *If P has no factor of degree 1, then for almost all values of x_0, the numbers b_i are all non-zero (for $i = 1, \ldots, n$).*

Proof. We consider the analytic function on U defined by $\mathcal{P} = \prod_{i=1}^n b_i(X) = \prod_{i=1}^n b(X, \varphi_i(X)) \in \mathbb{C}(X, \varphi_1(X), \ldots, \varphi_n(X))$.

$\mathcal{P}$ is a symmetric rational function of the $\varphi_i(X)$. Thus by definition of the $\varphi_i(X)$, we deduce that $\mathcal{P}$ is a rational function of the coefficients of P as a polynomial in Y. So $\mathcal{P}$ belongs to $\mathbb{C}(X)$.

If $\mathcal{P} \neq 0$ in $\mathbb{C}(X)$ then for almost all x_0 in $\mathbb{C}$, $\mathcal{P}(x_0) \neq 0$. Thus $b_i(x_0) = b(x_0, \varphi_i(x_0)) \neq 0$ for all i, and we are done.

If $\mathcal{P} = 0$ in $\mathbb{C}(X)$, we shall see that we get a contradiction.

As $\mathcal{P}(x) = \prod_{i=1}^n b(x, \varphi_i(x)) = 0$ in V and $b(x, \varphi_i(x))$ are analytic functions on U, there exists an index i_0 such that $b(x, \varphi_{i_0}(x)) = 0$ on U (because the ring of analytic functions on U is an integral domain).

This implies $\varphi_{i_0}''(x) = 0$ on U then $\varphi_{i_0}^{(r)}(x) = 0$ on U for every $r \geq 2$.

Thus $\varphi_{i_0}(x) = y_{i_0} + \varphi_{i_0}'(x_0)(x - x_0)$ on U.

We perform the Euclidean division of $P(X, Y)$ by

$$F(X, Y) = Y - y_{i_0}(x_0) - \varphi_{i_0}'(x_0)(X - x_0)$$

in $\mathbb{C}(X)[Y]$. Since $F(X, Y)$ is monic in Y, we get

$$P(X, Y) = A(X, Y)F(X, Y) + R(X),$$

with $A(X, Y) \in \mathbb{C}[X, Y]$ and $R(X) \in \mathbb{C}[X]$.

Therefore, for every $x \in U$ as $F(x, \varphi_{i_0}(x)) = 0$, we have $R(x) = 0$ then $R(X) = 0$ in $\mathbb{C}[X]$. This implies that P has a factor of degree 1, contrary to hypothesis.

Now we can prove the theorem:

Theorem 9.3.24. *Let P be an irreducible polynomial of degree n. Consider $f_\lambda(X, Y, \lambda) = P(X + \lambda Y, Y)$. Then for almost all specializations (x_0, λ_0) of (x, λ) in $\mathbb{C} \times \mathbb{Q}$, none of the sums $\sum_{i \in J} b_i$, for $J \subsetneq \{1, \ldots, n\}$, vanishes.*

Remark 9.3.25. In the previous statement, "almost all" means that we have to avoid a finite number of x_0 and a finite number of λ_0.

Proof. First we perform a change of coordinates, and we choose λ_0 such that the conclusion of the Affine Harris Theorem 9.3.20 is true.

We denote by $f_{\lambda_0}(X, Y)$ the polynomial $P(X + \lambda_0 Y, Y)$. Hereafter, the numbers b_i, $y_i(x_0)$ and the rational function b are related to f_{λ_0}. We set $m < n$.

Now we want to prove that for almost all x_0, we do not have $\sum_{i \in I} b_i = 0$ with $I \subsetneq \{1, \ldots, n\}$ and $|I| = m$. We consider the functions $\mathcal{B}_I = \sum_{i \in I} b_i(X)$, and $\mathcal{B} = \prod_{\mathcal{E}_m} \mathcal{B}_I$ where
$$\mathcal{E}_m = \{\{\sigma(1), \ldots, \sigma(m)\} | \sigma \in \mathcal{S}_n\}.$$
$\mathcal{B}$ is a rational function in $X, \varphi_1(X), \ldots, \varphi_n(X)$, and is symmetric with respect to these last n arguments. So as in Lemma 9.3.23, we have that $\mathcal{B}$ is a rational function of X. We want to prove that $\mathcal{B} \neq 0$ in $\mathbb{C}(X_0)$. For this, we suppose that $\mathcal{B} = 0$ in $\mathbb{C}(X_0)$, and show that we get a contradiction. Choose $x_0 \notin \Delta = Disc_Y(f_{\lambda_0}(X, Y))$ such that for $i = 1, \ldots n, b_i(x_0) \neq 0$ (∗) (this is possible by Lemma 9.3.23). Then for every x in U (the neighborhood where all the φ_i are defined),
$$\beta(x, \varphi_i(x)) = \frac{\partial f}{\partial Y}(x, \varphi_i(x)) \neq 0 \text{ for } i = 1, \ldots, n.$$
Therefore $b(x, \varphi_i(x))$ is well defined and analytic on U, thus $\mathcal{B}_I(x)$ is analytic on U. As $\mathcal{B} = 0$, there exists a set I_0 such that $\mathcal{B}_{I_0}(x) = 0$ on U. We can suppose $I_0 = \{1, 2, \ldots, m\}$, and we set $J_0 = \{2, 3, \ldots, m - 1, m, m + 1\}$.

By the Affine Harris Theorem 9.3.20, there exists a closed path γ such that $[\gamma]$ acts on $\{y_1(x_0), \ldots, y_n(x_0)\}$ as the transposition $(1 \quad m + 1)$. That is to say: if we denote by $\tilde{\gamma}_{y_i}$ the lifting of γ above y_i, so $\tilde{\gamma}_{y_i} = (\gamma, \delta_{y_i})$ where γ and δ_{y_i} are analytic (see Lemma 9.3.9) then
$$\begin{cases} \delta_{y_i}(0) = y_i \text{ for } i = 1, \ldots, n \\ \delta_{y_i}(1) = y_i \text{ if } i \neq 1 \text{ and } i \neq m + 1 \\ \delta_{y_i}(1) = y_{m+1} \text{ and } \delta_{y_{m+1}}(1) = y_1. \end{cases}$$
Now we set $H(t) = \sum_{i=1}^{m} b(\gamma(t), \delta_{y_i}(t))$. This is an analytic function on $]0, 1[$ such that $H(0) = \mathcal{B}_{I_0}(x_0)$ and $H(1) = \mathcal{B}_{J_0}(x_0)$.

As $H(t) = \mathcal{B}_{I_0}(\gamma(t))$ for $t \in \gamma^{-1}(U)$ and $\mathcal{B}_{I_0} = 0$ on U, we get $H = 0$ because H is analytic. Thus $\mathcal{B}_{I_0}(x_0) = \mathcal{B}_{J_0}(x_0)$. This implies $b_1 = b_{m+1}$. We can do the same thing for all the other indices, so we have $b_1 = \ldots = b_n$.

Finally the necessary condition $\sum_{i=1}^{n} b_i = 0$ gives $b_i = 0$ for $i = 1, \ldots, n$, and this leads to contradiction (see (∗)).

With the same method of proof, we get the following theorem in the reducible case:

Theorem 9.3.26. *Let P be a polynomial of degree n and let $Q(x, y, \lambda) = P(x + \lambda y, y)$. Then for almost any specialization (x_0, λ_0) of (x, λ), the sums $\sum_{i \in J} b_i$, with J in $\{1, \ldots, n\}$, vanishes if only if it corresponds to the union of roots of a family of factors of P.*

The factorization algorithm uses also another similar generic property of the number b_i: the number b_i (defined in Section 9.3.2) are all different. See

[Rup00].

Terms of higher degree in the expansion of φ_i

We have seen that after a generic change of coordinates, a vanishing sum of a subset of the numbers b_i (defined in Section 9.3.2) corresponds to a factor of P. We could also show the same result on other terms in the series expansion near a root y_i: if we write

$$\varphi_i(x) = y_i + a_i(x - x_0) + b_i(x - x_0)^2 + c_i(x - x_0)^3 + \cdots$$

a factor of P (generically) corresponds to a vanishing sum of the numbers c_i. This remains true for terms of higher order. This could be used to get a stronger certification on the partition of $\{1, \ldots, n\}$. However it would take time to compute these higher degree terms and their sums, and the condition on the numbers b_i is strong enough to discover the relation.

The Algorithm

This algorithm was implemented. The implementation made by D. Rupprecht, is written in C using the PARI library for multiprecision computation and computation in extensions of $\mathbb{Q}$. The algorithm takes as input 2 constants $prec_1$ and $prec_2$. The first one is used to test if a number is equal to 0 (if its absolute value is lower than 10^{-prec_1} then the number is 0). The other one $prec_2$ is the number of digits for computations.

Algorithm 9.3.27 THE GALLIGO-RUPPRECHT ALGORITHM

Input: $P(X, Y)$ a square-free polynomial in $\mathbb{Q}[X, Y]$, irreducible in $\mathbb{Q}[X, Y]$.

1. *After a generic linear change of coordinates $X \leftarrow x + \lambda_0 . y + x_0$, $Y \leftarrow y$ (with rational coefficients), we get a monic polynomial in y, denoted by $Q(x, y)$.*
2. *Compute numerically the roots $y_1, \ldots, y_n$ of $Q(0, y)$ and the second order coefficients b_i defined above.*
3. *Look for a minimal partition $I_1, \ldots, I_m$ of $\{1, \ldots, n\}$ such that $\mathrm{card}(I_1) = \cdots = \mathrm{card}(I_m)$ and $B_{I_k} = \sum_{j \in I_k} b_k = 0$ for $k = 1, \ldots, m$. Denote by R_k the polynomial $R_k = \prod_{j \in I_k}(y - y_j)$. We have $Q(0, y) = R_1 \cdots R_m$.*
4. *By the last theorem, the polynomials R_k correspond to the trace of the factors of Q for $x = 0$. Performing Hensel liftings on these polynomials, one obtains new polynomials Q_k such that*

$$\begin{cases} Q = Q_1 \cdots Q_m & \mod x^{n+1} \\ Q_k(0, y) = R_k(y) & \text{for } k = 1, \ldots, m. \end{cases}$$

5. *Performing the inverse change of coordinates, one obtains numerical polynomials $\tilde{P}_1, \ldots, \tilde{P}_m$ which provide a candidate for an absolute factorization of P.*
6. *The last step of the algorithm is to find an extension of $\mathbb{Q}$ and conjugate polynomials $P_1, \ldots, P_m$ where $\tilde{P}_k$ is a good approximation of P_k. Finally one can test if P_1 is a divisor of P.*

Output: $P_1(X, Y) \in \mathbb{Q}[\alpha][X, Y]$, an absolute factor of $P(X, Y)$, and the minimal polynomial $q(t)$ of α over $\mathbb{Q}$.

Zero-sums B_I

This problem is the difficult part of the algorithm. We have a set of complex numbers $b_1, \cdots, b_n$ and we are looking for vanishing sums of these numbers. This combinatorial problem could be solved by an extensive search among all the 2^n sums. For $n = 60$, we would have to compute more than 10^{18} sums (or keep in memory some of these sums). D. Rupprecht [Rup00] proposed several improvements for detecting vanishing sums ; then the complexity that he got for this step is $O(2^{n/4})$. With this we can easily factorize polynomials up to degree 80.

Zero-sums and the knapsack problem

We write the problem of the vanishing sums in the following way. Let $v_1 = (1, 0, \ldots, 0, \Re(b_1), \Im(b_1)), \ldots, v_i = (0, \ldots, 0, 1, 0, \ldots, 0, \Re(b_i), \Im(b_i)), \ldots, v_n = (0, \ldots, 0, 1, \Re(b_n), \Im(b_n))$ be n vectors of $\mathbb{R}^{n+2}$ ($\Re(z)$ is the real part of the complex number z and $\Im(z)$ is the imaginary part). We consider the lattice $\mathcal{L}$ generated by these vectors. A zero-sum $\sum_{i \in I} b_i = 0$ corresponds to the "small" vector of $\mathcal{L}$:

$$\sum_{i \in I} v_i = (\lambda_1, \ldots, \lambda_n, \sum_{i \in I} \Re(b_i), \sum_{i \in I} \Im(b_i)) = (\lambda_1, \ldots, \lambda_n, 0, 0)$$

where $\lambda_i = 1$ if i belongs to I and $\lambda_i = 0$ otherwise. We call these vectors zero-sum vectors, and the minimal zero-sum vector the vector corresponding to the minimal zero-sum relation (i.e. there is no subset $J \subsetneq I$ such that $\sum_{j \in J} b_j = 0$).

Thus, instead of computing all the 2^n sums, we can try to get a small vector of $\mathcal{L}$ in order to obtain a zero-sum vector. Now we remark that the matrix with rows corresponding to minimal zero-sums vectors is in a reduced row echelon form. Indeed its n first columns contain, as entries, one 1 and otherwise zero, moreover the last columns are identically zero.

Therefore we get the following method to obtain the zero-sum relations. Firstly, we compute a new basis $\{w_1, \ldots, w_n\}$ of $\mathcal{L}$ with the LLL algorithm. Secondly, we take all the w_i with small norm. More precisely, we set:

$$\mathcal{L}' = \{w_i \,|\, \|w_i\| < B\}$$

where B is a small real number. Therefore we compute the reduced row echelon form of the matrix whose i^{th} row is the ith vector of $\mathcal{L}'$. This gives, if B is not too large, a zero-sum vector.

This method is very close to the algorithm of van Hoeij which factorizes a polynomial $f(X)$ over $\mathbb{Z}[X]$ [vH02] and has been developed by Chèze in [Chè04]. One of the ideas of van Hoeij is to use 0-1 vectors, instead of using a vector of coefficients of a divisor of f which can have much larger coordinates.

9.3.3 Contribution of other authors

Monodromy and homotopy methods

The first use of a monodromy method to provide an algorithm for computing an absolute factorization was made by Bajaj et al. [BCGW93] in order to prove a complexity result. Their algorithm was never implemented because it amounts to considering all the loops around the points of the discriminant locus Δ.

A.J. Sommese, J. Verschelde and C.W. Wampler developed a geometric method, in a series of articles, to separate the components of an algebraic variety. Specialized to the case of a plane curve it provides a geometric algorithm for computing an absolute factorization. It is based on numerical computations and relies on so-called continuation or homotopy methods. This amounts to following, in $\mathbb{C}$ or $\mathbb{C}^2$, integral curves of some differential equation and avoiding singularities.

As in the last section, suppose the input is a reduced plane curve C given by a square-free polynomial $P(x, y)$ of (total) degree n and monic in y. They consider a generic smooth section, say for $x = 0$, which consists of n simple points. So the question is to find the partition of this fiber by the irreducible components of C.

The curve C is a ramified covering of degree n of $\mathbb{C}$. Let Δ be the discriminant locus of the projection ϕ. We supposed that 0 is not in Δ. As we have seen in a previous section, for any loop γ in $\mathbb{C} - \Delta$, (starting and ending in 0), following the roots of $P(\gamma(t), y)$ over such γ, one gets a permutation of the fiber $\phi^{-1}(x_0)$.

Sommese, Verschelde and Wampler made the following important observation. They considered a few random loops γ in $\mathbb{C} - \Delta$ (starting and ending at 0), and noted that in general they generate enough permutations of the fiber $\phi^{-1}(x_0)$ to recover the desired partition. So, in practice, they did not need to follow all the loops as in [BCGW93].

It is not at all an easy task to follow precisely such a path γ and the n paths over it to get the corresponding permutation of the fiber. Indeed, if the chosen time step is too large near a value x_0, then the computation can cause a confusion between the various roots of $P(x_0, y)$, and the obtained permutation could be false. Such scaling problems are really tough. They later improved their algorithm by using a criterion based on our Zero-sum method; see Chapter 8 in this book.

They demonstrate on a large problem, coming from an application in robotics, that their strategy and implementation are efficient. Moreover, in [CGKW02], R.M. Corless, A. Galligo, I.S. Kotsireas, and S.M. Watt proposed a combination of a homotopy method with the other two approaches, in order to diminish the potential risk of errors.

Zero-sum relations and the Japanese school

As above, the input is a square-free polynomial $P(x, y)$ of (total) degree n and monic in y. Consider the fiber over 0 and the corresponding factorization into n linear factors in $\mathbb{C}[[x]][y]$:

$$P = \prod (y - \varphi_i(x)).$$

In [SSKS91], the authors became the first to develop an algorithm based on Zero-sum relation, a concept that they introduced. Sasaki and his coworkers also proposed an algorithm which proceeds as follows. They consider the (integer) k powers of the $\varphi_i(x)^k$ of the $\varphi_i(x)$. Their sums are called Newton sums and are symmetric functions of the coefficients of P, hence are polynomials in x of bounded degrees. So if we denote by $|f|_l$ the sum of terms of degree $\geq l$ of f then we have, for some well chosen $d \geq n + 1$:

$$|\varphi_1(x)^k + \ldots + \varphi_n(x)^k|_d = 0.$$

Conversely, as this is true for any polynomial factor of P, by searching for the same kind of zero-sum relations among a subset of the n linear factor of P on the ring of power series, we can find a family $\varphi_{i_1}(x), \ldots, \varphi_{i_m}(x)$ such that $(y - \varphi_{i_1}(x)) \cdots (y - \varphi_{i_m}(x))$ is a polynomial factor of P.

Each zero-sum gives rise to many relations on the coefficients of the n series $\varphi_i(x)$. The first coefficients can be computed within some approximation, Sasaki and his coworkers derived a method using linear algebra for recognizing the indices i of a factor. The Japanese school has been successful in inventing original algorithms for that purpose. The algorithms of Sasaki et al. proceed by filling a matrix with numerical coefficients coming from k-powers of the series. Then they look for elements in the kernel of that matrix whose coefficients are only zero or one. They were able to provide fine error analysis for their method.

Remark 9.3.28. The above analysis has been recently completed in a work of A. Bostan, G. Lecerf, B. Salvy, E. Schost and B. Wiebelt presented at ISSAC'04 (see [BLS$^+$04]). In this work, they use logarithmic derivatives (as in Gao's algorithm), and zero-sum relations (as in Sasaki's algorithm). In the special case of absolute factorization, their algorithm studies the subspace L_σ of $\mathbb{C}^n \times \mathbb{C}[X, Y]_{n-1}$ given by:

$$L_\sigma = \{(l_1, \ldots, l_n, Q) / \sum_{i=1}^n l_i \frac{P(X, Y)}{(Y - \varphi(X))} = Q + \mathcal{O}(X^\sigma)\}.$$

The idea here is, as in van Hoeij's algorithm ([vH02]), to find 0-1 vectors $(e_1^{(i)}, \ldots, e_n^{(i)})$, such that $P_i = \prod_{k=1}^n (Y - \varphi(X))^{e_k^{(i)}}$. Since we have

$$\frac{P_i'}{P_i} P = \sum_{k=1}^n e_k^{(i)} \frac{(Y - \varphi_k(X))'}{(Y - \varphi_k(X))} P$$

(where P_i' denotes the first derivative of P_i with respect to Y), this leads to consider L_σ. We denote by $\pi(L_\sigma)$ the canonical projection of L_σ to $\mathbb{C}^n$. In their paper, they prove that if $\sigma \geq 3n - 2$, then with $\pi(L_\sigma)$ we can get the absolute factors of P, and furthermore that this method is equivalent to Sasaki's. Finally, this method only uses Hensel's lifting and linear algebra in order to get 0-1 vectors $(e_1^{(i)}, \ldots, e_n^{(i)})$. This method works in general for fields of characteristic zero or at least $n(n-1) + 1$.

9.4 Lecture 4: Reconstruction of the exact factors

Thanks to the results of the last section, we know how to compute an approximate absolute factorization. Thus here we are in the following situation:

We have an irreducible polynomial $P \in \mathbb{Q}[X, Y]$. Let us denote by $P = P_1 \cdots P_s$ its absolute factorization. Let $\mathbb{Q}[\alpha]$ be the smallest extension of $\mathbb{Q}$

which contains all the coefficients of the factor P_1. Let $P \approx \tilde{P}_1 \cdots \tilde{P}_s$ be an approximate absolute factorization of P. By this we mean that $\tilde{P}_i \in \mathbb{C}[X,Y]$ and the coefficients of $\tilde{P}_i$ are numerical approximations of the coefficients of P_i with a given precision ϵ. That is to say $\|P_i - \tilde{P}_i\|_\infty < \epsilon$, with respect to the norm $\|\sum_{i,j} a_{i,j} X^i Y^j\|_\infty = \max_{i,j} |a_{i,j}|$.

A natural question is: Can we get an exact factorization from the approximate one? If it is possible, how can we find the minimal polynomial of α over $\mathbb{Q}$, and how can we express the coefficients of P_1 in $\mathbb{Q}[\alpha]$?

We will answer positively if ϵ is small enough. As the coefficients of $\tilde{P}_1$ are given with an error ϵ, in order to find the minimal polynomial f_α of α, ($f_\alpha \in \mathbb{Q}[T]$), we should have to recognize its coefficients which are rational numbers from imprecise floating point numbers. David Rupprecht gave a preliminary study of this problem in [Rup00]. Here we present a complete and satisfactory answer. We closely follow the exposition given in [CG03].

9.4.1 Notations and elementary results

In all this lecture we have: $P \in \mathbb{Q}[X,Y]$ and $P = P_1 \cdots P_s$ in $\mathbb{C}[X,Y]$.

P_i are irreducible factors of P in $\mathbb{C}[X,Y]$. $\mathbb{K}$ is the smallest field which contains all the coefficients of P_1 ; this is a finite extension of $\mathbb{Q}$. By the primitive element theorem we can write $\mathbb{K} = \mathbb{Q}[\alpha]$. Let $x \in \mathbb{K}$, we denote by f_x the minimal polynomial of x over $\mathbb{Q}$. $\mathcal{O}_\mathbb{K}$ is the ring of algebraic integers in $\mathbb{K}$: if $x \in \mathcal{O}_\mathbb{K}$ then $f_x(T) \in \mathbb{Z}[T]$ and is monic.

Let $x \neq 0$ be an element of $\mathbb{K}$. We denote by m_x the homomorphism of multiplication by x in $\mathbb{K}$, by $P_{char}(x)(T)$ the characteristic polynomial of m_x and by $Tr_{\mathbb{K}/\mathbb{Q}}(x)$ the trace of m_x.

We recall that $P_{char}(x)(T) = f_x^k$ where $k = [\mathbb{K} : \mathbb{Q}(x)]$ is the degree of $\mathbb{K}$ over $\mathbb{Q}(x)$.

Let $[\mathbb{K} : \mathbb{Q}] = s$ and $(x_1, \ldots, x_s)$ be an element of $\mathbb{K}^s$. We define the discriminant $disc_{\mathbb{K}/\mathbb{Q}}(x_1, \ldots, x_s)$ to be the determinant of the matrix whose (i,j)-coefficient is $Tr_{\mathbb{K}/\mathbb{Q}}(x_i x_j)$ (for $i, j = 1, \ldots, s$).

For the special case $(1, \alpha, \alpha^2, \ldots, \alpha^{s-1})$, where α is a primitive element of $\mathbb{K}$ over $\mathbb{Q}$, we set

$$disc_{\mathbb{K}/\mathbb{Q}}(\alpha) = disc_{\mathbb{K}/\mathbb{Q}}(1, \alpha, \ldots, \alpha^{s-1})$$

and call this number the discriminant of α. If $f_\alpha(T) = T^n + a_{n-1} T^{n-1} + \ldots + a_0$ then we denote by $Disc(f_\alpha)$ the number satisfying

$$Res(f_\alpha, f'_\alpha) = (-1)^{n(n-1)/2} Disc(f_\alpha)$$

where Res is the resultant and f'_α is the derivative of f_α. When α is a primitive element of $\mathbb{K}$, we have

$$disc_{\mathbb{K}/\mathbb{Q}}(\alpha) = Disc(f_\alpha).$$

9.4.2 The strategy

Our aim is to compute the minimal polynomial of a primitive element α of $\mathbb{K}$ and then the coefficients of P_1 in $\mathbb{K}$.

Our strategy is based on the following observations.

Let $P_i(X, Y) = \sum_u \sum_v a_i^{(u,v)} X^u Y^v$. Then we have (by the fundamental lemma):

$$P_{char}(a_1^{(u,v)})(T) = \prod_{i=1}^{s}(T - a_i^{(u,v)}) = T^s + c_{s-1}T^{s-1} + \cdots + c_0.$$

Furthermore if $\gcd(P_{char}(a_1^{(u,v)}), \frac{\partial}{\partial T}P_{char}(a_1^{(u,v)})) = 1$, then $a_1^{(u,v)}$ is a primitive element of $\mathbb{K}$ and consequently $P_{char}(a_1^{(u,v)})(T) = f_{a_1^{(u,v)}}(T)$. So it is easy to obtain theoretically the minimal polynomial of a coefficient which is a primitive element of $\mathbb{K}$. But in our situation we do not have exact data $a_1^{(u,v)}, \ldots, a_s^{(u,v)}$; we only have approximations $a_1^{(u,v)} + \epsilon_1, \ldots, a_s^{(u,v)} + \epsilon_s$ and a bound ϵ on the errors: $|\epsilon_i| < \epsilon$. Expanding $\prod_{i=1}^{s}(T - a_i^{(u,v)} - \epsilon_i)$ we get

$$T^s + c_{s-1}(\epsilon)T^{s-1} + \cdots + c_0(\epsilon),$$

and we have to recognize c_i from $c_i(\epsilon)$.

However, since we do not have a bound on the denominators of the rational number c_i, this might be hard.

In order to avoid this difficulty, we show that we can restrict our study to a polynomial $P(X, Y) \in \mathbb{Z}[X, Y]$. Then we prove that the coefficients of P_i are algebraic integers over $\mathbb{Z}$. Therefore, the coefficients of the minimal polynomial will be integers, so it is easy to recognize them, and we can certify the result.

In Section 9.4.6 we propose a method to obtain the expression of the coefficients of P_1 in $\mathbb{K}$. We will use the fundamental lemma and adapted representations of these algebraic integers over $\mathbb{Z}$. The algorithm has been implemented. The last subsection provides illustrative examples.

9.4.3 Reduction to $\mathbb{Z}[X, Y]$

Let $Q(X, Y) = \sum_{i=0}^{n} \sum_{j=0}^{i} q_{j,n-i} X^j Y^{n-i}$ be an irreducible and monic polynomial in $\mathbb{Q}[X, Y]$ of total degree n. Let d be a common denominator of the coefficients of Q ; that is to say $dq_{j,n-i} \in \mathbb{Z}$. Then $d^n Q$ is irreducible in $\mathbb{Q}[X, Y]$ and $d^n Q(X, Y) = \sum_{i=0}^{n} \sum_{j=0}^{i} d^i q_{j,n-i} X^j (dY)^{n-i}$. Setting $Z = dY$ we define $P(X, Z) \in \mathbb{Z}[X, Z]$ by

$$d^n Q(X, Y) = d^n Q\left(X, \frac{Z}{d}\right) = Z^n + dq_{1,n-1}XZ^{n-1} + \cdots + d^n q_{0,0} = P(X, Z).$$

Since $d^n Q(X, Y)$ is irreducible in $\mathbb{Q}[X, Y]$, $d^n Q(X, \frac{Z}{d})$ is irreducible in $\mathbb{Q}[X, Z]$. Hence $P(X, Z)$ is monic in Z, irreducible in $\mathbb{Q}[X, Z]$ and belongs to $\mathbb{Z}[X, Z]$. We state two lemmas whose proofs are obvious.

Lemma 9.4.1. *Let $Q(X,Y)$ be a polynomial satisfying the hypothesis of the fundamental lemma and d a common denominator of the coefficients of Q. Let $Q(X,Y) = Q_1(X,Y) \cdots Q_s(X,Y)$ be an absolute factorization. We set:*

$$P(X,Y) = d^n Q\left(X, \frac{Y}{d}\right) = d^m Q_1\left(X, \frac{Y}{d}\right) \cdots d^m Q_s\left(X, \frac{Y}{d}\right).$$

Then $P(X,Y) \in \mathbb{Z}[X,Y]$ is irreducible in $\mathbb{Q}[X,Y]$ and monic relative to Y, and $P_i(X,Y) = d^m Q_i\left(X, \frac{Y}{d}\right) \in \mathbb{C}[X,Y]$ are the irreducible factors of P in $\mathbb{C}[X,Y]$.

Lemma 9.4.2. *Let $\mathbb{K}'$ be the smallest field generated by the coefficients of Q_1 and $\mathbb{K}$ the smallest field generated by the coefficients of P_1. Then $\mathbb{K}' = \mathbb{K}$.*

So for now on, we can suppose that our input polynomial belongs to $\mathbb{Z}[X,Y]$.

9.4.4 The coefficients of P_i are algebraic integers over $\mathbb{Z}$

Here we first prove a lemma, then we prove the following theorem.

Theorem 9.4.3. *Let $P \in \mathbb{Z}[X,Y]$ be monic in Y and irreducible in $\mathbb{Q}[X,Y]$. Then it admits a factorization $P_1 \cdots P_s$ in $\mathbb{C}[X,Y]$ which consists of polynomials whose coefficients are algebraic integers over $\mathbb{Z}$.*

Lemma 9.4.4. *Let α be an algebraic number of degree s over $\mathbb{Q}$ and let $p(X) \in \mathbb{Q}[\alpha][X]$ be an integer over $\mathbb{Z}[X]$. Then all the coefficients of $p(X)$ are integers over $\mathbb{Z}$.*

Proof. We denote by l the degree of $p(X)$ relative to X. We remark that $\mathbb{Q}(X)[\alpha]$ is an extension of $\mathbb{Q}(X)$ of degree s. Moreover,

($*$) all the conjugates of $p(X)$ belong to $\mathbb{C}[X]$ and have degree l.

As $\mathbb{Z}[X]$ is an integrally closed ring we deduce (see e.g. [Sam67] p. 45) that

($**$) the coefficients of the characteristic polynomial $P_{char}(p(X))$ of $p(X)$ over $\mathbb{Q}(X)$ are in $\mathbb{Z}[X]$.

Let $k = [\mathbb{Q}(X)[\alpha] : \mathbb{Q}(X)[p(X)]]$. We denote the conjugates of $p(X)$ over $\mathbb{Q}(X)$ by q_i where $i = 1, \ldots, \frac{s}{k}$ and $q_1 = p(X)$. Then

$$P_{char}(p(X))(Z) = \prod_{i=1}^{s/k} (Z - q_i)^k$$

is the characteristic polynomial of $p(X)$.

Now we prove by induction that all the coefficients of $p(X)$ are integers over $\mathbb{Z}$. We start by the leading term of $p(X)$. We have:

$$P_{char}(p(X))(Z) = Z^s + (\sum_i q_i)Z^{s-1} + \cdots + \prod_i q_i$$

$$= Z^s + c_{s-1}(X)Z^{s-1} + \cdots + c_0(X)$$

with $c_i(X) \in \mathbb{Z}[X]$ by $(**)$, and $\deg(c_{s-i}(X)) \leq il$, because $\deg(q_i) = deg(p(X)) = l$ by $(*)$. Thus $\deg(c_{s-i}(X)p(X)^{s-i}) = deg(c_{s-i}(X)) + (s-i)deg(p(X)) \leq ls$. As $P_{char}(p(X))(p(X)) = 0$ in $\mathbb{C}[X]$, considering the term of degree ls, we get: $\lambda_l^s + \sum_{i \in I} \text{lc}(c_{s-i})\lambda_l^{s-i} = 0$, where $\text{lc}(c_i)$ is the leading coefficient of $c_i(X)$, $\lambda_l = \text{lc}(p(X))$ and I is the set $I = \{i/deg(c_{s-i}(X)p(X)^{s-i}) = ls\}$.

The fact that all $\text{lc}(c_i)$ are integers implies that λ_l is an algebraic integer over $\mathbb{Z}$ and therefore $\lambda_l X^l$ is an algebraic integer over $\mathbb{Z}[X]$.

To prove the other steps of the induction, we remark that $p(X) - \lambda_l X^l$ belongs to $\mathbb{Q}[\alpha][X]$ and is an integer over $\mathbb{Z}[X]$, then we can repeat the same argument with $p(X) - \lambda_l X^l$ instead of $p(X)$.

Now we can prove the theorem.

Proof. As in the previous section, $\mathbb{Q}[\alpha]$ is the extension field generated by all the coefficients of P_1, and the degree of α over $\mathbb{Q}$ is s. By Steinitz's theorem, there exists an algebraically closed field $\mathcal{K}$ such that $\mathcal{K} \supset \mathbb{Q}(X) \supset \mathbb{Z}[X]$ and

$$P(X,Y) = Y^n + a_{n-1}(X)Y^{n-1} + \cdots + a_0(X) = \prod_{i=1}^n (Y - r_i(X))$$

where $r_i(X) \in \mathcal{K}$ and $r_i(x)$ is an algebraic integer over $\mathbb{Z}[X]$. To be more precise in the description of $r_i(X)$, there is an integer p such that $r_i(X) \in \mathbb{C}[[X^{1/p}]]$; see, e.g. [Eis95, p.300], Corollary 13.16.

Since $P_1(X,Y)$ is a factor of $P(X,Y)$, then we have

$$P_1(X,Y) = \prod_{i=1}^m (Y - r_i(X)) = Y^m + p_{m-1}(X)Y^{m-1} + \cdots + p_0(X)$$

where the $p_i(X)$ are in $\mathbb{Q}[\alpha][X]$ and are integers over $\mathbb{Z}[X]$ because they are polynomials in $r_i(X)$. Then we can apply the previous lemma to each $p_i(X)$.

9.4.5 Finding a primitive element

The coefficients of P_1 generate an extension $\mathbb{K}$ of $\mathbb{Q}$. We aim to get a primitive element of $\mathbb{K}$ which is an *algebraic integer over* $\mathbb{Z}$.

First we check if there is a primitive element among the coefficients of P_1. If this is not the case, we present a method which constructs a primitive element.

Recognition

We let $P_i(X, Y) = \sum_u \sum_v a_i^{(u,v)} X^u Y^v$. The fundamental lemma implies that $P_{char}(a_1^{(u,v)})(T) = \prod_{i=1}^s (T - a_i^{(u,v)})$. Furthermore, $a_i^{(u,v)} \neq a_j^{(u,v)}$ for all $i \neq j$ if and only if $a_1^{(u,v)}$ is a primitive element of $\mathbb{K}$. Thus,

$$a_1^{(u,v)} \text{ is a primitive element of } \mathbb{K}$$

if and only if

$$\gcd(P_{char}(a_1^{(u,v)}), \frac{\partial}{\partial T} P_{char}(a_1^{(u,v)})) = 1.$$

We derived the following characterization:

Lemma 9.4.5. *With the previous notations, we have:*

$$\gcd(P_{char}(a_1^{(u,v)}), \frac{\partial}{\partial T} P_{char}(a_1^{(u,v)})) = 1$$

if and only if
$$a_1^{(u,v)} \text{ is a primitive element of } \mathbb{K}.$$

In this case $P_{char}(a_1^{(u,v)})$ is the minimal polynomial $f_{a_1^{(u,v)}}$ of $a_1^{(u,v)}$ over $\mathbb{Q}$. Moreover $a_1^{(u,v)}$ is an algebraic integer over $\mathbb{Z}$.

This lemma allows us to recognize effectively a primitive element.

Construction

If all the coefficients of P_1 are not primitive, we construct with high probability a primitive element, which is integer over $\mathbb{Z}$. By Lemma 9.4.5 we can check if this constructed element is primitive.

We denote by σ_i $(1 \leq i \leq s)$ the s independent $\mathbb{Q}$-homomorphisms from $\mathbb{K}$ to $\mathbb{C}$ and by $a_1^{(u,v)}$ the coefficients of P_1. We recall that they generate $\mathbb{K}$.

For any pair (i, j) such that $i \neq j$, there exists a coefficient $a_1^{(u,v)}$ of P_1 such that $\sigma_i(a_1^{(u,v)}) \neq \sigma_j(a_1^{(u,v)})$. Thus the polynomial
$$H(\lambda_{(1,0)}, \ldots, \lambda_{(2,n-1)})$$
$$= \prod_{i<j} ((\sigma_i - \sigma_j)(a_1^{(0,0)}) + \lambda_{(1,0)}(\sigma_i - \sigma_j)(a_1^{(1,0)}) + \cdots + \lambda_{(2,n-1)}(\sigma_i - \sigma_j)(a_1^{(2,n-1)}))$$
is a nonzero polynomial in $\mathbb{C}[\lambda_{(i,j)}]$. So we can find $(\lambda_{(1,0)}, \ldots, \lambda_{(2,n-1)})$ with $\lambda_{(i,j)} \in \mathbb{Z}$ such that for $\forall i \neq j$:

$$\sigma_i(a_1^{(0,0)} + \lambda_{(1,0)} a_1^{(1,0)} + \ldots + \lambda_{(2,n-1)} a_1^{(2,n-1)})$$

differs from

$$\sigma_j(a_1^{(0,0)} + \lambda_{(1,0)}a_1^{(1,0)} + \ldots + \lambda_{(2,n-1)}a_1^{(2,n-1)}).$$

This means that $a_1^{(0,0)} + \lambda_{(1,0)}a_1^{(1,0)} + \ldots + \lambda_{(2,n-1)}a_1^{(2,n-1)}$ is a primitive element. We apply Lemma 9.2.15 (see exercise 9.2.14) to the polynomial

$$H(\lambda_{(1,0)}, \ldots, \lambda_{(2,n-1)}) \in \mathbb{C}[\lambda_{(i,j)}]$$

and get the following proposition:

Proposition 9.4.6. *Let P be a polynomial in $\mathbb{Z}[X,Y]$, which is monic and irreducible in $\mathbb{Q}[X,Y]$, and $P = P_1 \cdots P_s$ its irreducible decomposition in $\mathbb{C}[X,Y]$. Let $a_1^{(u,v)}$ denote the coefficients of P_1, and $\mathbb{K}$ the extension of $\mathbb{Q}$ they generate.*

Let S be a subset of $\mathbb{Z}$. Then we have the following estimation of probability:
$$\mathcal{P}\left(a_1^{(0,0)} + s_{(1,0)}a_1^{(1,0)} + \cdots + s_{(2,n-1)}a_1^{(2,n-1)} \text{ non primitive} \mid s_i \in S, 2 \le i \le r \right)$$
$$\le \frac{\binom{s}{2}}{|S|}.$$

The $a_1^{(u,v)}$ are integers over $\mathbb{Z}$, so $a_1^{(0,0)} + s_{(1,0)}a_1^{(1,0)} + \cdots + s_{(2,n-1)}a_1^{(2,n-1)}$ is an algebraic integer over $\mathbb{Z}$ and the probability that this element is non-primitive is less than $\frac{\binom{s}{2}}{|S|}$. So we can make the probability as small as we want. Moreover as it is possible to check the result with Lemma 9.4.5 ; this provides a method (efficient) and easy to implement.

Choice of the precision

In practice we can only compute an approximation of a minimal polynomial $f_\alpha(T)$, with $f_{\alpha+\epsilon}(T) = \prod_{k=1}^s (T - (\alpha_k + \epsilon_k))$.

We have perturbed roots and we want to know if the perturbation on the coefficients is smaller than 0.5 in order to recognize the polynomial f_α from $f_{\alpha+\epsilon}$. The following map describes the situation:

$$\varphi: \quad \mathbb{C}^s \longrightarrow \mathbb{C}^s$$
$$\begin{pmatrix} \alpha_1 \\ \vdots \\ \alpha_k \\ \vdots \\ \alpha_s \end{pmatrix} \longmapsto \begin{pmatrix} S_1(\alpha_1, \ldots, \alpha_s) = \alpha_1 + \alpha_2 + \cdots + \alpha_s \\ \vdots \\ S_k(\alpha_1, \ldots, \alpha_s) = \displaystyle\sum_{1 \le i_1 < \ldots < i_k \le s} \alpha_{i_1} \cdots \alpha_{i_k} \\ \vdots \\ S_s(\alpha_1, \ldots, \alpha_s) = \alpha_1 \times \cdots \times \alpha_s \end{pmatrix}.$$

We define $\|.\|_\infty$ by $\|(\alpha_1, \ldots, \alpha_s)\|_\infty = \max_{i=1,\ldots,s} |\alpha_i|$. We look for a condition on ϵ which implies $\|\varphi(\alpha + \epsilon) - \varphi(\alpha)\|_\infty < 0.5$. φ is a polynomial map such that the degree of each component is less than or equal to s and is of degree 1 in each variable. With the notation

$$\left[\sum_{i=1}^{s}\epsilon_i\frac{\partial\varphi}{\partial\alpha_i}(\alpha)\right]^{[k]} = \sum_{\substack{i_1+\cdots+i_s=k \\ i_j\in\{0,1\}}}\frac{k!}{i_1!\ldots i_s!}\epsilon_1^{i_1}\ldots\epsilon_s^{i_s}\frac{\partial^k\varphi}{\partial\alpha_1^{i_1}\ldots\partial\alpha_s^{i_s}}(\alpha),$$

the Taylor expansion of φ is

$$\varphi(\alpha+\epsilon) - \varphi(\alpha) = \left[\sum_{i=1}^{s}\epsilon_i\frac{\partial\varphi}{\partial\alpha_i}(\alpha)\right] + \frac{1}{2!}\left[\sum_{i=1}^{s}\epsilon_i\frac{\partial\varphi}{\partial\alpha_i}(\alpha)\right]^{[2]} + \cdots$$

$$+ \frac{1}{s!}\left[\sum_{i=1}^{s}\epsilon_i\frac{\partial\varphi}{\partial\alpha_i}(\alpha)\right]^{[s]}.$$

We introduce the constants ϵ and M such that

- $|\alpha_i| \le M$ for all $1 \le i \le s$
- $|\epsilon_i| < \epsilon < 1$.

Lemma 9.4.7. *With the previous notation, we have*

$$\|\varphi(\alpha+\epsilon) - \varphi(\alpha)\|_\infty \le \left(\sum_{k=1}^{s}\binom{s}{k}k!\max(1,\max_{j=k+1,\ldots,s}(\binom{s-k}{j-k}M^{j-k}))\right)\epsilon.$$

Proof. The total degree of the polynomial S_j is j, so we deduce

- If $k > j$ then $\dfrac{\partial^k S_j}{\partial\alpha_1^{i_1}\ldots\partial\alpha_s^{i_s}}(\alpha) = 0$.
- If $k = j$ then $\dfrac{\partial^k S_j}{\partial\alpha_1^{i_1}\ldots\partial\alpha_s^{i_s}}(\alpha) = 1$.

Moreover, we easily get the following upper bound, for $k < j$,

$$\left|\frac{\partial^k S_j}{\partial\alpha_1^{i_1}\ldots\partial\alpha_s^{i_s}}(\alpha)\right| \le \binom{s-k}{j-k}M^{j-k}.$$

As a result we obtain

$$\left\|\frac{\partial^k\varphi}{\partial\alpha_1^{i_1}\ldots\partial\alpha_s^{i_s}}(\alpha)\right\|_\infty \le \max(1,\max_{j=k+1,\ldots,s}(\binom{s-k}{j-k}M^{j-k})).$$

It follows that

$$\left\|\left[\sum_{i=1}^{s}\epsilon_i\frac{\partial\varphi}{\partial\alpha_i}(\alpha)\right]^{[k]}\right\|_\infty \le \sum_{\substack{i_1+\cdots+i_s=k \\ i_j\in\{0,1\}}}\frac{k!}{i_1!\ldots i_s!}|\epsilon_1|^{i_1}\ldots|\epsilon_s|^{i_s}\left\|\frac{\partial^k\varphi}{\partial\alpha_1^{i_1}\ldots\partial\alpha_s^{i_s}}(\alpha)\right\|_\infty$$

then

$$\left\|\left[\sum_{i=1}^{s}\epsilon_i\frac{\partial\varphi}{\partial\alpha_i}(\alpha)\right]^{[k]}\right\|_{\infty}\le\binom{s}{k}k!\epsilon^k\max(1,\max_{j=k+1,\dots,s}(\binom{s-k}{j-k}M^{j-k})).$$

Since $\epsilon < 1$, we deduce the required result.

Corollary 9.4.8. *With the previous notations, if the error ϵ on the roots is bounded by*

$$(*)\qquad \epsilon\le 0.5\left(\sum_{k=1}^{s}\binom{s}{k}k!\max(1,\max_{j=k+1,\dots,s}(\binom{s-k}{j-k}M^{j-k}))\right)^{-1}$$

then the error on the coefficient of $f_{\alpha+\epsilon}$ is smaller than 0.5.

So we have proven the following proposition.

Proposition 9.4.9. *We denote by $Digits1$ the number of significant digits used for the computation of the minimal polynomial. If $Digits1 \ge$*

$$E\left(\left|\log_{10}\left(0.5\left(\sum_{k=1}^{s}\binom{s}{k}k!\max(1,\max_{j=k+1,\dots,s}(\binom{s-k}{j-k}M^{j-k}))\right)^{-1}\right)\right|\right.$$
$$\left. + \log_{10}(\max_{k=1,\dots,s}(\binom{s}{k}M^k)))\right.$$

then we can recognize all the coefficients of $f_\alpha(T)$ from the coefficients of $f_{\alpha+\epsilon}(T)$.

To give an idea of the size of $Digits1$, we provide the following table.

s	M	$Digits1$
2	10^5	16
2	10^{10}	31
2	10^{20}	61
5	10^5	47
5	10^{10}	91
5	10^{20}	182

s	M	$Digits1$
10	10^5	97
10	10^{10}	192
10	10^{20}	382
15	10^5	147
15	10^{10}	292
15	10^{20}	582

9.4.6 A method to obtain the exact factorization

We start with a polynomial f_α of a primitive element α of $\mathbb{K}$, obtained as explained in the last section. We will use another canonical representation of the coefficients of P_1.

$f'_\alpha(\alpha)$ is a common denominator

We recall some classical results of algebraic number theory.

Definition 9.4.10. *Let $\mathbb{K}$ be a finite extension of $\mathbb{Q}$. Let M be a subset of $\mathbb{K}$. We set $M^* = \{x \in \mathbb{K} \mid \forall y \in M, Tr_{\mathbb{K}/\mathbb{Q}}(xy) \in \mathbb{Z}\}$ and call it the complementary set of M.*

Proposition 9.4.11. *(see [Rib01, page 242]) Let $\mathbb{K}$ be a finite extension of $\mathbb{Q}$, $\alpha \in \mathcal{O}_\mathbb{K}$ a primitive element of $\mathbb{K}$ and f_α its minimal polynomial. Then we have $\mathcal{O}_\mathbb{K} \subset \mathbb{Z}[\alpha]^* = \dfrac{1}{f'_\alpha(\alpha)}\mathbb{Z}[\alpha]$. This implies that all $a \in \mathcal{O}_\mathbb{K}$ can be written in the following way:*

$$a = \frac{z_0}{f'_\alpha(\alpha)} + \frac{z_1}{f'_\alpha(\alpha)}\alpha + \cdots + \frac{z_{s-1}}{f'_\alpha(\alpha)}\alpha^{s-1} \text{ with } z_i \in \mathbb{Z}.$$

Recognition of the coefficients of P_1

Having the denominator $f'_\alpha(\alpha)$, we only have to recognize the numerators. Let $a_1^{(u,v)}$ be a coefficient of P_1, so $a_1^{(u,v)}$ belongs to $\mathcal{O}_\mathbb{K}$. We have

$$a_1^{(u,v)} = \frac{z_0}{f'_\alpha(\alpha)} + \frac{z_1}{f'_\alpha(\alpha)}\alpha + \cdots + \frac{z_{s-1}}{f'_\alpha(\alpha)}\alpha^{s-1}.$$

Applying the $\mathbb{Q}$-homomorphism σ_i, we get

$$a_i^{(u,v)} = \frac{z_0}{f'_\alpha(\sigma_i(\alpha))} + \frac{z_1}{f'_\alpha((\sigma_i(\alpha))}\sigma_i(\alpha) + \cdots + \frac{z_{s-1}}{f'_\alpha(\sigma_i(\alpha))}\sigma_i(\alpha)^{s-1}, \text{ then}$$

$$(\star) \quad \begin{pmatrix} 1 & \sigma_1(\alpha) & \sigma_1(\alpha)^2 & \cdots & \sigma_1(\alpha)^{s-1} \\ 1 & \sigma_2(\alpha) & \sigma_2(\alpha)^2 & \cdots & \sigma_2(\alpha)^{s-1} \\ \vdots & \vdots & \vdots & \vdots & \vdots \\ 1 & \sigma_s(\alpha) & \sigma_s(\alpha)^2 & \cdots & \sigma_s(\alpha)^{s-1} \end{pmatrix} \begin{pmatrix} z_0 \\ z_1 \\ \vdots \\ z_{s-1} \end{pmatrix} = \begin{pmatrix} f'_\alpha(\sigma_1(\alpha))a_1^{(u,v)} \\ f'_\alpha(\sigma_2(\alpha))a_2^{(u,v)} \\ \vdots \\ f'_\alpha(\sigma_s(\alpha))a_s^{(u,v)} \end{pmatrix}.$$

We remark that in practice we do not have $a_i^{(u,v)}$ but $a_i^{(u,v)} + \nu_i$ and we do not have $\sigma_i(\alpha)$ but $\sigma_i(\alpha) + \epsilon_i$. So we need to solve the Vandermonde system and take the nearest integer of each component of the solution. Now we explain how to certify the result.

Choice of the precision

First we set some notation: $\mathcal{M}_{m,n}(\mathbb{C})$ is the ring of matrices with m rows and n columns, with coefficients in $\mathbb{C}$. If $\mathcal{M} = (m_{i,j})_{i,j=0}^{s-1}$ is a matrix of $\mathcal{M}_{s,s}(\mathbb{C})$, let $\|\mathcal{M}\|_\infty = \max_{i=0,\ldots,s-1} \sum_{j=0}^{s-1} |m_{i,j}|$. If $\mathbf{v}$ is a vector of $\mathbb{C}^s$ (with i-th coordinate equal to v_i), then $\|\mathbf{v}\|_\infty = \max_{i=0,\ldots,s-1} |v_i|$. With this notation we have $\|\mathcal{M}\mathbf{v}\|_\infty \leq \|\mathcal{M}\|_\infty \|\mathbf{v}\|_\infty$. Now we set: $\alpha_i = \sigma_i(\alpha)$, ϵ_i is the error on α_i,

ν_i is the error on $a_i^{(u,v)}$, e_i is the error on z_i, ϵ is a real number such that
$$\begin{cases} \forall 1 \le i \le s \; |\epsilon_i| < \epsilon < 1 \\ \forall 1 \le i \le s \; |\nu_i| < \epsilon < 1 \end{cases},$$

M is a real number such that: $\max\limits_{i,u,v} |a_i^{(u,v)}| \le M$,

$$M(\alpha) = \begin{pmatrix} 1 & \alpha_1 & \alpha_1^2 & \cdots & \alpha_1^{s-1} \\ 1 & \alpha_2 & \alpha_2^2 & \cdots & \alpha_2^{s-1} \\ \vdots & \vdots & \vdots & \vdots & \vdots \\ 1 & \alpha_s & \alpha_s^2 & \cdots & \alpha_s^{s-1} \end{pmatrix}^{-1} \begin{pmatrix} f_\alpha'(\alpha_1) & \cdots & 0 & \cdots & 0 \\ \vdots & \ddots & & & \vdots \\ 0 & \cdots & f_\alpha'(\alpha_k) & \cdots & 0 \\ \vdots & & & \ddots & \vdots \\ 0 & \cdots & 0 & \cdots & f_\alpha'(\alpha_s) \end{pmatrix},$$

$$\mathbf{z} = \begin{pmatrix} z_0 \\ \vdots \\ z_{s-1} \end{pmatrix}, \; \mathbf{a}^{(u,v)} = \begin{pmatrix} a_1^{(u,v)} \\ \vdots \\ a_s^{(u,v)} \end{pmatrix}, \; \mathbf{e} = \begin{pmatrix} e_1 \\ \vdots \\ e_s \end{pmatrix}, \; \boldsymbol{\epsilon} = \begin{pmatrix} \epsilon_0 \\ \vdots \\ \epsilon_{s-1} \end{pmatrix}, \; \boldsymbol{\nu} = \begin{pmatrix} \nu_0 \\ \vdots \\ \nu_{s-1} \end{pmatrix}.$$

Then we have the equality $\mathbf{z} + \mathbf{e} = M(\alpha + \epsilon)(\mathbf{a}^{(u,v)} + \boldsymbol{\nu})$.

Now, we are going to give an expression for the coefficients of $M(\alpha)$ as a function of α_i. We will deduce that $M(\alpha + \epsilon) = M(\alpha) + \epsilon N$, where N is a matrix with bounded coefficients and hence get the bound

$$\|\mathbf{e}\|_\infty \le (\|M(\alpha)\|_\infty + \|N\|_\infty (1 + M)) \, \epsilon.$$

Expression of the coefficients of $M(\alpha)$ and $M(\alpha + \epsilon)$

Lemma 9.4.12. *Let* $M(\alpha) = (m_{i,j}(\alpha))_{i,j=0}^{s-1}$ *then we have:*

$$m_{i,j}(\alpha) = (-1)^{s-i-1} S_{s-i-1}(\alpha_1, \dots, \alpha_j, \alpha_{j+2}, \dots, \alpha_s).$$

Proof. We denote by $V(\alpha)^{-1} = (w_{i,j}(\alpha))_{i,j=0}^{s-1}$ the inverse of the Vandermonde matrix.

The value of the polynomial $l_k(x) = \sum\limits_{j=0}^{s-1} w_{j,k} x^j$ is 1 when $x = \alpha_{k+1}$ and it is 0 when $x \in \{\alpha_1, \dots, \alpha_s\} \setminus \{\alpha_{k+1}\}$. Hence $l_k(x)$ is the Legendre polynomial and we get

$$l_k(x) = \prod_{\substack{i=1 \\ i \ne k+1}}^{s} \left(\frac{x - \alpha_i}{\alpha_{k+1} - \alpha_i} \right) = \prod_{\substack{i=1 \\ i \ne k+1}}^{s} (x - \alpha_i) \times \frac{1}{f_\alpha'(\alpha_{k+1})}.$$

Therefore $w_{j,k}(\alpha) = \dfrac{(-1)^{s-1-j} S_{s-1-j}(\alpha_1, \dots, \alpha_k, \alpha_{k+2}, \dots, \alpha_s)}{f_\alpha'(\alpha_{k+1})}$ where S_k is the symmetric polynomial (see Section 9.4.5), and we set $S_0 = 1$. The definition of $M(\alpha)$ gives $m_{i,j}(\alpha) = w_{i,j}(\alpha) f_\alpha'(\alpha_{j+1})$. Thus the claim is true.

Corollary 9.4.13. *There exists a matrix $N \in \mathcal{M}_{s,s}(\mathbb{C})$ such that*

$$M(\alpha + \epsilon) = M(\alpha) + \epsilon N$$

with $\|N\|_\infty \leq s \left(\sum_{k=1}^{s-1} \binom{s-1}{k} k! \max(1, \max_{j=k+1,\ldots,s-1} (\binom{s-1-k}{j-k} M^{j-k})) \right).$

Proof. Apply Lemma 9.4.7.

Upper bound for $\|\mathbf{e}\|_\infty$
    In the last paragraph we showed that $M(\alpha + \epsilon) = M(\alpha) + \epsilon N$. So, we deduce the following.

Lemma 9.4.14. *With the previous notations, we have*

$$\|\mathbf{e}\|_\infty \leq (\|M(\alpha)\|_\infty + \|N\|_\infty(1 + M))\epsilon.$$

Proof. The equality $\mathbf{z} + \mathbf{e} = M(\alpha + \epsilon)(\mathbf{a} + \boldsymbol{\nu})$ becomes $\mathbf{z} + \mathbf{e} = (M(\alpha) + \epsilon N)(\mathbf{a} + \boldsymbol{\nu})$. Then, $\mathbf{e} = M(\alpha)\boldsymbol{\nu} + \epsilon N\mathbf{a} + \epsilon N\boldsymbol{\nu}$. We deduce that:
$\|\mathbf{e}\|_\infty \leq \|M(\alpha)\|_\infty \epsilon + \|N\|_\infty \epsilon^2 + \|N\mathbf{a}\|_\infty \epsilon \leq (\|M(\alpha)\|_\infty + \|N\|_\infty + \|N\|_\infty M)\epsilon.$

Conclusion
    The results of the previous parts lead to the following.

Proposition 9.4.15. *If the error ϵ is such that*

$$\epsilon \leq 0.5 \left(\max_{i=0,\ldots,s-1} (s \binom{s-1}{s-i-1} M^{s-i-1}) + \right.$$
$$\left. s \left(\sum_{k=1}^{s-1} \binom{s-1}{k} k! \max(1, \max_{j=k+1,\ldots,s-1} (\binom{s-1-k}{j-k} M^{j-k})) \right) (1+M) \right)^{-1}$$

then, with the system $(\star)$ (see Section 9.4.6), we can recognize the exact coefficients of P_1.

Proof. Lemma 9.4.12 gives
$$|m_{i,j}(\alpha)| \leq S_{s-i-1}(|\alpha_1|, \ldots, |\alpha_j|, |\alpha_{j+2}|, \ldots, |\alpha_s|).$$
So $|m_{i,j}(\alpha)| \leq \sum_{1 \leq k_1 < .. < k_{s-i-1} \leq s-1} M^{s-i-1} \leq \binom{s-1}{s-i-1} M^{s-i-1}.$ It follows that

$$\sum_{j=0}^{s-1} |m_{i,j}(\alpha)| \leq \sum_{j=0}^{s-1} \binom{s-1}{s-i-1} M^{s-i-1} = s \binom{s-1}{s-i-1} M^{s-i-1}.$$

Thus $\|M(\alpha)\|_\infty \leq \max_{i=0,\ldots,s-1} (s \binom{s-1}{s-i-1} M^{s-i-1}).$ Together with Corollary 9.4.13 this implies

$$\|e\|_\infty \leq \left(\max_{i=0,\ldots,s-1} \left(s \binom{s-1}{s-i-1} M^{s-i-1} \right) \right.$$

$$\left. +s \left(\sum_{k=1}^{s-1} \binom{s-1}{k} k! \max(1, \max_{j=k+1,\ldots,s-1} \left(\binom{s-1-k}{j-k} M^{j-k} \right)) \right) (1+M) \right) \epsilon.$$

So we get the stated bound.

Proposition 9.4.16. *We denote by Digits2 the number of significant digits used for the step of recognition of the exact coefficients of P_1. If*

$$Digits2 \geq E \left(\left| \log_{10} \left(0.5 \left(\max_{i=0,\ldots,s-1} (s\, C_{s-1}^{s-i-1} M^{s-i-1}) \right. \right. \right. \right.$$

$$\left. \left. +s \left(\sum_{k=1}^{s-1} C_{s-1}^k k! \max(1, \max_{j=k+1,\ldots,s-1} (C_{s-1-k}^{j-k} M^{j-k})) \right) (1+M) \right)^{-1} \right) \right|$$

$$\left. + \log_{10} \left(\max_{i=0,\ldots,s-1} (s\, C_{s-1}^{s-i-1} M^{s-i-1}) M \right) \right)$$

then we can recognize the coefficients of P_1 from the solution of the system $(\star)$.

In order to give an idea of the size of *Digits2* we provide the following tables.

s	M	$Digits2$
2	10^5	16
2	10^{10}	31
2	10^{20}	61
5	10^5	48
5	10^{10}	93
5	10^{20}	183

s	M	$Digits2$
10	10^5	98
10	10^{10}	193
10	10^{20}	383
15	10^5	149
15	10^{10}	294
15	10^{20}	584

9.4.7 Conversion

Let $\beta \in \mathcal{O}_{\mathbb{K}}$. We have the following two representations:

$$\beta = \sum_{j=0}^{s-1} \frac{z_j}{f'_\alpha(\alpha)} \alpha^j = \sum_{i=0}^{s-1} q_i \alpha^i \text{ where } z_j \in \mathbb{Z} \text{ and } q_i \in \mathbb{Q}.$$

Let $B(\alpha)$ be the inverse of $f'_\alpha(\alpha)$, and set $\alpha^j B(\alpha) = \sum_{i=0}^{s-1} b_{i,j} \alpha^i$ where $b_{i,j} \in \mathbb{Q}$. It can be easily computed (once for all coefficients of P_1).

Lemma 9.4.17. *With the previous notations and with*

$$\mathbf{q} = \begin{pmatrix} q_0 \\ \vdots \\ q_{s-1} \end{pmatrix}, \quad \mathbf{z} = \begin{pmatrix} z_0 \\ \vdots \\ z_{s-1} \end{pmatrix}, \quad \text{and} \quad \mathcal{M}_B = (b_{i,j})_{i,j=0}^{s-1} \in \mathcal{M}_{s,s}(\mathbb{Q}), \text{ we}$$

have $\mathbf{q} = \mathcal{M}_B(\mathbf{z})$.

9.4.8 The algorithm

Algorithm 9.4.18 THE CHÈZE-GALLIGO ALGORITHM

Input: $P \in \mathbb{Z}[X,Y]$ *irreducible in* $\mathbb{Q}[X,Y]$, *monic in* Y.

1. *Compute an approximate absolute factorization of* P, *with a number of significant digits* = $Digits$.
   Compute $Digits1$ *and* $Digits2$. *If* $\max(Digits1, Digits2) \geq Digits$ *then go to step 1 with* $Digits = \max(Digits1, Digits2)$. *Else go to step 2.*
2. *Recognize all the primitive coefficients of* P_1 *and their minimal polynomial. [If no coefficients are primitives then construct a primitive element.] Choose a primitive element. We denote* f_α *its minimal polynomial.*
3. *Recognize the exact coefficients of* P_1 *by solving a Vandermonde system. Give for each coefficient of* P_i *its canonical expression in* $\mathbb{Q}[\alpha]$.

Output: The minimal polynomial of a primitive element of $\mathbb{K}$ *and* $P_1(X,Y) \in \mathbb{K}[X,Y]$, *an absolute factor of* P.

9.4.9 Description of the algorithm

Input: $P(X,Y) = Y^4 + 2Y^2X + 14Y^2 - 7X^2 + 6X + 47$.

Step 1)
Apply an approximate absolute polynomial factorization to P with $Digits = 4$, and get
$$\tilde{P}_1(X,Y) = Y^2 + 3.828X + 8.414,$$
$$\tilde{P}_2(X,Y) = Y^2 - 1.828X + 5.585.$$

We have $s = 2$ and we can take $M = 10$ (in fact we have to choose $M \geq 8.414$) $Digits1 = 4$, $Digits2 = 4$.

Step 2)
As before, we get
$$f_{a_1^{(0,0)}} = T^2 - 14T + 47, \text{ and } Disc(f_{a_1^{(0,0)}}) = 8,$$
$$f_{a_1^{(1,0)}} = T^2 - 2t - 7, \text{ and } Disc(f_{a_1^{(1,0)}}) = 32.$$

$\alpha = a_1^{(0,0)}$ is a primitive element of $\mathbb{K}$, and $f_\alpha(T) = T^2 - 14T + 47$.

Step 3)
$$\begin{pmatrix} 1 & 8.414 \\ 1 & 5.585 \end{pmatrix} \begin{pmatrix} \tilde{z}_0 \\ \tilde{z}_1 \end{pmatrix} = \begin{pmatrix} 2.828 \times 3.828 \\ -2.830 \times (-1.828) \end{pmatrix}.$$

This gives $\tilde{z}_0 = -5.989$ and $\tilde{z}_1 = 1.998$.

So $z_0 = -6$, $z_1 = 2$ and $a_1^{(1,0)} = \dfrac{-6}{f_\alpha'(\alpha)} + 2\dfrac{\alpha}{f_\alpha'(\alpha)}$.

We have $f_\alpha(T) = T^2 - 14T + 47$, $f_\alpha'(T) = 2T - 14$ and

$-\dfrac{1}{2}f_\alpha(T) + f_\alpha'(T)(\dfrac{1}{4}T - \dfrac{7}{4}) = 1$. This implies $\dfrac{1}{4}T - \dfrac{7}{4} = f_\alpha'(\alpha)^{-1}$.

Thus $a_1^{(1,0)} = \dfrac{-6}{f_\alpha'(\alpha)} + 2\dfrac{\alpha}{f_\alpha'(\alpha)} = -13 + 2\alpha$.

Output: $f_\alpha(T) = T^2 - 14T + 47$, $P_1(X, Y) = Y^2 + (-13 + 2\alpha)X + \alpha$.

Acknowledgment

Guillaume Chèze thanks ECOS-Sud for financial support which allowed him to participate in the CIMPA school in Buenos Aires and the First Latin American Workshop on Polynomial Systems, in July 2003.

References

[ABKR00] J. Abbott, A. Bigatti, M. Kreuzer, and L. Robbiano. Computing ideals of points. *J. Symbolic Computation*, 30:341–356, 2000.

[ABRW96] M.-E. Alonso, E. Becker, M.-F. Roy, and T. Wörmann. Zeros, multiplicities, and idempotents for zero-dimensional systems. In L. González-Vega and T. Recio, editors, *Algorithms in algebraic geometry and applications (Proc. MEGA'94), Santander, Spain*, number 143 in Progress in Math., pages 1–15, Basel, 1996. Birkhäuser.

[ACGH85] E. Arbarello, M. Cornalba, P.A. Griffiths, and J. Harris. *Geometry of algebraic curves. Vol. I*, volume 267 of *Grundlehren der Mathematischen Wissenschaften (Fundamental Principles of Mathematical Sciences)*. Springer-Verlag, New York, 1985.

[ACW89] D.C.S. Allison, A. Chakraborty, and L.T. Watson. Granularity issues for solving polynomial systems via globally convergent algorithms on a hypercube. *J. Supercomputing*, 3:5–20, 1989.

[AG90a] E.L. Allgower and K. Georg. *Numerical Continuation Methods, an Introduction*, volume 13 of *Computat. Math*. Springer-Verlag, 1990.

[AG90b] E.L. Allgower and K. Georg. *Numerical Path Following*. Springer-Verlag, 1990.

[AG93] E.L. Allgower and K. Georg. Continuation and path following. *Acta Numerica*, pages 1–64, 1993.

[AG97] E.L. Allgower and K Georg. Numerical Path Following. In P.G. Ciarlet and J.L. Lions, editors, *Techniques of Scientific Computing (Part 2)*, volume 5 of *Handbook of Numerical Analysis*, pages 3–203. North-Holland, 1997.

[AGR95] C. Alonso, J. Gutierrez, and T. Recio. An implicitization algorithm with fewer variables. *Comput. Aided Geom. Des.*, 12(3):251–258, 1995.

[AGZV85] V.I. Arnold, S.M. Gusein-Zade, and A.N. Varchenko. *Singularities of differentiable maps. Vol. I*, volume 82 of *Monographs in Mathematics*. Birkhäuser, Boston, 1985. Translated from the Russian.

[AK81] L.A. Aĭzenberg and A.M. Kytmanov. Multidimensional analogues of Newton's formulas for systems of nonlinear algebraic equations and some of their applications. *Sibirsk. Mat. Zh.*, 22(2):19–30, 1981.

[AL94] W. Adams and P. Loustaunau. *An Introduction to Gröbner Bases*. AMS, Providence R.I., 1994.

[ALMM99] P. Aubry, D. Lazard, and M. Moreno Maza. On the theories of triangular sets. *J. Symbolic Computation*, 28(1-2):105–124, 1999.

[AS88]   W. Auzinger and H.J. Stetter. An elimination algorithm for the computation of all zeros of a system of multivariate polynomial equations. In *Proc. Intern. Conf. on Numerical Math., Singapore 1988*,
number 86 in Intern. Series of Numerical Math., pages 11–30, Basel,
1988. Birkhäuser.

[AS89]   W. Auzinger and H.J. Stetter. A study of numerical elimination for the
solution of multivariate polynomial systems. Technical report, Technical Univ. Wien, 1989.

[AS96]   A. Adolphson and S. Sperber. Differential modules defined by systems
of equations. *Rend. Sem. Mat. Univ. Padova*, 95:37–57, 1996.

[AS01]   F. Aries and R. Senoussi. An implicitization algorithm for rational
surfaces with no base points. *J. Symbolic Computation*, 31:357–365,
2001.

[AV00]   P. Aubry and A. Valibouze. Using Galois ideals for computing relative
resolvents. *J. Symbolic Computation*, 30:635–651, 2000.

[AY83]   L.A. Aĭzenberg and A.P. Yuzhakov. *Integral representations and
residues in multidimensional complex analysis*, volume 58 of *Transla-
tions of Mathematical Monographs*. AMS, Providence, R.I., 1983.

[Bar94]   A.I. Barvinok. Computing the Ehrhart polynomial of a convex lattice
polytope. *Discr. and Comput. Geometry*, 12(1):35–48, 1994.

[BC]   L. Busé and M. Chardin. Implicitizing rational hypersurfaces using
approximation complexes. *J. Symbolic Computation*. To appear.

[BCD03]   L. Busé, D. Cox, and C. D'Andrea. Implicitization of surfaces in $\mathbb{P}^3$ in
the presence of base points. *J. Algebra & Appl.*, 2(2):189–214, 2003.

[BCGW93] C. Bajaj, J. Canny, T. Garrity, and J. Warren. Factoring rational
polynomials over the complex numbers. *SIAM J. Comput.*, 22(2):318–
331, 1993.

[BCRS96] E. Becker, J.P. Cardinal, M.-F. Roy, and Z. Szafraniec. Multivariate
Bezoutians, Kronecker symbol and Eisenbud-Levine formula. In L.
González-Vega and T. Recio, editors, *Algorithms in algebraic geometry
and applications (Proc. MEGA'94), Santander, Spain*, number 143 in
Progress in Math., pages 79–104. Birkhäuser, Basel, 1996.

[BCS97]   P. Bürgisser, M. Clausen, and M.A. Shokrollahi. *Algebraic com-
plexity theory. With the collaboration of T. Lickteig*, volume 315 of
Grundlehren der Mathematischen Wissenschaften. Springer-Verlag,
Berlin, 1997.

[BCSS98] L. Blum, F. Cucker, M. Shub, and S. Smale. *Complexity and real
computation*. Springer-Verlag, New York, 1998.

[Bec00]   M. Beck. Counting lattice points by means of the residue theorem.
Ramanujan J., 4(3):299–310, 2000.

[BEM00]   L. Busé, M. Elkadi, and B. Mourrain. Generalized resultants for unira-
tionnal algebraic varieties. *J. Symbolic Computation*, 59:515–526, 2000.

[BEM01]   L. Busé, M. Elkadi, and B. Mourrain. Residual resultant of complete
intersection. *J. Pure & Applied Algebra*, 164:35–57, 2001.

[BEM03]   L. Busé, M. Elkadi, and B. Mourrain. Using projection operators in
computer aided geometric design. In *Topics in Algebraic Geometry and
Geometric Modeling,*, pages 321–342. Contemporary Mathematics 334,
2003.

[Ber75]   D.N. Bernstein. The number of roots of a system of equations. *Funct. Anal. and Appl.*, 9(2):183–185, 1975.

[Ber84]   S.J. Berkowitz. On computing the determinant in small parallel time using a small number of processors. *Inf. Process. Lett.*, 18:147–150, 1984.

[Béz64]   E. Bézout. Recherches sur les degrés des équations résultantes de l'évanouissement des inconnues et sur les moyens qu'il convient d'employer pour trouver ces équations. *Hist. de l'Acad. Roy. des Sciences*, pages 288–338, 1764.

[BGV02]   E. Briand and L. Gonzalez-Vega. Multivariate Newton sums: identities and generating functions. *Comm. Algebra*, 30(9):4527–4547, 2002.

[BHH78]   G.E.P. Box, W.G. Hunter, and J.S. Hunter. *Statistics for Experimenters*. Wiley series in probability ans statistics. J. Wiley & Sons, New York, 1978.

[Bin96]   D. Bini. Numerical computation of polynomial zeros by means of Aberth's method. *Numerical Algorithms*, 13:179–200, 1996.

[BJ03]   L. Busé and J.-P. Jouanolou. On the closed image of a rational map and the implicitization problem. *J. Algebra*, 265:312–357, 2003.

[BKL98]   V. Bykov, A. Kytmanov, and M. Lazman. *Elimination methods in polynomial computer algebra*, volume 448 of *Mathematics and its Applications*. Kluwer Academic Publishers, Dordrecht, 1998.

[BKM90]   W. Bruns, A.R. Kustin, and M. Miller. The resolution of the generic residual intersection of a complete intersection. *J. Algebra*, 128:214–239, 1990.

[BLS⁺04]   A. Bostan, G. Lecerf, B. Salvy, E. Schost, and B. Wiebelt. Complexity issue in bivariate polynomial factorization. In J. Gutierrez, editor, *Proc. Annual ACM Intern. Symp. on Symbolic and Algebraic Computation*. ACM, 2004.

[BM82]   B. Buchberger and H.M. Möller. The construction of multivariate polynomials with preassigned zeros. In *Computer Algebra (Marseille, 1982)*, volume 144 of *Lect. Notes in Comp. Science*, pages 24–31. Springer-Verlag, Berlin New York, 1982.

[BM02]   V.V. Batyrev and E.N. Materov. Toric residues and mirror symmetry. *Mosc. Math. J.*, 2(3):435–475, 2002.

[BMP00]   D. Bondyfalat, B. Mourrain, and V.Y. Pan. Computation of a specified root of a polynomials system of equations using eigenvector. *Lin. Alg. & Appl.*, 319:193–209, 2000.

[BR90]   R. Benedetti and J.J. Risler. *Real algebraic and semi-algebraic sets*. Hermann, Paris, 1990.

[Bro87]   D. Brownawell. Bounds for the degrees in the Nullstellensatz. *Ann. Math. 2nd Series*, 126(3):577–591, 1987.

[Buc88a]   B. Buchberger. Applications of Gröbner bases in non-linear computational geometry. In *Trends in computer algebra*, volume 296 of *Lect. Notes in Comp. Science*, pages 52–80, New York, 1988. Springer-Verlag.

[Buc88b]   B. Buchberger. Applications of Gröbner bases in non-linear computational geometry. In J.R. Rice, editor, *Scientific Software*, volume 114 of *IMA Volumes in Math. & Appl.*, pages 59–87. Springer-Verlag, 1988.

[Bus01a]   L. Busé. *Étude du résultant sur une variété algébrique*. PhD thesis, Univ. de Nice Sophia-Antipolis, France, 2001.

[Bus01b]   L. Busé. Residual resultant over the projective plane and the implici-
            tization problem. In B. Mourrain, editor, *Proc. Annual ACM Intern.
            Symp. on Symbolic and Algebraic Computation*, pages 48–55. ACM,
            2001.

[Bus03]   L. Busé. Computing resultant matrices with Macaulay2 and Maple.
            Rapport de recherche 0280, INRIA, Sophia-Antipolis, France, 2003.

[BW93]   T. Becker and V. Weispfenning. *Gröbner bases*, volume 141 of *Graduate
            Texts in Mathematics*. Springer-Verlag, New York, 1993.

[BY99]   C.A. Berenstein and A. Yger. Residue calculus and effective Nullstel-
            lensatz. *Amer. J. Math.*, 121:723–796, 1999.

[Byr89]   C.I. Byrnes. Pole assignment by output feedback. In H. Nijmacher
            and J.M. Schumacher, editors, *Three Decades of Mathematical Systems
            Theory*, volume 135 of *Lect. Notes in Control and Inform. Sci.*, pages
            13–78. Springer-Verlag, 1989.

[Can90]   J. Canny. Generalised characteristic polynomials. *J. Symbolic Compu-
            tation*, 9:241–250, 1990.

[Car61]   H. Cartan. *Théorie élémentaire des fonctions analytiques d'une ou
            plusieurs variables complexes (Avec le concours de R. Takahashi)*. En-
            seignement des Sciences. Hermann, Paris, 1961.

[CARW93]   A. Chakraborty, D.C.S. Allison, C.J. Ribbens, and L.T. Watson. The
            parallel complexity of embedding algorithms for the solution of systems
            of nonlinear equations. *IEEE Trans. Parallel and Distributed Systems*,
            4(4):458–465, 1993.

[Cay48]   A. Cayley. On the theory of elimination. *Dublin Math. J.*, II:116–120,
            1848.

[CCD97]   E. Cattani, D. Cox, and A. Dickenstein. Residues in toric varieties.
            Compositio Math., 108(1):35–76, 1997.

[CCS99]   A. Cohen, H. Cuypers, and H. Sterk, editors. *Some Tapas of com-
            puter algebra*, volume 4 of *Algorithms and Computation in Mathemat-
            ics*. Springer-Verlag, Berlin, 1999.

[CD]   E. Cattani and A. Dickenstein. A note on the computation of sparse
            resultants. Manuscript, 2004.

[CD97]   E. Cattani and A. Dickenstein. A global view of residues in the torus.
            J. Pure & Applied Algebra, 117-118:119–144, 1997. Special Issue on
            Algorithms for algebra (Proc. MEGA'96).

[CD04]   E. Cattani and A. Dickenstein. Planar configurations of lattice vec-
            tors and GKZ-rational toric fourfolds in P^6. *J. Algebr. Combinatorics*,
            19(1):47–65, 2004.

[CdlOF$^+$94] P. Candelas, X. de la Ossa, A. Font, S. Katz, and D.R. Morrison. Mirror
            symmetry for two-parameter models. I. *Nuclear Phys. B*, 416(2):481–
            538, 1994.

[CDS96]   E. Cattani, A. Dickenstein, and B. Sturmfels. Computing multidimen-
            sional residues. In L. González-Vega and T. Recio, editors, *Algorithms
            in algebraic geometry and applications (Proc. MEGA'94), Santander,
            Spain*, number 143 in Progress in Math., pages 135–164. Birkhäuser,
            Basel, 1996.

[CDS98]   E. Cattani, A. Dickenstein, and B. Sturmfels. Residues and resultants.
            J. Math. Sci. Univ. Tokyo, 5(1):119–148, 1998.

[CDS01]   E. Cattani, A. Dickenstein, and B. Sturmfels. Rational hypergeometric
            functions. *Compositio Math.*, 128(2):217–239, 2001.

[CDS02]   E. Cattani, A. Dickenstein, and B. Sturmfels. Binomial residues. *Ann. Inst. Fourier (Grenoble)*, 52(3):687–708, 2002.

[CE93]   J. Canny and I. Emiris. An efficient algorithm for the sparse mixed resultant. In G. Cohen, T. Mora, and O. Moreno, editors, *Proc. Intern. Symp. Applied Algebra, Algebraic Algorithms and Error-Correcting Codes (AAECC, Puerto Rico)*, number 263 in Lect. Notes in Comp. Science, pages 89–104, Berlin, 1993. Springer-Verlag.

[CE00]   J.F. Canny and I.Z. Emiris. A subdivision-based algorithm for the sparse resultant. *J. ACM*, 47(3):417–451, May 2000.

[CG80]   J.A. Carlson and P.A. Griffiths. Infinitesimal variations of Hodge structure and the global Torelli problem. In *Journées de Géometrie Algébrique d'Angers / Algebraic Geometry, Angers, 1979*, pages 51–76. Sijthoff & Noordhoff, Alphen aan den Rijn, 1980.

[CG83]   A.L. Chistov and D.Y. Grigoryev. Subexponential-time solving systems of algebraic equations I, II. *Steklov Math. Institute, LOMI Preprints E-9-83, 0e-10-c83*, 1983.

[CG84]   A.L. Chistov and D.Y. Grigoryev. Complexity of quantifier elimination in the theory of algebraically closed fields. In *Proc. 11th Symp. Math. Foundations Comp. Science, Praha, Czechoslovakia*, number 176 in Lect. Notes in Comp. Science, pages 17–31, Berlin, 1984. Springer-Verlag.

[CG03]   G. Chèze and A. Galligo. From an approximate to an exact factorization. Submitted for publication, 2003.

[CGKW01]   R.M. Corless, M.W. Giesbrecht, I.S. Kotsireas, and S.M. Watt. Numerical implicitization of parametric hypersurfaces with linear algebra. In *Artificial intelligence and symbolic computation (Madrid, 2000)*, pages 174–183. Springer-Verlag, Berlin, 2001.

[CGKW02]   R.M. Corless, A. Galligo, I.S. Kotsireas, and S.M. Watt. A geometric-numeric algorithm for factoring multivariate polynomials. In T. Mora, editor, *Proc. Annual ACM Intern. Symp. on Symbolic and Algebraic Computation*, pages 37–45. ACM, 2002.

[CGT97]   R.M. Corless, P.M. Gianni, and B.M. Trager. A reordered Schur factorization method for zero-dimensional polynomial systems with multiple roots. In W. Küchlin, editor, *Proc. Annual ACM Intern. Symp. on Symbolic and Algebraic Computation*, pages 133–140, 1997.

[CGvH+01]   R.M. Corless, M.W. Giesbrecht, M. van Hoeij, I.S. Kotsireas, and S.M. Watt. Towards factoring bivariate approximate polynomials. In B. Mourrain, editor, *Proc. Annual ACM Intern. Symp. on Symbolic and Algebraic Computation*, pages 85–92. ACM, 2001.

[CGZ00]   D. Cox, R. Goldman, and M. Zhang. On the validity of implicitization by moving quadrics for rational surfaces with no base points. *J. Symbolic Computation*, 29(3):419–440, 2000.

[CH78]   N.R. Coleff and M.E. Herrera. *Les courants résiduels associés à une forme méromorphe*, volume 633 of *Lect. Notes in Math.* Springer-Verlag, Berlin, 1978.

[Cha95]   M. Chardin. Multivariate subresultants. *J. Pure & Applied Algebra*, 101(2):129–138, 1995.

[Chè04]   G. Chèze. Absolute polynomial factorization in two variables and the knapsack problem. In J. Gutierrez, editor, *Proc. Annual ACM Intern.*

              Symp. on Symbolic and Algebraic Computation, pages 87–94. ACM,
              2004.

[Chi86]   A.L. Chistov. Algorithm of polynomial complexity for factoring poly-
              nomials and finding the components of varieties in subexponential time.
              J. Sov. Math, 34(4):1838–1882, 1986.

[CJ98]   R.M. Corless and D.J. Jeffrey. Graphing elementary Riemann surfaces.
              SIGSAM Bulletin, 32(1):11–17, 1998.

[CK90]   C.C. Chang and H.J. Keisler. *Model theory*. Number 73 in Studies
              in Logic and the Foundations of Mathematics. North-Holland, Amster-
              dam, 3rd edition, 1990.

[CK99]   D.A. Cox and S. Katz. *Mirror symmetry and algebraic geometry*, vol-
              ume 68 of *Mathematical Surveys and Monographs*. AMS, Providence,
              RI, 1999.

[CKL89]   J.F. Canny, E. Kaltofen, and Y. Lakshman. Solving systems of non-
              linear polynomial equations faster. In *Proc. Annual ACM Intern. Symp.
              on Symbolic and Algebraic Computation*, pages 121–128. ACM, 1989.

[CKW02]   R.M. Corless, E. Kaltofen, and S.M. Watt. Hybrid methods. In J.
              Grabmeier, E. Kaltofen, and V. Weispfenning, editors, *Computer Alge-
              bra Handbook*, pages 112–125. Springer-Verlag, 2002.

[CLO97]   D. Cox, J. Little, and D. O'Shea. *Ideals, varieties, and algorithms*.
              Undergraduate Texts in Mathematics. Springer-Verlag, New York, 2nd
              edition, 1997.

[CLO98]   D. Cox, J. Little, and D. O'Shea. *Using algebraic geometry*, volume 185
              of *Graduate Texts in Mathematics*. Springer-Verlag, New York, 1998.

[CM92]   J.F. Canny and D. Manocha. Implicit representation of rational para-
              metric surfaces. *J. Symbolic Computation*, 13(5):485–510, 1992.

[CMPY79]   S.N. Chow, J. Mallet-Paret, and J.A. Yorke. A homotopy method for
              locating all zeros of a system of polynomials. In H.O. Peitgen and H.O.
              Walther, editors, *Functional differential equations and approximation
              of fixed points*, volume 730 of *Lect. Notes in Math.*, pages 77–88, New
              York, 1979. Springer-Verlag.

[Col67]   G.E. Collins. Subresultants and reduced polynomial remainder se-
              quences. *J. ACM*, 14:128–142, 1967.

[Cox]   D. Cox. What is the multiplicity of a basepoint? Lecture notes available
              at http://www.amherst.edu/~dacox.

[Cox95]   D. Cox. The homogeneous coordinate ring of a toric variety. *J. Alge-
              braic Geom.*, 4(1):17–50, 1995.

[Cox96]   D.A. Cox. Toric residues. *Ark. Mat.*, 34(1):73–96, 1996.

[CP93]   J. Canny and P. Pedersen. An algorithm for the Newton resultant.
              Technical Report 1394, Comp. Science Dept., Cornell University, 1993.

[CR91]   J. Canny and J.M. Rojas. An optimal condition for determining the ex-
              act number of roots of a polynomial system. In S.M. Watt, editor, *Proc.
              Annual ACM Intern. Symp. on Symbolic and Algebraic Computation*,
              pages 96–101. ACM, 1991.

[CR97]   M. Caboara and L. Robbiano. Families of ideals in statistics. In W.
              Küchlin, editor, *Proc. Annual ACM Intern. Symp. on Symbolic and
              Algebraic Computation*, pages 404–409. ACM, 1997.

[CR01]   M. Caboara and L. Robbiano. Families of estimable terms. In B.
              Mourrain, editor, *Proc. Annual ACM Intern. Symp. on Symbolic and
              Algebraic Computation*, pages 56–63, 2001.

[CSTU02] O. Cormier, M.F. Singer, B.M. Trager, and F. Ulmer. Linear differential operators for polynomial equations. *J. Symbolic Computation*, 34(5):355–398, 2002.

[CU02] M. Chardin and B. Ulrich. Liaison and Castelnuovo-Mumford regularity. *Amer. J. Math.*, 124(6), 2002.

[D'A01] C. D'Andrea. Resultants and moving surfaces. *J. Symbolic Computation*, 31:585–602, 2001.

[D'A02] C. D'Andrea. Macaulay-style formulas for the sparse resultant. *Trans. AMS*, 354:2595–2629, 2002.

[DD01] C. D'Andrea and A. Dickenstein. Explicit formulas for the multivariate resultant. *J. Pure & Applied Algebra*, 164(1-2):59–86, 2001. Effective methods in algebraic geometry (Proc. MEGA'00).

[DE01a] C. D'Andrea and I.Z. Emiris. Computing sparse projection operators. In *Symbolic Computation: Solving Equations in Algebra, Geometry, and Engineering*, volume 286 of *Contemporary Mathematics*, pages 121–139, Providence, R.I., 2001. AMS.

[DE01b] C. D'Andrea and I.Z. Emiris. Hybrid resultant matrices of bivariate polynomials. In B. Mourrain, editor, *Proc. Annual ACM Intern. Symp. on Symbolic and Algebraic Computation*, pages 24–31. ACM, 2001.

[DE01c] D. Daney and I.Z. Emiris. Robust parallel robot calibration with partial information. In *Proc. IEEE Intern. Conf. Robotics & Automation*, pages 3262–3267, Seoul, S. Korea, 2001.

[DE03] A. Dickenstein and I.Z. Emiris. Multihomogeneous resultant formulae by means of complexes. *J. Symbolic Computation*, 36(3-4):317–342, 2003.

[Dem87] J.W. Demmel. On condition numbers and the distance to the nearest ill-posed problem. *Numer. Math.*, 51(3):251–289, 1987.

[DES98] P. Diaconis, D. Eisenbud, and B. Sturmfels. Lattice walks and primary decomposition. In B.E. Sagan and R.P. Stanley, editors, *Mathematical Essays in Honor of Gian-Carlo Rota*, volume 161 of *Progress in Math.*, pages 173–193. Birkhäuser, 1998.

[DF88] M.E. Dyer and A.M. Frieze. On the complexity of computing the volume of a polyhedron. *SIAM J. Comput.*, 17(5):967–974, 1988.

[DGH98] M. Dyer, P. Gritzmann, and A. Hufnagel. On the complexity of computing mixed volumes. *SIAM J. Comput.*, 27(2):356–400, 1998.

[DGP99] W. Decker, G.-M. Greuel, and G. Pfister. Primary decomposition: algorithms and comparisons. In *Algorithmic algebra and number theory (Heidelberg, 1997)*, pages 187–220. Springer-Verlag, Berlin, 1999.

[Die98] P. Dietmaier. The Stewart-Gough platform of general geometry can have 40 real postures. In *Advances in robot kinematics: analysis and control (Salzburg, 1998)*, pages 7–16. Kluwer Academic Publishers, Dordrecht, 1998.

[Dix08] A.L. Dixon. The eliminant of three quantics in two independent variables. *Proc. London Math. Soc*, 6:49–69, 1908.

[DK] C. D'Andrea and A. Khetan. Macaulay style formulas for toric residues. *Compositio Math.* To appear.

[DKK03] Y. Dai, S. Kim, and M. Kojima. Computing all nonsingular solutions of cyclic-n polynomials using polyhedral homotopy continuation methods. *J. Comput. Appl. Math.*, 152(1-2):83–97, 2003.

400   References

[DMPT03]   O. Devillers, B. Mourrain, F.P. Preparata, and P. Trebuchet. Circular cylinders by four or five points in space. *Discr. and Comput. Geometry*, 29:83–104, 2003.

[Dre77]   F.J. Drexler. Eine Methode zur Berechnung sämtlicher Lösungen von Polynomgleichungssystemen. *Numer. Math.*, 29(1):45–58, 1977.

[DRMRT02] G. Dos Reis, B. Mourrain, R. Rouillier, and Ph. Trébuchet. An environment for symbolic and numeric computation. In *Proc. Intern. Conf. Math. Software 2002*, World Scientific, pages 239–249, 2002.

[DS91]   A. Dickenstein and C. Sessa. An effective residual criterion for the membership problem in $\mathbf{C}[z_1, \cdots, z_n]$. *J. Pure & Applied Algebra*, 74(2):149–158, 1991.

[DS02]   A. Dickenstein and B. Sturmfels. Elimination theory in codimension two. *J. Symbolic Computation*, 34:119–135, 2002.

[DT89]   R. Dvornicich and C. Traverso. Newton symmetric functions and the arithmetic of algebraically closed fields. In *Proc. Intern. Symp. Applied Algebra, Algebraic Algorithms and Error-Correcting Codes (AAECC, Menorca, 1987)*, volume 356 of *Lect. Notes in Comp. Science*, pages 216–224. Springer-Verlag, Berlin, 1989.

[Dum06]   G. Dumas. Sur quelques cas d'irréductibilité des polynomes coefficients rationnels. *J. Math. Pures et Appliquées*, 1906.

[DvH01]   B. Deconinck and M. van Hoeij. Computing Riemann matrices of algebraic curves. *Physica D*, 152:28–46, 2001.

[DY93]   J.-P. Dedieu and J.-C. Yakoubsohn. Computing the real roots of a polynomial by the exclusion algorithm. *Numerical Algorithms*, 4:1–24, 1993.

[EC95]   I.Z. Emiris and J.F. Canny. Efficient incremental algorithms for the sparse resultant and the mixed volume. *J. Symbolic Computation*, 20(2):117–149, 1995.

[Edw04]   H.M. Edwards. *Essays in Constructive Mathematics*. Springer-Verlag, New York, 2004.

[EGH96]   D. Eisenbud, M. Green, and J. Harris. Cayley-Bacharach theorems and conjectures. *Bull. AMS (N.S.)*, 33(3):295–324, 1996.

[Ehr67]   E. Ehrhart. Sur un problème de géométrie diophantienne linéaire ii. *J. reine angewandte Math.*, 227:25–49, 1967.

[EHV92]   D. Eisenbud, C. Huneke, and W. Vasconcelos. Direct methods for primary decomposition. *Invent. Math.*, 110:207–235, 1992.

[Eis95]   D. Eisenbud. *Commutative algebra*, volume 150 of *Graduate Texts in Mathematics*. Springer-Verlag, New York, 1995.

[EK03]   I.Z. Emiris and I.S. Kotsireas. Implicitization with polynomial support optimized for sparseness. In V. Kumar, M.L. Gavrilova, C.J.K. Tan, and P. L'Ecuyer, editors, *Proc. Intern. Conf. Comput. Science & Appl. 2003, Montreal, Canada*, volume 2669 of *LNCS*, pages 397–406. Springer-Verlag, 2003.

[EL02]   T. Ekedahl and D. Laksov. Splitting algebras, symmetric functions, and Galois theory. Available as math.AC/0211125, 2002.

[EM]   M. Elkadi and B. Mourrain. *Introduction à la Résolution des Systémes d'Équations Polynomiales*. To appear.

[EM96]   M. Elkadi and B. Mourrain. Approche effective des résidus algébriques. Rapport de recherche 2884, INRIA, Sophia-Antipolis, France, 1996.

[EM98]   M. Elkadi and B. Mourrain. Some applications of Bezoutians in effec-
            tive algebraic geometry. Rapport de recherche 3572, INRIA, Sophia-
            Antipolis, France, 1998.

[EM99a]   M. Elkadi and B. Mourrain. A new algorithm for the geometric decom-
            position of a variety. In S. Dooley, editor, *Proc. Annual ACM Intern.
            Symp. on Symbolic and Algebraic Computation.* ACM, 1999.

[EM99b]   I.Z. Emiris and B. Mourrain. Computer algebra methods for study-
            ing and computing molecular conformations. *Algorithmica,* 25:372–402,
            1999. Special Issue on Algorithms for Computational Biology.

[EM99c]   I.Z. Emiris and B. Mourrain. Matrices in elimination theory. *J. Sym-
            bolic Computation,* 28:3–44, 1999. Special Issue on Elimination.

[EM00]   M. Elkadi and B. Mourrain. Algorithms for residues and Lojasiewicz
            exponents. *J. Pure & Applied Algebra,* 153:27–44, 2000.

[Emi96]   I.Z. Emiris. On the complexity of sparse elimination. *J. Complexity,*
            12:134–166, 1996.

[Emi97]   I.Z. Emiris. A general solver based on sparse resultants: Numerical
            issues and kinematic applications. Technical Report 3110, INRIA,
            Sophia-Antipolis, France, 1997.

[EP02]   I.Z. Emiris and V.Y. Pan. Symbolic and numeric methods for exploiting
            structure in constructing resultant matrices. *J. Symbolic Computation,*
            33:393–413, 2002.

[ER94]   I.Z. Emiris and A. Rege. Monomial bases and polynomial system solv-
            ing. In *Proc. Annual ACM Intern. Symp. on Symbolic and Algebraic
            Computation,* pages 114–122. ACM, July 1994.

[ES96]   D. Eisenbud and B. Sturmfels. Binomial ideals. *Duke Math. J.,* 84(1):1–
            45, 1996.

[ESW03]   D. Eisenbud, F.-O. Schreyer, and J. Weyman. Resultants and Chow
            forms via exterior syzygies. *J. AMS,* 16(3):537–579, 2003.

[EV99]   I.Z. Emiris and J. Verschelde. How to count efficiently all affine roots
            of a polynomial system. *Discrete Applied Mathematics,* 93(1):21–32,
            1999.

[Ewa96]   G. Ewald. *Combinatorial convexity and algebraic geometry,* volume 168
            of *Graduate Texts in Mathematics.* Springer-Verlag, New York, 1996.

[Fau93]   O. Faugeras. *Three-Dimensional Computer Vision: a Geometric View-
            point.* MIT press, 1993.

[Fau99]   J.-C. Faugère. A new efficient algorithm for computing Gröbner bases
            (F_4). *J. Pure & Applied Algebra,* 139(1-3):61–88, 1999. Effective meth-
            ods in algebraic geometry (Proc. MEGA'98), Saint-Malo, France.

[FG90]   N. Fitchas and A. Galligo. Nullstellensatz effectif et conjecture de
            Serre (théorème de Quillen-Suslin) pour le calcul formel. (an effective
            Nullstellensatz and the Serre conjecture (Quillen-Suslin theorem) for
            the computer algebra). *Math. Nachr.,* 149:231–253, 1990.

[FGLM93]   J.-C. Faugère, P. Gianni, D. Lazard, and T. Mora. Efficient computa-
            tion of zero-dimensional Groebner bases by change of ordering. *J. Sym-
            bolic Computation,* 16(4):329–344, 1993.

[FGM90]   N. Fitchas, A. Galligo, and J. Morgenstern. Precise sequential and
            parallel complexity bounds for quantifier elimination over algebraically
            closed fields. *J. Pure & Applied Algebra,* 67(1):1–14, 1990.

[FGS95]   N. Fitchas, M. Giusti, and F. Smietanski. Sur la complexité du théorème des zéros. In J. Gudatt, editor, *Proc. 2nd Intern. Conf. Approximation & Optimization, Havana, Cuba, 1993*, number 8 in Approximation Optimization, pages 274–329. Peter Lang, Frankfurt, 1995.

[Fuh96]   P.A. Fuhrmann. *A polynomial approach to linear algebra.* Universitext. Springer-Verlag, New York, 1996.

[Ful93]   W. Fulton. *Introduction to toric varieties. The 1989 William H. Roever lectures in geometry*, volume 131 of *Annals of Mathematics Studies.* Princeton University Press, Princeton, NJ, 1993.

[Ful98]   W. Fulton. *Intersection theory*, volume 2 of *Ergebnisse der Mathematik und ihrer Grenzgebiete. 3. Folge. A Series of Modern Surveys in Mathematics.* Springer-Verlag, Berlin, 2nd edition, 1998.

[Gal99]   A. Galligo. Real factorization of multivariate polynomials with integer coefficients. *Zap. Nauchn. Sem. S.-Peterburg. Otdel. Mat. Inst. Steklov. (POMI)*, 258(4):60–70, 1999.

[Gao01]   S. Gao. Absolute irreducibility of polynomials via Newton polytopes. *J. Algebra*, 237(2):501–520, 2001.

[Gao03]   S. Gao. Factoring multivariate polynomials via partial differential equations. *Math. Comp.*, 72(242):801–822, 2003.

[GCMT02] O. Grellier, P. Comon, B. Mourrain, and Ph. Trébuchet. Analytical blind channel identification. *IEEE Trans. Signal Processing*, 50(9):2196–2207, 2002.

[GD93]   M. Griffis and J. Duffy. Method and apparatus for controlling geometrically simple parallel mechanisms with distinctive connections. US Patent 5,179,525, 1993.

[GH78]   P. Griffiths and J. Harris. *Principles of algebraic geometry.* Pure and Applied Mathematics. Wiley-Interscience, New York, 1978.

[GH91]   M. Giusti and J. Heintz. Algorithmes - disons rapides - pour la décomposition d'une variété algébrique en composantes irréductibles et équidimensionelles. (So-called fast algorithms for the decomposition of an algebraic variety into irreducible and equidimensional components). In *Effective methods in algebraic geometry (Proc. MEGA'90), Castiglioncello, Italy*, volume 94 of *Progess in Math.*, pages 169–194, 1991.

[GH93]   M. Giusti and J. Heintz. La détermination des points isolés et de la dimension d'une variété algébrique peut se faire en temps polynomial. (the determination of isolated points and of the dimension of an algebraic variety can be done in polynomial time). In *Computational algebraic geometry and commutative algebra (Cortona, Italy, 1991)*, Proc. Symp. Math., XXXIV, pages 216–256. Cambridge University Press, Cambridge, 1993.

[GHH⁺97] M. Giusti, J. Heintz, K. Hägele, J.E. Morais, L.M. Pardo, and J.L. Montana. Lower bounds for diophantine approximations. *J. Pure & Applied Algebra*, 117-118:277–317, 1997.

[GHKM01] D. Geiger, D. Heckerman, H. King, and C. Meek. Stratified exponential families: graphical models and model selection. *Ann. Statist.*, 29(2):505–529, 2001.

[GHM⁺98] M. Giusti, J. Heintz, J.E. Morais, J. Morgenstern, and L.M. Pardo. Straight-line programs in geometric elimination theory. *J. Pure & Applied Algebra*, 124(1-3):101–146, 1998.

[GHS93]   M. Giusti, J. Heintz, and J. Sabia. On the efficiency of effective Nullstellensätze. *Comput. Complexity*, 3(1):56–95, 1993.

[GKK+04]   T. Gunji, S. Kim, M. Kojima, A. Takeda, K. Fujisawa, and T. Mizutani.
PHoM – a polyhedral homotopy continuation method for polynomial
systems. *Computing*, 73:55–77, 2004.

[GKZ91]   I.M. Gelfand, M.M. Kapranov, and A.V. Zelevinsky. Discriminants of
polynomials in several variables and triangulations of Newton polytopes. *Leningrad Math. J.*, 2(3):449–505, 1991.

[GKZ94]   I.M. Gelfand, M.M. Kapranov, and A.V. Zelevinsky. *Discriminants,
Resultants and Multidimensional Determinants*. Birkhäuser, Boston,
1994.

[GL80]   C.B. Garcia and T.Y. Li. On the number of solutions to polynomial
systems of equations. *SIAM J. Numerical Analysis*, 17(4):540–546,
1980.

[GL00]   T. Gao and T.Y. Li. Mixed volume computation via linear programming. *Taiwan J. of Math.*, 4:599–619, 2000.

[GL03]   T. Gao and T.Y. Li. Mixed volume computation for semi-mixed systems. *Discr. and Comput. Geometry*, 29(2):257–277, 2003.

[GLS01]   M. Giusti, G. Lecerf, and B. Salvy. A Gröbner free alternative for
polynomial system solving. *J. Complexity*, 17(1):154–211, 2001.

[GLW99]   T. Gao, T.Y. Li, and X. Wang. Finding isolated zeros of polynomial
systems in C^n with stable mixed volumes. *J. Symbolic Computation*,
28:187–211, 1999.

[GM83]   M. Goresky and R. MacPherson. Stratified Morse theory. In *Proc.
Symp. Pure Math.*, volume 40, pages 517–533. AMS, Providence, R.I.,
1983.

[Gou97]   F.Q. Gouvêa. *p-adic numbers*. Universitext. Springer-Verlag, Berlin,
2nd edition, 1997.

[GP02]   G.-M. Greuel and G. Pfister. *A Singular introduction to commutative
algebra*. Springer-Verlag, Berlin, 2002.

[GPS01]   G.-M. Greuel, G. Pfister, and H. Schönemann. Singular 2.0. A
computer algebra system for polynomial computations, Centre for
Computer Algebra, University of Kaiserslautern, 2001. Available at
http://www.singular.uni-kl.de.

[GR01]   A. Galligo and D. Rupprecht. Semi-numerical determination of irreducible branches of a reduced space curve. In B. Mourrain, editor,
*Proc. Annual ACM Intern. Symp. on Symbolic and Algebraic Compu-
tation*, pages 137–142. ACM, 2001.

[GR02]   A. Galligo and D. Rupprecht. Irreducible decomposition of curves.
J. Symbolic Computation, 33(5):661–677, 2002.

[Gri86]   D.Y. Grigoryev. Factorization of polynomials over finite field and the
solution of systems of algebraic equations. *J. Sov. Math*, 34(4):1762–
1803, 1986.

[Gri87]   D.Y. Grigoryev. The complexity of the decision problem for the first
order theory of algebraically closed fields. *Math. USSR, Izv.*, 29:459–
475, 1987.

[GS]   D.R. Grayson and M.E. Stillman. Macaulay 2, a software system for research in algebraic geometry. Available at
http://www.math.uiuc.edu/Macaulay2.

[GSS]   L.D. Garcia, M. Stillman, and B. Sturmfels. Algebraic geometry of
           bayesian networks. arXiv:math.AG/0301255.

[GTZ88]   P. Gianni, B. Trager, and G. Zacharias. Gröbner bases and primary
           decomposition of polynomial ideals. *J. Symbolic Computation*, 6:149–
           167, 1988.

[GV97]   L. Gonzalez-Vega. Implicitization of parametric curves and surfaces by
           using multidimensional Newton formulae. *J. Symbolic Computation*,
           23(2-3):137–151, 1997.

[GVL83]   G.H. Golub and C.F. Van Loan. *Matrix Computations*. The Johns
           Hopkins University Press, 1983.

[GVRR99]   L. Gonzalez-Vega, F. Rouillier, and M.-F. Roy. Symbolic recipes for
           polynomial system solving. In *Some Tapas of computer algebra*, vol-
           ume 4 of *Algorithms Comput. Math.*, pages 34–65. Springer-Verlag,
           Berlin, 1999.

[GW97]   A. Galligo and S. Watt. A numerical absolute primality test for bi-
           variate polynomials. In W. Küchlin, editor, *Proc. Annual ACM Intern.
           Symp. on Symbolic and Algebraic Computation*, pages 217–224. ACM,
           1997.

[GZ79]   C.B. Garcia and W.I. Zangwill. Finding all solutions to polynomial sys-
           tems and other systems of equations. *Math. Programming*, 16(2):159–
           176, 1979.

[GZK89]   I.M. Gelfand, A.V. Zelevinsky, and M.M. Kapranov. Hypergeometric
           functions and toric varieties. *Funktsional. Anal. i Prilozhen.*, 23, 1989.

[Haa96]   U. Haagerup. Orthogonal maximal abelian *-algebras of the n x n
           matrices and cyclic n-roots. In *Operator Algebras and Quantum Field
           Theory*, pages 296–322. International Press, 1996.

[Haa02]   B. Haas. A simple counterexample to Kouchnirenko's conjecture.
           *Beitraege zur Algebra und Geometrie/Contributions to Algebra and
           Geometry*, 43(1):1–8, 2002.

[Har66]   R. Hartshorne. *Residues and duality. Lecture notes of a seminar on the
           work of A. Grothendieck, given at Harvard 1963/64. With an appendix
           by P. Deligne.*, volume 20 of *Lect. Notes in Math.* Springer-Verlag,
           Berlin, 1966.

[Har80]   J. Harris. The genus of a space curve. *Math. Ann.*, 249:191–204, 1980.

[Har83]   R. Hartshorne. *Algebraic geometry*, volume 52 of *Graduate Texts in
           Mathematics*. Springer-Verlag, New York-Heidelberg-Berlin, 1983.

[Har95]   J. Harris. *Algebraic geometry*, volume 133 of *Graduate Texts in Math-
           ematics*. Springer-Verlag, New York, 1995.

[Hei83]   J. Heintz. Definability and fast quantifier elimination in algebraically
           closed fields. *Theor. Comput. Sci.*, 24:239–277, 1983.

[Her26]   G. Hermann. Zur Frage der endlich vielen Schritte in der Theorie der
           Polynomideale. *Math. Ann.*, 95:736–788, 1926.

[HK00]   M.L. Husty and A. Karger. Self-motions of Griffis-Duffy type parallel
           manipulators. In *Proc. IEEE Intern. Conf. Robotics & Automation*,
           2000. San Francisco, Calif.

[HKP+00]   J. Heintz, T. Krick, S. Puddu, J. Sabia, and A. Waissbein. Deformation
           techniques for efficient polynomial equation solving. *J. Complexity*,
           16(1):70–109, 2000.

[HMW01]   J. Heintz, G. Matera, and A. Waissbein. On the time-space complexity of geometric elimination procedures. *J. Applic. Algebra in Engin. Communic. & Computing*, 11(4):239–296, 2001.

[Hof89]   C.M. Hoffmann. *Geometric and solid modeling: an introduction*. Morgan Kaufmann, 1989.

[HP00]   H. Hirukawa and Y. Papegay. Motion planning of objects in contact by the silhouette algorithm. In *Proc. IEEE Intern. Conf. Robotics & Automation*, San Francisco, Calif., 2000. IEEE.

[HRS00]   B. Huber, J. Rambau, and F. Santos. The Cayley trick, lifting subdivisions and the Bohne-Dress theorem on zonotopal tilings. *J. Eur. Math. Soc.*, 2(2):179–198, 2000.

[HS80]   J. Heintz and M. Sieveking. Lower bounds for polynomials with algebraic coefficients. *Theor. Comput. Sci.*, 11:321–330, 1980.

[HS81]   J. Heintz and M. Sieveking. Absolute primality of polynomials is decidable in random polynomial time in the number of variables. In *Automata, languages and programming (Akko, 1981)*, volume 115 of *Lect. Notes in Comp. Science*, pages 16–28. Springer-Verlag, Berlin, 1981.

[HS82]   J. Heintz and C.-P. Schnorr. Testing polynomials which are easy to compute. In *Logic and algorithmic Intern. Symp. (Zürich, 1980)*, volume 30 of *L'Enseign. Math.*, pages 237–254. Geneva, Switzerland, 1982.

[HS95]   B. Huber and B. Sturmfels. A polyhedral method for solving sparse polynomial systems. *Math. Comp.*, 64(212):1541–1555, 1995.

[HS97a]   V. Hribernig and H. J. Stetter. Detection and validation of clusters of polynomial zeros. *J. Symbolic Computation*, 24:667–681, 1997.

[HS97b]   B. Huber and B. Sturmfels. Bernstein's theorem in affine space. *Discr. and Comput. Geometry*, 17(2):137–142, March 1997.

[HS00]   S. Hoşten and J. Shapiro. Primary decomposition of lattice basis ideals. *J. Symbolic Computation*, 29(4-5):625–639, 2000.

[HSS98]   B. Huber, F. Sottile, and B. Sturmfels. Numerical Schubert calculus. *J. Symbolic Computation*, 26(6):767–788, 1998.

[Hus96]   M.L. Husty. An algorithm for solving the direct kinematics of general Stewart-Gough platforms. *Mechanism Machine Theory*, 31(4):365–380, 1996.

[HV98]   B. Huber and J. Verschelde. Polyhedral end games for polynomial continuation. *Numerical Algorithms*, 18(1):91–108, 1998.

[HV00]   B. Huber and J. Verschelde. Pieri homotopies for problems in enumerative geometry applied to pole placement in linear systems control. *SIAM J. Control Optim.*, 38(4):1265–1287, 2000.

[HW75]   J. Heintz and R. Wüthrich. An efficient quantifier elimination algorithm for algebraically closed fields. *SIGSAM Bull.*, 9:11, 1975.

[HW89]   S. Harimoto and L.T. Watson. The granularity of homotopy algorithms for polynomial systems of equations. In G. Rodrigue, editor, *Parallel processing for scientific computing*, pages 115–120. SIAM, 1989.

[HWSZ00]   Y. Huang, W. Wu, H.J. Stetter, and L. Zhi. Pseudofactors of multivariate polynomials. In C. Traverso, editor, *Proc. Annual ACM Intern. Symp. on Symbolic and Algebraic Computation*, pages 161–168. ACM, 2000.

[Ier89]   D. Ierardi. Quantifier elimination in the theory of an algebraically-closed field. *J. ACM*, pages 138–147, 1989.

[IS03]   I. Itenberg and E. Shustin. Viro theorem and topology of real and complex combinatorial hypersurfaces. *Israel Math. J.*, 133:189–238, 2003.

[IV96]   I. Itenberg and O. Viro. Patchworking algebraic curves disproves the Ragsdale conjecture. *The Math. Intelligencer*, 18(4):19–28, 1996.

[Jac30]   C. G. J. Jacobi. De resolutione aequationum per series infinitas. *J. Reine Angew. Math*, 6:257–286, 1830.

[JKSS04]   G. Jeronimo, T. Krick, J. Sabia, and M. Sombra. The computational complexity of the Chow form. *Found. of Comput. Math.*, 4(1):41–117, 2004.

[Jou79]   J.-P. Jouanolou. Singularités rationnelles du résultant. In *Algebraic geometry (Proc. Summer Meeting, Copenhagen, Copenhagen, 1978)*, volume 732 of *Lect. Notes in Math.*, pages 183–213. Springer-Verlag, Berlin, 1979.

[Jou83]   J.-P. Jouanolou. *Théorèmes de Bertini et applications*, volume 42 of *Progress in Math*. Birkhäuser, Boston, Mass., 1983.

[Jou91]   J.-P. Jouanolou. Le formalisme du résultant. *Adv. Math.*, 90(2):117–263, 1991.

[Jou97]   J. P. Jouanolou. Formes d'inertie et résultant: un formulaire. *Adv. Math.*, 126(2):119–250, 1997.

[JS02]   G. Jeronimo and J. Sabia. Effective equidimensional decomposition of affine varieties. *J. Pure & Applied Algebra*, 169(2-3):229–248, 2002.

[Kal85a]   E. Kaltofen. Fast parallel absolute irreducibility testing. *J. Symbolic Computation*, 1(1):57–67, 1985.

[Kal85b]   E. Kaltofen. Polynomial-time reductions from multivariate to bi- and univariate integral polynomial factorization. *SIAM J. Comput.*, 14(2):469–489, 1985.

[Kal90]   E. Kaltofen. Polynomial factorization 1982–1986. In *Computers in mathematics (1986)*, volume 125 of *Lect. Notes in Pure & Appl. Math.*, pages 285–309. Dekker, New York, Stanford, Calif., 1990.

[Kal91]   M. Kalkbrener. Implicitization of rational parametric curves and sur- faces. In *Proc. Intern. Symp. Applied Algebra, Algebraic Algorithms and Error-Correcting Codes (AAECC-8)*, volume 508 of *Lect. Notes in Comp. Science*, pages 249–259, 1991.

[Kal00]   E. Kaltofen. Challenges of symbolic computation: my favorite open problems. *J. Symbolic Computation*, 29(6):891–919, 2000.

[Kea90]   R.B. Kearfott. *Interval arithmetic techniques in the computational so- lution of nonlinear systesm of equations: Introduction, examples and comparisons*, pages 337–357. Lectures in Applied Mathemetics. AMS, Providence, R.I., 1990.

[Kem93]   G.R. Kempf. *Algebraic Varieties*, volume 172 of *London Mathematical Society Lecture Note Series*. Cambridge University Press, 1993.

[Khe]   A. Khetan. Exact matrix formula for the unmixed resultant in three variables. *J. Pure & Applied Algebra*. To appear.

[Khe03]   A. Khetan. The resultant of an unmixed bivariate system. *J. Symbolic Computation*, 36(3-4):425–442, 2003.

[Kho78a]   A.G. Khovanskii. Newton polyhedra and the Euler-Jacobi formula. *Uspekhi Mat. Nauk*, 33(6):237–238, 1978.

[Kho78b] A.G. Khovanskii. Newton polyhedra and the genus of complete inter-sections. *Functional Anal. Appl.*, 12(1):38–46, 1978. Translated from the Russian, *Funktsional. Anal. i Prilozhen.* 12(1),51–61,1978.

[KK87] M. Kreuzer and E. Kunz. Traces in strict frobenius algebras and strict complete intersections. *J. reine angewandte Math.*, 381:181–204, 1987.

[KK93] J.M. Kantor and A.G. Khovanskii. Une application du théorème de Riemann-Roch combinatoire au polynôme d'Ehrhart des polytopes en-tier de $\searrow^n$. *C.R. Acad. Sci. Paris, Series I*, 317:501–507, 1993.

[KK03a] A. Kehrein and M. Kreuzer. Characterizations of border bases. preprint, 2003.

[KK03b] S. Kim and M. Kojima. Numerical stability of path trac-ing in polyhedral homotopy continuation methods. Research re-port b-390, Tokyo Institute of Technology, 2003. Available at http://www.is.titech.ac.jp/~kojima/sdp.html.

[Kle98] S.L. Kleiman. Bertini and his two fundamental theorems. *Rend. Circ. Mat. Palermo (2) Suppl.*, 55:9–37, 1998. Studies in the history of mod-ern mathematics, III.

[KM03] E. Kaltofen and J. May. On approximate irreducibility of polynomials in several variables. In *Proc. Annual ACM Intern. Symp. on Symbolic and Algebraic Computation*, pages 161–168, 2003.

[Knu96] D.E. Knuth. Theory and Practice II. In *Selected Papers on Computer Science*, pages 129–139. Cambridge University Press, 1996.

[Kol88] J. Kollár. Sharp effective Nullstellensatz. *J. AMS*, 1(4):963–975, 1988.

[KP96] T. Krick and L.M. Pardo. A computational method for diophantine approximation. In L. González-Vega and T. Recio, editors, *Algorithms in algebraic geometry and applications (Proc. MEGA'94), Santander, Spain*, number 143 in Progress in Math., pages 193–253. Birkhäuser, Basel, 1996.

[KR00] M. Kreuzer and L. Robbiano. *Computational Commmutative Algebra 1*. Springer-Verlag, Heidelberg, 2000.

[Kro82] L. Kronecker. Grundzüge einer arithmetischen Theorie der algebrais-chen Grössen. *J. reine. amgew. Math.*, 92:1–122, 1882.

[Kro31] L. Kronecker. *Leopold Kronecker's Werke, Volumes II and III*. B.G. Teubner, Leipzig, Volume II 1897, Volume III 1899 and 1931. Reprint by Chelsea, New York, 1968.

[Kru13] E. Kruppa. Zur Ermittlung eines Objektes aus zwei Perspektiven mit innere Orientierung. *Sitz.-Ber. Akad. Wiss., Wien, Math.-Naturw. Kl.*, Abt. IIa(122):1939–1948, 1913.

[KSS97] T. Krick, J. Sabia, and P. Solern'o. On intrinsic bounds in the Null-stellensatz. *J. Applic. Algebra in Engin. Communic. & Computing*, 8(2):125–134, 1997.

[KSZ92] M.M. Kapranov, B. Sturmfels, and A.V. Zelevinsky. Chow polytopes and general resultants. *Duke Math. J.*, 67(1):189–218, 1992.

[Kun85] E. Kunz. *Introduction to commutative algebra and algebraic geometry*. Birkhäuser, Boston, 1985.

[Kun86] E. Kunz. *Kähler differentials*. Advanced Lectures in Mathematics. Friedr. Vieweg & Sohn, Braunschweig, 1986.

[Kur23] J. Kurschak. Irreduzible Formen. *J. Reine Angew. Math.*, 152:180–191, 1923.

[Kus75]   A.G. Kushnirenko. The Newton polyhedron and the number of solutions of a system of k equations in k unknowns. *Uspekhi Mat. Nauk.*, 30:266–267, 1975.

[Kus76]   A.G. Kushnirenko. Newton Polytopes and the Bézout Theorem. *Functional Anal. Appl.*, 10(3):233–235, 1976. Translated from the Russian, *Funktsional. Anal. i Prilozhen.* 10(3),82–83,1976.

[KX94]   R.B. Kearfott and Z. Xing. An interval step control for continuation methods. *SIAM J. Numerical Analysis*, 31(3):892–914, 1994.

[Lan83]   S. Lang. *Fundamentals of Diophantine geometry*. Springer-Verlag, New York, 1983.

[Lau96]   S.L. Lauritzen. *Graphical Models*. Oxford University Press, 1996.

[Laz77]   D. Lazard. Algèbre linéaire sur $k[x_1, \ldots, x_n]$ et élimination. *Bull. Soc. Math. Fr.*, 105:165–190, 1977.

[Laz81]   D. Lazard. Résolution des systèmes d'equations algébriques. *Theor. Comput. Sci.*, 15:77–110, 1981.

[Laz92]   D. Lazard. Stewart platform and Gröbner basis. In *Proc. ARK*, pages 136–142, 1992.

[Laz93]   D. Lazard. Generalized Stewart platform: How to compute with rigid motions? In *IMACS-SC'93*, 1993.

[Lec00]   G. Lecerf. Computing the equidimensional decomposition of an algebraic variety by means of geometric resolutions. In C. Traverso, editor, *Proc. Annual ACM Intern. Symp. on Symbolic and Algebraic Computation*, pages 209–216. ACM, 2000.

[Lec02]   G. Lecerf. Quadratic Newton iteration for systems with multiplicity. *Found. of Comput. Math.*, 2(3):247–293, 2002.

[Lec03]   G. Lecerf. Computing the equidimensional decomposition of an algebraically closed set by means of lifting fibers. *J. Complexity*, 19(4):564–596, 2003.

[Lee91]   C.W. Lee. Regular triangulations of convex polytopes. In P. Gritzmann and B. Sturmfels, editors, *Applied Geometry and Discrete Mathematics - The Victor Klee Festschrift*, volume 4 of *DIMACS Series*, pages 443–456. AMS, Providence, R.I., 1991.

[Lef53]   S. Lefschetz. *Algebraic Geometry*. Princeton University Press, 1953.

[Li83]   T.Y. Li. On Chow, Mallet-Paret and Yorke homotopy for solving systems of polynomials. *Bulletin of the Institute of Mathematics. Acad. Sin.*, 11:433–437, 1983.

[Li87]   T.Y. Li. Solving polynomial systems. *The Mathematical Intelligencer*, 9(3):33–39, 1987.

[Li97]   T.Y. Li. Numerical solution of multivariate polynomial systems by homotopy continuation methods. *Acta Numerica*, 6:399–436, 1997.

[Li03]   T.Y. Li. Numerical solution of polynomial systems by homotopy continuation methods. In F. Cucker, editor, *Handbook of Numerical Analysis*, volume XI. Special Volume: Found. of Comp. Math., pages 209–304. North-Holland, 2003.

[Lip87]   J. Lipman. *Residues and traces of differential forms via Hochschild homology*, volume 61 of *Contemporary Math.* AMS, Providence, R.I., 1987.

[LL91]   Y. N. Lakshman and D. Lazard. On the complexity of zero-dimensional algebraic systems. In *Effective Methods in Algebraic Geometry (Proc.*

MEGA'90), Castiglioncello, Italy, volume 94 of *Progress in Math.,* pages 217–225. Birkhäuser, 1991.

[LL01]   T.Y. Li and X. Li. Finding mixed cells in the mixed volume computation. *Found. of Comput. Math.,* 1(2):161–181, 2001. Software available at http://www.math.msu.edu/~li.

[LN83]   R. Lidl and H. Niederreiter. *Finite Fields.* Number 20 in Encyclopedia Math. & Appl. Addison-Wesley, Reading, Mass., 1983.

[LRD00]   H. Lombardi, M.-F. Roy, and M. Safey El Din. New structure theorem for subresultants. *J. Symbolic Computation,* 29(4-5):663–690, 2000.

[LS87]   T.Y. Li and T. Sauer. Regularity results for solving systems of polynomials by homotopy method. *Numer. Math.,* 50(3):283–289, 1987.

[LSY87a] T.Y. Li, T. Sauer, and J.A. Yorke. Numerical solution of a class of deficient polynomial systems. *SIAM J. Numerical Analysis,* 24(2):435–451, 1987.

[LSY87b] T.Y. Li, T. Sauer, and J.A. Yorke. The random product homotopy and deficient polynomial systems. *Numer. Math.,* 51(5):481–500, 1987.

[LSY89]   T.Y. Li, T. Sauer, and J.A. Yorke. The cheater's homotopy: an efficient procedure for solving systems of polynomial equations. *SIAM J. Numerical Analysis,* 26(5):1241–1251, 1989.

[Lut86]   E. Lutwak. Volume of mixed bodies. *Trans. AMS,* 294(2):487–500, 1986.

[LW91]   T.Y. Li and X. Wang. Solving deficient polynomial systems with homotopies which keep the subschemes at infinity invariant. *Math. Comp.,* 56(194):693–710, 1991.

[LW92]   T.Y. Li and X. Wang. Nonlinear homotopies for solving deficient polynomial systems with parameters. *SIAM J. Numerical Analysis,* 29(4):1104–1118, 1992.

[LW96]   T.Y. Li and X. Wang. The BKK root count in C^n. *Math. Comp.,* 65(216):1477–1484, 1996.

[LWW96]   T.Y. Li, T. Wang, and X. Wang. Random product homotopy with minimal BKK bound. In J. Renegar, M. Shub, and S. Smale, editors, *The Mathematics of Numerical Analysis,* volume 32 of *Lectures in Applied Mathematics,* pages 503–512, Providence, R.I., 1996. AMS. Proc. ACM-SIAM Summer Seminar in Applied Math. Park City, Utah, 1995.

[LWW02]   T.Y. Li, X. Wang, and M. Wu. Numerical Schubert calculus by the Pieri homotopy algorithm. *SIAM J. Numerical Analysis,* 20(2):578–600, 2002.

[Mac02]   F.S. Macaulay. Some formulae in elimination. *Proc. London Math. Soc.,* 1(33):3–27, 1902.

[Mac94]   F.S. Macaulay. *The Algebraic Theory of Modular Systems,* volume 19 of *Cambridge tracts in Math. & Math. Physics.* Cambridge University Press, 1994. Reprint, original edition: 1916.

[MC92]   D. Manocha and J. Canny. Algorithms for implicitizing rational parametric surfaces. *Computer Aided Geometric Design,* 9(1):25–50, 1992.

[MC00]   T. Michiels and R. Cools. Decomposing the secondary cayley polytope. *Discr. and Comput. Geometry,* 23, 2000.

[McD02]   J. McDonald. Fractional power series solutions for systems of equations. *Discr. and Comput. Geometry,* 27(4):501–529, 2002.

[Mig92]   M. Mignotte. *Mathematics for computer algebra.* Springer-Verlag, New York, 1992.

410   References

[MMM93] M. Marinari, H. M. Möller, and T. Mora. Gröbner bases of ideals defined by functionals with an application to ideals of projective points. *J. Applic. Algebra in Engin. Communic. & Computing*, 4:103–145, 1993.

[MMM96] M. Marinari, H. M. Möller, and T. Mora. On multiplicities in polynomial system solving. *Trans. Am. Math. Soc.*, 348:3283–3321, 1996.

[Möl93] H.M. Möller. Systems of algebraic equations solved by means of endomorphisms. In G.Čohen et al., editor, *Proc. Intern. Symp. Applied Algebra, Algebraic Algorithms and Error-Correcting Codes (AAECC-10)*, pages 43–56, 1993.

[Mon02] C. Monico. Computing the primary decomposition of zero-dimensional ideals. *J. Symbolic Computation*, 34:451–459, 2002.

[Mor83] A.P. Morgan. A method for computing all solutions to systems of polynomial equations. *ACM Trans. Math. Softw.*, 9(1):1–17, 1983.

[Mor87] A. Morgan. *Solving polynomial systems using continuation for engineering and scientific problems*. Prentice-Hall, 1987.

[Mou93] B. Mourrain. The 40 generic positions of a parallel robot. In M. Bronstein, editor, *Proc. Annual ACM Intern. Symp. on Symbolic and Algebraic Computation*, pages 173–182. ACM, July 1993.

[Mou96] B. Mourrain. Enumeration problems in Geometry, Robotics and Vision. In L. González and T. Recio, editors, *Algorithms in Algebraic Geometry and Applications*, volume 143 of *Progress in Math.*, pages 285–306. Birkhäuser, Basel, 1996.

[Mou98] B. Mourrain. Computing isolated polynomial roots by matrix methods. *J. Symbolic Computation*, 26(6):715–738, December 1998.

[Mou99] B. Mourrain. A new criterion for normal form algorithms. In M. Fossorier, H. Imai, S. Lin, and A. Poli, editors, *Proc. Intern. Symp. Applied Algebra, Algebraic Algorithms and Error-Correcting Codes (AAECC-13, Honolulu, 1999)*, volume 1719 of *Lect. Notes in Comp. Science*, pages 440–443. Springer-Verlag, 1999.

[MP00] B. Mourrain and V.Y. Pan. Multivariate polynomials, duality and structured matrices. *J. Complexity*, 16(1):110–180, 2000.

[MR02] B. Mourrain and O. Ruatta. Relation between roots and coefficients, interpolation and application to system solving. *J. Symbolic Computation*, 33(5):679–699, 2002.

[MS87a] A. Morgan and A. Sommese. Computing all solutions to polynomial systems using homotopy continuation. *Appl. Math. Comput.*, 24(2):115–138, 1987. Errata: *Appl. Math. Comput.* 51 (1992), p. 209.

[MS87b] A.P. Morgan and A.J. Sommese. A homotopy for solving general polynomial systems that respects *m*-homogeneous structures. *Appl. Math. Comput.*, 24(2):101–113, 1987.

[MS89] A.P. Morgan and A.J. Sommese. Coefficient-parameter polynomial continuation. *Appl. Math. Comput.*, 29(2):123–160, 1989. Errata: *Appl. Math. Comput.* 51:207(1992).

[MS90] A.P. Morgan and A.J. Sommese. Generically nonsingular polynomial continuation. In E.L. Allgower and K. Georg, editors, *Computational Solution of Nonlinear Systems of Equations*, pages 467–493, Providence, R.I., 1990. AMS.

[MS95]   H.M. Möller and H.J. Stetter. Multivariate polynomial systems with multiple zeros solved by matrix eigenproblems. *Numer. Math.*, 70:311–329, 1995.

[MSW91]   A.P. Morgan, A.J. Sommese, and C.W. Wampler. Computing singular solutions to nonlinear analytic systems. *Numer. Math.*, 58(7):669–684, 1991.

[MSW92a]   A.P. Morgan, A.J. Sommese, and C.W. Wampler. Computing singular solutions to polynomial systems. *Adv. Appl. Math.*, 13(3):305–327, 1992.

[MSW92b]   A.P. Morgan, A.J. Sommese, and C.W. Wampler. A power series method for computing singular solutions to nonlinear analytic systems. *Numer. Math.*, 63:391–409, 1992.

[MSW95]   A.P. Morgan, A.J. Sommese, and C.W. Wampler. A product-decomposition bound for Bézout numbers. *SIAM J. Numerical Analysis*, 32(4):1308–1325, 1995.

[MT00]   B. Mourrain and Ph. Trébuchet. Solving projective complete intersection faster. In C. Traverso, editor, *Proc. Annual ACM Intern. Symp. on Symbolic and Algebraic Computation*, pages 231–238, New York, 2000. ACM.

[MT01]   H.M. Möller and R. Tenberg. Multivariate polynomial system solving using intersections of eigenspaces. *J. Symbolic Computation*, 32:513–531, 2001.

[MT02]   B. Mourrain and Ph. Trébuchet. Algebraic methods for numerical solving. In *Proc. 3rd Intern. Workshop Symbolic & Numeric Algorithms for Scientific Computing'01 (Timisoara, Romania)*, pages 42–57, 2002.

[Mul87]   K. Mulmuley. A fast parallel algorithm to compute the rank of a matrix over an arbitrary field. *Combinatorica*, 7:101–104, 1987.

[MV99]   T. Michiels and J. Verschelde. Enumerating regular mixed-cell configurations. *Discr. and Comput. Geometry*, 21(4):569–579, 1999.

[MVY02]   B. Mourrain, M. Vrahatis, and J.C. Yakoubsohn. On the complexity of isolating real roots and computing with certainty the topological degree. *J. Complexity*, 18(2):612–640, 2002.

[MW83]   D.W. Masser and G. Wuestholz. Fields of large transcendence degree generated by values of elliptic functions. *Invent. Math.*, 72:407–464, 1983.

[Nag02]   K. Nagasaka. Towards certified irreducibility testing of bivariate approximate polynomials. In T. Mora, editor, *Proc. Annual ACM Intern. Symp. on Symbolic and Algebraic Computation*, pages 192–199, 2002.

[Nie93]   H. Niederreiter. A new efficient factorization algorithm for polynomials over small finite fields. *J. Applic. Algebra in Engin. Communic. & Computing*, 4(2):81–87, 1993.

[Ore23]   O. Ore. Zur theorie der irreduzibilitätskriterien. *Math. Zeit.*, 18:277–288, 1923.

[Ore24a]   O. Ore. Zur theorie der algebraischen körper. *Acta Math.*, 44:219–314, 1924.

[Ore24b]   O. Ore. Zur theorie der der eisensteinschen gleichungen. *Math. Zeit.*, 20:267–279, 1924.

[Pan96]   V.Y. Pan. Optimal and nearly optimal algorithms for approximating polynomial zeros. *Comp. & Math. (with Appl.)*, 31:97–138, 1996.

412   References

[Pan97]   V.Y. Pan. Solving a polynomial equation: Some history and recent progress. *SIAM Rev.*, 39(2):187–220, 1997.

[Poi87]   H. Poincaré. Sur les résidus des intégrales doubles. *Acta Math.*, 9:321–380, 1887.

[Pom93]   J.E. Pommersheim. Toric varieties, lattice points and Dedekind sums. *Math. Ann.*, 295(1):1–24, 1993.

[PRS93]   P.S. Pedersen, M.-F. Roy, and A. Szpirglas. Counting Real Zeros in the multivariate Case. In A. Galligo and F. Eyssette, editors, *Effective Methods in Algebraic Geometry (Proc. MEGA'92), Nice, France*, Progress in Math., pages 203–223. Birkhäuser, 1993.

[PRW00]   G. Pistone, E. Riccomagno, and H.P. Wynn. *Algebraic Statistics: Computational Commutative Algebra in Statistics*, volume 89 of *Monographs on Statistics and Applied Probability*. CRC Press, Boca Raton, 2000.

[PS83]   C. Peters and J. Steenbrink. Infinitesimal variations of Hodge structure and the generic Torelli problem for projective hypersurfaces (after Carlson, Donagi, Green, Griffiths, Harris). In *Classification of algebraic and analytic manifolds (Katata, 1982)*, volume 39 of *Progrress in Math.*, pages 399–463. Birkhäuser, Boston, Mass., 1983.

[PS96]   P. Pedersen and B. Sturmfels. Mixed monomial bases. In L. González-Vega and T. Recio, editors, *Algorithms in algebraic geometry and applications (Proc. MEGA'94), Santander, Spain*, number 143 in Progress in Math., pages 307–316, Basel, 1996. Birkhäuser.

[PS98]   S. Puddu and J. Sabia. An effective algorithm for quantifier elimination over algebraically closed fields using straight line programs. *J. Pure & Applied Algebra*, 129(2):173–200, 1998.

[PZ89]   M. Pohst and H. Zassenhaus. *Algorithmic Algebraic Number Theory*. Cambridge University Press, Cambridge, 1989.

[Rag93]   M. Raghavan. The Stewart platform of general geometry has 40 configurations. *ASME J. Mech. Design*, 115:277–282, 1993.

[Rag97]   J.F. Ragot. *Sur la factorisation absolue des polynmes*. PhD thesis, Univ. Limoges, 1997.

[Ram01]   J. Rambau. TOPCOM, 2001. www.zib.de/rambau/TOPCOM.

[Rel27]   T. Rella. Ordnungsbestimmungen in integritätsbereichen und newtonsche polygone. *J. fr die Reine und Angew. Math.*, 158:33–48, 1927.

[Ren92]   J. Renegar. On the computational complexity and geometry of the first-order theory of the reals. *J. Symbolic Computation*, 13(3):255–352, 1992.

[Rib01]   P. Ribenboim. *Classical theory of algebraic numbers*. Universitext. Springer-Verlag, New York, 2001.

[Roba]   L. Robbiano. Cocoa, Computational Commutative Algebra. Available at http://cocoa.dima.unige.it/.

[Robb]   L. Robbiano. Zero-dimensional ideals or the inestimable value of estimable terms. Constructive Algebra and Systems Theory, Proc. Acad. Coll., 2000, Amsterdam. To appear.

[Rob98]   L. Robbiano. Gröbner bases and statistics. In F. Winkler B. Buchberger, editor, *Gröbner Bases and Applications*, volume 251 of *London Math. Soc LNS*, pages 179–204. Cambridge University Press, 1998.

[Roj94]   J.M. Rojas. A convex geometric approach to counting the roots of a polynomial system. *Theor. Comput. Sci.*, 133(1):105–140, 1994.

[Roj99a]   J.M. Rojas. Solving degenerate sparse polynomial systems faster. *J. Symbolic Computation*, 28:155–186, 1999.

[Roj99b]   J.M. Rojas. Toric intersection theory for affine root counting. *J. Pure & Applied Algebra*, 136(1):67–100, 1999.

[Ros94]   J. Rosenthal. On dynamic feedback compensation and compactifications of systems. *SIAM J. Control and Optimization*, 32(1):279–296, 1994.

[Rot88]   J.J. Rotman. *An introduction to algebraic topology*, volume 119 of *Graduate Texts in Mathematics*. Springer-Verlag, New York, 1988.

[Rou99]   F. Rouillier. Solving zero-dimensional polynomial systems through rational univariate representation. *J. Applic. Algebra in Engin. Communic. & Computing*, 9(5):433–461, 1999.

[Roy96]   M.F. Roy. Basic algorithms in real algebraic geometry: from Sturm theorem to the existential theory of reals. In *Lectures on Real Geometry in memoriam of Mario Raimondo*, volume 23 of *Exposition in Mathematics*, pages 1–67, 1996.

[RR95]   M. Raghavan and B. Roth. Solving polynomial systems for the the kinematic analysis of mechanisms and robot manipulators. *ASME J. Mech. Design*, 117(2):71–79, 1995.

[RR98]   L. Robbiano and M.P. Rogantin. Full factorial designs and distracted fractions. In F. Winkler B. Buchberger, editor, *Gröbner Bases and Applications*, volume 251 of *London Math. Soc LNS*, pages 473–482. Cambridge University Press, 1998.

[RRW96]   M.S. Ravi, J. Rosenthal, and X. Wang. Dynamic pole placement assignment and Schubert calculus. *SIAM J. Control and Optimization*, 34(3):813–832, 1996.

[RRW98]   M.S. Ravi, J. Rosenthal, and X. Wang. Degree of the generalized Plücker embedding of a quot scheme and quantum cohomology. *Math. Ann.*, 311:11–26, 1998.

[RS98]   M.-F. Roy and A. Szpirglas. Bezoutiens et résidus, affines, projectifs et dans le tore. Prépublication IRMAR, 1998.

[RSV02]   G. Reid, C. Smith, and J. Verschelde. Geometric completion of differential systems using numeric-symbolic continuation. *SIGSAM Bulletin*, 36(2):1–17, 2002.

[Rup99]   W.M. Ruppert. Reducibility of polynomials $f(x, y)$ modulo p. *J. Number Theory*, 77(1):62–70, 1999.

[Rup00]   D. Rupprecht. *Elements de géométrie algébrique approchée: Etude du pgcd et de la factorisation*. PhD thesis, Univ. de Nice Sophia-Antipolis, France, 2000.

[Rup04]   D. Rupprecht. Semi-numerical absolute factorization of polynomials with integer coefficients. *J. Symbolic Computation*, 37:557–574, 2004.

[RV95]   F. Ronga and T. Vust. Stewart platforms without computer? In *Real Analytic and Algebraic Geometry. Proc. Intern. Conf., (Trento, 1992)*, pages 196–212. Walter de Gruyter, 1995.

[RW96]   J.M. Rojas and X. Wang. Counting affine roots of polynomial systems via pointed Newton polytopes. *J. Complexity*, 12:116–133, 1996.

[RW99]   J. Rosenthal and J.C. Willems. Open problems in the area of pole placement. In V.D. Blondel, E.D. Sontag, M. Vidyasagar, and J.C. Willems, editors, *Open Problems in Mathematical Systems and Control Theory*, pages 181–191. Springer-Verlag, 1999.

[RZ03]   F. Rouillier and P. Zimmermann. Efficient isolation of a polynomial real roots. *J. Comput. Appl. Math.*, 162(1):33–50, 2003.

[SAG84]   T.W. Sederberg, D.C. Anderson, and R.N. Goldman. Implicit representation of parametric curves and surfaces. *Computer Vision, Graphics and Image Processing*, 28:72–84, 1984.

[Sam67]   P. Samuel. *Théorie algébrique des nombres*. Hermann, Paris, 1967.

[Sas01]   T. Sasaki. Approximate multivariate polynomial factorization based on zero-sum relations. In B. Mourrain, editor, *Proc. Annual ACM Intern. Symp. on Symbolic and Algebraic Computation*, pages 284–291. ACM, 2001.

[SC95]   T.W. Sederberg and F. Chen. Implicitizing using moving curves and surfaces. In *Proc. ACM Symp. SIGGRAPH*, pages 301–308, 1995.

[Sch78]   C.P. Schnorr. Improved lower bounds on the number of multiplications/divisions which are necessary to evaluate polynomials. *Theor. Comput. Sci.*, 7:251–261, 1978.

[Sch80]   J.T. Schwartz. Fast probabilistic algorithms for verification of polynomial identities. *J. ACM*, 27:701–717, 1980.

[Sch93]   R. Schneider. *Convex Bodies: The Brunn-Minkowski Theory*. Cambridge University Press, Cambridge, 1993.

[Sch03a]   H. Schenck. *Computational Algebraic Geometry*, volume 58 of *London Mathematical Society Student Texts*. Cambridge University Press, Cambridge, 2003.

[Sch03b]   E. Schost. Computing parametric geometric resolutions. *J. Applic. Algebra in Engin. Communic. & Computing*, 14(1):349–393, 2003.

[Sha77]   I. R. Shafarevich. *Basic algebraic geometry*. Springer-Verlag, Berlin, 1977.

[Sha90]   R. Y. Sharp. *Steps in Commutative Algebra*. Cambridge University Press, Cambridge, 1990.

[Som97]   M. Sombra. Bounds for the Hilbert function of polynomial ideals and for the degrees in the Nullstellensatz. *J. Pure & Applied Algebra*, 117–118:565–599, 1997.

[Sot97a]   F. Sottile. Enumerative geometry for real varieties. In J. Kollár, R. Lazarsfeld, and D.R. Morrison, editors, *Algebraic Geometry (Santa Cruz, 1995)*, volume 62, Part I of *Proc. Symp. Pure Math.*, pages 435–447, Providence, R.I., 1997. AMS.

[Sot97b]   F. Sottile. Pieri's formula via explicit rational equivalence. *Can. J. Math.*, 49(6):1281–1298, 1997.

[Sot03]   F. Sottile. Enumerative real algebraic geometry. In S. Basu and L. Gonzalez-Vega, editors, *Algorithmic and Quantitative Real Algebraic Geometry*, pages 139–180. AMS, Providence, R.I., 2003. Web-based survey available at http://www.math.umass.edu/~sottile.

[SS75]   G. Scheja and U. Storch. Uber Spurfunktionen bei vollständigen Durchschnitten. *J. Reine Angew. Math.*, 278:174–190, 1975.

[SS79]   G. Scheja and U. Storch. Residuen bei vollständigen durchschnitten. *Math. Nachr.*, 91:157–170, 1979.

[SS93]   M. Shub and S. Smale. On the complexity of Bezout's theorem I – geometric aspects. *J. AMS*, 6(2):459–501, 1993.

[SS95]   J. Sabia and P. Solern'o. Bounds of traces in complete intersections and degrees in the Nullstellensatz. *J. Applic. Algebra in Engin. Communic. & Computing*, 6(6):353–376, 1995.

[SS00]   R. Settimi and J.Q. Smith. Geometry, moments and conditional independence trees with hidden variables. *Ann. Statist.*, 28(4):1179–1205, 2000.

[SS01]   F. Sottile and B. Sturmfels. A sagbi basis for the quantum grassmannian. *J. Pure & Applied Algebra*, 158(2-3):347–366, 2001.

[SSKS91]   T. Sasaki, M. Suzuki, M. Kolář, and M. Sasaki. Approximate factorization of multivariate polynomials and absolute irreducibility testing. *Japan J. Indust. Appl. Math.*, 8(3):357–375, 1991.

[SST00]   M. Saito, B. Sturmfels, and N. Takayama. *Gröbner deformations of hypergeometric differential equations*, volume 6 of *Algorithms and Computation in Mathematics*. Springer-Verlag, Berlin, 2000.

[Ste04]   H.J. Stetter. *Numerical polynomial algebra*. SIAM, Philadelphia, 2004.

[Stu93]   B. Sturmfels. Sparse elimination theory. In *Computational algebraic geometry and commutative algebra (Cortona, Italy, 1991)*, Proc. Symp. Math., XXXIV, pages 264–298. Cambridge University Press, Cambridge, 1993.

[Stu94a]   B. Sturmfels. On the Newton polytope of the resultant. *J. Algebr. Combinatorics*, 3:207–236, 1994.

[Stu94b]   B. Sturmfels. On the number of real roots of a sparse polynomial system. In A. Bloch, editor, *Hamiltonian and Gradient Flows: Algorithms and Control*, pages 137–143, Providence, R.I., 1994. AMS.

[Stu94c]   B. Sturmfels. Viro's theorem for complete intersections. *Annali della Scuola Normale Superiore di Pisa*, 21(3):377–386, 1994.

[Stu98]   B. Sturmfels. Introduction to resultants. In *Applications of computational algebraic geometry*, volume 53 of *Proc. Symp. Appl. Math.*, pages 25–39. AMS, Providence, RI, 1998.

[Stu02]   B. Sturmfels. *Solving Systems of Polynomial Equations*. Number 97 in CBMS Regional Conference Series in Math. AMS, Providence, R.I., 2002.

[SV00]   A.J. Sommese and J. Verschelde. Numerical homotopies to compute generic points on positive dimensional algebraic sets. *J. Complexity*, 16(3):572–602, 2000.

[SVW]   A.J. Sommese, J. Verschelde, and C.W. Wampler. Homotopies for intersecting solution components of polynomial systems. *SIAM J. Numerical Analysis*. To appear.

[SVW01a]   A.J. Sommese, J. Verschelde, and C.W. Wampler. Numerical decomposition of the solution sets of polynomial systems into irreducible components. *SIAM J. Numerical Analysis*, 38(6):2022–2046, 2001.

[SVW01b]   A.J. Sommese, J. Verschelde, and C.W. Wampler. Numerical irreducible decomposition using projections from points on the components. In E.L. Green, S. Hoşten, R.C. Laubenbacher, and V. Powers, editors, *Symbolic Computation: Solving Equations in Algebra, Geometry, and Engineering*, volume 286 of *Contemporary Mathematics*, pages 37–51, Providence, R.I., 2001. AMS.

[SVW01c]   A.J. Sommese, J. Verschelde, and C.W. Wampler. Using monodromy to decompose solution sets of polynomial systems into irreducible components. In C. Ciliberto, F. Hirzebruch, R. Miranda, and M. Teicher, editors, *Application of Algebraic Geometry to Coding Theory, Physics and Computation*, pages 297–315. Kluwer Academic Publishers, 2001. Proc. NATO Conf., Eilat, Israel.

[SVW02a]   A.J. Sommese, J. Verschelde, and C.W. Wampler. Advances in polyno-
           mial continuation for solving problems in kinematics. In *Proc. ASME
           Design Engineering Technical Conf.*, Montreal, Quebec, 2002. Paper
           DETC2002/MECH-34254.

[SVW02b]   A.J. Sommese, J. Verschelde, and C.W. Wampler. A method for track-
           ing singular paths with application to the numerical irreducible decom-
           position. In M.C. Beltrametti, F. Catanese, C. Ciliberto, A. Lanteri,
           and C. Pedrini, editors, *Algebraic Geometry, a Volume in Memory of
           Paolo Francia*, pages 329–345. Walter de Gruyter, 2002.

[SVW02c]   A.J. Sommese, J. Verschelde, and C.W. Wampler. Symmetric functions
           applied to decomposing solution sets of polynomial systems. *SIAM
           J. Numerical Analysis*, 40(6):2026–2046, 2002.

[SVW03]   A.J. Sommese, J. Verschelde, and C.W. Wampler. Numerical irre-
           ducible decomposition using PHCpack. In M. Joswig and N. Takayama,
           editors, *Algebra, Geometry, and Software Systems*, pages 109–130.
           Springer-Verlag, 2003.

[SVW04]   A.J. Sommese, J. Verschelde, and C.W. Wampler. Numerical factoriza-
           tion of multivariate complex polynomials. *Theor. Comput. Sci.*, 315(2–
           3):651–669, 2004. Special Issue on Algebraic and Numerical Algorithms.

[SW96]   A.J. Sommese and C.W. Wampler. Numerical algebraic geometry. In
           J. Renegar, M. Shub, and S. Smale, editors, *The Mathematics of Nu-
           merical Analysis*, volume 32 of *Lectures in Applied Mathematics*, pages
           749–763, Providence, R.I., 1996. AMS. Proc. AMS-SIAM Summer Sem-
           inar in Applied Math. Park City, Utah, 1995.

[SWS96]   M. Sosonkina, L.T. Watson, and D.E. Stewart. Note on the end game
           in homotopy zero curve tracking. *ACM Trans. Math. Softw.*, 22(3):281–
           287, 1996.

[SY96]   T. Shimoyama and K. Yokoyama. Localization and primary decompo-
           sition of polynomial ideals. *J. Symbolic Computation*, 22(3):247–277,
           1996.

[SZ94]   B. Sturmfels and A. Zelevinsky. Multigraded resultants of Sylvester
           type. *J. Algebra*, 163(1):115–127, 1994.

[Tar51]   A. Tarski. *A decision method for elementary algebra and geometry*.
           University of California Press, Berkeley, 2nd edition, 1951.

[TKF02]   A. Takeda, M. Kojima, and K. Fujisawa. Enumeration of all solutions
           of a combinatorial linear inequality system arising from the polyhe-
           dral homotopy continuation method. *J. Operations Research Society of
           Japan*, 45(1):64–82, 2002.

[Tré02]   Ph. Trébuchet. *Vers une résolution stable et rapide des équations
           algébriques*. PhD thesis, Université P. et M. Curie, 2002.

[Tsi92]   A.K. Tsikh. *Multidimensional residues and their applications*, volume
           103 of *Translations of Mathematical Monographs*. AMS, Providence,
           R.I., 1992. Translated from the Russian (1988).

[Tsü94]   W.W. Tsün. *Mechanical Theorem Proving in Geometries: Basic Prin-
           ciples*. Texts and Monographie in Symbolic Computation. Springer-
           Verlag, 1994. Translated from the Chinese.

[TY84]   A.K. Tsikh and A.P. Yuzhakov. Properties of the complete sum of
           residues with respect to a polynomial mapping, and their applications.
           Sibirsk. Mat. Zh., 25(4):207–213, 1984.

[Usp48]   J.Y. Uspensky. *Theory of equations*. Mac-Graw Hill, 1948.

[VC93]   J. Verschelde and R. Cools. Symbolic homotopy construction. *J. Applic. Algebra in Engin. Communic. & Computing*, 4(3):169–183, 1993.

[VC94]   J. Verschelde and R. Cools. Symmetric homotopy construction. *J. Comput. Appl. Math.*, 50:575–592, 1994.

[vdW49]   B.L. van der Waerden. *Modern algebra*, volume I. Frederick Ungar Publishing Co., New York, 1949.

[Ver99a]   J. Verschelde. Algorithm 795: PHCpack: A general-purpose solver for polynomial systems by homotopy continuation. *ACM Trans. Math. Softw.*, 25(2):251–276, 1999. Software available at http://www.math.uic.edu/~jan.

[Ver99b]   J. Verschelde. Polynomial homotopies for dense, sparse and determinantal systems. Technical Report 1999-041, Mathematical Sciences Research Institute, 1999. Available at http://www.math.uic.edu/~jan.

[Ver00]   J. Verschelde. Toric Newton method for polynomial homotopies. *J. Symbolic Computation*, 29(4–5):777–793, 2000.

[VG95]   J. Verschelde and K. Gatermann. Symmetric Newton polytopes for solving sparse polynomial systems. *Adv. Appl. Math.*, 16(1):95–127, 1995.

[VGC96]   J. Verschelde, K. Gatermann, and R. Cools. Mixed-volume computation by dynamic lifting applied to polynomial system solving. *Discr. and Comput. Geometry*, 16(1):69–112, 1996.

[VH93]   J. Verschelde and A. Haegemans. The *GBQ*-Algorithm for constructing start systems of homotopies for polynomial systems. *SIAM J. Numerical Analysis*, 30(2):583–594, 1993.

[vH02]   M. van Hoeij. Factoring polynomials and the knapsack problem. *J. Number Theory*, 95(2):167–189, 2002.

[VVC94]   J. Verschelde, P. Verlinden, and R. Cools. Homotopies exploiting Newton polytopes for solving sparse polynomial systems. *SIAM J. Numerical Analysis*, 31(3):915–930, 1994.

[VW02]   J. Verschelde and Y. Wang. Numerical homotopy algorithms for satellite trajectory control by pole placement. In *Proc. Math. Theory Networks & Systems (MTNS)*, Notre Dame, Indiana, 2002. CDROM.

[vzG85]   J. von zur Gathen. Irreducibility of multivariate polynomials. *J. Comput. System Sci.*, 31(2):225–264, 1985. Special Issue: Symp. Found. of Comp. Science (1983).

[Wal62]   R.J. Walker. *Algebraic curves*. Dover Publications, Inc., New York, 1962.

[Wam92]   C.W. Wampler. Bezout number calculations for multi-homogeneous polynomial systems. *Appl. Math. Comput.*, 51(2–3):143–157, 1992.

[Wam96]   C.W. Wampler. Forward displacement analysis of general six-in-parallel SPS (Stewart) platform manipulators using soma coordinates. *Mechanism Machine Theory*, 31(3):331–337, 1996.

[Wan95]   D. Wang. Elimination procedures for mechanical theorem proving in geometry. *Annals of Mathematics and Artificial Intelligence*, 13:1–24, 1995.

[Wat86]   L.T. Watson. Numerical linear algebra aspects of globally convergent homotopy methods. *SIAM Rev.*, 28(4):529–545, 1986.

[Wat89]   L.T. Watson. Globally convergent homotopy methods: a tutorial. *Appl. Math. Comput.*, 31:369–396, 1989. Special Issue.

[Wat02]   L.T. Watson. Probability-one homotopies in computational science. *J. Comput. Appl. Math.*, 140(1-2):785–807, 2002.

[WBM87]   L.T. Watson, S.C. Billups, and A.P. Morgan. Algorithm 652: HOM-PACK: a suite of codes for globally convergent homotopy algorithms. *ACM Trans. Math. Softw.*, 13(3):281–310, 1987.

[WMS90]   C.W. Wampler, A.P. Morgan, and A.J. Sommese. Numerical continuation methods for solving polynomial systems arising in kinematics. *ASME J. Mech. Design*, 112(1):59–68, 1990.

[WMS92]   C.W. Wampler, A.P. Morgan, and A.J. Sommese. Complete solution of the nine-point path synthesis problem for four-bar linkages. *ASME J. Mech. Design*, 114(1):153–159, 1992.

[Wri85]   A.H. Wright. Finding all solutions to a system of polynomial equations. *Math. Comp.*, 44(169):125–133, 1985.

[WSM+97]   L.T. Watson, M. Sosonkina, R.C. Melville, A.P. Morgan, and H.F. Walker. HOMPACK90: A suite of Fortran 90 codes for globally convergent homotopy algorithms. *ACM Trans. Math. Softw.*, 23(4):514–549, 1997. Available at http://www.cs.vt.edu/~ltw.

[WSW00]   S.M. Wise, A.J. Sommese, and L.T. Watson. Algorithm 801: POL-SYS_PLP: a partitioned linear product homotopy code for solving polynomial systems of equations. *ACM Trans. Math. Softw.*, 26(1):176–200, 2000.

[WZ94]   J. Weyman and A. Zelevinsky. Multigraded formulae for multigraded resultants. *J. Algebr. Geom.*, 3(4):569–597, 1994.

[YNT92]   K. Yokoyama, M. Noro, and T. Takeshima. Solutions of systems of algebraic equations and linear maps on residue class rings. *J. Symbolic Computation*, 14:399–417, 1992.

[Zip79]   R. Zippel. Probabilistic algorithms for sparse polynomials. In *Proc. Symbolic and algebraic computation, EUROSAM Intern. Symp., Marseille*, number 72 in Lect. Notes in Comp. Science, pages 216–226, 1979.

[Zip93]   R. Zippel. *Effective polynomial computation*. Kluwer Academic Publishers, 1993.

[Zul88]   W. Zulehner. A simple homotopy method for determining all isolated solutions to polynomial systems. *Math. Comp.*, 50(181):167–177, 1988.

Index